University Chemistry

Based on Common Core Syllabus of Universities of Andhra Pradesh

Volume II

Syllabus of B.Sc. II Year (Theory and Practicals)

Series Editor

C. Parameswara Murthy

Authors

C. Parameswara Murthy
M.Sc., Ph.D
Professor of Chemistry
University College of Engineering
Osmania University, Hyderabad (A.P.)

Syed Fazal Mehdi Ali
M.Sc:, Ph.D.
Principal
Rishi Junior and Degree College
Hyderabad (A.P.)

Pramod Kumar Dubey
M.Sc., Ph.D.
Reader in Chemistry
JNTU College of Engineering
Hyderabad (A.P.)

D. Ashok
M.Sc., Ph.D.
P.G. College of Science
O.U. Saifabad
Hyderabad (A.P.)

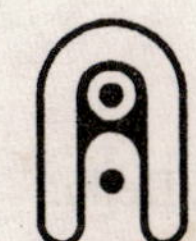

NEW AGE INTERNATIONAL (P) LIMITED, PUBLISHERS

New Delhi • Bangalore • Chennai • Cochin • Guwahati • Hyderabad
Jalandhar • Kolkata • Lucknow • Mumbai • Ranchi
Visit us at www.newagepublishers.com

Published by New Age International (P) Ltd., Publishers
First Edition: 1996
Reprint: 2008

Branches:

- 36, Malikarjuna Temple Street, Opp. ICWA, Basavanagudi, **Bangalore**. © (080) 26677815
- 26, Damodaran Street, T. Nagar, **Chennai**. © (044) 24353401
- Hemsen Complex, Mohd. Shah Road, Paltan Bazar, Near Starline Hotel, **Guwahati**. © (0361) 2543669
- No. 105, 1st Floor, Madhiray Kaveri Tower, 3-2-19, Azam Jahi Road, Nimboliadda, **Hyderabad**. © (040) 24652456
- RDB Chambers (Formerly Lotus Cinema) 106A,1st Floor, S.N. Banerjee Road, **Kolkata**. © (033) 22275247
- 18, Madan Mohan Malviya Marg, **Lucknow**. © (0522) 2209578
- 142C, Victor House, Ground Floor, N.M. Joshi Marg, Lower Parel, **Mumbai**. © (022) 24927869
- 22, Golden House, Daryaganj, **New Delhi**. © (011) 23262370, 23262368

ISBN : 978-81-224-0955-0

Rs. 395.00

C-08-06-2651

Printed in India at Mohanlal Printers, Delhi.
Typeset at Hitek Computer Graphix.

PUBLISHING FOR ONE WORLD
NEW AGE INTERNATIONAL (P) LIMITED, PUBLISHERS
4835/24, Ansari Road, Daryaganj, New Delhi-110002
Visit us at **www.newagepublishers.com**

Foreword

Chemistry is a fast trac science and every year we read and hear many developments in the subject. This explosion of knowledge often makes it difficult to cope with and normally every university adopts its own pattern and this brings many difficulties for students who seek transfer from one university to other university. So educationalists of A.P. Universities have strived for a common syllabus and recently they even revised the syllabus. This series is written on the lines of this revised common core syllabus.

As I went through the book I found many creative aspects both in the authors write-up and publishers efforts in bringing a good quality book. Firstly the cover is apealing and I liked the idea of putting C_{60} molecule, a new innovation of chemists, on the cover page. The printing and paper are of high quality and clubbed with computer generated organic structures will definitely give a feeling of reading a book of international standard. Coming to the textual matter, all the three branches of chemistry are well written. Inclusion of practical chemistry section in a text book is a novel idea.

This will serve the new generation students, who for want of proper books or for want of sufficient desk time, normally neglect the details and settle for guides.

The authors have taken great pains in bringing entire syllabus of second year in one single text book. This is highly commandable job.

For a student who can devote sufficient time there are ample number of interesting questions followed by answers at the end of every chapter. The self test questions will not only develop the objective knowledge of the subject but also cater to the needs of entrance examination oriented students.

As a good effort needs a real good pat, I congratulate the authors, series editor and the publishers for their sincere effort to bring a very useful text book that caters to the needs of undergraduate students. I am also proud of them because this good academic persuit mainly is from colleagues in the university.

Prof. M. BHAGVANTH RAO
Ph.D., F.I.I.Ch.E.
Registrar
Osmania University
Hyderabad - 500 007.

Preface

The over whelming response from teaching colleagues and students to our first volume of University Chemistry has given impetus in presenting this second volume in a better form. The topics of this volume are those prescribed for the second year B.Sc. course of the Universities of Andhra Pradesh.

The Authors have attempted to write this volume in a style that the students could easily absorb the pulse of the subject and additional information is hoped to give required direction for the students in exploring the subject.

In inorganic section a considerable number of descriptive details with logical consequences of the principles, specially in the chapter of Nuclear Chemistry, are given. Although the topic of alloys appear in the third year B.Sc. syllabus of A.P. Universities, it is put along with metallurgy chapter in view of its easy miscibility.

In organic chemistry sufficient care is taken to provide the mechanitic details of organic reactions involved and list a number of solved and unsolved examples at the end of each chapter. For the first time a novel attempt to explain the organic reactions with the help of potential energy profiles was made.

In physical chemistry section a large number of numerical problems were solved systematically with step by step details to infuse numerical skills in the students. The unsolved problems and self test questions incorporated at the and of each chapter will help the student in understanding the theoretical concepts.

The unique feature of this book could be inclusion of separate section for practical study of second year B.Sc. course. Normally the theoritical background of lab work is given a second fiddle and the students carryout the experiments in a mechanical way. Each semimicro chart is followed by corresponding chemical reactions and similarly gravimetric analytical methods include a brief account of their theory and calculations.

All the care has been taken by the authors and publishers to minimize the printers devils and present a useful book to the students and teachers. The biggest room in the world is the room for improvement. As a critic is always a good friend who helps in furnishing this room, I humbly request the readers to pointout the short comings that no doubt will pave way for further improvement of the book.

I take this opportunity to thank Prof. M. Bhagavantha Rao, Registrar, Osmania University for according permission to take up this work and readily accepting my request and writing the forward. I also thank Prof. D.C.Reddy, Principal, University College of Engineering for his constant encouragement in discharging this academic activity.

It is shocking to know that Mr. Asang Machwe, the M.D. of New Age International (P) Ltd., Publishers and a well balanced man with a vision, has expired at a young age, during the production of this volume. I express my deep sense of gratitude for his detailed discussions that helped in the development of this series.

I will be failing my duty if I don't record the inspiration given by my parents Ms Subba Lakshmi and Dr. Lakshmi Narayana, service rendered by my wife Ms. Sarvani and the omni patience shown by my children Kalyani and Srinivas and other family members in my disgusting and depressing hours.

C. PARAMESWARA MURTHY

Acknowledgements

We have a great pleasure in expressing our sincere thanks to all those teachers and students who readily encouraged the Volume-I of this series and sent their constructive comments.

First, we record our deep sense of gratitude to our family members who were real source of inspiration and without whose cooperation we would not have completed this venture.

We are greatful to Mr. Machwe (Late), Ms. Supriya Bhale Rao, Mr. Govindan, Mr. Siva Kumar and other executives of New Age International for their constant support.

We thank Mr. V.Venkatesh, Mr. G. Srinivas and Mr. M.V. Venugopala Rao for the DTP work and Mr.A. Satyanarayana for his neat drawings.

Finally we express our sincere thanks to our friends Prof. C. Ratnam, Andhra University, Dr. G. Subba Rao, Principal, SGS College, Jaggayyapeta, and all our colleagues, mainly Dr. G. Veerabhadram, Dr. M. Vittal and Dr. Padma Reddy for their thought provoking ideas and criticisms, proof reading and providing question papers and other materials.

C. PARAMESWARA MURTHY

PROMOD KUMAR DUBEY

SYED FAZAL MEHDI ALI

D. ASHOK

COMMON SYLLABUS FOR II YEAR B.Sc., CHEMISTRY

A. INORGANIC CHEMISTRY

1. BONDING IN METALS

General properties of metals, crystal structure of metals. Free electron and valence bond theory. Band theory of solids, metals, semi conductors and insulators. General principles of extraction of metals, metallurgy of nickel, uranium and thorium.

2. *d*-BLOCK ELEMENTS

Electronic configuration and comparative study of elements of first transition series with reference to atomic and ionic radii, ionisation potential, redox potential. Oxidation states, Oxidation state diagrams on the basis of radox potentials, magnetic properties, complex formation, metallic nature, catalytic activity, colour and spectral behaviour with reference to d^1 and d^9.

Trends in physical and chemical properties in passing from the first to the second and to the third series.

3. *f*-BLOCK ELEMENTS

Electronic configuration-General properties, Ionic size, Oxidation states-complexes, magnetic properties, lanthanide contration-comparative study of lanthanous and actinous based on their ionic, covalent radii and exidations states.

4. NUCLEAR CHEMISTRY

Artificial transmutation, nuclear reactions, spellation, nuclear fission and fusion. Radio active isotopes, Tracer Chemistry. Carbon dating, some typical applications in industry, agriculture, medicine and biochemistry, theraptic uses of isotopes.

B. ORGANIC CHEMISTRY

1. ALKYNES

Preparation and properties, mechanism of electrophilic addition, Acidity of acetylenic hydrogens, nucleophilic addition of acetylenes.

2. ARENES

Coal-Source of aromatic compounds, Structure of benzene, concept of aromaticity, mechanism of electrophilic substitution reactions, orientation effects. Alkyl benzenes, naphthalene, Anthracene.

3. HALOGEN COMPOUNDS

Preparation of alkyl, aryl and aralkyl halides. Physical and chemical properties-reactivity of alkyl halides - mechanism and stereochemistry of S_N1 and S_N2 mechanisms. E_1 and E_2 reactions, Saytzeffs elimination, relative reactivity of alkyl and aryl halides - Polyhalogen compounds.

4. HYDROXY COMPOUNDS

Preparation of alcohols and phenols. Physical properties and Hydrogen bonding. Chemical properties, comparison of reactivity of alcohols and phenols. Polyhydroxy compounds; Ethylene glycol and glycerol.

5. ETHERS

Preparation, properties, functional isomerism, cyclic ethers.

6. ALDEHYDES AND KETONES

Preparation, properties, mechanism of nucleophilic addition Aldol condensation, Cannizzaro reaction, Benzoin condensation, Grignard reaction, Perkin reaction, Addition of $NaHSO_3$, HCN, differences in the reactivity of aliphatic and aromatic aldehydes.

7. MONOCORBOXYLIC ACIDS AND THEIR DERIVATIVES

Preparation and properties, Acidic nature of Carboxylic group. Mechanism of esterification and ester hydrolysis. Properties and reactivity of acid chlorides, anhydrides, esters. Acidic nature of Carboxylic group.

8. ORGANIC SYSNTHESIS BASED ON CARBANIONS

Sources and generation of Carbanions, Claisens condensation. Keto-enol tautomerism, synthetic application of acetoacetic ester and diethyl malonate.

9. NITROGEN COMPOUNDS

Nitroalkanes and alkyl nitrites, Nitrobenzene. Amines-Classification, preparation, basic character, reactivity and separation, Aryl diazonium salts and their synthetic applications. Cyanides and isocyanides.

C. PHYSICAL CHEMISTRY

1. SOLIDS

Laws of symmetry- Crystal systems - Internal structure of crystals - space lattice - lattice planes and unit cells. Defects of crystals - Frankel and Schottky's defects - Semi conductors and their usefulness in photo electro - chemical cells.

2. PHASE RULE

Statement and explanation of terms - construction of phase diagrammes - detailed application to water, silver-lead and salt-water systems.

3. COLLOIDS

Distinguishing features - classification - mechanical, electrical and optical properties and uses of colloids - stability, coagulation and protection of colloids - Gold number.

Emulsions and gels - Donnan membrane equilibrium. Origin of charge on colloids - zetapotential.

Adsorption : Types of adsorption - Factors affecting adsorption - Freundlich and Langmuir adsorption isotherms.

4. MACROMOLECULES

Number average and weight average molecular weights - Determination of molecular weight by osmotic pressure method.

D. PRACTICAL CHEMISTRY

Semimicro analysis of a mixture containing two anions and two cations without interference

Anions : Acetate, borate, halides, carbonate, sulphide, sulphite, nitrate, sulphate and phosphate (microinterfering)

Cations of : Ag, Pb, Hg, Bi, Cd, Cu, Sb, Sn, Al, Cr, Fe, Zn, Ni, Ca, Ba, Sr, Mg, K and NH_4.

Contents

Part I - INORGANIC CHEMISTRY

9. TRANSITION ELEMENTS - THE *d*-BLOCK ELEMENTS ***3***

* 9.1 Introduction *3* * 9.2 Electronic configuration *4* * 9-3 Shapes of *d*-Orbitals *6* * 9-4 General group trends of *d*-Block Elements *8* * 9-5 Short and essay type questions *21* * 9-6 Self test questions 23

10. LANTHANIDES AND ACTINIDES *f*-BLOCK ELEMENTS ***25***

* 10-1 Introduction *3* * 10-2 Shapes of *f*-orbitals *26* * 10-3 Lanthanides *26* * 10-4 General Characteristics of Lanthanides *33* * 10-5 Actinides *37* * 10-6 Electronic Configuration of Lanthanides & Actinides *39* * 10-7 Self Test Questions *42* * 10-8 Short and essay type questions *43*

11. METALLURGY; BONDING IN METALS AND ALLOYS ***45***

* 11-1 Introduction *45* * 11-2 Metallurgical Processes *47* * 11-3 Uranium *65* * 11-4 Thorium *69* * 11-5 Nickel *70* * 11-6 Zirconium *73* * 11-7 Chromium *76* * 11-8 Cadmium *78* * 11-9 Metallic Bonding *79* * 11-10 General Properties of Metals *11-11 Crystal Structures of Metals *85* * 11-12 Band Theory of Metals *87* * ALLOYS : 11-13 Introduction *90* * 11-14 Self test questions 98 * 11-15 Short and essay type questions *100*

12. NUCLEAR CHEMISTRY ***102***

* 12-1 Introduction *102* * 12-2 Forces in Nucleus *104* * 12-3 Properities of Nuclides *105* * 12-4 Stability of the Nucleus *107* * 12-5 Mass Defect, Binding Energy and Packing Fraction *115* *12-6 Magic Numbers *120* *12-7 Nuclear Models *122* *12-8 Radioactive Emanations *124* * 12-9 Nuclear Reactions *131* * 12-10 Rate of Disintegration, Half Life Period and Radioactive Equilibrium *133* *12-11 Radioactive Series *137* * 12-12 Nuclear Fission *141* * 12-13 Nuclear Fusion *143* * 12-14 Uses of Radio Isotopes *144* *12-15 Self test questions *14* * 12-16 Numerical Problems *150* * 12-17 Short and essay type questions *151*

Part II - ORGANIC CHEMISTRY

6. ALKYNES *157*

* 6-1 Introduction *157* * 6-2 Electronic Structure of Alkynes *157* * 6-3 Nomenclature *158* * 6-4 Isomerism *159* * 6-5 Methods of Preparation *159* * 6-6 Physical Properties *163* * 6-7 Chemical Properties *163* *6-8 Industrial Significance of Acetylene *172* * 6-9 Self test questions *174* * 6-10 Short and essay type questions *174*

7. ARENES *177*

* 7-1 Introduction *177* * 7-2 Sources of Aromatic Compounds *177* * 7-3 Physical Properties of Benzene *180* * 7-4 Structure of Benzene *180* * 7-5 Huckel's (4n+2) π *189* * 7-6 Nomenclature of Benzene Derivatives *191* * 7-7 Methods of Preparation of Benzene *192* * 7-8 Uses of Benzene *193* * 7-9 Chemical reactions of Benzene *193* * 7-10 Alkylbenzenes *213* * 7-11 Biphenyls *221* * 7-12 Fused Ring Systems of Benzene, Napthalene *224* * 7-13 Anthracence and Phenanthrene *231* * 7-14 Self test questions *235* * 7-15 Short and essay type questions *236*

8. HALOGEN COMPOUNDS *244*

* 8.1 Introduction *244* * 8.2 Nomenclature *244* * 8.3 Classification *245* * 8.4 Alkyl Halides *246* * 8.5 Aryl Halides *262* * 8.6 Self test questions *268* * 8.7 Short and essay type questions *269*

9. HYDROXY COMPOUNDS *271*

* 9.1 Introduction *271* * 9.2 Classification *272* * 9.3 Nomenclature *272* * 9.4 Preparation of Alcohols *273* * 9.5 Physical and Chemical Properties *287* * 9.6 Industrial Significance of Hydroxy compounds *316* * 9.7 Self-test questions *319* * 9.8 Short and essay type questions *320*

10. ETHERS *324*

* 10.1 Definition *324* * 10.2 Nomenclature *324* * 10.3 Structue of Ethers *324* * 10.4 Functional Isomerism *325* * 10.5 Preparation of Ethers *325* * 10.6 Physical Properties of Ethers *327* * 10.7 Chemical Properties *327* * 10.8 Uses of Ethers *328* * 10.9 Self test questions *329* * 10.10 Short and essay type questions *330*

11 ALDEHYDES AND KETONES *334*

* 11.1 Introduction *334* * 11.2 Clasification *334* * 11.3 Nomenclature *335*

* 11.4 Isomerism in Carbonyl Compounds *335* * 11.5 Strcture of Carbonyl Compounds *336* * 11.6 General Methods of Preparation of Aldehydes and Ketones *337* * 11.7 Preparation of Aromatic Aldehydes and Ketones *342* * 11.8 Physical Properties of Aldehydes and Ketones *345* * 11.9 Chemical Properties of Aldehydes and Ketones *346* * 11.10 Self test questions *368* * 11.11 Short and essay type questions *369*

12 MONOCARBOXYLIC ACIDS AND THEIR DERIVATIVES ***375***

* 12.1 Introduction *375* * 12.2 Nomenclature *376* * 12.3 Structure *378* * 12.4 Preparation of Carboxylic Acids *378* * 12.5 Physical Properties *382* * 12.6 Chemical Properties *386* * 12.7 Self test questions *394* * 12.8 Essay and short type questions *395*

13. CARBANIONS ***397***

* 13.1 Introduction *397* * 13.2 Structure *397* * 13.3 Formation and Stability *398* * 13.4 Claisen Condensation *401* * 13.5 Acylated Esters *404* * 13.6 Acetoacetic ester synthesis *405* * 13.7 Problems and exercises *415* * 13.8 Short and essay type question *416* * 13.9 solutions to problems *418*

14. NITROGEN COMPOUNDS ***420***

* 14.1 Alkyl Nitrites *420* * 14.2 Nitroalkanes *421* * 14.3 Nitrobenzenes *428* * 14.4 Amines *432* * 14.5 Aromatic Diazonium Salts *451* * 14.6 Cyanides and Isocyanides *455* * 14.7 Self test questions *459* * 14.8 Essay and short type question *461*

Part III - PHYSICAL CHEMISTRY

5. THE SOLID STATE ***569***

* 5.1 Introduction *569* * 5.2 Classfication of Solids *470* * 5.3 Characteristic Geometry of Crystalline Solid *470* * 5.4 Laws of Crystallography *473* * 5.5 Crystal Systems *478* * 5.6 Difraction of X-Rays *482* * 5.7 Crystal Types and Properties *487* * 5.8 Defects in Crystals *493* * 5.9 Metallic Bonding *497* * 5.10 Photovoltaic Cell *502* * 5.11 Solved Problems *504* * 5.12 Short and essay type questions *506*

6. PHASE RULE ***508***

* 6.1 Homogeneity and Inhomogeneity of Systems *508* * 6.2 Phase Rule *508* * 6.3 Phase Diagram *513* * 6.4 One Component System *513* * 6.5 Two Component System *517* * 6.6 short and essay type questions *525*

7. COLLOIDS AND ADSORPTION ***527***

COLLOIDS : * 7.1 Introduction *527* * 7.2 Types of Colloidal Dispersion

529 * 7.3 Classification of Collioids 530 * 7.4 Methods of Preparation of Colloidal Solutions 532 * 7.5 Purification of Colloidal Dispersion 535 * 7.6 Properties of Colloids 536 * 7.7 Emulsion 548 * 7.8 Gels 550 * 7.9 Applications of Colloid Chemistry 551 * ADSORPTION : 7.10 Introduction 553 * 7.11 Types of Adsorption 555 * 7.12 Adsorption Isotherms 557 * 7.13 Applications of Adsorption 562 * 7.14 Worked out Numerical Problems 563 * 7.15 Short and Essay Type Questions 564

8. MACROMOLECULES **567**

* 8.1 Introduction 567 * 8.2 Classification 570 * 8.3 Physical Properties of Macfromolecules 577 * 8.4 Determination of Molecular Weight 582 * 8.5 Numerical Problemes 586 * 8.6 Self test questions 590 * 8.7 Short and essay type questions 591

Part IV - PRACTICAL CHEMISTRY

1. SEMI-MICRO ANALYSIS **595**

* 1.1 Introduction 595 * 1.2 Spot Tests 596 * 1.3 Semimicro Analysis 597 * 1.4 Analysis of Acid Radicals 598 * 1.5 Analytical Classification of the Basic Radical 602 * 1.6 Ammonium Radical test 603 * 1.7 Analysis of Basic Radicals 603 * 1.8 Important Points to be Remembered 610

2. GRAVIMETRIC ANALYSIS **611**

* 2.1 A Brief Theory of Gravimetric Analysis 611 * 2.2 Estimation of Barium as (a) Chromate (b) Sulphate 612 * 2.3 Estimation of silver or Chloride 614 * 2.4 Estimation of Iron as it's Oxide 615 * 2.5 Estimation of Nickel as Nickel Dimethyl Glyoximate 616 * 2.6 Estimation of Aluminium as Aluminium Oxinate 617

APPENDIX **618**

* A) Names, symbols and definitions of certain SI derived units 618 * B) General physical constants and their values 618 * C) Data of the strength of Aqueous solutions of the common acids and of Aqueous Ammonia 619 * D) Dissociation constants of organic acids and bases 619

QUESTION PAPERS OF PREVIOUS YEARS **620**

Part I

Inorganic Chemistry

9

TRANSITION ELEMENTS - THE *d*-BLOCK ELEMENTS

9-1 INTRODUCTON

The *d*-block is placed in the periodic table between the *s*- and *p*- blocks, thus forming a transition between these two sets of representative elements. Hence these *d*-block elements acquired the name transition elements. A transition element is defined *as that element in which an electron is filled in the d-orbital and all the valence orbitals are filled to their maximum capacity.* A transition series is that in which the *d*-electron starts and completes its filling. There are four transition series in the periodic table each comprising of ten elements. The fourth series is an incomplete series. The first transition series comprises of elements from scandium (21) to zinc (30), the second series from yttrium (39) to cadmium (48) and the third series from lanthanum (57) to mercury (80) excluding fourteen elements from cerium (58) to lutetium (71), referred to as lanthanides and belong to the inner transition series of the *f*-block of the periodic table. The fourth transition series is an incomplete series containing actinium (89) along with the newly discovered kurchtorium/rutherfordium/unnilquadium, (104th element with different names as given by the Russains/Americans/IUPAC respectively) and a host of new as well as predicted elements till the atomic number 112.

According to the IUPAC regulations the new elements that follow the 103 element are named as follows:

1. All the elements will end in the letters-ium.
2. The name will be constructed from the following numerical roots:

   0 = nil; 1 = un; 2 = bi; 3 = tri; 4 = quad; 5 = pent; 6 = hex; 7 = sept; 8 = oct; 9 = enn

3. The symbol will consist of three letters, instead of one or two, derived from the numerical roots above.

For example, 106th element is

1	0	6
Un	nil	hexium

Unnilhexium(Unh)

Hence according to IUPAC nomenclature the elements from 104 to 109 are named as follows:

Atomic number	Nomenclature	Symbol	Expected Isotopic weight
104	Unnilpentium	Unp	261.11
105	Unnilquadium	Unq	262.114
106	Unnilhexium	Unh	263.118
107	Unnilseptiium	Uns	262.12
108	Unniloctium	Uno	
109	Unnilennium	Une	

Although the properties of these elements are not yet known, but they ought to be the *d*-block elements possesing the characteristic properties of the transition elements. The isotopic weights of 108,109 elements are yet to be determined.

Between the elements actinium (89) and unnilquadium (104) lies the second inner transition series - the actinides, belonging to the f-block of the periodic table.

9-2 ELECTRONIC CONFIGURATION

The general electronic configuration of the transition metals is $ns^2(n\text{-}1)d^{1\text{-}10}$ for the first and the second series. In the third series between La and Hf lies the lanthanide series, and the fourth series between Ac and Unq lies the actinide series, in these series the $4f$ and $5f$ orbitals are filled, hence the general electronic configuration becomes $ns^2(\text{n-1})d^{1\text{-}10}\ (n\text{-}2)f^{14}$.

A number of elements of the *d*-block like Cr, Cu, Mo, Ag, etc., show deviations from the standard rules, of electronic configuration , by promoting the 4s or 5s electrons to 3d or 4d orbitals respectively for attributing either a d^5 or a d^{10} configuration. The electronic configuration of the metals is given in the following sub-heads.

(A) First transition series (3*d* series)

Scandium	(Sc-21)	$1s^2\ 2s^2\ 2p^6\ 3s^2\ 3p^6\ 4s^2\ 3d^1$
Titanium	(Ti-22)	$1s^2\ 2s^2\ 2p^6\ 3s^2\ 3p^6\ 4s^2\ 3d^2$
Vanadium	(V-23)	$1s^2\ 2s^2\ 2p^6\ 3s^2\ 3p^6\ 4s^2\ 3d^3$

Chromium (Cr-24) $1s^2$ $2s^2$ $2p^6$ $3s^2$ $3p^6$ $4s^2$ $3d^4/4s^1$ $3d^5$
Manganese (Mn-25) $1s^2$ $2s^2$ $2p^6$ $3s^2$ $3p^6$ $4s^2$ $3d^5$
Ferrum (Fe-26) $1s^2$ $2s^2$ $2p^6$ $3s^2$ $3p^6$ $4s^2$ $3d^6$
Cobalt (Co-27) $1s^2$ $2s^2$ $2p^6$ $3s^2$ $3p^6$ $4s^2$ $3d^7$
Nickel (Ni-28) $1s$ $2s^2$ $2p^6$ $3s^2$ $3p^6$ $4s^2$ $3d^8$
Copper (Cu-29) $1s$ $2s^2$ $2p^6$ $3s^2$ $3p^6$ $4s^2$ $3d^9/4s^1$ $3d^{10}$
Zinc (Zn-30) $1s^2$ $2s^2$ $2p^6$ $3s^2$ $3p^6$ $4s^2$ $3d^{10}$

(B) Second transition series (4*d* series)

Yttrium (Y-39) $1s^2$ $2s^2$ $2p^6$ $3s^2$ $3p^6$ $4s^2$ $3d^{10}$ $4p^6$ $5s^2$ $4d^1$
Zirconium (Zr-40) $1s^2$ $2s^2$ $2p^6$ $3s^2$ $3p^6$ $4s^2$ $3d^{10}$ $4p^6$ $5s^2$ $4d^2$
Niobium (Nb-41) $1s^2$ $2s^2$ $2p^6$ $3s^2$ $3p^6$ $4s^2$ $3d^{10}$ $4p^6$ $5s^2$ $4d^3$
Molybdenum (Mo-42) $1s^2$ $2s^2$ $2p^6$ $3s^2$ $3p^6$ $4s^2$ $3d^{10}$ $4p^6$ $5s^2$ $4d^4/5s^1$ $4d^5$
Technetium (Tc-43) $1s^2$ $2s^2$ $2p^6$ $3s^2$ $3p^6$ $4s^2$ $3d^{10}$ $4p^6$ $5s^2$ $4d^5$
Ruthenium (Ru-44) $1s^2$ $2s^2$ $2p^6$ $3s^2$ $3p^6$ $4s^2$ $3d^{10}$ $4p^6$ $5s^2$ $4d^6$
Rhodium (Rh-45) $1s^2$ $2s^2$ $2p^6$ $3s^2$ $3p^6$ $4s^2$ $3d^{10}$ $4p^6$ $5s^2$ $4d^7$
Palladium (Pd-46) $1s^2$ $2s^2$ $2p^6$ $3s^2$ $3p^6$ $4s^2$ $3d^{10}$ $4p^6$ $5s^2$ $4d^8$
Silver (Ag-47) $1s^2$ $2s^2$ $2p^6$ $3s^2$ $3p^6$ $4s^2$ $3d^{10}$ $4p^6$ $5s^2$ $4d^9/5s^1$ $4d^{10}$
Cadmium (Cd-48) $1s^2$ $2s^2$ $2p^6$ $3s^2$ $3p^6$ $4s^2$ $3d^{10}$ $4p^6$ $5s^2$ $4d^{10}$

(C) Third transition Series (5*d* series)

Lathanum (La-57) $1s^2$ $2s^2$ $2p^6$ $3s^2$ $3p^6$ $4s^2$ $3d^{10}$ $4p^6$ $5s^2$ $4d^{10}$ $5p^6$ $6s^2$ $5d^1$
Hafnium (Hf-72) $1s^2$ $2s^2$ $2p^6$ $3s^2$ $3p^6$ $4s^2$ $3d^{10}$ $4p^6$ $5s^2$ $4d^{10}$ $5p^6$ $6s^2$ $4f^{14}$ $5d^2$
Tantalum (Ta-73) $1s^2$ $2s^2$ $2p^6$ $3s^2$ $3p^6$ $4s^2$ $3d^{10}$ $4p^6$ $5s^2$ $4d^{10}$ $5p^6$ $6s^2$ $4f^{14}$ $5d^3$
Tungsten (W-74) $1s^2$ $2s^2$ $2p^6$ $3s^2$ $3p^6$ $4s^2$ $3d^{10}$ $4p^6$ $5s^2$ $4d^{10}$ $5p^6$ $6s^2$ $4f^{14}$ $5d^4$
Rhenium (Re-75) $1s^2$ $2s^2$ $2p^6$ $3s^2$ $3p^6$ $4s^2$ $3d^{10}$ $4p^6$ $5s^2$ $4d^{10}$ $5p^6$ $6s^2$ $4f^{14}$ $5d^5$
Osmium (Os-76) $1s^2$ $2s^2$ $2p^6$ $3s^2$ $3p^6$ $4s^2$ $3d^{10}$ $4p^6$ $5s^2$ $4d^{10}$ $5p^6$ $6s^2$ $4f^{14}$ $5d^6$
Iridium (Ir-77) $1s^2$ $2s^2$ $2p^6$ $3s^2$ $3p^6$ $4s^2$ $3d^{10}$ $4p^6$ $5s^2$ $4d^{10}$ $5p^6$ $6s^2$ $4f^{14}$ $5d^7$
Platinum (Pt-78) $1s^2$ $2s^2$ $2p^6$ $3s^2$ $3p^6$ $4s^2$ $3d^{10}$ $4p^6$ $5s^2$ $4d^{10}$ $5p^6$ $6s^2$ $4f^{14}$ $5d^8$
Gold (Au-79) $1s^2$ $2s^2$ $2p^6$ $3s^2$ $3p^6$ $4s^2$ $3d^{10}$ $4p^6$ $5s^2$ $4d^{10}$ $5p^6$ $6s^2$ $4f^{14}$ $5d^9$
Mercury (Hg-80) $1s^2$ $2s^2$ $2p^6$ $3s^2$ $3p^6$ $4s^2$ $3d^{10}$ $4p^6$ $5s^2$ $4d^{10}$ $5p^6$ $6s^2$ $4f^{14}$ $5d^{10}$

The Common features exhibited due to electronic configuration of the transition metals is that:

(i) the incoming electron is added to the penultimate shell, hence all the *d*-electrons may or may not participate in bonding to an extent of determining the observed variable oxidation states;

(ii) the electron distribution in the *d*-orbitals of transitional elements is affected by the proximity ligand electrons;

(iii) the electrons tend to occupy vacant *d*-orbitals singly, when available, because it takes extra pairing energy to enter into half filled orbital for a joint occupancy with another electron; and

(iv) if the *d*-orbitals are unequal in energy, an electron may find the next vacant orbital at a higher energy over half filled orbital, in which case it prefers to pair.

Properties specially affected by the electron distribution among penultimate *d*-orbitals are absorption spectra and magnetic susceptibility (see section 19-4-5).

When the five *d*-orbitals contain zero, five or ten electrons, the stabilization energy is zero because the less stable pairing electrons cancel the advantage of the more stable half filled electrons. Intermediate number of electrons, however, represent more stabilization energy. Consequently compounds having *d*-electrons other than zero, five or ten are more stable than would otherwise be expected, but by an amount often difficult to determine for solids is determined by the absorption spectra.

Another additional feature which may be due to the electronic distribution and still remains unexplained is that, the d-block elements cannot form stable sigma (σ) bonds with carbon or hydrogen as the other metals of the representative blocks do. The other most important property being the electronic configuration of Cr, Cu, Mo and Ag. The electron from the 4*s* or 5*s* gets promoted to the 3*d* or 4*d* orbitals respectively to provide stability to the *d*-orbital. Cr and Mo attain a half filled orbital stability and Cu & Ag attain a completely filled stability. Instead the electrons in, Zn, Cd and Hg thus gets filled in *s*-orbital. Therefore these elements are not true transition metals and are therefore referred to as **pseudo transition metals.**

9-3 SHAPES OF *d*-ORBITALS

For each principal quantum number there are *s*-orbitals, which are spherically symmetrical. For each prinicpal quantum number greater than one there are also three degenerate *p*-orbitals directed mutually at right angles to each other along the *x,y* and *z* axis, but similar in shape, Fig.9-1.

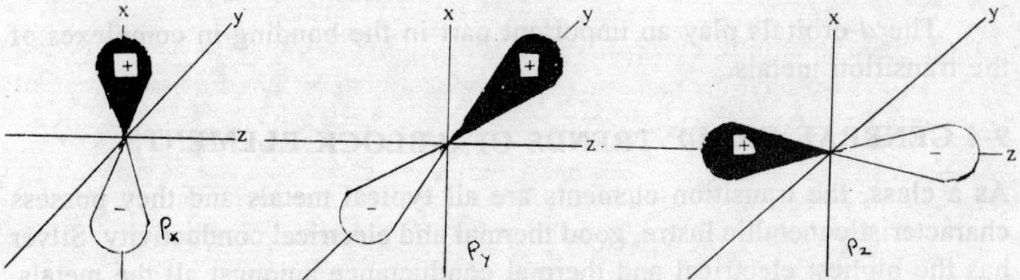

Fig. 9-1 Shapes of p_x, p_y and p_z orbitals

For each principal quantum number greater than 2 there are also five degenerate *d*-orbitals, which are not, however, have similar shape. In a set of three, the lobes of the orbitals, figure 9-2, lie along lines making angles of 45° to the *x*-, *y*- and *z*- axes, i.e., the lobes fall between the axes.

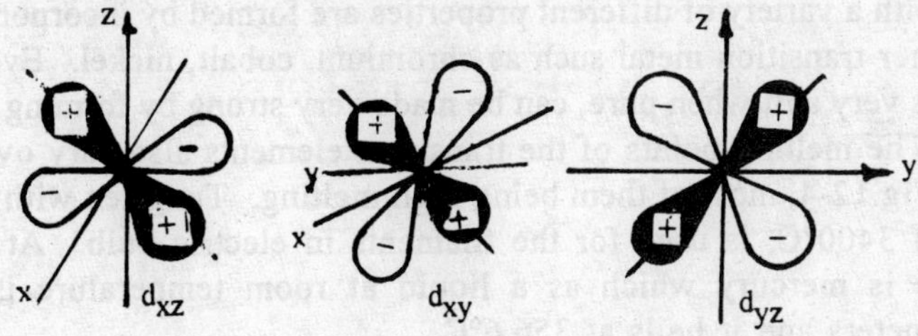

Fig. 9-2 Shapes of t_{2g} orbtals (d_{xy}, d_{yz}, d_{zx})

These three orbitals are called d_{xy}, d_{yz} and d_{zx} according to the plane in which the orbitals lie. This set of three orbitals is known collectively as the t_{2g} or dε orbitals.

In the second set of two orbitals the lobes lie either along the *x*-and *y*-axes or along the *z*-axis, Fig. 9-3, the first of these orbitals is referred to as the $d_{x^2-y^2}$ orbital and second as the d_{z^2} orbital.

The two orbitals together are known as e_g or dγ orbitals.

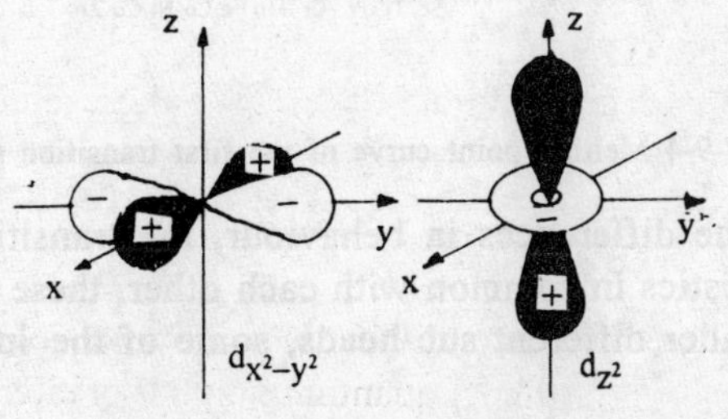

Fig. 9-3 Shapes of the e_g-orbitals (d_x, d_y and d_z)

The *d*-orbitals play an important part in the bonding in complexes of the transition metals.

9-4 GENERAL GROUP TRENDS OF *d*-BLOCK ELEMENTS

As a class, the transition elements are all typical metals and they possess characteristic metallic lustre, good thermal and electrical conductivity. Silver has the highest electrical and thermal conductance amongst all the metals, followed closely by copper and aluminium. Owing to atmospheric stability and much lower cost in comparison to silver, copper is used widely in electrical appliances and wiring. Silver is preferred for coating of mirrors because of its high reflectivity. The chemical and physical properties of the transition elements cover a wide range and account for the larger variety of uses. Some of these metals like iron are very hard and strong and are used as structural metals, either in the pure state or as alloys. Steel, an iron alloy, with a variety of different properties are formed by incorporating iron with other transition metal such as chromium, cobalt, nickel. Even copper which is very soft when pure, can be made very strong by forming a suitable alloy. The melting points of the transition elements also vary over a wide range, Fig.12-4, most of them being high melting. Tungsten with a melting point of 3400°C, is used for the filaments in electric bulb. At the other extreme is mercury which as a liquid at room temperature is used in thermometers and it boils at 356.6°C.

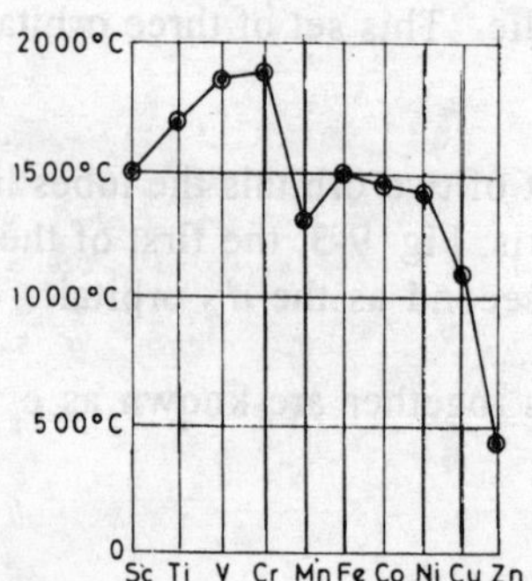

Fig. 9-4 Melting point curve of the first transition series

Despite some differences in behaviour, the transition elements have several characteristics in common with each other, these characteristics are dealt in detail under different sub-heads, some of the important ones are:

1. Most of these metals have multiple oxidation states, with only a few exceptions (section 9-4-1).

2. Many of the compounds of *d*-block elements are paramagnetic. Because the transition elements tend to have partially completed d-subshells in both the free state and in their compounds, the metal atoms and ions often possess unpaired electrons. This imparts the property of paramagnetism (section 9-4-5).

3. Many (if not most) of their compounds are coloured. The origin of the colours of complex ions of the transition elements is due to d-d transitions (section 9-4-4).

4. They have a strong tendency to form complexes. As a group, these elements form a large number of complex ions of varying degrees of complexities (section 9-4-7).

9-4-1 Variable Oxidation States

The transition elements commonly exhibit more than one oxidation state. In general, these states are formed because electrons are lost not only from the outer *s*-subshell but from the underlying *d*-subshell as well. The oxidation states of the transition elements of all the three series are given in Table 9-1.

Table 9-1 : Oxidation states of transition elements

[A] Oxidation states of first transition series:

IIIA	IVA	VA	VIA	VIIA		VIII		IB	IIB
Sc	Ti	V	Cr	Mn	Fe	Co	Ni	Cu	Zn
		+1						+1	
	+2	+2	+2	**+2**	**+2**	**+2**	**+2**	**+2**	**+2**
+3	+3	+3	**+3**	+3	**+3**	+3	+3	+3	+3
	+4	**+4**	+4	+4	+4	+4	+4		
		+5	+5	+5	+5	+5			
			+6	+6	+6				
				+7					

[B] Oxidation states of second transition series:

Y	Zr	Nb	Mo	Tc	Ru	Rh	Pd	Ag	Cd
						+1		**+1**	
	+2	+2	+2	+2	+2	+2	**+2**	+2	**+2**
+3	+3	+3	+3	+3	+3	+3	+3	+3	
	+4	+4	**+4**	**+4**	**+4**	**+4**	**+4**		
		+5	+5	+5	+5	+5			
			+6	+6	+6	+6			
				+7					
					+8				

[C] Oxidation states of third transition series :

La	Hf	Ta	W	Re	Os	Ir	Pt	Au	Hg
						+1		+1	+1
		+2	+2		+2	+2	+2		+2
+3	+3	+3	+3	**+3**	+3	**+3**	+3	**+3**	
	+4	+4	+4	**+4**	**+4**	**+4**	**+4**		
		+5	**+5**	+5	+5	+5	+5		
			+6	+6	+6	+6	+6		
				+7					
					+8				

The stable oxidation states are in ***bold*** *print.*

With regards to the trend in the stabilities of oxidation states it is found that;

(i) ***Elements at the left in a transition series prefer higher oxidation state. As we move to the right, the lower oxidation states become increasingly more stable relative to the higher ones, moreover as, expected the highest oxidation state is equal to the group number;***

(ii) ***As we go down a group among the d-block elements the higher oxidation states become progressively more stable relative to the lower ones.***

These trends are helpful in making comparisons of the strengths of oxidizing agents. A transition series shows an increasing tendency towards a stable +2 oxidation state and very few higher oxidation states, let us compare the relative stabilities of Fe^{2+} and Fe^{3+} with Ni^{2+} and Ni^{3+}. The horizontal trends suggests that Ni^{3+} has a greater tendency to become Ni^{2+} than the tendency of Fe^{3+} to become Fe^{2+}. In other words Ni^{3+} has a greater tendency to gain an electron than Fe^{3+}. Since these species are functioning as oxidizing agents when they gain electrons, we conclude that Ni^{3+} is a stronger oxidizing agent than Fe^{3+} On comparing vertically the permanganate (MnO_4^-) ion with perrhenate (ReO_4^-) ion, becuase Mn is vertically above Re in the d-block, the higher oxidation states become more stable as we go down, thus making ReO_4^- a weaker oxidizing agent than MnO_4 .

Similarly, we can also compare any element of the transition series with any other element of the same series. Let us examine the oxidizing capabilities of titanium dioxide (TiO_2) with managnese dioxide (MnO_2)belonging to the first series. In this series titanium is placed left of manganese. Since the higher oxidation states loose their stability as we move from left to right, Mn(IV) should be relativley less stable than Ti(IV) towards reduction. Thus Mn(IV) should have a greater tendency to get reduced than Ti(IV), thereby MnO_2 should be a stronger oxidizing agent than TiO_2. TiO_2 is a stable white pigment used in paints while MnO_2 is a commonly employed as an oxidizing agent.

9-4-2 Atomic & Ionic Radii

In the transition elements, the change in size from one element to the next is not as large as observed in the representative elements. This is because the electrons added to the atoms enter the penultimate *d*-orbital without disturbing the outermost shell which effectively shield or insulate the outer shell from the inner nuclear charge.

So only a small increase in effective nuclear charge is felt by the outer electrons with the increase in atomic number, resulting in a small decrease in size. The atomic radii of all the transition elements is graphically represented in Fig.9-5. It is observed that there is a minimum change in

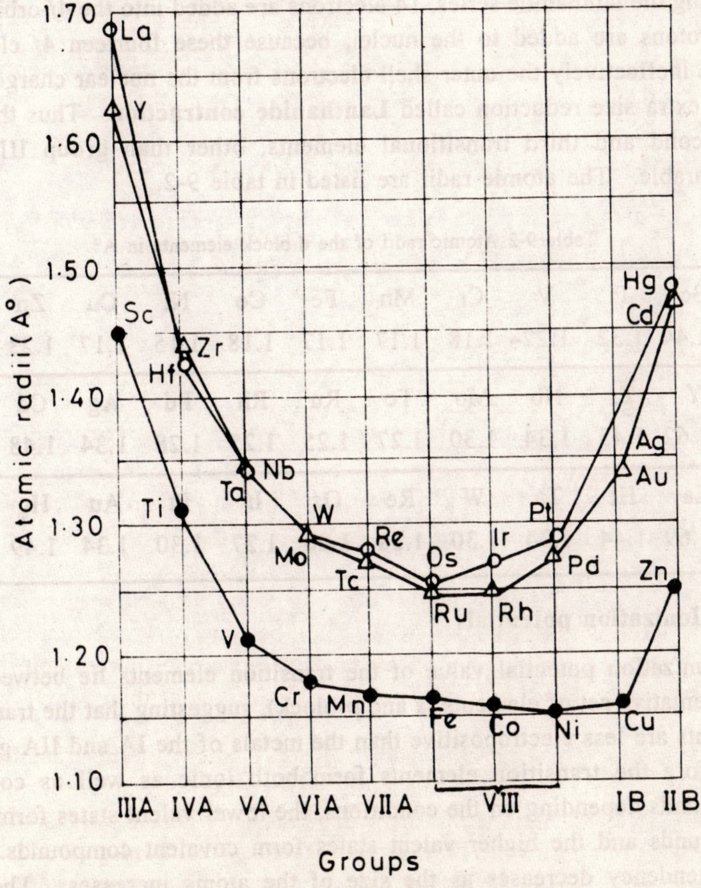

Fig. 9-5 Size variations among the transtion elements

atomic radii that occurs near the middle of each transition series. The addition of first five electrons pose no problem of electronic repulsion as they are placed in empty orbitals. In addition their shielding of outer most atomic orbitals help in shrinking of the atomic size. But once *d*-subshell becomes half filled, addition of electrons force the *d*-orbitals to expand in size due to an increase in the inter electron repulsions.

From Fig. 9-5 it is seen, for the elements in group IIIA viz., Sc, Y and La, there is a gradual increase in the size as expected to be in a group. However, for the elements in the second transition series, although there is substantial increase in atomic size when compared to the first series, virtually we observe no change in size on going from the second series to the third. This can be understood by analyzing the role played by intervening *f*-block electrons/elements in the third series. In this series the elements between La(57) and Hf(71), called lanthanides, cause an extra shrinking in size. In crossing the lanthanide series, 14 electrons are added into the 4*f*-orbital and 14 protons are added to the nuclei, because these fourteen 4*f* electrons shield ineffectively the outer shell electrons from the nuclear charge, there is an extra size reduction called **Lanthanide contraction**. Thus the radii of second and third transitional elements, other than group IIIA, are comparable. The atomic radii are listed in table 9-2.

Table 9-2 Atomic radii of the d-block elements in A°

Sc	Ti	V	Cr	Mn	Fe	Co	Ni	Cu	Zn
1.44	1.32	1.22	1.18	1.17	1.17	1.18	1.15	1.17	1.23
Y	**Zr**	**Nb**	**Mo**	**Tc**	**Ru**	**Rh**	**Pd**	**Ag**	**Cd**
1.62	1.45	1.34	1.30	1.27	1.25	1.25	1.28	1.34	1.48
La	**Hf**	**Ta**	**W**	**Re**	**Os**	**Ir**	**Pt**	**Au**	**Hg**
1.69	1.44	1.34	1.30	1.28	1.26	1.27	1.30	1.34	1.49

9-4-3 Ionization potentials

The ionization potential value of the transition elements lie between the representative set of elements (*s* and *p*-block), suggesting that the transition elements are less electropositive than the metals of the IA and IIA groups. Therefore the transition elements form both ionic as well as covalent compounds depending on the conditions, the lower valent states form ionic compounds and the higher valent states form covalent compounds. The ionic tendency decreases as the size of the atoms increases. The first ionization potential of all the elements lie between 5.5 to 10.5 eV. The first ionization potential do not differ greatly in the first and second transition series families of transition elements. The ionization potentials are given in table 9-3.

Table 9-3 Ionization potential values in electron volls (eV) of the three transition series

Sc	Ti	V	Cr	Mn	Fe	Co	Ni	Cu	Zn
6.5	6.9	6.8	6.8	7.4	7.9	7.9	7.6	7.7	9.4
Y	Zr	Nb	Mo	Tc	Ru	Rh	Pd	Ag	Cd
6.6	6.9	6.8	7.2	7.3	7.5	7.7	8.3	7.6	9.0
La	Hf	Ta	W	Re	Os	Ir	Pt	Au	Hg
5.6	5.5	6.0	8.0	7.9	8.7	9.2	9.0	9.2	10.5

9-4-4 Colour

Transition metal ions and their compounds containing unpaired d-electrons are coloured both in solid as well as in liquid states. The colour of transition metal ion is associated with:

(i) An incomplete *d*-level (between 1 and 9 *d*-electrons)

(ii) The nature of the ligands (NH_3, H_2O, Cl^-, CN^- etc.) sorrounding the metal ion.

It is suggested that the property of colour of these metal ions is due to the excitation of electrons from one *d*-level to another, which is generally referred to as *d-d* transition. The *d*-orbitals in an atom are closely spaced energetically. This small energy differences allow the excitation of electrons present in one *d*-orbital to another *d*-orbital by mere absorption of visible light (an electromagnetic radiation) corresponding to the energy gap. This removal of a characteristic colour component from the visible light leaves the other colours unattended. Thus $CuSO_4$, which selectively absorbs red colour leaves blue colour unabsorbed, thus imparting blue colour to the crystals and its solution.

The colour of a particular transition metal ion depends upon the nature of the ligands bonded to the metal ion. The colourless anhydrous $CuSO_4$, turns pale blue on addition of even a slight trace of moisture, and changes to dark blue in the presence of ammonia. The colors are explained by the strength of coordination linkages with various ligands, viz., H_2O, NH_3, and Cl^-. Hence ligands play an importanct role in imparting colour to the metal ions as seen below.

$$[Cu(CN)_4]^{2-}$$

Prussian Blue

↑HCN

$$\underset{\textit{Dark blue}}{\{Cu(NH_3)_4)]^{2+}} \xleftarrow{NH_3} \underset{\textit{(Colorless)}}{CuSO_4} \xrightarrow{H_2O} \underset{\textit{Blue}}{CuSO_4 6H_2O}$$

↓HCl

$$[CuCl_4]^{2-}$$

Green

Another concept is, since the five *d*-orbitals are arranged differently in the space, the electron/electrons which are close to a ligand will be repelled raising the energy of such orbital, when compared to the others. For example in the case of Co^{2+}, Fig. 9-6, the splitting of the *d*-orbitals destroys the degeneracy of the 3*d*-orbital, and the distribution of electrons is different when it combines with CN^-, compared to the distribution when it combines with neutral H_2O molecule, Cyanide ion being a strong field ligand the t_{2g} level are filled completely first and then the lone seventh electron goes in the e_g level, but in the case of H_2O, which being a weak field ligand

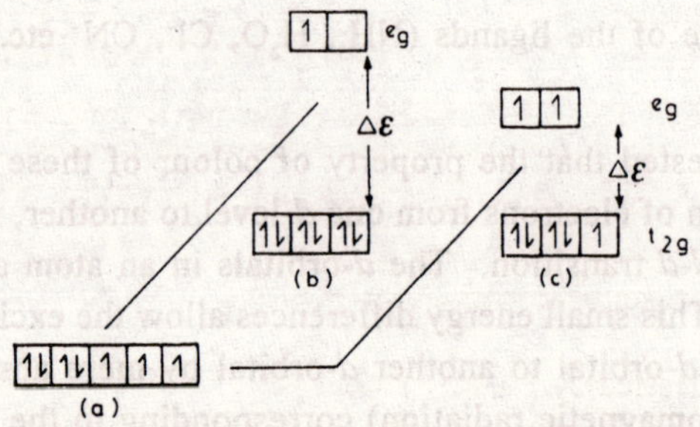

Fig.9-6(a). Degenerated *d*-orbital of Co^{2+}; (b) The splitted *d*-orbital of Co^{2+} ion in $[Co(CN)_6]^{4-}$; (C) The splitted *d*-orbital of Co^{2+} ion in $[Co(H_2O)_6]^{2+}$.

shows the filling to be t_{2g}^5 and e_g^2. Therefore the *d-d* transition and the corresponding energy gap $\Delta\varepsilon$ will actually depend on the nature of ligand and correspond to differrent frequencies of radiation in the visible spectrum, thus imparting different colours to these compounds.

The same metal in various oxidation states exhibit different colours because of similar reasons mentioned above. The different colours of some of the metal ions in different oxidation states are given in table 9-4.

Table 9-4 Transition metal ions show different colours in different oxidation states

	$V^{II}O$ *Lavender*	$Cr^{II}Cl_2$ *Yellow*	$Mn^{II}O$ *Pink*	$Fe^{II}(aq)$ *Green*	$Co^{II}(aq)$ *Pink*	$Ni^{II}(aq)$ *Green*	$Cu^{II}(aq)$ *Blue*
$Ti^{III}(aq)$ *Purple*	$V^{III}(aq)$ *Green*	$Cr^{III}(aq)$ *Violet*	$Mn_2{}^{III}O_3$ *Red*	$Fe^{III}(aq)$ *Yellow*	$Co^{III}(aq)$ *Dark blue*		
	$V^{IV}O_2$ *Blue*		$Mn^{IV}O_2$ *Grey*				
	$V_2{}^{V}O_5$ *Orange*						
		$Cr^{VI}O_4{}^{2-}$ *Yellow*	$Mn^{VI}O_4{}^{2-}$ *Dark green*				
		$Cr_2{}^{VI}O_7{}^{2-}$ *Orange*					
			$Mn^{VII}O_4{}^{-}$ *Violet*				

9-4-5 Magnetic properties

The magnetic properties of atoms and molecules are dependent on the total spins of their electrons. As electrons are charged particles, the unpaired spinning electrons present in an atom produce a magnetic field, and an atom behaves like a tiny bar magnet. They effectively interfere with the applied magnetic field and orient along the magnetic field. This phenomenon of attraction between the metal atom or ion and the applied magnetic field, owing to the presence of unpaired electrons is known as **paramagnetism.**

Characteristically, many transition elements and their compounds possess partially filled *d*-subshell, therefore, many of them exhibit the phenomenon of paramagnetism. The prediction of magnetic properties of transition metal compounds is therefore dependent on the theories of bonding.

Related to the property of paramagnetism is the phenomenon called **ferromagnetism,** observed for the three pure elements-iron, cobalt and nickel. Ferromagnetic materials like paramagnetic ones, are also attracted to a magnetic field, however, the magnitude of the interaction for a ferromagnetic substance is approximately a million times stronger than it is with other paramagnetic materials. Before we proceed to see how does the transition metals behave and what are their paramagnetic values let us see how ferromagnetism can be illustrated. The origin of ferromagnetism is the same as paramagnetism that is the existence of unpaired electrons in the ferromagnetic atoms with their atomic magnets all lined up in the same direction as shown in Fig 9-7. Or binarily these domains are randomly oriented in a ferromagnetic solid, so evenly, that, each domain behaves as relatively a large magnet by itself and hence combined effect gets cancelled. When the ferromagnetic material is placed in a magnetic field, the domains tend to become aligned much the same as the atomic magnets of a paramagnetic substance.

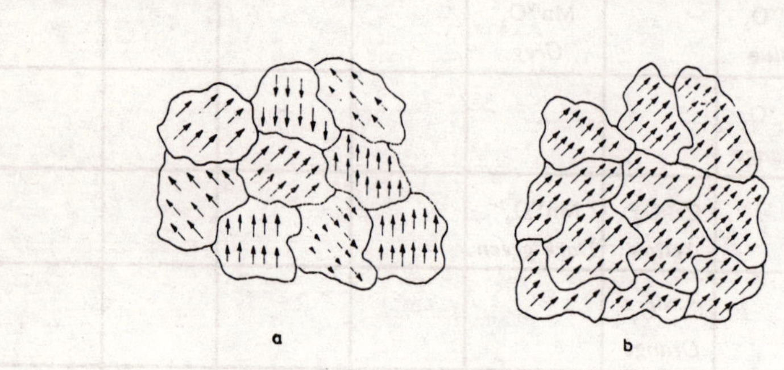

Fig. 9-7. Domains in a ferromagnetic field.

Alignment of million (9-7b) of such tiny atomic magnets bring the property of ferromagnetism. As a result the interaction between the ferromagnetic solid and the magnetic field is very much larger than that experienced by paramagnetic substances.

When the magnetic field is removed from a paramagnetic substance, its atomic magnets very quickly become randomly oriented (9-7a) and no permanent magnetism is induced. For a ferromagnetic substance, however, the domains tend to remain in the orientation in which they are found themselves when the external magnetic field was present, giving permanent magnetic property to the substance. Any piece of iron can be then permanently magnetized simply by striking it with another permanent magnet.

Paramagnetism is exhibited by all the transition metals in their elemental state and in all their ions except those whose oxidation states is equal to the group number. For example Sc^{3+}, Ti^{4+}, V^{5+}, Cr^{6+}, Mn^{7+}, Cu^{+}, Zn^{2+} whose group numbers are equal to their oxidation state are **diamagnetic.** Diamagnetism is exhibited by such ions, atoms or molecules of a compound which do not posses unpaired electron but instead all the electrons are paired. Such species repell in the magnetic field instead of getting attracted.

When a species behaves paramagnetically then its **magnetic moment,** μ can be determined based on the number of lone electrons (n) present by the expression.

$$\mu = \sqrt{n(n+2)}$$

The unit for the **paramagnetic moment, also called magnetic susceptibility, μ is Bhor magneton.** Thus the expected moment for unpaired electrons would be 1.73 B.M., 2.83 B.M., 3.87 B.M. 4.90 B.M., 5.92 B.M and 6.93 B.M. respectively for 1,2,3,4,5 and 6 electrons. The magnetic moment, μ, of a species may be calculated if the number of unpaired electrons is known or conversely the observed moment may be used to indicate the number of unpaired electrons present. Such a procedure is well adapted to the transition metals as shown in Fig. 9-8. In this figure a comparison

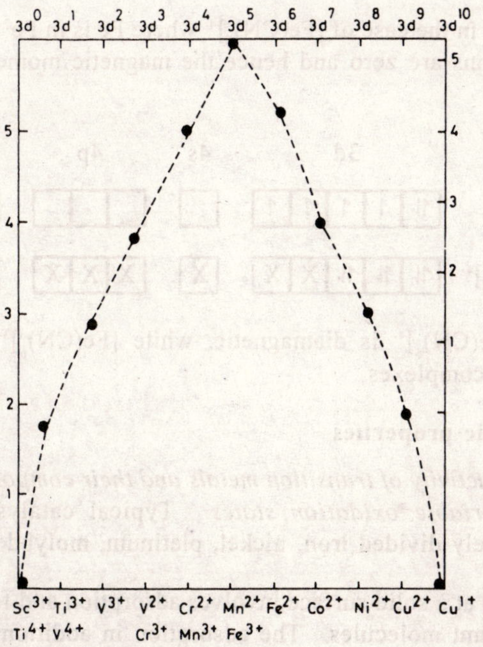

Fig. 9-8. The magnetic susceptibilities (in Bohr magnetons) of transition metals ions.

between calculated and observed moments is established for ions from the first transition series. It is observed that when the oxidation state changes the magnetic moments also change. The different oxidation states will have different electronic configuration, therefore there will be different number of lone electrons.

The magnetic moment can be determined for the complex species, depending upon the nature of ligand. Complex species containing strong field ligands, the magnetic moment will be different from the complex species containing a weak field ligand for the same metal ion. For example in the case of $[Fe(CN)_6]^{3-}$ there is only one lone electron, whereas in $[FeF_6]^{3-}$ there are five lone electrons; thus the magnetic moment of the former complex is 1.73 B.M., and the later has a magnetic moment of 5.92 B.M.

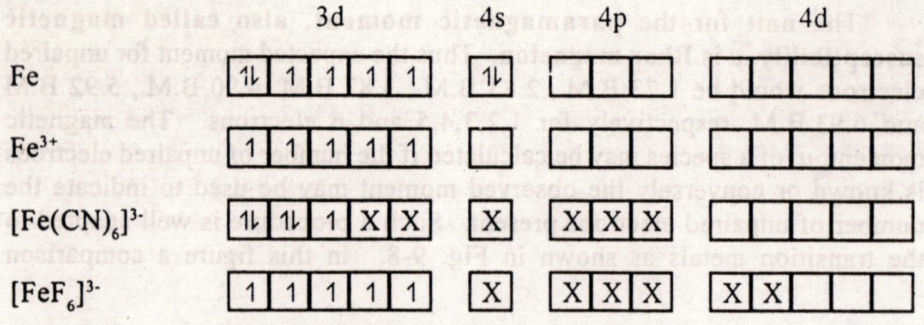

X indicate lone pair of electrons from the ligand.

Similarly in the case of $[Fe(CN)_6]^{4-}$ where Fe is in Fe^{2+} state, the number of lone electrons are zero and hence the magnetic moment is found to be zero.

	3d	4s	4p	4d
Fe^{2+}	↑↓ ↑ ↑ ↑ ↑			
$[Fe(CN)_6]^{4-}$	↑↓ ↑↓ ↑↓ X X	X	X X X	

Thus $[Fe(CN)_6]^{4-}$ is diamagnetic, while $[Fe(CN)_6]^{3-}$ and $[FeF_6]^{3-}$ are paramagnetic complexes.

9-4-6 Catalytic properties

The catalytic activity of transition metals and their compounds is associated with their variable oxidation states. Typical catalysts are vanadium pentoxide, finely divided iron, nickel, platinum, molybdenum etc.

Catalysis at a solid surface involves adsorption and formation of bonds between reactant molecules. The adsorption in addition to increasing the concentration of the reactants at the catalyst surface also weakens the bonds in the reactant molecules, resulting in the lowering of the activation energy.

Transition metal ions that have the flexibility of changing the oxidation state also function as catalysts. For example; in the iron catalysed reaction of persulphate and iodide, the Fe^{3+} ion reduces itself to Fe^{2+}, effecting a simultaneous oxidation of iodide ion, but gets oxidize again to Fe^{3+} by the oxidation by persulphate.

$$2Fe^{3+} + 2I^- \longrightarrow 2Fe^{2+} + I_2$$

$$2Fe^{2+} + S_2O_8^{2-} \longrightarrow 2Fe^{3+} + 2SO_4^{2-}$$

$$2I^- + S_2O_8^{2-} \xrightarrow{Fe^{3+}/Fe^{2+}} I_2 + 2SO_4^{2-}$$

9-4-7 Complex formation

It is an important characteristic of the transition elements to form numerous coordination compounds, in sharp contrast to the *s*-and *p*-block elements because of small, highly charged ions and vacant *d*-orbitals of appropriate energy to accept lone pairs of electrons donated by other groups or ligands. The valence bond, crystal field and ligand field theories of bonding explain the formation of complexes very clearly. These theories are explained in detail in the coordination chemistry chapter of volume III, but for our understanding a brief account of these theories will suffice.

Valence bond theory postulated by Pauling explains the stereochemistry and magnetic properties of coordination compounds. Consider for example the complex $[Cr(NH_3)_6]^{3+}$ In this complex all the six ligands are electron donors. In order to accept these six electron pairs the metal ion provides a set of six equivalent σ orbitals with their lobes directed towards the six apexes of an octahedron, such a set of orbitals is constructed by hybridizing the following group of orbitals.

$$d_{x^2-y^2},\ d_{z^2},\ s,\ p_x,\ p_y \text{ and } p_z$$

It can be represented as shown below.

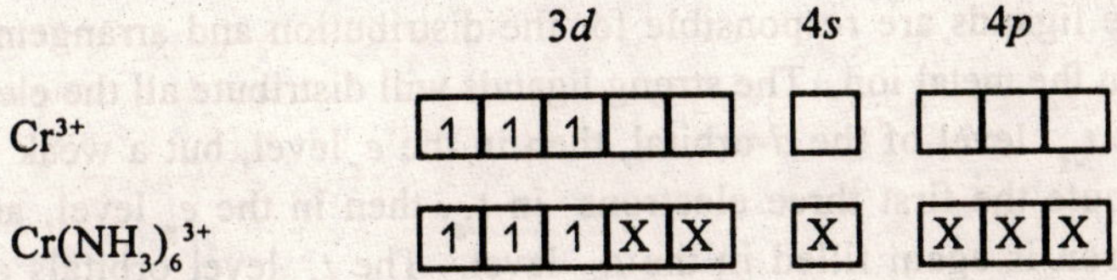

A hybridization of d^2sp^3 type occurs and the formed complex is paramagnetic with three lone unpaired electrons with a magnetic moment of μ = 3.87 B.M.

Now, considerr the forrmation of $[Co(CN)_6]^{3-}$ by Co^{3+}, a d^6 system. To provide the required d^2sp^3 hybrid orbitals, it is necessary to clear two of these electrons by promoting them to higher orbitals above 4p orbitals or by pairing within the 3d orbitals. The former is not energetically permitted so the latter is favoured as shown below:

	3*d*					4*s*	4*p*		
Co^{3+}	⇅	1	1	1	1				
$[Co(CN)_6]^{3-}$	⇅	⇅	⇅	X	X	X	X	X	X

This complex is diamagnetic with d^2sp^3 hybridization. But the fluoro complex of Co^{3+}, $[Co(F)_6]^{3-}$, having four unpaired electrons and sp^3d^2

hybridized is paramagnetic in nature whereas in the fluorocomplex instead of 3*d* ($d_{x^2-y^2}$ and d_{z^2}) orbital, the 4*d* (d_{xy}, d_{yz}) orbital is involved in the bond formation as shown below.

	3*d*					4*s*	4*p*			4*d*				
$[CoF_6]^{3-}$	⇅	1	1	1	1	X	X	X	X	X	X			

The reason for this cannot be explained by valence bond theory, hence the crystal field theory, comes to the rescue.

Crystal field theory suggests that, the attraction between the central metal ion and ligands is purely electrostatic. The ligands are considered as the point charges. The d-orbital of the metal are all of the same energy in their free state or degenerated state, but their degeneracy is destroyed by the ligands when a complex is formed. The common ligands when arranged in order of their strength (Δ_o) of destroying the degeneracy, known as **spectrochemical series** are:

$$I^- < Br^- < Cl^- < NO_3^- < F^- < OH^- < H_2O < NH_3 < CN^-$$

(Weak ligands) (Strong ligands)

These ligands are responsible for the distribution and arrangement of electrons in the metal ion. The strong ligands will distribute all the electrons first in the t_{2g} level of the *d*-orbital, then in the e_g level, but a weak ligand will distribute the first three electrons in t_{2g}, then in the e_g level, and the sixth electron is again filled in the t_{2g} level. The t_{2g} level orbitals are (in an octahedral field) d_{xy}, d_{yz}, d_{zx} and the e_g level has d_{x2-y2} and d_{z^2} orbitals in it, each orbital having a capacity of two electrons. The splitting of *d*-orbitals is seen in Fig. 9-6. The detail study is done as mentioned before in volume III

The ligand field theory was developed by van Vleck and it is a modified form of crystal field theory. It refers to the understanding of electronic properties, magnetic, spectroscopic, etc. of complexes of the transition metals. This theory suggests that, not only the atomic orbitals of metal but the ligational orbitals also get mixed up to form a complex, by overlapping either with a σ bond or a π bond. Hence, depending upon the formation of the complex, the solubility, colour and magnetic properties of the same metal ion changes from one complex to another.

9-4-8 Non-stoichiometry

An interesting feature of the transition elements is the existence of nonstoichiometry in their compounds; i.e., compounds of indefinite structure and composition. For example the iron (II) oxide has a varying formula of $Fe_{0.94}O$ and $Fe_{0.84}O$ instead of FeO, indicating that the Fe : O ratio is not exactly 1 : 1. Similarly CdO is $CdO_{0.89}$, vanadium forms a series of compounds ranging from $VSe_{0.98}$ to VSe_2; TiSe is $TiSe_{0.95}$ or $Ti_{0.9}Se$; TiO_2 is $TiO_{1.75}$; MO_3 is found to be $MO_{2.875}$ to $MO_{2.929}$ (where M is either Mo or W). The non stoichiometry is due to the variability of the oxidation states among transition metals and also due to the defects in the solid structures. The intermediate phases formed by the defects in an inner-transition element like praseodymium -oxygen system is given in table 12-5. This example is shown as it is one of the best non-stoichiometric examples which shows non-stoichiometric changes below 1000°C for Pr_nO_{2n-2} with n=4, 7, 9, 10, 11, 12 and ∞. The 'y' is proportionate amount of oxygen with respect to one atom of praseodymium.

Table 12-5 Non stoichiometry in praseodymium oxides

n	Formula Pr_nO_{2n-2},	y in PrO_y	Non-Stoichiometric limits at T°C	T°C
4	Pr_2O_3	1.500	1.500-1.503	1000
7	Pr_7O_{12}	1.714	1.713-1.719	700
9	Pr_9O_{16}	1.778	1.776-1.778	500
5	Pr_5O_9	1.800	1.799-1.801	450
11	$Pr_{11}O_{20}$	1.818	1.817-1.820	430
12	Pr_6O_{11}	1.833	1.831-1.836	400
∞	PrO_2	2.000	1.999-2000	>400

The non-stoichiometric oxide phases are of great importance in semiconductor devices, in heterogeneous catalysis and in understanding photoelectric, thermoelectric, magnetic and diffusion properties of solids. They have been in use in thermistors, photoelectric cells, rectifiers, transistors, phosphors, and computer components. They are crucially implicated in reactions at electrode surfaces, the performance of batteries, the tarnishing and corrosion of metals and in many other significant catalytic reactions.

9-5 SHORT AND ESSAY TYPE QUESTIONS

9-5-1 Discuss the ability of d-block elements (i) to from complexes (ii) to act as Catalytic agents.

9-5-2 Write notes on ability of d-block elements to form complexes.
(A.U.Dec. 1989) & (SKU., Apr. 1990)

9-5-3 Write about the ionization potentials of d-block elements. (A.U., Dec. 1989)

9-5-4 Write about catalytic properties of d-block elements. (A.U., Jun. 1990)

9-5-5 Explain the variable oxidation state, colour and magnetic properties of d-block element. (N.U. Mar., 96) (A.U., Jun. 1990)

9-5-6 Discuss the following aspects of 3-d elements.
(A) Ionization potential (B) Magnetic Properties (O.U., Oct 1990)

9-5-7 (a) Why are d-block elements called, transition elements? Name all the elements present in the 3d-series.

(b) Write a note on the variable oxidation states of the 3-d elements. (O.U., Apr 1990)

9-5-8 Describe the catalytic properties of transition elements. (O.U., Apr 1989)

9-5-9 Discuss the position of d-block elements in the periodic table.

9-5-10 What are d-d transitions? How are they responsible for imparting colour to a transition metal ions.

9-5-11 Discuss the ionization potentials of the first transition series.

9-5-12 Define the following terms:

(a) Paramagnetism (b) Diamagnetism (c) Ferromagnetism

How are these magnetic properties applied to study the magnetic criterion of the transition elements.

9-5-13 How is the paramagnetic susceptibility calculated for a paramagnetic substance? Calculate the magnetic susceptibility of the following:

(a) Fe^{+2} (b) Fe^{+3} (c) Cr^{+3} (d) Co^{+2} (e) Mn^{+5}

9-5-14 Write a note on valence bond theory and its application to transition metal complexes.

9-5-15 What is a spectrochemical series how it helps in the spliting of the *d*-orbitals?

9-5-16 Write the electronic configuration of chromium (24), Copper (29), Molybdenum [42], Silver [47].

9-5-17 Draw the shapes of the d-orbitals.

9-5-18 Write in detail about

(a) Variable oxidation states,

(b) Magnetic properties and

(c) colour of d-block elements. (A.U., Jun 1989) & (SKU., Apr 1990)

9-5-19 What are transition elements. Describe their oxidation states, magnetic properties and capability of forming complexes.

9-5-20 (a) What are transition elements? Give their general electronic configuration.

(b) What are the trends in ionization potential of transition elements? How do you explain the variable oxidation states of 3*d* transition elements.

(c) What are complexes? Explain the suitability of transition elements to form complexes. (N.U. Mar., 95) (K.U., Apr. 1989)

9-5-21 Justify how Zn, Cd & Hg can be included in *d*-block elements (O.U. Apr 94)

9-5-22 Mention any two metals of *d*-block elements or their compounds which exhibit catalytic properties. Write the manufacturing processes also in which they are used. (O.U. April 94)

9-5-23 Discuss the following properties of transitional elements. (a) Variable oxidation states (b) Formation of colour ions (c) Paramagnetism (d) Formation of complexes (A.U. Nov.94)

9-5-24 Discuss the magnetic properties of d-block elements (A.U. March 94)

9-5-25 Why *d*-block elements are known as transition elements? Transition metal ions are coloured, paramagnetic in nature and exhibit variable valency - Explain (O.U. Oct. 93)

9-5-26 Write the electronic configuration of Zn (At.No. 30), Cd(At. No. 48) and Hg (At. No. 80) (O.U. Oct.93)

9-5-27 a) Discuss why transition elements form complexes

b) Cu^{2+} are coloured and paramagnitic while Zn^{2+} ions are colourless and diamagnetic explain

c) Why ferric salts are more stable than ferrous salts. (O.U. March 93.

9-5-28 What are transition elements? Give their general electronic configuration. Discuss (a) variable oxidation state (b) Colour (c) Magnetic properties of transition elements with reference to 3*d* - elements. (N.U. Mar., 95) (O.U. Oct 92)

9-5-29 Explain the ionization protential and catalytic properties of 3d series of transition elements. (O.U. Oct 92)

9-5-30 Discuss the general characteristics of transitional elements with reference to the following

1. Magnetic properties 2. Colour 3. Variable valency (O.U. March 92)

9-5-31 a) What are transition elements ? Give their general electronic configuration.

b) Write notes on various oxidation states of 3*d* elements (A.U. Sept.93)

9-5-32 Write about the catalytic properties of *d*-block elements (A.U.Sept.93)

9-5-33 What are transitional elements? Write in detail about their colour, variable oxidation states magnetic and catalytic properties (A.U. March 93) (A.U. Sept. 93)

9-5-34 How do the transition elements form complexes (A.U. March 93)

9-5-35 Explain the anamolous electronic configuration of chromium (A.U. Sept. 93)

9-5-36 What are transitional elements? Discuss the extraction of nickel from its sulphide ores (N.U. Oct 92)

9-5-37 Explain the ionisation fotentional and catalytic properties of 3d series of transition elements. (O.U. Oct.92)

9-6 SELF TEST QUESTIONS

9-6-1 Out of the following pairs, Which one is not isoelectronic,
(A) Sc^{3+} and Pt^{4+} (B) V^{5+} and Cr^{6+} (C) Fe^{2+} and $Ni^{2}+$ (D) V^{3+} and Ca

9-6-2 The electronic configuration of chronium is
(A) $4s^2 3d^5$ (B) $4s^1$ $3d^4$ (C) $4s^1$ $3d^5$ (D) $4s^0$ $3d^4$

9-6-3 Zinc can exhibit the following oxidation states.
(A) +4, +3, +2, (B) Only +2 (C) Only +4 (D) Only +3

9-6-4 Transition metal ions
(A) Exhibit diamagnetism (B) Undergo inert pair effect
(C) Inert gas (D) Non - transition elements.

9-6-5 The oxidation state common for all the transition elements.
(A) +5 (B) +3 (C) +2 (D) +1

9-6-6 Which one of the following is a coloured ion
(A) Fe^{++} (B) Mg^{++} (C) Zn^{++} (D) Ba^{++}

9-6-7 The second transition series lies between
(A) Ca and Ga (B) Sr and In (C) Mg and Al (D) Ba and In

9-6-8 Transition elements can form complexes due to
(A) Small sized ions (B) Large size and low charge ions
(C) Small and highly charged ions (D) Large sized ions

9-6-9 The catalyst used in the manufacture of ammonia
(A) V_2O_5 (B) Pt (C) Fe (D) $Ni(CO)_4$

ANSWERS:

9-6-1 A, 9-6-2 C, 9-6-3 A, 9-6-4 A, 9-6-5 C, 9-6-6 A,
9-6-7 B, 9-6-8 C, 9-6-9 C

10

LANTHANIDES AND ACTINIDES - *f*-BLOCK ELEMENTS

10-1 INTRODUCTION

When Mandeleeff propounded his periodic classification of elements, he was unable to give a proper place to the lanthanides. He suggested that as these fourteen elements are similar to yttrium and lanthanum of III A group of the periodic table, and as they all possess much similarity in their properties, they cannot be separated from each other and therefore be kept separately below the main periodic table. It was a major defect of the Mendeleevian periodic table as he was unable to justify, in keeping these elements separately below the main periodic table. Bohr, Werner and Bury gave their modern periodic table, called the long form of the periodic table, which was developed on the basis of electronic configuration of the elements. Electronic configuration is written based on the four quantum orbital numbers (*s,p,d,f*) and Aufbau principle. It is interesting to see the arrangement of the long form of the periodic table, Fig. 10-1, and note that the lanthanides still occupy the same position as assigned by Mendeleeff, only difference being the explanation part of it. The lanthanides and actinides according to Aufbau principle have an electronic configuration of $ns^2 (n-1)d^1$ or $(n-2) f^{1-14}$. As the 4f orbital starts and completes its filling from atomic number 57 to 71, it covers all the elements from cerium to lutecium. From the atomic number 90 to 103. i e., from element thorium to lawrencium, the electron filling takes place in the 5*f* orbital. The 4*f* and the 5*f* elements together occupy the

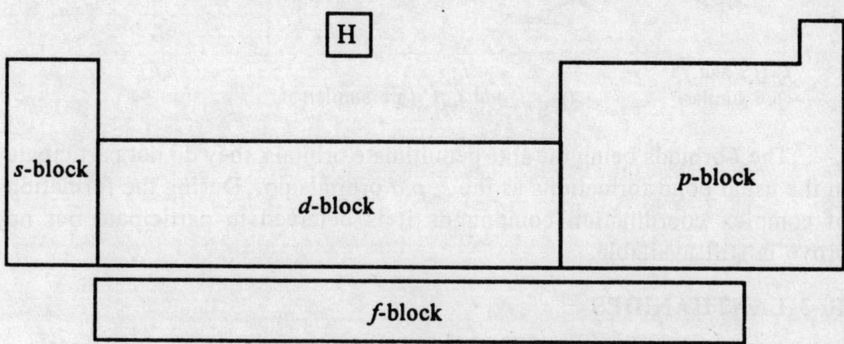

Fig 10-1 Skeleton of the long form of periodic table

f-block of the periodic table. Thus the placing of lanthanides and actinides, forming the f-block, separately is in full confirmity with the division of periodic table into *s*-, *p*-, *d*- and *f*-blocks.

Tomorrow, when the *g*-block elements are discovered they will find, most probably, a place below the *f*-block. Hence, there is no question at present of any block being placed above or below. Each block of elements are placed in the periodic table according to the convenience and the electronic distribution in the elements. These elements are also referred to as inner transition series as the incoming electrons of this series of elements always find place in the antepenultimate shell. For the sake of convenience and to get a clear understanding of both the series they are discussed separately.

10-2 SHAPES OF *f*-ORBITALS

The shapes of the *f*-orbitals are difficult to be represented on a plain surface because of their complicated three dimensional structures. The *f*-orbitals are important in accounting for the behaviour of elements with atomic number greater than 57 ie., the lanthanides and actinides. The f_{x^3}, f_{y^3} and f_{z^3} are the least stable and are similar to each other depending upon the axis on which the orbitals lie. The intermediately stable, $f_{x(y^2-z^2)}$, $f_{y(z^2-x^2)}$ and $f_{z(x^2-y^2)}$ are similar to each other depending upon, between which two axes these are lying. The most stable *f*-orbital is the f_{xyz} in which the orbitals lie in between the three axes. The shapes of the orbitals are given below.

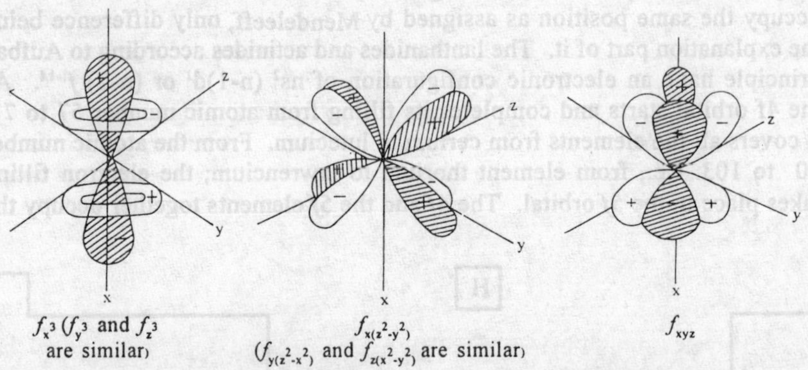

f_{x^3} (f_{y^3} and f_{z^3} are similar) $f_{x(z^2-y^2)}$ ($f_{y(z^2-x^2)}$ and $f_{z(x^2-y^2)}$ are similar) f_{xyz}

The *f*-orbitals being the ante penultimate orbitals, they do not participate in the usual bond formations as the *s, p,d* orbitals do. During the formation of complex coordination compounds it is beleived to participate but no prove is still available.

10-3 LANTHANIDES

Lanthanides or lanthanons are the elements: cerium, praseodymium, neodymium, samarium; europim, gadolinium, terbium, dysporsium, holmium, erbium, thulium, ytterbium and lutecium, including the artificially prepared promethium. This family is also called the rare earths, which also includes yttrium and scandium, because of the similarity in their properties with the lanthanides. The physical data of these elements is given in table 10-1.

Table 10-1 Some physical Properties of Lanthanides

Element	Atomic radii A°	Ionic radii A°	Oxidation states	Ionization potential	Electron negativity	Electrode potential M^{3+}/M volts
Y	1.801	1.06	3+	6.377	1.11	-2.37
La	1.877	1.22	3+	5.61	1.08	-2.52
Ce	1.820	1.18	3+, 4+	6.91	1.08	-2.48
Pr	1.828	1.16	3+ (4+)	5.78	1.07	-2.44
Nd	1.821	1.15	3+	6.3	1.07	-2.44
Pm^{3+}	0.279	1.14	3+	—	1.07	-2.42
Sm	1.802	1.13	2+, 3+	5.67	1.01	-2.41
Eu	2.042	1.13	2+, 3+	5.6	1.07	-2.41
Gd	1.802	1.11	3+(4+)	6.16	1.11	-2.40
Td	1.782	1.09	3+	6.74	1.10	-2.39
Dy	1.773	1.07	3+	6.82	1.10	-2.35
Ho	1.766	1.05	3+	—	1.10	-2.32
Er	1.757	1.04	3+	—	1.11	-2.30
Tm	1.746	1.04	3+	—	1.11	-2.38
Yb	1.940	1.00	2+, 3+	6.22	1.06	-2.27
Lu	1.734	0.99	3+	6.15	1.14	-2.25

The electronic configuration is discussed along with the actinides as a comparative study of the two series. The elements are called rare earths because of their very low concentrations in the earth's crust and other stellar bodies of the solar system. The spectral data of the giant Ba II star and Hd 46407 shows the presence of La , Ce, Pr Nd, Sm, Gd and Yb, in them. Substantial amounts of these elements are present in different ores but to a very small percentage. The abundances of the rare earths in different sources is shown in table 10-2.

Table 10-2 Abundance of the Rare earths from different sources

Element	Igneous Rocks g / ton	Monazite %	Onas ppm	Sediment ppm	Earths crust %
Y	28.1	2.8×10^{-3}	63	—	2.8×10^{-3}
La	18.3	1.8×10^{-3}	34	6	1.8×10^{-3}
Ce	46.1	4.0×10^{-3}	85	10	4.5×10^{-3}
Pr	5.53	5.5×10^{-4}	17	1	7.0×10^{-4}
Nd	23.9	2.4×10^{-3}	69	3	2.5×10^{-3}
Pm	—	—	—	—	—
Sm	6.47	6.5×10^{-4}	17	0.8	7.0×10^{-4}
Eu	1.06	1.0×10^{4}	2.6	—	1.2×10^{-4}
Gd	6.36	6.5×10^{-3}	17	0.6	1.0×10^{-3}
Tb	0.91	9×10^{-5}	3.5	—	1.5×10^{-4}
Dy	4.47	4.5×10^{-9}	13	0.4	4.5×10^{-4}
Ho	1.15	1.1×10^{-4}	3.5	0.1	1.3×10^{-4}
Er	2.47	2.5×10^{-4}	8.7	0.5	4.0×10^{-4}
Tm	0.20	2.0×10^{-5}	1.8	0.1	8.0×10^{-5}
Yb	2.66	2.7×10^{-4}	8.8	0.1	3.0×10^{-4}
Lu	0.75	7.5×10^{-5}	3.5	—	1×10^{-4}

10-3-1 Extraction of the metals

Monazite is the principle ore from which the rare earths-lanthanides are extracted. It is an orthophosphate of the lanthanides and thorium, containing 3-10% of thorium oxide and about sixty percent of the rare earth-lanthanide oxides. The lanthanides are classified mainly into three groups:

the cerium group: containing the ligher lanthanides, with yttria as the main source,

the gadolinite group: having the middle elements and

the ytterbia group: with the heavy lanthanides in it. All the three group elements are extracted from monazite sand, although monazite is particularly rich in cerium group rare earths. The other group of rare earths are present in monazite in amounts varying between 0.1 and 4 percent.

1. Extraction from monazite Two methods are usually employed to effect the break down of the rare earth metals from monazite sand:

A. The sulphuric acid breakdown of monazite and

B. The caustic soda break down of monazite.

A. The Sulphuric acid Breakdown of Monazite: The massive minerals and some coarse sands are first crushed and reduced to 40 mesh size for improved reactivity of sulphuric acid with monazite sand. The sulphuric acid used is about 0.6% of the weight of the mineral monazite. First the thorium content is precipitated by diluting the solution with 6 to 7 parts of water and neutralized with ammonia to a pH of 1.05. The amount of thorium thus recovered is 98-99% monazite. Leaching liquors with electrolytically generated alkali, a more selective precipitation of thorium is obtained. Rare earths are then precipitated from the filtrate by careful acidification to pH 2.3. At pH 6, uranium is obtained. The rare earths found in all the three fractions, can be separated by the solvent extraction and other methods described later in this section.

The other sulphuric acid digestion process, that allows elegant recovery of thorium, uranium, cerium and cerium free rare earths also involves monazite. The monazite sulphate liquors are diluted to about 15 g lit^{-1} of rare earth oxides and neutralized to pH 1.5 with ammonia. The rare earths and thorium are then treated with sodium hydroxide and the sodium oxalate solution is recycled. The resulting oxides are dissolved in 8N nitric acid and extracted by tributylphosphate (T.B.P.). Cerium is removed from the solvent by scrubbing the organic phase with a 0.1 N sodium nitrite solution. Thorium is recollected as its oxalate with T.B.P. in the final stage.

B. The Caustic Soda Breakdown of Monazite: This is a relatively recent process where the monazite sand is attacked by alkalies. This procedure has the advantage of removing phosphate more readily than the sulphuric acid procedure. In this procedure the sand is dissolved in hydrochloric acid, diluted with water and then neutralized with caustic soda or ammonia to pH 5.8. The Eimoo vacuum filters, the filter cake being continuously removed and repulped. The filtrate and the washings contain together about 7 g lit^{-1} of the rare earths which are removed as hydroxides. The rare earths are then extracted from adopting techniques given for minerals like **bastnasite** and **yttroparasite** discussed below.

2. Extraction from Bastnasite. Bastnasite is a fluorocarbonate of cerium group rare earths with traces of heavy rare earths and yttrium. The process starts from the crude ore containing about 10% of rare earth. 20-25% of calcium and 25-30% of barium sulphate. The ore is crushed and calcined in a rotary kiln at 800-900°C. The calcined ore is ground to 10 mesh and digested with 57% nitric acid. A violent exothermic reaction takes place. The resulting slurry is diluted with 30-40% of water and the insoluble berite and silica are filtered out. The contents of the rare earths are recovered from the filtrate by the solvent extraction with T.B.P. is 98%.

3. Extraction from Yttroparasite. Yttroparasite is a complex of fluorocarbonate from U.S.S.R. which occurs associated with several other rare earth minerals (Monazite, Fluocerite and Xenotime). This contained equal amount of cerium and yttrium group rare earths. Yttroparasite is decomposed by baking with ammonium sulphate in the ratio of 1 : 2 at a temperature of 650°C. The decomposition is done for an hour or two and the sulphates of rare earths are recovered with cold water.

4. Extraction from Gadolinite. It is a silicate of the rare earth containing 40-45% of rare earth oxides it is leached with nitric acid or sulphuric acid. The iron and beryllium impurities are separated as their oxalates. The leaching is done with 80% sulphuric acid. The rare earths formed as their sulphates are separated by their leaching with water from the other constituents. By basicity fractionation under led pH control gadolinite can also be fused with caustic soda-sodium peroxide mixutre at 700°C to allow more complete recovery of rare earths than the acid leaching. The melt is taken up with water, rare earth hydroxides are filtered off. Dissolved mineral acids, are then recovered as oxalates which are further calcined to get their respective oxides.

10-3-2 Separation of the lanthanides

Separation may be affected by one of the following general procedures:

1. Fractional crystallization of isomorphous salts.
2. Fractional precipitation from the solution.
3. Fractional thermal decomposition of salts.
4. Ion exchange.
5. Solvent extraction.
6. Selective oxidation or reduction.

Of all the mentioned procedures, except oxidation or reduction process, they are fractional in character. Separation results of fractionation, procedures 1-5 are satisfactory only if that step is repeated a number of times. The efficiency of a fractionation procedure is given by the magnitude of its separation factor (α). No. separation results if α is equal to unity, but greater the departure from unity, the more efficient the process will be. If two rare earths (RE) and (RE') being changed from an intial concentration (a) to a final concentration (b), the separation factor (α) may be defined as

$$\alpha = \frac{C_{RE\,(b)} / C_{RE'(b)}}{C_{RE\,(a)} / C_{RE'(a)}}$$

or

$$\alpha = \frac{C_{RE\,(b)} \times C_{RE'\,(a)}}{C_{RE\,(a)} \times C_{RE'\,(b)}}$$

Among all the above mentioned methods the recently developed ion exchange, solvent extraction and oxidation - reduction methods are employed today for the separation of the rare earths (lanthanides). So only these modern techniques are discussed here.

The Ion Exchange separation. The ion exchange separation of metal ions in solution involves exchange of heavy metal ions (M^{3+}) with protons (H^+) present on a solid cation resin exchange as shown in the following equilibrium process.

$$3RH^+ + M^{3+}(aq) \longrightarrow RM^{3+} + 3H^+(aq)$$

where RH^+ represents a proton adsorbed on to a resin surface, RM^{3+} is a tripositive lanthanide ion bounded to the resin, and M^{3+} (aq) and H^+(aq) are the aquated ions. The equilibrium constant, K, expression can be written as:

$$K = \frac{[H^+(aq)]^3\,[RM^{3+}]}{[RH^+]^3\,[M^{3+}(aq)]}$$

The constant K decreases with increase in atomic number of the series because the ratio RM^{3+}/M^{3+} (aq) decreases. So the ion exchange separation is based upon the fact that a steady decrease in size of lanthanide cation and consequent decrease in the basicity means an increase in the K value and vice versa. The lanthanide (III) ions increase with increasing atomic number. Since the binding of cations to the anionic sites of the exchange resin seems to involve it is found that the larger the hydratea radius of the ion, the less strongly it is bound. Thus this effect gives a separation of different Ln(III) ions. When a solution containing several of them is passed slowly through a column of a cation exchange resin, the heavier members of the family like lutecium comes out first and lanthanum is the last metal that passes through the citrate system at pH 2.5 - 3.2 proved effective on tracer scale. Use of 0.1% ammonium citrate at pH prominent, k however leads to a rapid separation with the development of sharpy 5% ammonium citrate system. Both the systems proved to be suitable and lanthanides of 99.99% purity are obtained.

The Solvent Extraction Technique: The solvents employed in this method of the extraction of the lanthanides are usually tri n-butyl phosphate (TBP) and di(2-ethylhexyl phosphoric acid (HDGP).

The extraction by n-butyl phosphate (TBP) is generally used in nitrate medium, the equilibrium of the following type is established.

$$Ln^{3+}(aq) + 3NO_3^-(aq) + 3TBP\,(org) \longrightarrow Ln\,(NO_3)_3\,(TBP)_3\,(org)$$

where (org) represents the organic phase. Distribution between the two phases is then described as

$$\lambda = CLn\,(No_3)_3(TBP)_3(org)/\,CLn^{3+}(aq)$$

and for two lanthanides the separation factor becomes

$$\alpha = \frac{\lambda'}{\lambda}$$

$$= \frac{[Ln' (NO_3)_3 (TBP)_3 (org)] / [Ln'^{3+} (aq)]}{[Ln (NO_3)_3 (TBP)_3 (org)] / [Ln^{3+} (aq)]}$$

It was reported by Pappard and his collaborators that an average separation factor has a value of 1.5 for the adjacent lanthanides in 15.8M HNO_3 and 100% in TBP system.

The extraction of the lanthanides by di(2-ethyl hexyl) phosphoric acid is more resistant to hydrolytic decomposition when in contact with aqueous nitric acid than TBP. In this case the average separation factor reported for the adjacent lanthanides is 2.5. The solvent which is an ester exists in dimeric form in the nonaqueous phase. Extraction equilibrium in this case is

$$Ln^{3+} (aq) + 3(HDGP)_2 (org) \longrightarrow Ln [H(DGP)_2]_3 (org) + 3H^+ (aq)$$

and its distribution ratio is

$$\lambda = K [(HDGP)_2 (org)]^3 / [H^+(aq)]^3$$

The separation of the lanthanides by the solvent extraction process is of less use when compared to that of ion exchange. Its major use is in the purification of cerium, thorium and lanthanum.

Selective Oxidation - Reduction. A change in oxidation state is accompanied by a profound change in the properties of the element. Oxidation and Reduction separates the specific lanthanides. For example oxidation of Ce^{3+} by bromite or permanganate gives Ce^{4+} which completely precipitates by suitable change in the pH.

Electrochemical Separation: The separation of rare earth elements by electrolysis with a mercury cathode depends on the relative distribution of the rare earths between mercury and the electrolyte. It was noted that in lithium citrate solution samarium, europium and ytterbium can be separated from other rare earths (whose deposition at the mercury cathode requires more severe conditions) and also from one another if they are present in different concentration. They can be separated at an initial ratio of Eu: Sm of 1:62 (0.447 M Sm and 7.2 X 10^{-3} M Eu). The concentration of the Europium impurity in Sm can be reduced by a thousand fold by electrolysis. Neodymium and praseodymium can be separated at a current density of 7.7 mA cm^{-2} and a tartarate concentration of 0.34M. This can be in the presence of lithium tartarate and the separation factor was 58. The conditions investigated for the formation of iridium and gadolinium, at mercury cathode of the acetate - citrate solutions in the presence of lithium ions and also binary mixtures of lanthanide elements were found to be : 6:8 for Gd:Ir, 10:40 for Gd;Ho, 5:10 for Ir:Dy, 10:30 for Ir:Ho, 20:40 for Ir:Er, 30:60 for Ir:Th. The yield of the lanthanide elements as a rule decrease with their concentration.

10-4 GENERAL CHARACTERISTICS OF LANTHANIDES

The lanthanides are silvery white metals, the lighter lanthanides retain their lustre at room temperature but the higher lanthanides get tarnished due to oxidation. The lanthanides are not good conductors of heat and electricity, and their mechanical characteristics are also very poor.

All the elements of the family have a very low electronegative character as their ions exist in a stable Ln^{3+} state. However it is suggested that each successive 4*f* electron does less than a through job of cancelling the nuclear charge, so that the effective nuclear charge increase slowly through the series. Because the lone electrons are present in the antepenultimate (4*f*) orbit, the ionization potential values of the lanthanides is quite high and as the electrons are removed from the same orbit the ionization potential values are found to be almost the same for all the lanthanides. The electrode potential values for M^{3+}/M system shows that the lanthanides are strong reducing agents and hence are used for thermite type reactions. A mixture containing mostly Ce and La with small amounts of other lanthanides and iron, known as **misch metal** is used for such purposes. It is pyrophoric and when alloyed with more iron the hardness increases and it is used for flints in lighters.

10-4-1 Oxidation States

The most common oxidation state exhibited by the lanthanides is the +3 state, in which the M^{3+} ion contains an outer shell containing eight electrons and an underlying layer containing upto fourteen 4*f* electrons as the $6s^2$ and $5d^1$ electrons are involved in exhibiting this state. The La^{3+}, Gd^{3+} and Lu^{3+} containing respectively an empty, a half filled and a completely filled 4*f* levels are especially very stable. Cerium, exhibit an oxidation state of +4 in which it is isoelectronic with La^{3+} similarly Tb^{4+} is isoelectronic to Gd^{3+}, the former pair has empty 4*f* level ($4f^0\ 5s^2\ 5p^6$) and the latter pair has half filled 4*f* level ($4f^7\ 5s^2\ 5p^6$).

Europium, ytterbium and samarium exhibit a +2 state. The Eu^{2+} is isoelectronic with Gd^{3+} with a half filled 4*f* level of $4f^7\ 5s^2\ 5p^6$, whereas Yb^{2+} is isoelectronic with Lu^{3+} with a completely filled 4*f* level. The lanthanides which contain one, two, six and thirteen 4*f* electrons are Pr^{4+}, Nd^{4+}. Sm^{2+} and

Tm^{2+}. Hence it is seen that a few lanthanides exhibit an oxidation state of +4 while some show an oxidation state of +2 and +3 is only to attain the most stable electronic configuration. Ce^{4+} is a strong oxidizing agent and Sm^{3+} a strong reductant respectively, only indicate the stability of the 3+ oxidation state.

$$Ce^{4+} + Fe^{2+} \longrightarrow Ce^{3+} + Fe^{3+}$$

$$2Sm^{2+} + 2H_2O \longrightarrow 2Sm^{3+} + H_2 + 2OH^-$$

10-4-2 Colour

Many of lanthanide ions are coloured. The colour of these ions can be attributed to the *4f* transitions but it is difficult to interpret. But it is observed that the colour is dependent of the number of lone electrons present in the 4f orbital, the colour of the lanthandies based on this property is shown in table 10-3.

Table 10-3 Colours of Lanthanides

Ion	No. of lone electrons	Colour	Ion	No.of Lone electrons	Colour
La	0	Colourless	Lu	0	Colourless
Ce	1	Colourless	Yb	1	Colourless
Pr	2	Green	Tm	2	Green
Nd	3	Reddish	Er	3	Red
Pm	4	Pink yellow	Ho	4	Orange
Sm	5	Yellow	Dy	5	Yellow
Eu	6	Pink	Tb	6	Colourless
Gd	7	Colourless	Gd	7	Colourless

10-4-3 Lanthanide contraction, Atomic Radii and Ionic Radii

In a group of elements it is found that the atomic radii always increases on going down the group and in a period on passing from right to left. This property of periodicity of elements is very common for the elements. In the case of lanthandies something very interesting is observed. Each succeeding lanthanide differs from its immedicate predecessor in having an electron extra in the *4f* orbital and one extra proton in the nucleus of the atom. The

4*f* electrons constitute the antepenultimate shell and are hence ineffective in the screening of the nucleus. Thus there is a gradual increase in the attraction of the nucleus for the peripheral electrons as the nuclear charge increases, and a consequent contraction in atomic and ionic radii of the trivalent lanthanides.

It is surprising to note, that the non-lanthanide, scandium has a radius of 0.83Å and yttrium 1.06Å a difference of radii equal to (0.23Å), whilst in passing from lanthanum to lutetium there is a total contraction of 0.23Å.

The lanthanide contraction influences the chemistry of the *d*- and *p*-block elements which succeed the lanthanides in the periodic table. For instance the atomic radii of zirconium (1.45Å) and hafnium (1.44Å) are almost same and the chemistry of these elements is also similar, which make their separation from each other very difficult. In general it is observed that the second and third rows of transition elements resemble each other more closely than do the first and second rows.

10-4-4 Spectra and Magnetic Properties

All the lanthanide ions show an absorption in the visible or the near ultra violet region of the spectrum, except for Lu^{3+} which has a full-shell.

The electronic absorption spectra are produced when electromagnetic radiation promote the ions from their ground state to the excited states. For the lanthanides the most common of such transitions involve excited states which are the components of the group state or else belong to excited terms which arise from the same $4f^n$ configuration as the ground state. In both of these cases the transtions involve only a redistribution of electrons within the 4*f* orbitals (creating a *f*-*f* transition) and these vibrations within the orbitals can be recorded through an electronic spectra. The absorption bands are found to be very sharp in the *f*-*f* transitions.

Paramagnetism is associated with the presence of unpaired electrons in an element or its compound or in its ions. Majority of the lanthanide ions exhibit paramagnetism, since according to Hund's rule each of the 4*f* orbitals must be filled singly before the pairing takes place. Those ions which do not show paramagnetism are those with no 4*f* electrons like La^{3+} and Ce^{4+} or with fully filled 4*f* level like that in Yb^{2+} and Lu^{3+}, which are diamagnetic. It is interesting to note in the lanthanides Gd^{3+} eventhough has the highest number of unpaired electrons (7 in number) is not the most paramagnetic element but Dy^{3+} having five unpaired electrons is the most

Table 10-4a Paramagnetism susceptibilities of lanthanides

Lanthanides (Ln^{3+} ions)	No.of lone electrons	μ (BM)(Van Vleck)
Cerium	$1(4f^1)$	2.54
Praseodymium	$2(4f^2)$	3.58
Neodymium	$3(4f^3)$	3.62
Promethium	$4(4f^4)$	2.68
Samarium	$5(4f^5)$	1.85
Europium	$6(4f^6)$	3.40
Gadolinium	$7(4f^7)$	7.94
Terbium	$6(4f^8)$	9.72
Dysprosium	$5(4f^9)$	10.65
Holmium	$4(4f^{10})$	10.60
Erbium	$3(4f^{11})$	9.58
Thulium	$2(4f^{12})$	7.56
Ytterbium	$1(4f^{13})$	4.54
Lutetium	$0(4f^{14})$	0

paramagnetic. This unusual feature shows that the paramagnetism in the lanthanides is not dependent on the number of lone '*f* electrons. The

Ion	Unpaired electrons	Theoretical	Observed in $Ln_2(SO_4)8H_2O$
La^{3+}	0	0	--
Ce^{3+}	1	2.56	--
Pr^{3+}	2	3.62	3.47
Nd^{3+}	3	3.68	3.52
Pm^{3+}	4	2.83	--
Sm^{3+}	5	1.60	1.58
Eu^{3+}	6	3.45	3.54
Gd^{3+}	7	7.94	7.90
Tb^{3+}	6	9.70	9.60
Dy^{3+}	5	10.60	10.40
Ho^{3+}	4	10.60	10.40
Er^{3+}	3	9.60	9.40
Tm^{3+}	2	7.60	7.00
Yb^{3+}	1	4.50	4.30
Lu^{3+}	0	0	0

Table 10-4b Paramagnetic moments of tripositive lanthanides ions

paramagnetic susceptibility value are given in table 10-4a and the paramagnetic moments of the tripositive lanthanide ions are depicted in Table 10-4b.

10-4-5 Complex formation

The lanthanides are capable of forming complexes. In the previous chapter we have seen how the transition elements form complexes utilizing the empty *d*- and the *s*- and *p*- orbitals after suitable hybridization. The 4*f* orbital of the lanthanides is an antepenultimate orbit hence its involvement is generally thought to be less in the complex formation. It is true that the *role played by the f-orbitals is not that important as that played by the d-orbitals in the complex formation.*

Much work on the coordination chemistry of these elements is being conducted world over and many new concepts and findings are being brought to light about the lanthanides. In all these complexes it is clearly observed that only *s*-, *p*-, *d*- orbitals are involved. The early work on the coordination chemistry of these elements, by Moeller showed that these complexes are quite stable, specially when the lanthanides are in their trivalent state. Some of the important complexes of these elements, apart from the double salts like $M^{III} M^{I}_3(SO_4)_3$ X H_2O, where M^{III} being tripositve lanthanide ion and M^I being univalent ion like Na^+, K^+, Tl^+, NH^+_4, etc; the others being acetylacetonate, dihydroxy benzoates, ethylene diamminetetraacetates, tartarates, citrates and even mixed ligand complexes with acetyl acetonates with triphenyl phosphine and arisine oxides. In these complexes the lanthanides ion gets coordinate through oxygen of the ligand, and in complexes with triethyl phosphite it is found to coordinate through phosphorous which is very uncommon. Nitrogen donor ligands like pyridine are also reported to complex with lanthanides.

10-5 ACTINIDES

Actinides or **Actinons** or **Actinoids** are the fourteen elements following actinium (as mentioned before) and these are thorium, protoactinium, uranium, neptunium, plutonium, americium, curium, berkelium, californium, einsteinium, fermium, mendelevium, nobelium, and lawrencium. The elements thorium, protoactinium and uranium are naturally occurring elements whereas the rest of the eleven elements which are often called as the **transuranic elements** are artificially prepared and the mother nuclide for all of them is uranium or one of its daughter nucleides. Table 10-5 gives a brief account of the year of discovery and origin of the name of these elements. All these elements were discovered by Seaborg and his collaborators.

Table 10-5 Year of discovery and the origin of the name of the transuranic elements.

Atomic number	Name	syn ol	Date	Origin of the name
93	Neptunium	(Np)	1940	The planet neptune
94	Plutonium	(Pu)	1940	The planet pluto
95	Americium	(Am)	1944	In honour of America
96	Curium	(Cm)	1944	In honour of Marie & Perri Curie
97	Berkelium	(Bk)	1949	Berkeley is the county where Seaborg's laboratory is present
98	Californium	(Cf)	1950	Location of the laboratory in the state of California
99	Einsteinium	(Es)	1952	In honour of Albert Einstein
100	Fermium	(Fm)	1952	In honour of Fermi who made nuclear reactor
101	Mandeleevium	(Md)	1955	In honour of Dimitri Mandelleeff
102	Nobelium	(No)	1958	In honour of Alfred Nobel
103	Lawrencium	(Lw)	1961	In honour of Ernest Lawrence who developed the cyclotron

It is clear from Table 10-5 that all the transuranic elements were discovered after 1940 only.

The actinides are obtained from the naturally occurring ores like monzite sand and as mentioned in the previous section, they get separated during the separation of the lanthanides.

The chemical properties of Th, Pa and U are similar to those of Ti, Vand Cr group of elements of the transition series.

10-6 ELECTRONIC CONFIGURATION OF LANTHANIDES & ACTINIDES

The most characteristic property of the *f*-block elements is the electronic configuration. The electronic configuration of the lanthanides is expected to be different from what it is actually. The theoritical electronic configuration of the lanthanides is $4f^{1-14}$ $5d^1$ $6s^2$, but it was observed to be different. The $4f^1$ is not found to be present either in La or Ce, instead, the $5d^1$ electron gets promoted to $4f$ orbital, thus giving the configuration of $4f^2$ $5d^0$ $6s^2$ to Ce. The $5d^1$ electron is not to be found in any of the elements of lanthanide family except in La, Gd and Lu, with a completely vacant *f*-orbital, a half filled *f*-orbital and completely filled *f*-orbital respectively. The electronic configuration of the lanthanides is shown in table 10-6.

Table 10-6 Electronic configuration of lanthanides

Atomic number	Symbol	Electronic configuration	
		Theoretical	Observed
57	La	[Xe] $5d^1$ $6s^2$	[Xe]$4f^0$ $5d^1$ $6s^2$
58	Ce	—$4f^1$ $5d^1$ $6s^2$	—$4f^2$ $6s^2$
59	Pr	—$4f^2$ $5d^1$ $6s^2$	—$4f^3$ $6s^2$
60	Nd	—$4f^3$ $5d^1$ $6s^2$	—$4f^4$ $6s^2$
61	Pm	—$4f^4$ $5d^1$ $6s^2$	—$4f^5$ $6s^2$
62	Sm	—$4f^5$ $5d^1$ $6s^2$	—$4f^6$ $6s^2$
63	Eu	—$4f^6$ $5d^1$ $6s^2$	—$4f^7$ $6s^2$
64	Gd	—$4f^7$ $5d^1$ $6s^2$	—$4f^7$ $5d^1$ $6s^2$
65	Tb	—$4f^8$ $5d^1$ $6s^2$	—$4f^9$ $6s^2$
66	Dy	—$4f^9$ $5d^1$ $6s^2$	—$4f^{10}$ $6s^2$
67	Ho	—$4f^{10}$ $5d^1$ $6s^2$	—$4f^{11}$ $6s^2$
68	Er	—$4f^{11}$ $5d^1$ $6s^2$	—$4f^{12}$ $6s^2$
69	Tm	—$4f^{12}$ $5d^1$ $6s^2$	—$4f^{13}$ $6s^2$
70	Yb	—$4f^{13}$ $5d^1$ $6s^2$	—$4f^{14}$ $6s^2$
71	Lu	—$4f^{14}$ $5d^1$ $6s^2$	—$4f^{14}$ $5d^1$ $6s^2$

On making a comparative study of the electronic configuration of actinides with the lanthanides it is found that the actinides do have the $6d^1$ electron in all the elements according to Seaborg, with the f-electrons being filled regularly as they ought to be. But Dawson suggests that Ac has a $5f^0$ $6d^1$ $7s^2$ electronic configuration with Th having $5f^0$ $6d^2$ $7s^2$ configuration, with the subsequent filling of the d^3 and d^4 in Pa and U respectively. The Seaborg and Dawson electronic configurations of actinides is shown in Table 19-7.

Table 10-7 The Seaborg and Dawson electronic configurations of actinides

Element	Atomic Number	Electronic configuration					
		Seaborg's [Rn core]			Dawson's [Rn core]		
Ac	89	$5f^0$	$6d^1$	$7s^2$		$6d^1$	$7s^2$
Th	90	$5f^1$	$6d^1$	$7s^2$		$6d^2$	$7s^2$
Pa	91	$5f^2$	$6d^1$	$7s^2$		$6d^3$	$7s^2$
U	92	$5f^3$	$6d^1$	$7s^2$		$6d^4$	$7s^2$
Np	93	$5f^4$	$6d^1$	$7s^2$	$5f^4$	$6d^1$	$7s^2$
Pu	94	$5f^5$	$6d^1$	$7s^2$	$5f^5$	$6d^1$	$7s^2$
Am	95	$5f^7$		$7s^2$	$5f^6$	$6d^1$	$7s^2$
Cm	96	$5f^7$	$6d^1$	$7s^2$	$5f^7$	$6d^1$	$7s^2$
Bk	97	$5f^8$	$6d^1$	$7s^2$	$5f^8$	$6d^1$	$7s^2$
Cf	98	$5f^9$	$6d^1$	$7s^2$	$5f^9$	$6d^1$	$7s^2$
Es	99	$5f^{10}$	$6d^1$	$7s^2$	$5f^{10}$	$6d^1$	$7s^2$
Fm	100	$5f^{11}$	$6d^1$	$7s^2$	$5f^{11}$	$6d^1$	$7s^2$
Md	101	$5f^{12}$	$6d^1$	$7s^2$	$5f^{12}$	$6d^1$	$7s^2$
No	102	$5f^{14}$		$7s^2$	$5f^{13}$	$6d^1$	$7s^2$
Lw	103	$5f^{14}$	$6d^1$	$7s^2$	$5f^{14}$	$6d^1$	$7s^2$

From this data we are unable to establish whether the $5f$ electron enters at thorium and gives an actinide series as thought by Seaborg or it enters at neptunium and gives a uranide series as proposed by Dawson. This concept is still undecided and more so due to the $5f$ and $6d$ orbitals being identical in their energies.

Although the general resemblance in electronic configuration between the actinides and the lanthanides is clear, there are several ways in which the later series differs from the former. Firstly, it is only with the second element (^{91}Pa) after actinium that the ground state of the atoms contain f-electrons; the atom for which the energy of the $5f$ orbitals is lower than $6d$ - orbitals. In case of lanthanides, the very first element that follws lanthanum, i.e., ^{58}Ce, starts filling its $4f$-orbital. In the first few elements of the actinide series the energies of the $5f$, $6d$, $7s$ and $7p$ orbitals are comparable, and therefore covalent bonding can take place with any of these orbitals. The fact that the $5f$ orbitals have greater spatial extension than the $4f$ orbitals is proved by the electron split resonance spectrum (esr) of UF_3 in a CaF_2 lattice. The *esr* spectrum of UF_3 shows a fine the U^{3+} ion with the fluorine nuclei, while NdF_3, the corresponding compound in the lanthanide series, shows no such effect.

The range of oxidation states exhibited by the actinides (particularly the lighter ones) is very much greater than that found among the lanthanides, and high oxidation states involving covalent bonding to oxygen are well established for uranium, neptunium, plutonium and americium. In the second half of the series, the tripositive state becomes the dominant oxidation state. However, the last two elements before the last actinide, i.e., mendelevium and nobelium (the homologues of thulium and ytterbium) both have dipositive states. In the case of nobelium this is, indeed, the most stable oxidation state. All oxidation states known at the present time are shown in Table 10-8.

The lanthanides are found to have a stable oxidation state of +3 uniformly from La to Lu, but in the case of actinides only +3 oxidation state is seen with the increasing atomic number, while in less heavier elements like thorium, protoactinium, uranium, neptunium, plutonium and americium alongwith +3 oxidation state, the stable oxidation states of +4, +5, +6, are also seen as in table 10-8.

Table 10-8 Different oxidation states of actinides
(Stable oxidation states are shown by bold lettering)

Element	Atomic number	Oxidation states
Ac	89	3
Th	90	3 4
Pa	91	3 4 5
U	92	3 4 5 6
Np	93	3 4 5 6
Pu	94	3 4 5 6
Am	95	3 4 5 6
Cm	96	3 4
Bk	97	3 4
Cf	98	3
Es	99	3
Fm	100	3
Md	101	2 3
No	102	2 3
Lw	103	3

The electronic spectra of the actinide ions are sharp like those of the lanthanides, but their intensities are some what greater. Though the spectrum are less influenced by ligand fields than those of ordinary transition metal ions, complexing does affect them even slightly, so there are small crystal field stabilization energy terms for actinide ions. The magnetic properties of the actinide ions are extremely difficult to interpret. The general shape of the curves of magnetic susceptibility against the number of *f*-electrons is, however, similar to the corresponding curves for the lanthanides, though the oxidation states involved are quite different. For the actinides, the f^1 configuration, for example, is represented by neptunium (VI) or protactinium (IV) rather than thorium (III). Thorium (III) is highly unstable oxidation state.

The crystal chemistry of the actinides, in the limited range of compounds that can be studied, show a general resemblance to the lanthanides. A similar contraction in ionic size take place along both the series. The difference in the radius between successive elements becomes smaller towards the end of the series.

The distribution of different oxidation states between the members of the actinide series is quite uneven and investigation techniques employed for the earlier and later members of the series are quite different. It is unsatisfactory or review their chemistry in terms of oxidation states as done in case of lanthanides.

10-7 SELF TEST QUESTIONS

Mark the correct answer

10-7-1 In the *f*-block; the extra electron enters into the
(A) Ultimate (B) Penultimate (C) Ante penultimate (D) None of these orbitals

10-7-2 In the *f*-block the next electron is added to
(A) *(n-1)d* nf orbital (B) *(n-2)d(n-1)f* orbital
(C) *(n-2)d(n-1)f* orbital (D) *(n-1)d(n-2)f* orbital.

10-7-3 The lanthanides and actinides are the elements belonging to
(A) *d*-block (B) *s*-block (C) *f*-block (D) *p*-block of the periodic table.

10-7-4 The transuranic elements are
(A) 10 (B) 12 (C) 11 (D) 14 in number.

10-7-5 In lanthanide contraction the size of
(A) M^{3+} ion increases (B) M^{4+} ion increases
(C) M^{3+} ion decreases (D) M^{4+} ion decreases

10-7-6 With the increase in atomic number of atomic radii
(A) increases (B) remains the same
(C) decreases (D) increases, decreases and again increases

10-7-7 The common oxidation state of lanthanides is
(A) +2 (B) +3 (C) +4 (D) +1

10-7-8 The common oxidation state of actinides is
(A) +2 (B) +3 (C) +4 (D) +1

10-7-9 The pink coloured pair of lanthanides ions are
(A) Nd^{3+}/Er^{3+} (B) La^{3+}/Lu^{3+} (C) Pr^{3+}/Im^{3+} (D) Sm^{3+}/Dy^{3+}

10-7-10 Lanthanides lie between

(A) Y and Zr (B) La and Hf (C) La and la (D) La and W

10-7-11 The number of elements present in the *f*-block are

(A) 14 (B) 28 (C) 24 (D) 32

10-7-12 The *f*-block elements are called

(A) transition elements (B) representative elements

(C) inner transition elements (D) transuranic elements

10-7-13 The element not belonging to *f*-block.

(A) Mo (B) Pr (C) Tb (D) Lu

10-7-14 The *f*-block elements show a unique characteristic known as

(A) Inert pair effect (B) Godolinium break

(C) Effective contraction in atomio size (D) Transuranic

13-7-15 Lanthanide contraction is responsible for similarities of the properties of elements with atomic number

(A) 41 and 73 (B) 11 and 19 (C) 42 and 81 (D) 22 and 40

13-7-16 An element with an atomic number 60 is

(A) a representative element (B) 4*f* element

(C) 5*d* element (D) 5*f* element:

10-7-17 An element with an atomic number 94 is

(A) A representative element (B) transition element

(C) Inner transition element (D) A transurnic element

10-7-18 The electronic configuration : $6s^2\ 5d^1\ 4f^n$ is of

(A) Ce (B) Pm (C) Sm (D) Nd

10-8 SHORT AND ESSAY TYPE QUESTIONS

10-8-1 Discuss the oxidation states and colour of lanthanides.

10-8-2 Write notes on lanthanide contraction (A.U., Dec. 1989) & O.U., Jun. 1989)

10-8-3 What is the element having atomic number 57? What are the elements which follow this element called? (A.U., Jun. 1990)

10-8-4 What are the oxidation states exhibited by lanthanides? (A.U., Jun. 1990)

10-8-5 (a) What are the inner transition elements? Given their general eléctronic configuration.

(b) Explain the term lanthanide contraction and discuss its effects.
(N.U. Mar., 96) (A.U., Oct. 1990) & (A.U., Apr. 1989)

10-8-6 (a) What is an inner transition element? Give two examples.

(b) What iś lanthanide contraction? Mention its consequences.

(c) Most of the transition metal ions are coloured and Paramagnetic, explain why?

(d) All the lanthanide ions exist in +3 oxidation state - Give reasons for this
(O.U., Apr. 1990)

10-8-7 What are the important consequences of lanthanide contraction? (O.U., Apr. 1984)

10-8-8 Discuss the oxidation states of lanthaoas. (O.U., Apr. 1990)

10-8-9 Write the electronic configuration of the lanthanides.

10-8-10 Discuss the position of the *f*-block elements in the periodic table based on the modern periodić law.

10-8-11 Discuss the various methods applied for the extraction of the lanthanides.

10-8-12 Write the note on the solvent extraction technique used in the extraction of lanthanides.

10-8-13 Discuss the following properties of lanthanides:

(a) Colour

(b) Spectral and Magnetic properties

(c) Complex formation.

10-8-14 Discuss the oxidation states of the actinides.

10-8-15 What is lanthanide contraction? What are its effects, discuss the anamolous oxidation states of lanthanides.

10-8-16 Write the differences and similarities between lanthanide and actinide elements. (SKU., Apr. 1990)

10-8-17 What are lanthanous? Write breifly about their electronic configuration and oxidation states what is lanthanide contiaction and how do you account for it (O.U., Apr. 1995) (N.U., Oct. 1992)

10-8-18 Discuss the position of actinides in the periodic table (O.U., Apr. 1995)

10-8-19 Explain lanthanide contraction and its impact on the properties of lanthenides (O.U. Mar., 96) (A.U., Apr.1995)

10-8-20 Discuss the position of actinides in the periodic table, compare & Contrast lanthanides with activides (N.U., Apr.1995)

10-8-21 Explain what are lanthanides and lanthanide contraction (N.U., Apr. 1995)

10-8-22 Write short notes on lanthanide contraction (A.U., Sept.1993) & (O.U., March 1992)

10-8-23 Write the differences & Similarities between lanthanide and actinide elements (A.U., Sept.1993)

10-8-24 Why are lanthanons called inner transition elements (A.U., March 1993)

10-8-25 Discuss the oxidation states and colour of lanthanous (A.U., March 1993)

10-8-26 Write a comparative account of lanthanous and actinous based on their ionic and covalent radii and oxidation states. (A.U., Sept.1993)

ANSWERS

10-7-1 C, 10-7-2 D, 10-7-3 C, 10-7-4 C, 10-7-5 C, 10-7-6 C,

10-7-7 B, 10-7-8 B, 10-7-9 A, 10-7-10 B, 10-7-11 B, 10-7-12 C,

10-7-13 A, 10-7-14 C, 10-7-15 A, 10-7-16 B, 10-7-17 D, 10-7-18 B.

11

METALLURGY; BONDING IN METALS AND ALLOYS

11-1 INTRODUCTION

The metals obtained from the earth's crust are always found to be in the combined state with many other elements or their compounds as impurities. The metals in their combined state are known as **minerals**. Those minerals from which the metals can conveniently and economically be obtained are called **ores**. All ores are minerals but not all minerals are ores. The metals like silver, gold, platinum, mercury etc., are present in free native state. The pure native metals with some washable impurities like clay or sand are present as lumps in the earth's crust, as well as on the sea bed as **nodules**. These lumps or nodules are commonly called **nuggets**.

Most of the metals, because of their high chemical reactivity , are present in the combined state with moisture, air, nitrogen, oxygen, phosphorous, carbon dioxide, sulphur, arsenic, halogens etc., and sometimes as their sulphates, phosphates, carbonates. Based on the nature of the combination, the common metal ores are classified as follows:

11-1-1 Arsenide Ores

The Arsenide ores are formed when the naturally occuring arsenic combines with metals like nickel. The arsenic containing ores of nickel are kupfernickel-NiAs, Nickel glance - NiAsS etc.

11-1-2 Halide ores

Metallic halides are very few in nature. The most common halide ores are those of chlorides of sodium, potassium, magnesium, etc. Some of the halide ores are common salt or Rock Salt -NaCl, Fluorspar - CaF_2, Horn Silver - AgCl, Carnallite - $KClMgCl_26H_2O$, Cryolite - AgF_3. 3NaF, Syerite - KCl etc. Sea water is a rich source of metal ions like Mg^{++}, Ca^{++}, Na^+, K^+ in the form of their chlorides, bromides and iodides.

11-1-3 Oxide Ores

The metal ores of this type are present in the form of either their oxides or oxysalts like carbonates, phosphates, silicates, sulphates etc. Some of the important oxide ores are Pyrolusite- MnO_2, Haematite - Fe_2O_3, Magnetite Fe_3O_4, Cuprite - Cu_2O, Zincite - ZnO, Bauxite - $Al_2O_32H_2O$, Corundum - Al_2O_3, Cassiterite - SnO_2, etc.

11-1-4 Carbonate Ores

Metal oxides present in earth's crust when get exposed to carbondioxide form carbonate ores. Some of the carbonate ores are Lime Stone/Marble/ Calcite - $CaCO_3$, Smithsonite -$ZnCO_3$, Siderite - $FeCO_3$, Cerrusite - $PbCO_3$, Magnesite $MgCO_3$, Dolomite - $MgCO_3CaCO_3$, Malachite - $CuCO_3Cu(OH)_2$, Witherite - $BaCO_3$ etc.

11-1-5 Phosphate ores

Phosphatic ores are not very common. The important phosphate ores are Rock Phosphate - $Ca_5(PO_4)_3OH$ and Pyromorphite- $3Pb_3(PO_4)_2$ $PbCl_2$.

11-1-6 Silicate ores

Almost every mineral ore contains silica in the form of sand as an impurity. But there are many minerals which contain the silicate groups in them. It is a very difficult task to remove the silicate to get the pure metal, hence in most cases the silicates ores are made use of in their original form itself. The common silicate ores are Mica -$KAl_3H_2Si_3O_{12}$, Asbestos - $CaMg_3Si_4O_{12}$, Felspar-K_2O, Al_2O_3, $6SiO_2$, Hemimorphite-$2ZnOSiO_2H_2O$, Beryl - $Be_3Al_2Si_6O_{18}$ etc.

11-1-7 Sulphate Ores

Sulphate ores are the oxidation products of the sulphide ores. The important sulphate ores which find various applications in the industry, in their original state itself, are Epsomite-$MgSO_4$, $7H_2O$ Barytes-$BaSO_4$, Anhydrite-$CaSO_4$, Anglesite ($PbSO_4$), Celestite- $SrSO_4$.

11-1-8 Sulphide Ores

The sulphide ores are also referred to as sulpherised ores as they are formed by the combination of the metals (usually the heavier ones) with the native sulphur of the soil . The commonly found sulphide ores are Argentite-Ag_2S, Chalcocite-Cu_2S, Copper Pyrites/Glance - CuS, Pyrites/Ironpyrites - FeS, Cinnabar - HgS, Gelana - PbS, Sphalerite - ZnS etc.

11-1-9 Some other Important Ores

The hydroxide ores like Talc-$Mg_3(Si_4O_{10})$ $(OH)_2$, Brucite-$Mg(OH)_2$ etc., and the chromate ore like Crocoite - $PbCrO_4$, are important. Further more, vast areas of the ocean floor are covered with manganese nodules, which are made up mostly of manganese along with iron, nickel, copper and cobalt.

Type	Minerals
Native Metals	Ag, Au, Bi, Cu, Pd, Pt.
Carbonates	$BaCO_3$ (lithrite) $CaCo_3$ (calcite), $MgCO_3$ (magnesite), $PbCO_3$ (cerussite), $ZnCO_3$ (smithsonite), $CaCO_3MgCO_3$ (dolomite)
Chromate	$PbCrO_4$ (crocite)
Halides	CaF_2 (fluorite), NaCl (halite), KCl (sylvite), Na_3AlF_6 (cryolite)
Hydroxides	$Mg(OH)_2$ (brucite), $Mg_3Si_4O_{10}(OH)_2$ (talc)
Oxides	$Al_2O_3.2H_2O$ (bauxite), Al_2O_3 (corundum), Fe_2O_3 (hematite), Fe_3O_4 (magnetite), Cu_2O (cuprite), MnO_2 (pyrolusite), SnO_2 (cassiterite), TiO_2 (rutile), ZnO_2 (zincite)
Phosphates	$Ca_3(PO_4)_2$ (phosphorite), $Ca_3(PO_4)_2CaF_2$ (flurapalite), $Ca_3(PO_4)_2CaF_2$ (chlorapatite).
Silicates	$Be_3Al_2Si_6O_{18}$ (beryl), $ZrSiO_4$ (zircon), $NaAlSi_3O_8$ (albite).
Sulphides	Ag_2S (argentite), CdS (greenockite), Cu_2S (Chalcocite), FeS_2 (pyrite), HgS (cinnabar), PbS (gelena), ZnS (sphalerite).
Sulphates	$BaSO_4$ (barite), $CaSO_4$ (anhydrite),, $PbSO_4$ (angelsite) $SrSO_4$ (celesite) $MgSO_4.7H_2O$ (epsomite).

Fig. 11-1 Classification of metals according to their best known minerals.

The figure 11-1, shows a classification of metals according to their best known minerals. From a close observation we note that lithium containing metal is spodumene - Li $AlSi_2O_6$ and the beryllium containing mineral is beryl. The minerals containing the rest of the alkaline earth metals are generally carbonates and sulphates. The minerals of Sc, Y, La and the lanthanides are usually phosphates, present in combined state in monazite, along with U and Th.

Some metals have more than one type of important mineral. For example in addition to sulphide, iron is also known to be found as the oxides hematite- Fe_2O_3 and magnetite - Fe_3O_4 and aluminium in addition to the oxide is found in beryl - $Be_3Al_2Si_6O_{18}$, as silicate.

11-2 METALLURGICAL PROCESSES

The metals have to be obtained from the ores in their purest form and the process usually involves three principal steps

i) Preparation of the ore
ii) Production of the metal and
iii) Purification of the metal.

The extraction of every metal adopts a different technology depending on its chemical reactivity. Th· science and technology of separating metals from their ores and compounding of alloys is known as **Metallurgy.**

Before going into the details of the principal steps involved in metallurgy let us first look at the general principles involved for the isolation of metals depending upon the chemical composition of the ore.

11-2-1 Principles involved in Metallurgy :

Every mineral ore needs a different type of treatment before it is subjected to the principles of preparation of the ore. The reactions below indicate principles involved for different chemical classes of ores.

The carbonate ores need strong heating, so that they form the respective metallic oxides. The oxide is then reduced by different processes to get the pure metal.

$$CaCO_3 \xrightarrow{\Delta} CaO + CO_2$$
$$ZnCO_3 \xrightarrow{\Delta} ZnO + CO_2$$

The halide ores which are usually water soluble, ionic compounds, such as those of the alkali and alkaline earth metals are electrolysed. These ores which are mostly chlorides of Li, Na, K, Rb, Cs, Mg, Ca, Sr, and Ba are very stable usually (cannot be decomposed by heating) therefore are thermally electrolyse in their fused state

$$KCl \rightleftharpoons K^+ + Cl^-$$
$$\text{at anode} \quad 2Cl \longrightarrow Cl_2 + 2e^-$$
$$\text{at cathode} \quad K^+ + e^- \longrightarrow K$$

Insoluble chlorides like AgCl are converted into soluble coordination compounds and the metal is obtained by the displacement of the central metal ion.

$$AgCl + 2NaCN \longrightarrow Na[Ag(CN)_2] + NaCl$$

After making the complex alkaline by the addition of NaOH, it is reduced either with Zinc or aluminium, which discharges pure silver metal into the solution.

$$2Na[Ag(CN)_2] + Zn \longrightarrow Na_2[Zn(CN)_4] + 2Ag$$

For recycling the NaCN, the cyanozincate complex is treated with NaOH.

$$Na_2[Zn(CN)_4] + NaOH \longrightarrow Na_2ZnO_2 + 4NaCN.$$

The oxide ores of metals like zinc, tin, iron, manganese, aluminium etc., are usually reduced by heating with charcoal to their metallic state.

$$ZnO + C \xrightarrow{\Delta} Zn + CO$$
$$SnO_2 + 2C \xrightarrow{\Delta} Sn + 2CO$$
$$Fe_2O_3 + 3C \xrightarrow{\Delta} 2Fe + 3CO$$

In case of metals like chromium and manganese, the oxide is reduced with aluminium, following Goldschmidt - thermite process. Such reduction is required because these metals have a tendency to react with carbon, to give carbides at the reduction temperature.

$$Cr_2O_3 + 2\,Al \xrightarrow{\Delta} 2\,Cr + Al_2O_3$$
$$3Mn_3O_4 + 8Al \xrightarrow{\Delta} 9\,Mn + 4\,Al_2O_3$$

Oxides of tungstun cannot be reduced by the above processes. Therefore they are reduced in a current of hydrogen

$$WO_3 + 3H_2 \longrightarrow W + 3H_2O$$

When none of these chemical reduction processes work, as in the case of alumina electrolytic reduction is carried out to get pure metal. The electrolytic reduction of alumina is performed in the presence of molten cryolite Na_3AlF_6.

$$\underset{\text{(molten)}}{Al_2O_3} \rightleftharpoons 2\ Al^{3+} + 3O^{2-}$$

at cathode : $2\ (2Al^{3+} + 6e^- \longrightarrow 2\ Al\ )$

at anode : $6\ O^{2-} \longrightarrow 3\ O_2 + 12e^-$

Amongst the sulphate minerals barytes, It is the most important, its powdered ore is heated strongly between 600-800°C with charcoal to get the sulphide which is further subjected to different other processes to get the metal. The conversion of sulphate to sulphide can be represented by the following reactions

$$BaSO_4 + 4\ C \longrightarrow BaS + 4\ CO$$
$$BaSO_4 + 2\ C \longrightarrow BaS + 2\ CO_2$$

The sulphide ores are usually converted to their oxides and the oxides are reduced with either carbon or aluminium to get the pure metal. For example, gelena - PbS is heated over 500°C for its partial conversion to the yellow plumbic oxide - PbO and to plumbous sulphate.

$$2PbS + 3O_2 \longrightarrow 2\ PbO + 2\ SO_2$$
$$PbS + 2\ O_2 \longrightarrow PbSO_4$$

PbO and $PbSO_4$ combines with PbS at very high temperatures in the absence of air to form lead metal.

$$PbS + 2PbO \longrightarrow 3\ Pb + SO_2$$
$$PbS + PbSO_4 \longrightarrow 2\ Pb + 2\ SO_2$$

Zinc blende - ZnS is heated in a free supply of air to yield zinc oxide which is readyly reduced to zinc metal in the presence of carbon.

$$2\ ZnS + 3O_2 \longrightarrow 2\ ZnO + 2\ SO_2$$
$$ZnO + C \longrightarrow Zn + CO$$

Copper glance/cuprous sulphide Cu_2S obtained from copper pyrites $CuFeS_2$, is heated in a current of air to Cu_2O, which combines further with Cu_2S to get pure copper.

$$2\ Cu_2S + 3\ O_2 \longrightarrow 2\ Cu_2O + 2\ SO_2$$
$$2\ Cu_2O + Cu_2S \longrightarrow 6\ Cu + SO_2$$

$$Cu_2S + O_2 \longrightarrow 2\ Cu + SO_2$$

11-2-3 Preparation of the Ore

In the preliminary treatment of an ore the desired mineral is crushed and pulverised. The crushing is done in the mechanical jaw crushers. These crushers consists of a set of stationary plates on which moves another set of plates with jaws, the ore is passed through these plates to be crushed into tiny pieces as seen the fig 11-2.

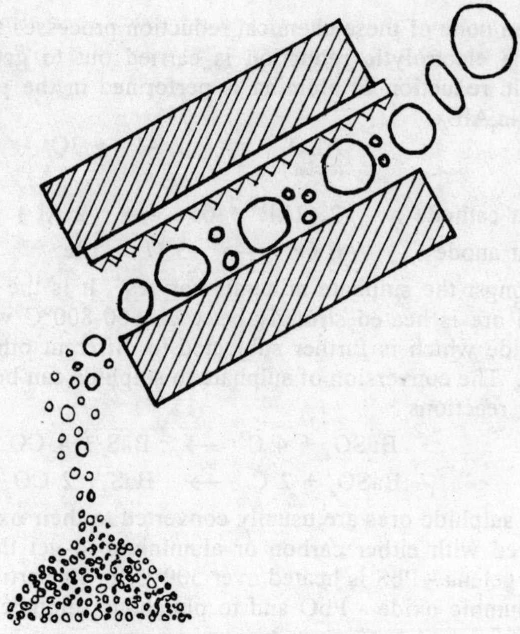

Fig 11-2 Mechanical Jaw Crusher

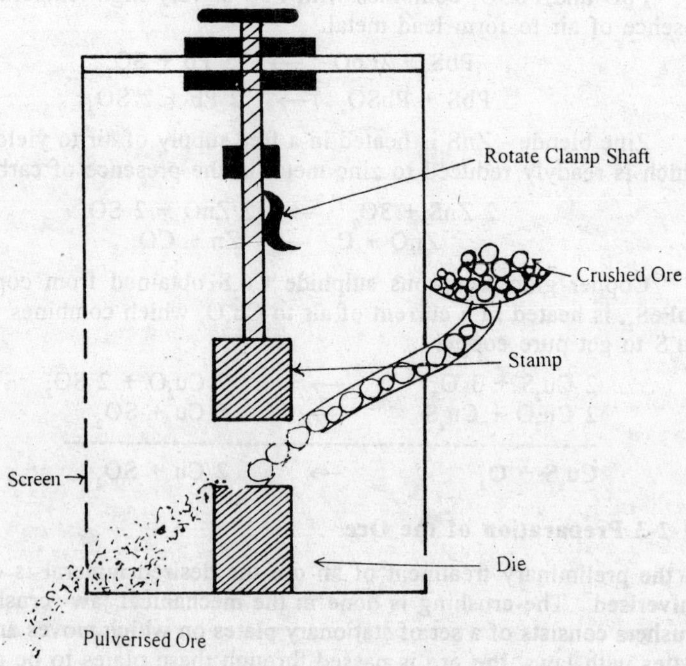

Fig 11.3 Rotating Stamp Crusher

The crushed ore is then subjected to pulverisation, usually in a stamp mill or a ball mill. The crushed ore is passed over a hard metallic die on which a rotating stamp hits with a regular interval, fig-11-3. To pulverise the ore in a ball mill the ore is placed in a huge steel cylinder containing iron balls, the cylinder is rotated in a fast motion, which helps the ore pulverise due to its contact with the ball and the barrel.

Now the pulverised ore is subjected to Concentration or ore dressing. The desired mineral is separated from waste materials - usually clay and silicate minerals - which are collectively called **gangue** or **matrix**. One very useful method for carrying out such a separation is called flotation. In this process, the ore is finely ground and added to water containing oil and

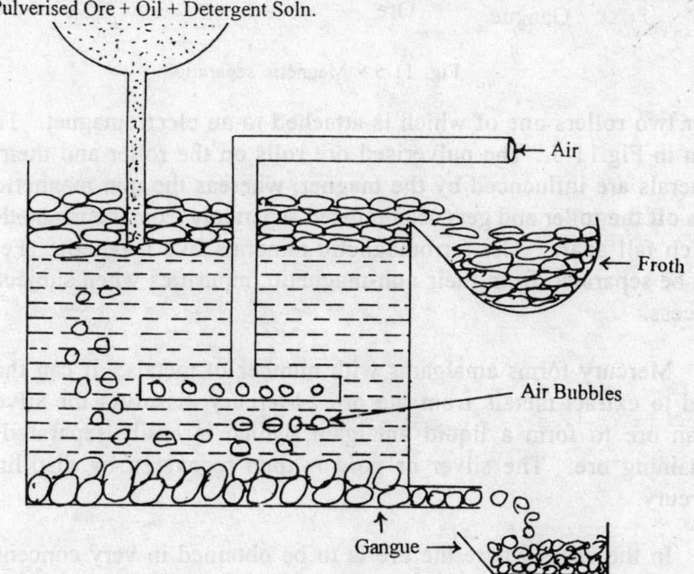

Fig. 11-4 - Froth Floatation process

a detergent. The liquid mixture is then beaten or a current of air is circulated to form a froth. The oil preferentially wets the mineral particles which are carried to the top in the form of a froth, while the gangue settles to the bottom. The froth is skimmed off, collected and allowed to subside and dry. This method is suitable for the low grade sulphides of zinc blende (ZnS), copper pyrites ($CuFeS_2$), etc. The froth flotation process can be visualised in Fig 11-4. The gangue particles settle at the bottom from where these are drawn off.

Another physical separation process depends upon the magnetic properties of certain minerals. The minerals possessing magnetic properties, can be separated from the nonmagnetic gangue by using a strong electromagnet, the pulverised ore is made to fall on to a leather belt moving

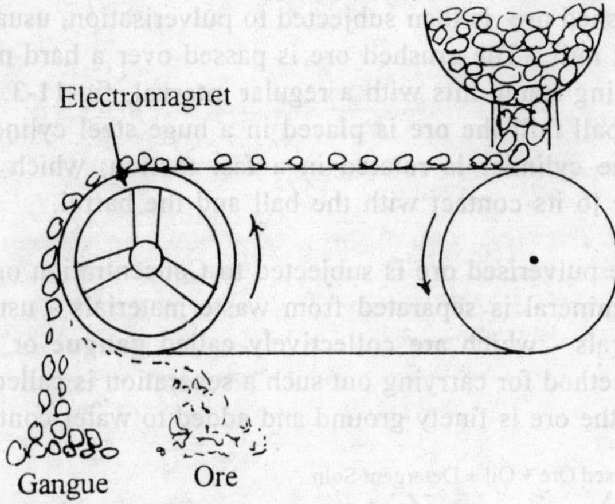

Fig. 11-5 - Magnetic separation

over two rollers one of which is attached to an electromagnet. This can be seen in Fig.11-5. The pulverised ore rolls on the roller and their magnetic minerals are influenced by the magnet, whereas the non magnetic particles flies off the roller and gets accumulated separately from the magnetic particles which fall apart. The ferromagnetic minerals like magnetite (Fe_3O_4), etc., can be separated from their non-magnetic impurities when subjected to this process.

Mercury forms amalgams with number of metals. It can therefore be used to extract metals from the ore. Mercury dissolves the silver or gold in an ore to form a liquid amalgam, which is easily separated from the remaining ore. The silver or gold is then recovered by distilling off the mercury.

In the cases where the ore is to be obtained in very concentrated and pure forms, chemical method is made use of. In particular, the extraction of aluminium is done by this process. Bauxite ($Al_2O_3.2H_2O$) which is the chief ore of aluminium contains oxides of silicon, iron, titanium etc., in it. these impurities are removed by digesting the pulverised ore with an aqueous solution of Sodium hydroxide around 150°C under pressure. The Al_2O_3 present in the ore dissolves in NaOH whereas the other oxides remain undissolved and are filtered off. The filtrate rich in sodium aluminate is diluted with water to get aluminium hydroxide, which is filtered and ignited to give aluminium oxide.

$$\underset{\text{(Bauxite)}}{Al_2O_3} + 6\ NaOH \longrightarrow 2\ Na_3AlO_3 + 3\ H_2O$$

$$Na_3AlO_3 + 3\ H_2O \longrightarrow Al(OH)_3 + 3\ NaOH$$

$$2\ Al(OH)_3 \xrightarrow{\Delta} Al_2O_3 + 3\ H_2O$$

11-2-4 Production of the metal

As metals in their combined form always have positive oxidation numbers, the production of a free metal is always a reduction process, preliminary operations may be necessary to convert the ore to a chemical state more suitable for reduction. An ore has either to be **calcined** or **roasted** to drive off volatile impurities and at the same time to convert the carbonates and sulphides to their corresponding oxides, which can be reduced more easily to yield the pure metals. *Calcination involves heating the ore in the absence or in a limited supply of air below the melting point of the ore where as roasting is done in the presence of excess air below the melting point of the ore.*

Calcination:

i) $ZnCO_3 \xrightarrow{\Delta} ZnO + CO_2$

ii) $Al_2O_3.2H_2O \xrightarrow{\Delta} Al_2O_3 + 2\ H_2O$

iii) $CaCO_3 \xrightarrow{\Delta} CaO + CO_2$

Roasting:

iv) $2\ ZnS + 3\ O_2 \xrightarrow{\Delta} 2\ ZnO + 2\ SO_2$

v) $PbS + 2\ O_2 \xrightarrow{\Delta} PbSO_4$

The equation (iv) shows the conversion of sulphide to sulphur dioxide which is a major air pollutant, hence usually the sulphide ores are heated in a limited quantity of oxygen to get a sulphate following equation (v). The developments in the recent years allow the convertion of sulphurdioxide to sulphuric acid. This has not only helped to reduce the gas emission that polute the atmosphere but convert them into useful biproducts. How a pure metal is obtained by reduction from its combined form depends on the standard reduction potentials of the metal.

Table 11-1 Standard Reduction Potentials of Metals at 25°C

Half Reaction	E°C=(V)	Half Reaction	E°C=(V)
$Li^+_{(aq)} + e^- \longrightarrow Li_{(s)}$	- 3.05	$Cr^{3+}_{(aq)} + 3e^- \longrightarrow Cr_{(s)}$	- 0.74
$K^+_{(aq)} + e^- \longrightarrow K_{(s)}$	- 2.93	$Fe^{2+}_{(aq)} + 2e^- \longrightarrow Fe_{(s)}$	- 0.44
$Ba^{2+}_{(aq)} + 2e^- \longrightarrow Ba_{(s)}$	- 2.90	$Cd^{2+}_{(aq)} + 2e^- \longrightarrow Cd_{(s)}$	- 0.40
$Sr^{2+}_{(aq)} + 2e^- \longrightarrow Sr_{(s)}$	- 2.89	$Co^{2+}_{(aq)} + 2e^- \longrightarrow Co_{(s)}$	- 0.28
$Ca^{2+}_{(aq)} + 2e^- \longrightarrow Ca_{(s)}$	- 2.87	$Ni^{2+}_{(aq)} + 2e^- \longrightarrow Ni_{(s)}$	- 0.25
$Na^+_{(aq)} + e^- \longrightarrow Na_{(s)}$	- 2.71	$Sn^{2+}_{(aq)} + 2e^- \longrightarrow Sn_{(s)}$	- 0.14
$Mg^{2+}_{(aq)} + 2e^- \longrightarrow Mg_{(s)}$	- 2.37	$Pb^{2+}_{(aq)} + 2e^- \longrightarrow Pb_{(s)}$	- 0.13
$Be^{2+}_{(aq)} + 2e^- \longrightarrow Be_{(s)}$	- 1.85	$Hg^{2+}_{(aq)} + 2e^- \longrightarrow Hg_{(s)}$	- 0.85
$Al^{3+}_{(aq)} + 3e^- \longrightarrow Al_{(s)}$	- 1.66	$Ag^+_{(aq)} + e^- \longrightarrow Ag_{(s)}$	-0.799
$Mn^{2+}_{(aq)} + 2e^- \longrightarrow Mn_{(s)}$	- 1.18	$Pt^{2+}_{(aq)} + 2e^- \longrightarrow Pt_{(s)}$	- 0.98
$Zn^{2+}_{(aq)} + 2e^- \longrightarrow Zn_{(s)}$	- 0.76	$Au^+_{(aq)} + e^- \longrightarrow Au_{(s)}$	- 1.68

From the standard reduction potential values, the type of reduction can be suggested as given in the following table 11-2.

Table 11-2 Reduction Processes for some common metals

	Metal	Type of Reduction Potental	Reduction Process
Decreasing activity of the metal ↓	Li, Ca, Na, Mg	Quite High	Most difficult to reduce chemically - Electrolytic reduction of the molten halide ore.
	Al	High	Difficult to reduce chemically - Electrolytic reduction of anhydrous oxide (in molten cryolite).
	Mn, Zn, Cr, Fe, V	Medium	Easily reduced - Reduction of the metal oxide with a more electropositive metal or reduction with coke and CO_2.
	Hg, Ag, Pt, Cu, Au	Low	Most easily reduced - These metals occur in the free (uncombined) state or can be obtained by roasting.

Another major metallurgical process is carried out at high temperature by a procedure called **pyrometallurgy**. For instance a more electropositive metal is used as a reducing agent to separate a less electro positive metal from its compounds at high temperatures :

$$Cr_2O_3 + 2\,Al \xrightarrow{\Delta} 2\,Cr + Al_2O_3$$

$$3\,Mn_3O_4 + 8\,Al \xrightarrow{\Delta} 9\,Mn + 4\,Al_2O_3$$

In some cases even molecular hydrogen (E°=0) can be used for reduction:

$$WO_3 + 3\,H_2 \xrightarrow{\Delta} W + 3\,H_2O$$

Apart from these **chemical reduction** processes, the other suitable method employed specially for the highly electropositive metals is the **electrolytic reduction**, this is employed for the reduction of potassium, sodium, magnesium, aluminium etc. The process is usually carried out on the anhydrous molten oxide or halide of the metal.

$$2\,MO_{(l)} \longrightarrow 2\,M_{(\text{at cathode})} + O_{2\,(\text{at anode})}$$

$$2\,MCl_{(l)} \longrightarrow 2\,M_{(\text{at cathode})} + Cl_{2\,(\text{at anode})}$$

On calcination as well as on roasting the ore becomes porous so that it undergoes easier reduction in the next step of metallurgy. Calcination/ roasting is performed in a **reverberatory furnace**. This furnace, Fig.11-6, consists of a fire box and a huge hearth, lined inside with a suitable flux. Air is circulated through vents. The charge which is kept on the hearth is heated by directing the flames from the fire place. The concave roof of the furnace helps in reverberating the hot flames on to the charge kept on the hearth. The charge is stirred mechanically through locking doors fixed on the hearth.

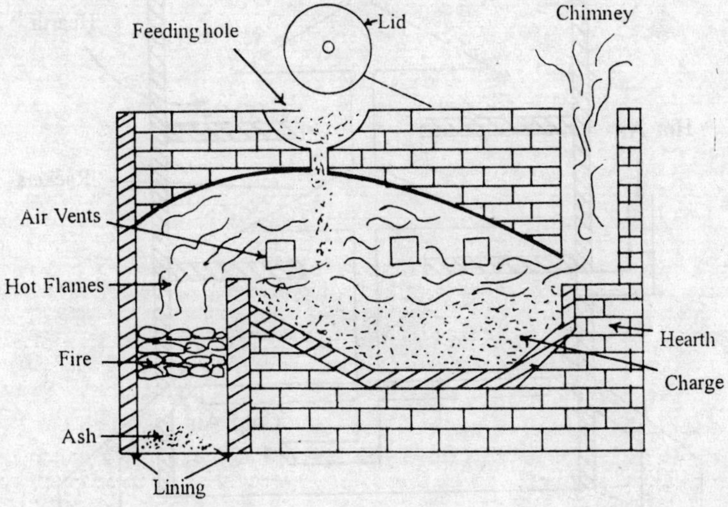

Fig 11-6-Reverberatory furnace.

The modern multiple hearth furnace has replaced the reverberatory furnace. The charge is dropped from the top which spreads in thin layers on the revolving mechanical rockers attached to the central shaft. These rockers are moving around the five to seven hearths present, as seen in Fig 11-7. Slowly the charge gets spread and is subjected to heating, producer gas is used as a fuel. The central shaft and the attached racks are usually hollow and a circulation of water/oil keeps them cool during the process. This is a continuous and an efficient process. The waste gases like SO_2 are captured and converted to SO_3 to be used in the manufacture of sulphuric acid.

After calcination or roasting the ore is subjected to reduction to get the pure metal. The reduction can be done either by chemical or electrolytic processes. The common methods involved are (A) **Smelting** (B) **Electrolytic Reduction** and (C) **Reduction by Precipitation Processes/ Hydrometallurgy.**

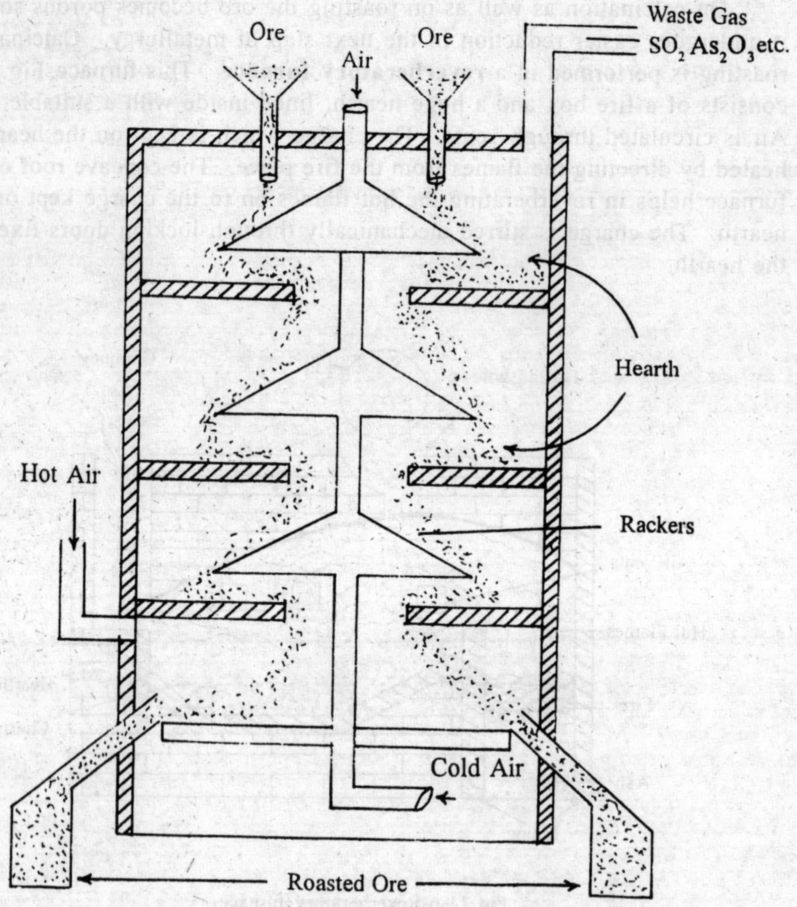

Fig 11-7-Multiple Hearth Furnace.

(A) ***Smelting*** *is the process in which the metal is extracted in a fused molten state, usually through reduction.* By this process the metals from the concentrated ores are obtained in more or less pure condition by one of the following methods.

i) *Reduction of the oxide with hydrogen in the case of ιtungstem and molybdenum.*

ii) *Reduction of the oxide with carbon or carbon monoxide in a blast furnace in the case of iron, zinc, tin, cobalt, copper, lead, magnesium, etc.*

iii) *Reduction of the oxide with aluminium - thermite process - in the case of chromium and manganese.*

iv) *Autoreduction (Oxidation) of the sulphide for example in the extraction of copper, lead, mercury, etc.*

v) *Reduction of the sulphide with iron, for example in antimony extraction.*

vi) *Reduction with water gas is done in the extraction of nickel.*

vii) *Reduction of tetra chlorides of titanium and vanadium with sodium or magnesium.*

i) ***Reduction*** of tungstun and molybdenum oxide ore is done at a temperature range of 1000-1200°C by passing a current of hydrogen over the metal oxide.

$$WO_3 + 3H_2 \longrightarrow W + 3H_2O$$
$$MoO_2 + 2H_2 \longrightarrow Mo + 2H_2O$$

ii) ***On reducing the oxide*** with carbon or carbon monoxide, free molten metal is obtained, these reactions are conducted in the blast furnace, fig. 11-8.

$$Fe_2O_3 + 3\ C \longrightarrow 2\ Fe + 3\ CO$$
$$Fe_2O_3 + 3\ CO \longrightarrow 2\ Fe + 3\ CO_2$$
$$ZnO + C \longrightarrow Zn + CO$$
$$SnO + CO \longrightarrow Sn + CO_2$$
$$CuO + C \longrightarrow Cu + CO$$
$$Co_3O_4 + 4\ C \longrightarrow 3\ Co + 4\ CO$$

In this process some amount of gangue may still remain with the molten metal hence to remove it another impurity in the form of **flux** is added. The flux combines with gangue to form **slag**. The slag formed being lighter than the molten metal floats on the surface and is drawn out of the blast furnace. Choósing of the right flux is very essential, for a basic gangue (CaO, MgO, etc.) an acidic flux (SiO_2) is added, and for an acidic gangue (SiO_2) a basic flux ($CaCO_3$, $MgCO_3$ etc.) is added. The reactions which follow on the addition of the flux are as follows;

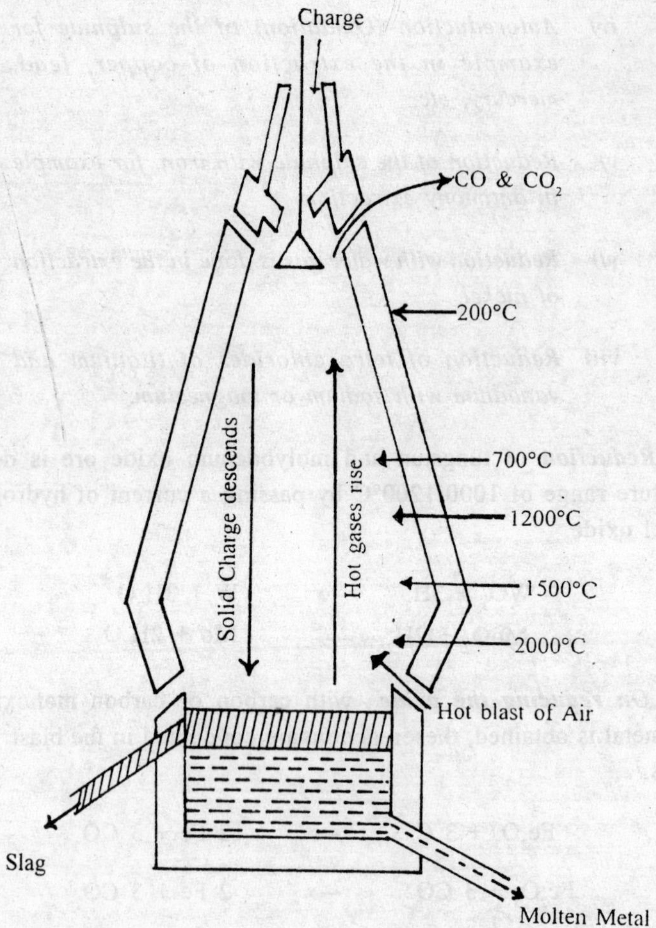

Fig 11-8-Blast furnace.

$$\underset{\text{(Gangue)}}{MgO} + \underset{\text{(Flux)}}{SiO_2} \longrightarrow \underset{\text{(Slag)}}{MgSiO_3}$$

$$\underset{\text{(Gangue)}}{SiO_2} + \underset{\text{(Flux)}}{CaCO_3} \longrightarrow \underset{\text{(Slag)}}{CaSiO_3} + CO_2$$

In to the blast furnace, the concentrated ore, the flux (eg. lime stone: $CaCO_3$) and coke are charged from the top. A blast of preheated air is forced simultaneously up the furnace from the bottom - hence the name - blast furnace. The oxygen of the air reacts with the carbon in the coke to form mostly carbon monoxide and some carbon dioxide. These reactions are highly exothermic, and as the hot carbon monoxide and carbondioxide gases rise, they react with the metal oxides at different temperature zones as shown in the fig 11-8. The key steps, for example in the case of the metallurgy of iron, are

$$3\ Fe_2O_{3(s)} + CO_{(g)} \longrightarrow 2\ Fe_3O_{4(s)} + CO_2$$
$$Fe_3O_{4(s)} + CO_{(g)} \longrightarrow 3\ FeO_{(s)} + CO_{2(g)}$$
$$FeO_{(s)} + CO_{(g)} \longrightarrow Fe_{(m)} + CO_2$$

The molten iron ($Fe_{(m)}$) is collected at the bottom and is drawn out. The limestone decomposes in the furnace as follows:

$$CaCO_{3(s)} \longrightarrow CaO_{(s)} + CO_{2(g)}$$

The calcium oxide then reacts with the impurities in the iron, which are mostly sand (SiO_2) and aluminium oxide (Al_2O_3)

$$CaO_{(s)} + SiO_{2(s)} \longrightarrow CaSiO_{3(l)}$$
$$CaO_{(s)} + Al_2O_{3(s)} \longrightarrow Ca(AlO_2)_{2(l)}$$

$CaSiO_3$ and $Ca(AlO_2)_2$ are the molten slags.

By the time the ore works its way down to the bottom of the furnace, most of it has already been reduced to iron. The temperature of the lower part of the furnace is above the melting point of impure iron, and so the molten iron at the lower level can be run off to a receiver and the slag being less dense forms the top layer above the molten iron and can be skimed off at that level. The iron obtained in the process is known as **Pig iron,** because of many impurities like silicon, phosphorous, manganese, sulphur and about 5% carbon, it is a granular type of iron. For various other type of iron, other procedures are made use of .

iii. Thermite Process- Reduction by Active Metals : Certain oxides are not satisfactorily reduced by carbon. In such cases the active metals like aluminium, magneseium, etc., are used as reducing agents. A mixture of the metallic oxide and aluminium powder is called *thermite* and hence the process involved is named *thermite process* or is also termed as *Goldschmidt's Alumino thermite process.* The thermite is ignited in a closed crucible, (fig.11-9), by means of a lighted magnesium ribbon. When the flash subsides, the molten metal is drawn from the bottom of the crucible. Metals like chromium and manganese are extracted by thermite process.

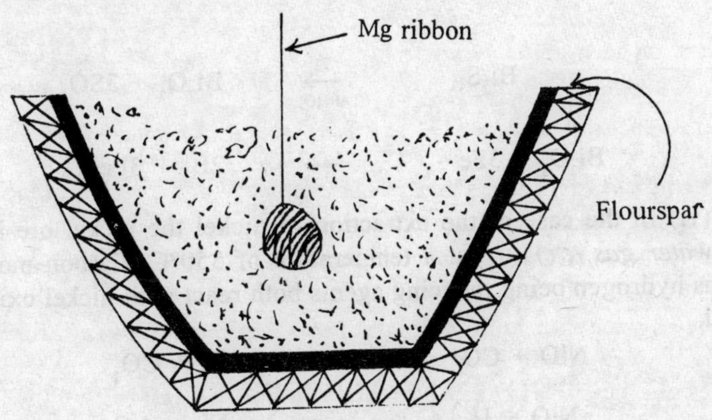

Fig. 11-9 Alumino Thermite Process

$$Cr_2O_3 + 2Al \longrightarrow Al_2O_3 + 2Cr$$

$$3Mn_3O_4 + 8Al \longrightarrow 4Al_2O_3 + 9Mn$$

The Reduction of tetra chlorides of titanium and vanadium *with active metal* can also be classed under these type of reactions.

$$TiCl_4 + 4Na \longrightarrow Ti + 4NaCl$$

$$VCl_4 + 2Mg \longrightarrow V + 2MgCl_2$$

(iv) Autoreduction of the sulphide ores of mercury and lead is performed at high temperature. For example:

$$2HgS + 3O_2 \longrightarrow 2HgO + 2SO_2$$

$$2HgO + HgS \longrightarrow 3Hg + SO_2$$

$$PbS + PbSO_4 \longrightarrow 2Pb + 2SO_2$$

At moderate heats (500-700°C) if the ore is heated in a limited supply of air then partial reduction takes place

$$2PbS + 3O_2 \longrightarrow 2PbO + 2SO_2$$

$$PbS + 2PbO \longrightarrow 3Pb + SO_2$$

(v) Reduction of Sulphide with iron, for example, in the case of antimony can be done, by heating the sulphide ore at 550°-600°C where in the ore which contains about 40-60% antimony in it gets melted. The molten ore, Sb_2S_3 is then treated with scrap iron to remove the sulphide.

$$Sb_2S_3 + 3Fe \longrightarrow 2Sb + 3FeS$$

In the case of bismuth, the sulphide ores are oxidised to their oxide form and then reduced by iron.

$$Bi_2S_3 \xrightarrow[\text{Air [O]}]{\Delta} Bi_2O_3 + 3SO_2$$

$$Bi_2O_3 + 3Fe \longrightarrow 2Bi + 3FeO$$

(vi) In the case of the extraction of nickel the nickel ore is *reduced with water gas* $(CO+H_2)$ at a temperature of 350°C. Carbon monoxide as well as hydrogen being reducing agents both react with nickel oxide to give nickel.

$$NiO + CO \longrightarrow Ni + CO_2$$

$$NiO + H_2 \longrightarrow Ni + H_2O$$

(B) Electrolytic Reduction/Electro metallurgy can be performed for very active metals of the alkali and alkaline earth groups and aluminium. These metals are extracted by the electrolysis of their oxides, hydroxides or chlorides in a fused state. The metal is liberated at the cathode and the gases like oxygen, chlorine etc. are collected at the anode. The electrolytic reactions which takes place to form some active metals are given below in Table 11-3.

Table 11-3 : Electrolytic reduction of active metals.

Compound Electrolysed	Metal at Cathode	Gases at Anode	Electrolytic Chemical Reactions
NaOH	Na alongwith H_2	O_2	$2\ NaOH \rightarrow 2Na + H_2 + O_2$
NaCl	Na	Cl_2	$2\ NaCl \rightarrow 2\ Na + Cl_2$
Al_2O_3	Al	O_2	$2\ Al_2O_3 \rightarrow 4\ Al + 3\ O_2$
$MgCl_2$	Mg	Cl_2	$MgCl_2 \rightarrow Mg + Cl_2$
$CaCl_2$	Ca	Cl_2	$CaCl_2 \rightarrow Ca + Cl_2$

The metals of the lanthanide family as well as Sc and Y can be obtained by the electrolysis of their fused chlorides.

(C) Hydrometallurgy : It is the process of bringing the metai into solution by the action of suitable chemical reagents, followed by recovery of the metal by electrolysis or by the use of proper precipitating agents, hence this process is known as **wet process of reduction by precipitation** process. For example, the sulphide ore of copper is treated with air to obtain copper sulphate which is then extracted by water and the copper is recovered either by electrolysis or by precipitation with the addition of scrap iron.

$$CuSO_4 + Fe \longrightarrow FeSO_4 + Cu$$

Ores of silver are kept in dilute solutions of sodium cyanide and air is blown into the solution. The silver goes into solution to form an argento-cyanide complex.

$$4Ag + 8\ NaCN + O_2 + 2H_2O \rightleftharpoons 4\ Na\ [Ag(CN)_2] + 4NaOH$$

or

$$Ag_2S + 4\ NaCN \rightleftharpoons 2\ Na[Ag(CN)_2] + Na_2S$$

The solution containing the complex is treated with sodium hydroxide to make it alkaline. It is then treated with zinc turnings, by which silver precipitates out.

$$2Na[Ag(CN)_2] + Zn \longrightarrow Na_2\ [Zn\ (CN)_4] + Ag$$

11-2-5 Purification of the metal

Metals prepared by Reduction usually need furthur treatment to remove various impurities. The extent of purification, of course, depends on the use to be made of the metal. Below we describe few common purfication procedures.

(a) Distillation : Metals that have low boiling points, such as mercury, magnenium and zinc, may be separated from other metals by fractional distillation. One well known method of fractional distillation is the Mond process for the purification of nickel. Carbon monoxide gas is passed over the impure nickel metal of about 70°C to form the volatile ttetracarbonyl nickel (b.p. 43°C), a highly toxic substance, which is then separated from the less volatile impurities by distillation.

$$Ni_{(s)} + 4CO_{(g)} \longrightarrow Ni(CO)_{4(g)}$$

Pure nickel is then recovered at 200°C by distillation of $Ni(CO)_4$

$$Ni(CO)_{4(g)} \xrightarrow{\Delta} Ni_{(s)} + 4CO_{(g)}$$

The carbon monoxide released is recycled back into the process.

(b) Electrolysis : Electrolysis is another important purification technique, the metals like Cu, Ag, Au, Cr, Zn, Sn, Pb, Ni, etc., are refined electrolytically. For example the copper metal obtained by roasting copper

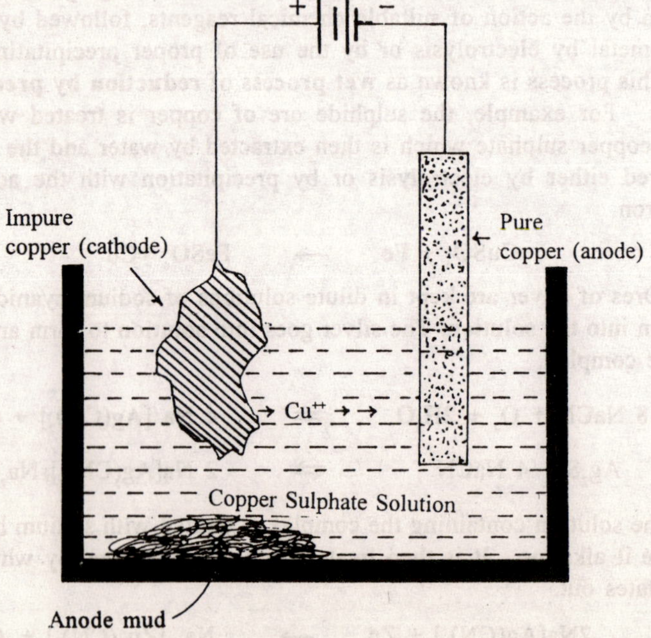

Fig. 11-10 Electrolytic refining

sulphide usually contains a number of impurities such as zinc, iron, silver and gold. The more electropositive metals are removed by an electrolysis process in which the impure copper acts as the anode and pure copper acts as a cathode. Fig 11-10 the reactions which proceed in the electrolytic bath are:

$$Cu_{(s)} \longrightarrow Cu^{2+} + 2e^- \text{ (Anode - oxidating)}$$

$$Cu^{2+} + 2e^- \longrightarrow Cu_{(s)} \text{ (Cathode - reduction)}$$

Iron and zinc and other more reactive metal impurities present in the copper anode get oxidised at the anode and enter the solution as Fe^{++} and Zn^{++}, but they are not reduced at the cathode. However, the less electropositive metals, such as gold and silver, are not oxidised at the anode. Eventually, as the copper anode dissolves these metals fall to the bottom of the cell, as **anode mud**. Thus the net result is the transfer of copper from the anode to the cathode. The copper obtained by electrolysis is 99.99% pure, hence electrolysis is a better way of refining a metal. The anode mud is then treated to separate precious Au and Ag.

(c) Zone Refining : Another method often used obtaining extremely pure metals is zone refining. It is based on the principle that melting point of a substance is lowered by the presence of impurities. Consequently, when an impure molten metal in cooled, crystals of pure metal separate out and the impurities remain in the molten metal. In this process a metal containing few impurities is drawn in the form of a bar. A circular heater is fitted around this bar and is slowly moved langitudinally from one end to other. At the heated zone the bar melts and as the heater moves on, pure metal crystallizes while the impurities pass into the adjacent molten portion. In this way the impurities are swept from one end of the bar to the other. As the molten rod from the heating coil, it cools and the pure metal crystallizes, leaving the impurities in the molten metal portion. When the molten zone carrying the impurites, now at increased concentration, reaches the end of the rod, it is allowed to cool and is then cut off. Repeating this procedure a number of times, results in the metal with a purity greater than 99.99 percent.

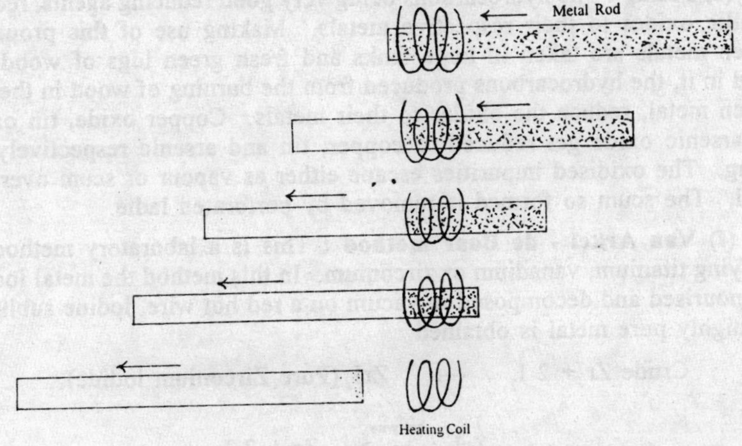

Fig. 11-11 Zone refining technique.

By repeating the process several times ultra pure metal can be obtained. This method is popularly employed in the preparation of metals such as Si, Te, Ge, etc., used in the semicunductors and solar panels.

(d) Liquation : Low melting metals like tin and lead are refined by this process. The impure metal is placed on a sloping hearth (Fig 11-12)

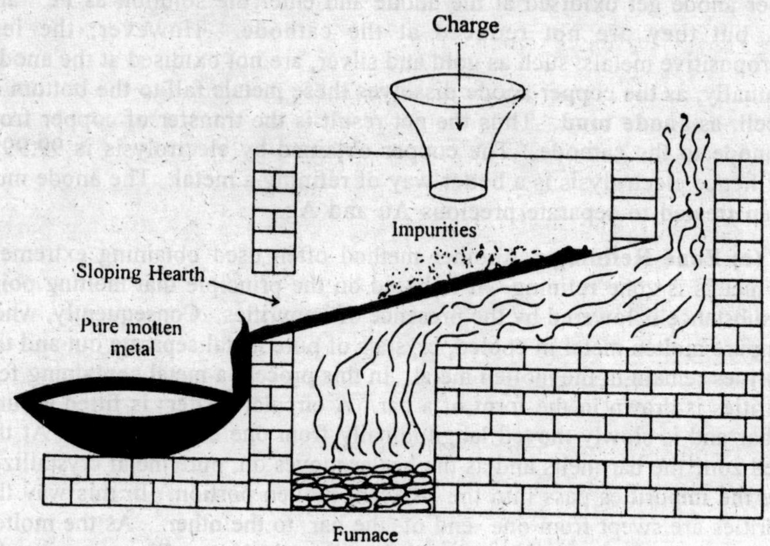

Fig 11-12 Liqution furnace

of a furnace and is heated gently. The metal in pure form flows down, leaving behind the infusible impurities called **dross**.

(e) Poling : The hydrocarbons being very good reducing agents, reduce metallic oxides to their respective metals. Making use of this property, molten metals are taken in huge tanks and fresh green logs of wood are stired in it, the hydrocarbons produced from the burning of wood in the hot molten metal, reduce the oxides to their metals. Copper oxide, tin oxide and arsenic oxide get reduced to copper, tin and arsenic respectively by poling. The oxidised impurities escape either as vapour or scum over the metal. The scum so formed is removed by perforated ladle.

(f) Van Arkel - de Boar method : This is a laboratory method of purifying titanium, vanadium or zirconium. In this method the metal iodide is vapourised and decomposed in vacum on a red hot wire, iodine sublimes and highly pure metal is obtained

$$\text{Crude Zr} + 2\,I_2 \longrightarrow ZrI_4 \text{ (Pure Zirconium iodide)}$$

$$ZrI_4 \xrightarrow[\text{vacum}]{\text{hot wire}} Zr + 2\,I_2$$

EXTRACTION OF METALS AND THEIR IMPORTANCE

11-3 URANIUM

11-3-1 Introduction

Uranium is the highest atomic number (92) and atomic weight (238) bearing and naturally occurring metal. This metal finds application for both destructive and constructive purposes. Uranium was discovered by Klaproth (1789) in a sample of the ore, pitchblende. Peligot (1840) suggested that the properties of this new element is due to its oxide UO_2, which also appeared like the metal. Peligot (1842) produced uranium by the reduction of UCl_4 with potassium metal in a closed platinum crucible. Madame Curie (1898) discovered radium from pitchblende and concluded that all uraninum minerals contain radium in small quantities as a result of radioactive equilibrium.

Uranium is widely distributed in nature and the average uranium ontent in the earth's crust is about 3.98 x 10^{-6}g per gram of rock. The principle ores of uranium are : pitchblende containing mainly U_3O_8 and carnolite containing $2K(UO_2)\ VO_4.3H_2O$. Uranium is present in phosphate rocks to an extent of 0.01%. In traces it is present in ores like fergnsonite, uronophane, autonite, smarskite and bituminous shales.

11-3-2 Metallurgy

Pitchblende and carnolite are two major sources which have been processed extensively. As the ores are different in composition, different procedures are made use of for the extraction of the metal. The processes usually employed are;

(a) The ores are brought into solution by treating them with a mineral acid like sulphuric acid, and sometimes, alkaline fusion using Na_2CO_3 to disintegrate the ore.

(b) Uranium is converted to a soluble complex carbonate and at the same time iron, aluminium, manganese etc., present in major proportions in the ore are removed.

(c) Lead, Copper etc. are precipitated as sulphides.

(d) Uranium is finally obtained as ammonium diuranate$(NH_4)_2U_2O_7$. This on igniting yields uranium oxide U_3O_8.

The process is given in Flow Sheet 11-1. The oxide obtained is now subjected to either chemical reduction or electrolysis or thermal decomposition.

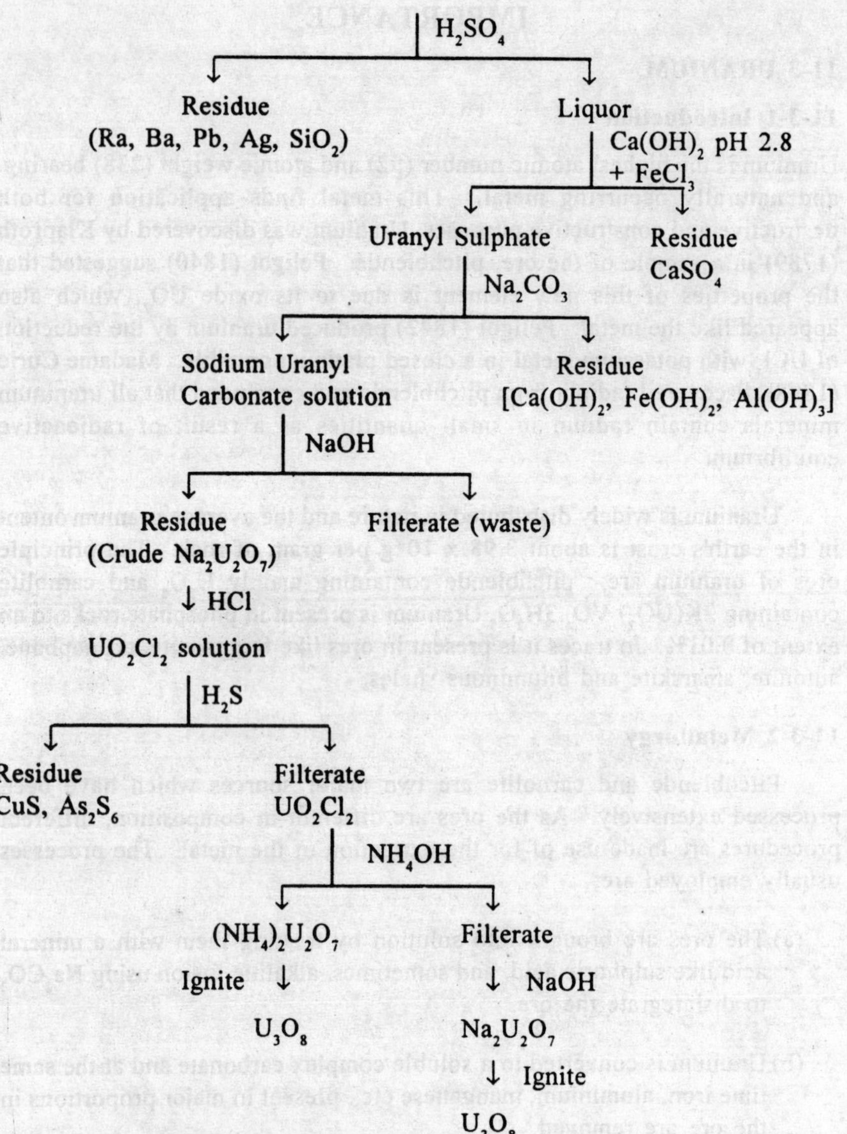

Flow Sheet 11-1 Extraction of Uranium

(A) Chemical reduction

(i) Reduction of the oxide with carbon at the electric arc temperature.

(ii) Reduction of U_3O_8 with aluminium, calcium or magnesium at a very high temperature.

(iii) Reduction of uranium halides with alkali or alkaline earth metals in steel bomb.

(B) Electrolysis of double salt, KUF_5 or UF_4 in fused salt baths of $CaCl_2$(80%) and NaCl(20%) at 900°C in a graphite crucible, which also serves as anode and a molybdenum cathode gives 99.9% of pure uranium.

(C) Thermal decomposition of uranium halides on a hot tungsten filament below 11 33°C (the melting point of U)also produces U.

Extremely pure uranium can be obtained by the reduction of UO_3 with hydrogen at a temperature of 650°C

$$UO_3 + H_2 \longrightarrow UO_2 + H_2O$$

The dioxide formed is digested in hydrofluoric acid to get UF_4, which is reduced with calcium to get the pure metal.

$$UO_2 + 4HF \longrightarrow UF_4 + 2H_2O$$

$$UF_4 + 2Ca \longrightarrow U + 2CaF_2$$

KLAPROTH'S PROCESS: In the Klaproth's process the pitchblende is first treated with nitric acid and the nitrate is evaporated to dryness. The dry mass is then extracted with hydrochloric acid. This solution is then subjected to react with hydrogen sulphide. Arsenic, lead, cadmium present in the ore get precipitated as sulphides and are removed by filtration. The filtrate is then treated with nitric acid followed by neutralisation of the acid with ammnonium carbonate. The solution is boiled to get the precipitates of zinc, cobalt and uranium as their oxides. On treating these oxides with hydrogen chloride, zinc oxide and cobalt oxide go into the solution as their soluble chlorides and uranium oxide being insoluble remains in the solution which is removed by filtration and is subjected to one of the above methods employed to get the pure metal as shown in flow sheet 11-2.

11-3-3 Importance of Uranium

The most important application of uranium is in producing nuclear energy. In the defence organizations uranium finds application in the making of atom bombs and hydrogen bombs. Uranium is employed in research reactors to produce plutonium, a nuclear fuel of fast breeder reactors and radio isotopes for research and industrial needs. Nuclear energy is converted into readily usable electrical energy in the nuclear reactors.

Uranium salts are used in the textile industry, ceramic industry as well as in medicine. Uranium hexafluoride is used in the separation of U^{235} from

U^{238}, Uranyl nitrate and acetate are used as reagents in the estimation of phosphate by volumetric titration. Uranyl acetate, $UO_2(C_2H_3O_2)_2.2H_2O$, is also used for the detection of sodium, it gets precipitated as its double salt in the presence of zinc or magnesium. This product has the formula of $NaMg[UO_2(C_2H_3O_2)_3].9H_2O$.

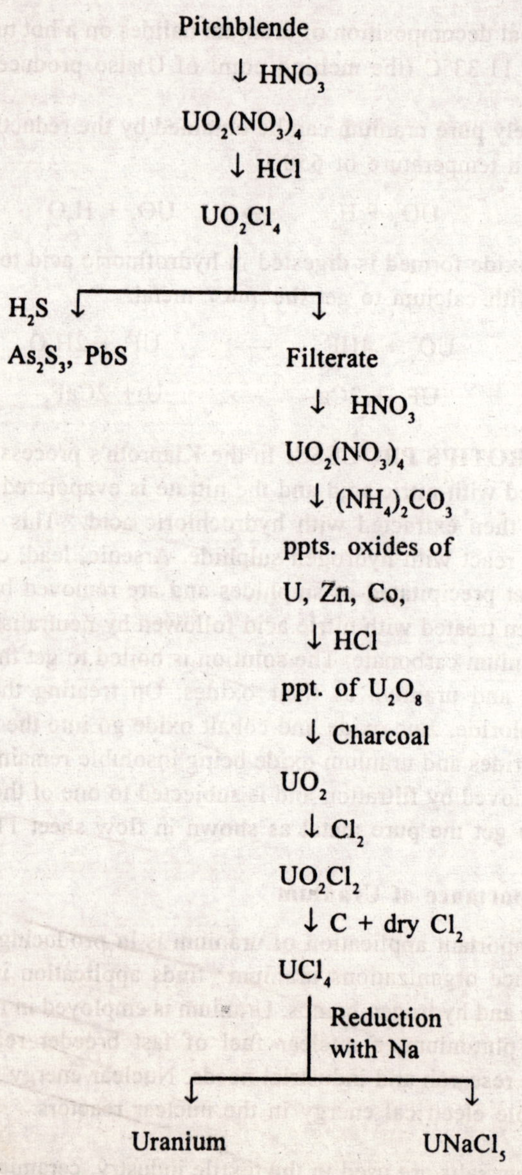

Flow Sheet 11-2 : Extraction of Uranium from Pitch blende

11-4 THORIUM

11-4-1 Introduction

Thorium was discovered by Berzelius (1828) . Madame Curie (1898) found, that, thorium exhibits radioactivity. Large deposits of thorium are present in monazite sand to an extent of about 15% in the form of ThO_2 (thoria). In India monazite is present on the shores of Travancore-Cochin beaches. It is a phosphate of cerium and lanthanum with thoria. It is very difficult to extract thorium from monazite (Th, Ln) PO_4 as it has many impurities in the form of zirconium, titanium and the rare earths and these metals interfere with the normal extraction procedures.

The other ores of thorium are thorianite ($ThO_2.UO_2$), thorite ($ThSiO_4$), cheralite (Th,Ca,Ce silicates and phosphates), gadolinite (a mixed silicate of the lanthanides, iron and berylium - [(Be,Fe)$Ln_2Si_2O_{10}$).

11-4-2 Metallurgy

Monazite

↓ H_2SO_4,MgO

$Th_3(PO_4)_4$

↓ $NaCO_3$

ThO_2 (Crude)

↓ Oxalic acid

Th $(C_2O_4)_2$

↓ Ignited

ThO_2

↓ HNO_3

$Th(NO_3)_4$

↓ Calcined

ThO_2

↓ $COCl_2$

$ThCl_4$

↓ Reduction with Mg/Ca/Na

Th (pure)

↓ I_2

ThI_4

↓ Vapourization

Th (High purity)

Flow Sheet 11-3
The extraction of thorium

Monazite when dissolved in concentrated sulphuric acid gets the thorium phosphate precipitated along with magnesium oxide. The phosphate is washed and heated with a solution of sodium carbonate to get crude thoria, ThO_2, which is then converted to the soluble oxalates, the lanthanide oxalates being insoluble remain as precipitates which are filtered and removed. The thorium oxalate is ignited to get once again ThO_2 which is converted to nitrate and recalcined to get pure ThO_2.

The thoria is then treated in presence of dry hydrogen chloride with phosgene to get $ThCl_4$, this on reducion with either calcium, magnesium or sodium gives the pure metal.

$$ThO_2 + 2COCl_2 \xrightarrow{HCl} ThCl_4 + 2CO_2$$

$$ThCl_4 + 2Mg \longrightarrow 2MgCl_2 + Th$$

As it is difficult to get a sample of high purity, the metal obtained by the above process is converted to volatile ThI_4, by heating the metal in iodine vapour. The ThI_4 vapour gets decomposed, on a hot filament; at a temperature of 1100 to 1200°C; into throium and iodine, the liberated iodine is recycled to use again to react with the impure metal. The extraction process of thorium is shown in flow sheet 11-3.

11-4-3 Importance of Thorium

Thorium is the most important source of nuclear fuel in fast breeder reactors, where it produces U^{233} a fissionable isotope of uranium which is used in normal reactors.

$$_{80}Th^{232} \xrightarrow{_0n^1} {}_{80}Th^{233} \xrightarrow{\beta^-} {}_{81}Pa^{233} \xrightarrow{\beta^-} {}_{82}U^{233}$$

Thoria is used as a catalyst in organic synthesis. Thoriated tungsten electrodes are used in welding and arc melting of the transition metals. Thorium nitrate is used for the titration of fluorides using Alizarin -S as indicator. The end point is observed by the appearance of a red colour due to the reaction of excess of thorium nitrate present in the titration mixture, with the indicator.

11-5 NICKEL

11-5-1 Introduction

The element is present in the earth's crust to an extent of 10^{-2} percent and its principle ore is pentlandite (NiS) having FeS as the main impurity. The other important ores of nickel are nicollite (Kupfer nickel) which is NiAs; nickel glance/blende NiAsS, garnierite which is (MgNi) H_2SiO_4, millerite NiS, and pyrrhotite (Fe_nS_{n+1}) that contains 3-5% Ni. Elemental nickel is also found alloyed with iron in many meteor remains.

11-5-2 Metallurgy

The metallurgy of nickel is complicated but in general the ore is transformed into Ni_2S_3 which is roasted in air to give NiO. This is then reduced with carbon to give the metal. High purity nickel is made by the **carbonyl process** which is referred to as **Mond's process**. This process involves the following steps.

Concentration: The ore employed for this process is pentlandite. This sulphide ore is first crushed into a fine powder and is subjected to the froth floatation process using pine oil, water and frothing agents like potassium ethyl xanthates. The froth carries all the fine ore mainly which contains sulphides of iron, copper, and nickel.

Roasting: The concentrated ore obtained then roasted to convert any iron sulphide present to iron oxde and sulphur dioxide.

$$2FeS + 3O_2 \longrightarrow 2FeO + 2SO_2 \uparrow$$

The copper and nickel sulphides remain unaffected.

Smeltng: Roasted ore is sent for smelting in a blast furnace after mixng with coke, lime and quartz. Any unoxidized FeS left in the roasted mixture is totally oxidized here and the FeO present readily combines with quartz and is removed as slag.

$$FeO + SiO_2 \longrightarrow FeSiO_3(slag)$$

The heavy Ni and Cu sulphide mother liquor called **matte** is drawn from the blast furnace and sent to Bessemer converter.

Bessemerizaton: The iron impurty present combines here also with the SiO_2 to be removed as slag and the bessemerized matte now contains about 56% of Ni, 25-30% of Cu, 15-16% of sulphur and 0.2 to 0.4% of iron.

Ni can be extracted from the bessemerized matte by one of the following processes; (i) Mond's Process (ii) Orford's Process (iii) Electrolysis (iv) German process.

Mond's process is simple and is based on the following two principles:

(a) Out of the components of bessemerized matte only nickel forms volatile carbonyl, when CO at 50°C is passed over matte.

(b) The nickel carbonyl, $Ni(CO)_4$,readily decomposes to give the metal on heating to 180°C.

Mond's process can be discussed as follows:

Oxidation: The matte from the bessemer converter is roasted to convert the Cu and Ni sulphides to their oxides.

$$2CuS + 3O_2 \longrightarrow 2CuO + 2SO_2$$

$$2NiS + 3O_2 \longrightarrow 2NiO + 2SO_2$$

In this step if any free sulphur is present it is also totally oxdized to sulphur doixide.

Leaching: The oxides of Cu and Ni formed are leached with H_2SO_4 to remove the CuO from the mixture. CuO gets dissolved in H_2SO_4, NiO being insoluble in H_2SO_4 is filtered off.

Reduction: The insoluble portion is now dried and subjected to reduction, in a current of water gas ($CO+H_2$) to obtain crude Ni,

$$NiO + H_2 \longrightarrow Ni + H_2O$$

$$NiO + CO \longrightarrow Ni + CO_2$$

The crude Ni contains some amount of copper and iron as impurities.

Carbonylation: The curde nickel is now placed in a volatilizer or volatizing tower maintained at 50°C and carbon monoxide is passed, forming $Ni(CO)_4$. The vapours of the nickel tetracarbonyl are sent into a decomposing tower.

Decomposition: The decomposing tower is filled with nickel pellets at a temperature of 150-180°C, where the carbonyl gets decomposed to give pure nickel and carbon monoxide is recycled.

$$Ni(CO)_4 \longrightarrow Ni + 4CO\uparrow$$

The nickel obtained is 93% pure and further refining is done by electrolytic process.

Electrolysis: The anode is made of a nickel compound and cathode is of iron, the electrolysis is conducted using nickel ammonium sulphate solution as the electrolyte, nickel gets deposited on the cathode.

The extraction of nickel is shown in flow sheet 11-4.

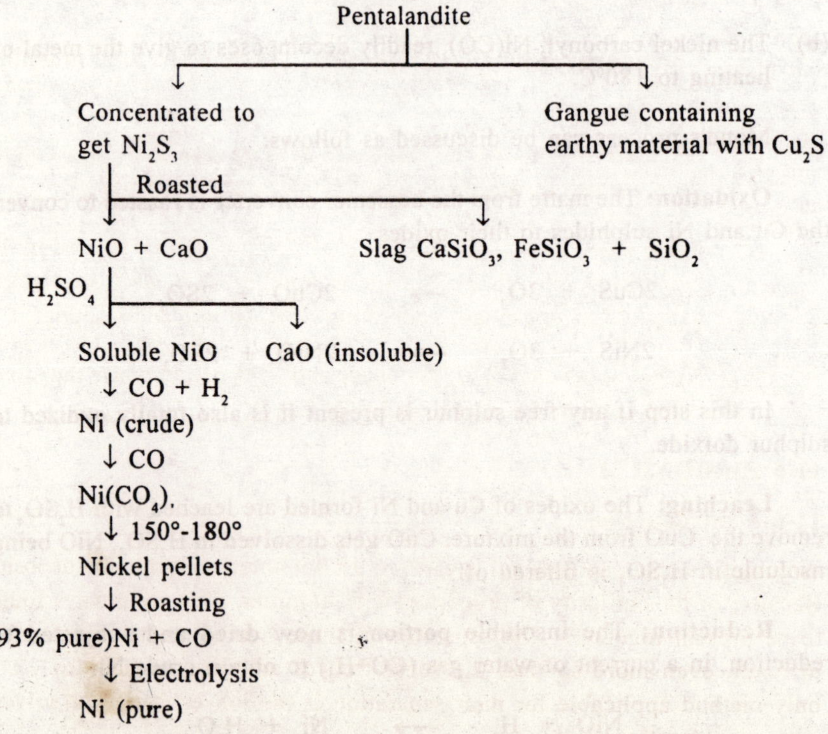

Flow sheet 11-4 Extraction of nickel from pentalandite

11-5-3 Importance of nickel

Nickel is one of the most important and widely used metal.

(i) Nickel is used in its finely divided form in the manufacture of vanaspati ghee. In this it acts as a hydrogenating catalyst for the hydrogenation of oils.

(ii) It is also used as a catalyst in many other chemical processes.

(iii) Many nickel alloys find importance in industry, because of their hardness.

(a) Its alloy with iron, called **nickel steel**, with 2.5 - 5 percent of nickel is used in armor plates, aeroplanes and automobile parts.

(b) **Invar**, contains 3.5% nickel with extremely low coefficient of expansion is used in making clock pendulums and other sophisticated, precison instruments.

(c) **Monel metal** with 60% nickel with extremely low coefficent of expansion is used in making clock pendulums and other sophisticated, precision instruments.

(d) **Cupronickel** with 80% copper and 20% nickel is used in bullets sheaths.

(e) **German silver** with 20% nickel, 20% of zinc and 60% copper finds its use in domestic utensils and decorative works.

(f) **Nichrome** containing 60% Ni, 25% Fe and 15% Cr is used in the making of electrical resistances, another alloy used in electric gadgets is Contsantan, containing 40% Ni and 60% copper.

(g) Nickel coins contain 75% copper and 25% nickel while the silver coins contain 5% nickel and 95% silver.

11-6 ZIRCONIUM

11-6-1 Introduction

Zirconium is abundantly found in earths crust but not in concentrated deposits. Because of its similarity of properties with hafnium, they are always found to be present together. The separaton of these two elements is extremely difficult, even more so than the separation of adjacent lanthanides. The only method applicable for their separation is the ion exchange or solvent extraction fractionation.

Separation of Zrconium and Hafnium: Hafnium is always in combination with zirconium and is difficult to be separated by normal techniques employed for other metals. For the separation the ore is first dissolved in concentrated hydrochloric acid to get $ZrCl_4$ and $HfCl_4$. This is then treated with a solution of ammonium thiocyanate. By subjecting this mixture to solvent extraction using iso-butyl ketone, hafnium chloride gets in to the organic solvent. The aqueous layer is diluted with water and boiled with sulphuric acid at 180°C, Zirconium sulphate gets precipitated. On hydrolysis with liquor ammonia, zirconium hydroxide is separated out as precipitate, which on incineration gives zirconiumdioxide. This on reduction with carbon in the presence of a current of chlorine, gives a pure sample of zirconium tetrachloride $ZrCl_4$. On reduction with magnesium by thermite process zirconium, free from hafnium is obtained.

The major minerals of zirconium are zirconia or baddeylite containing about 80% ZrO_2 and zircon or hyacynth, $ZrSiO_4$, with 85% ZrO_2. ZrO_2 has a monoclinic crystalline structure. The Zr-O distances range from 2.04 to 2.26 A° and the O-Zr-O angles vary from 72° to 91°

The metal is produced commercially by the Kroll process but the van Arkel - de Boer process is also useful when it is especially important to get 100% pure zirconium.

11-6-2 Metallurgy

According to the Van Arkel - de Boer Method, zircon is first reduced at 200°C in the presence of iodine when ZrI_4 a volatile compound is formed. These vapours are passed over a tungsten or zirconium filament electrically heated to about 1300°C. This decomposes the ZrI_4 and pure zirconium is deposited on the filament.

The extraction of zirconium on a large scale is shown in flow sheet 11-5. Zircon is fused with sodium hydroxide to get sodium zirconate and sodium silicate, on boiling with water sodium silicate goes in solution, leaving zirconium dioxide as residue. On treating the contaminated zirconium oxide with hydrochloric acid, zirconium oxychloride is formed with the liberation of silica. The solution is heated to dryness and silica is removed when zirconium oxychloride is extracted with hydrochloric acid. This is oxidised by treating it with ammonia to get hydrated zirconium oxide, which is then dehydrated by ignition. The oxide is heated in a current of chlorine in the presence of carbon to get zirconium tetrachloride, which on reduction with molten magnesium gives a mixture of zirconium with traces of hafnium. The spongy mixture is obtained in an atmosphere of argon to get pure sample of the mixture of zirconium and hafnium, which are virtually identical and their association in nature parallels their very close chemical similarity.

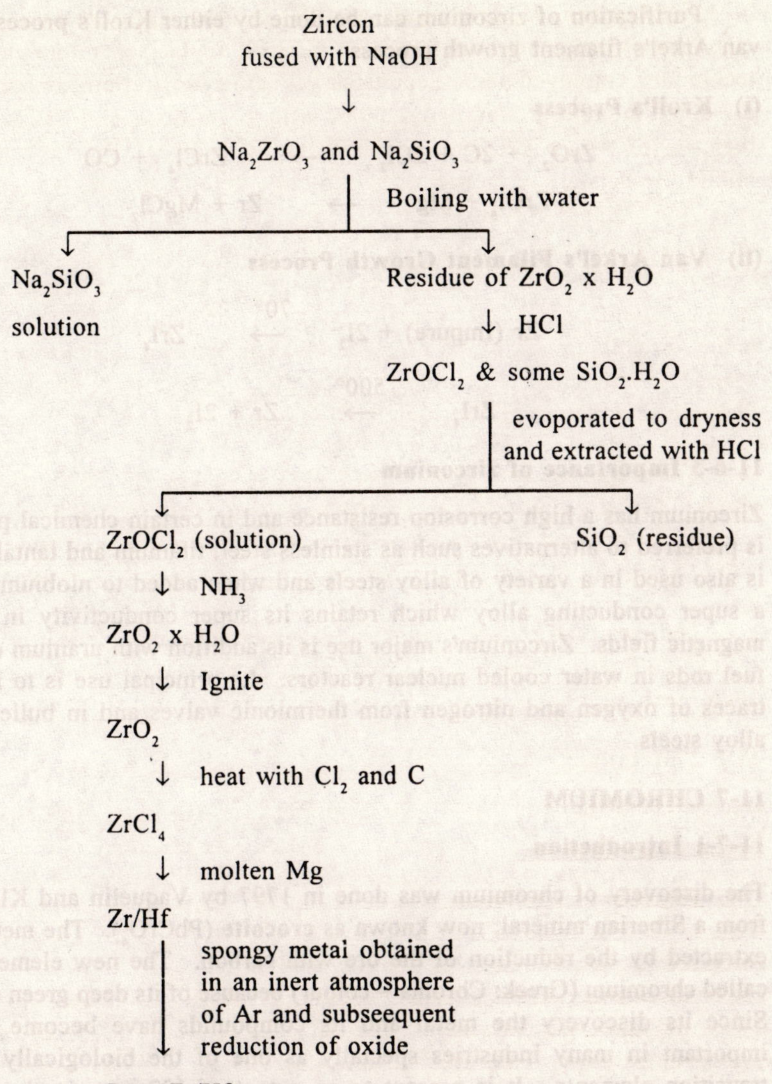

Flow sheet 11-5 Extraction of zirconium

In an alternate method the ore is finely powdered and heated with carbon in an electric furnace to a temperature of 2100°C at this temperature zircon gets decomposed forming ZrO_2, SiO_2, being volatile is suctioned away, ZrO_2 gets converted to $ZrCl_4$, which on reduction with magnesium by thermite process produces pure zirconium.

$$ZrSiO_4 \longrightarrow ZrO_2 + SiO_2$$

$$ZrO_2 + 2C + 2Cl_2 \longrightarrow ZrCl_4 + 2CO$$

$$ZrCl_4 + 2Mg \longrightarrow Zr + 2MgCl_2$$

Purification of zirconium can be done by either Kroll's process or by van Arkel's filament growth process

(i) Kroll's Process

$$ZrO_2 + 2C + 2Cl_2 \longrightarrow ZrCl_4 + CO$$

$$ZrCl_4 + Mg \longrightarrow Zr + MgCl_2$$

(ii) Van Arkel's Filament Growth Process

$$Zr\ (Impure) + 2I_2 \xrightarrow{70^\circ} ZrI_4$$

$$ZrI_4 \xrightarrow{500^\circ} Zr + 2I_2$$

11-6-3 Importance of zirconium

Zirconium has a high corrosion resistance and in certain chemical plants it is preferred to alternatives such as stainless steel, titanium and tantalum. It is also used in a variety of alloy steels and when added to niobium, forms a super conducting alloy which retains its super conductivity in strong magnetic fields. Zirconium's major use is its addition with uranium dioxide fuel rods in water cooled nuclear reactors. Its principal use is to remove traces of oxygen and nitrogen from thermionic valves and in bullet proof alloy steels.

11-7 CHROMIUM

11-7-1 Introduction

The discovery of chromium was done in 1797 by Vaquelin and Klaproth, from a Siberian mineral, now known as **crocoite** ($PbCrO_4$). The metal was extracted by the reduction of the ore with carbon. The new element was called chromium (Greek: Chroma = colour) because of its deep green colour. Since its discovery the metal and its compounds have become vitally important in many industries specially as one of the biologically active transition elements. It is present to an extent of 122 ppm in the earth's crust.

The commercially important ore is chromite, $FeCr_2O_4$. In India chromite is found in Bihar, Karnataka, Maharashtra, Orissa and Tamil Nadu. The other ores are crocoite, $PbCrO_4$; chromeochre, Cr_2O_3. Many gem stones like emerald, ruby, sapphire etc., owe their colours to traces of chromium in them.

11-7-2 Metallurgy

Chromium is produced from chromite, $FeCr_2O_4$. Iron oxide is mainly separated by gravitational and magnetic techniques. The processes adopted in the extraction of chromium (III) oxide, Cr_2O_3 are;

(i) The oxide is reduced by carbon in an electric furnace.

(ii) From the alumino-thermite process.

Process (I) The reduction process satisfies the requirement of ferro-chrome alloy industries, where chromic oxide can have iron impurities. In this the chromite ore is reduced with carbon in an electrical furnace affording ferrocromium alloys containing carbon.

$$FeCr_2O_4 + 4C \longrightarrow Fe + 2Cr + 4CO$$

Process (II) The alumino-thermite process affords pure chromic oxide. The chromite ore is roasted with potassium carbonate in the presence of lime and is leached with water to get crude potassium chromate. The solution is acidified with suphuric acid to get potassium dichromate. The potassium dichromate is then mixed with ammonium chloride and heated strongly to get pure Cr_2O_3. The sequence is represented by the following equations.

$$4FeCr_2O_4 + 7O_2 \longrightarrow 2Fe_2O_3 + 8CrO_3$$

$$CrO_3 + K_2CO_3 \longrightarrow K_2CrO_4 + CO_2$$

$$H_2SO_4 + 2K_2CrO_4 \longrightarrow K_2Cr_2O_7 + K_2SO_4 + H_2O$$

$$K_2Cr_2O_7 + 2NH_4Cl \longrightarrow (NH_4)_2Cr_2O_7 + 2KCl$$

$$(NH_4)_2Cr_2O_7 \longrightarrow Cr_2O_3 + N_2 + 4H_2O.$$

The oxide is washed with water and completely dried before it is subjected to alumino-thermite reduction.

The alumino thermite process is carried out in a fire clay crucible kept in sand bath. The charged ore is covered. with a ribbon containing aluminiumbarium peroxide. Magnesium powder is inserted into the ignition mixture setting an exothermic reaction on. The mixture melts and a displacement reaction takes place between aluminium and chromium oxide, producing molten chromium of 98.5% purity.

$$Cr_2O_3 + 2Al \longrightarrow 2Cr + Al_2O_3$$

Alumina produced floats on the surface as slag and is skimmed off.

11-7-3 Importance of Chromium

The main use of chromium metal is in the production of nonferrous alloys. The use of pure chromium is limited, because of its low ductility at ordinary trempeerature. Alternatively, the Cr_2O_3 can be dissolved in sulphuric acid to give the electrolyte used in chromium electroplating, which is both protective and decorative because of its shiny lustre. The sodium chromate produced in the isolation of chromium is itself the basis for the manufacture of all industrially important chromium chemicals. Potassium chromate and potassium dichromate find a very important position among the laboratory reagents.

11-8 CADMIUM

11-8-1 Introduction

Cadmium was discovered by Stromeyer. It is present to an extent of 0.16% ppm of the earth's crust.

Cadmium belongs to the group of chalcophiles. Chalcophiles are the elements which are present in the earth's crust in their sulphide form. Some chalcophiles are given in table 11-4. It's important ore is greenockite, CdS, but is commercially obtained from zinc ores which generally have 0.2 to 0.4%.

Table 11-4 Chalcophiles

															S		
							Fe			Cu	Zn	Ga		As	Se		
					Mo					Ag	Cd	In		Sb	Te		
											Hg	Ti	Pb	Bi			

11-8-2 Metallurgy

Cadmium is found to be present always with zinc. Cadmium and its oxide being more volatile than zinc and its oxides respectively, the process of fractional distillation is applied for its separation from zinc and its isolation in pure form. It is usually obtained during the extraction of zinc as a by product. It is collected in the receiver along with the distillate of zinc dust. But the zinc dust will be present with cadmium to some extent, therefore it is mixed with charcoal and the mixture is subjected to fractional distillation at 800-900°C till cadmium of 99.5 percent purity is collected.

11-8-3 Importance of Cadmium

Cadmium is used in food processing. The food to be preserved is stored in cadmium plated utensils, cadmium containing glazes are used for pottery and galvanized equipment. Extremely pure cadmium is used in standard cells, it is prepared by subliming the metal in vacuum or in hydrogen or by repeated electrolysis of an acidified concentrated solution of cadmium sulphate.

It is used as a deoxidizer in aluminium castings, in bronze and in telephone trans-way wires. Cadmium amalgam is used in Weston standard cell. The wood's metal is an important alloy containing 12.5% of tin, 25% of lead and 50% of bismuth. Cadmium sulphate is used as a medicine in eye ailments.

11-9 METALLIC BONDING

11-9-1 Introduction

We are very much familiar with the term chemical bonding, which in simple terms mean the affinity present between two or more number of chemical elements present in their atomic, ionic or molecular forms. We also know about the different types of bondings present. The electrovalent or ionic bonding is formed by the complete transference of electrons from one atom to another, thereby forming an ion of each of them which are attracted to each other owing to the opposite charges. The covalent bond is formed, by the mutual sharing of electrons by those atoms or molecules which are short of electrons and tend to complete their octets. In this process not only a simple electron is shared to form a single bond, but also two or three electrons are shared to form (multiple) double and triple bonds, as found in molecules of oxygen and nitrogen respectively. The single covalent bond is said to be of σ type whereas a double bond has one σ (sigma) and one π (pi) bond, the triple bond is constituted of one σ and two π bonds. The σ bond is formed by the endon overlapping of the orbitals, π is formed by the sideon overlapping. Then there is another type of bonding known as Δ (delta) bonding as found in ferrocene where there is an overlapping of two orbitals over each other, this type of bonding is not very common. The fourth type of chemical bonding is the coordinate covalent or the dative bonding, in which an atom, ion or molecule donates and electron pair to another atom or molecule that has an electron pair short of its octet or duplet, in all the cases stability is reached.

The chemical bonding in different simple and complex molecules is explained based on different theories like atomic orbital, molecular orbital, valence bond theory, crystal field theory, ligand field theory etc.

11-9-2 Theory of Metallic Bonding

Metals are characterized by lustre, malleability, ductility, further, metals also crystallize as structures with high coordination numbers (number of nearest neighbours in crystal) such as 8 or 12. Attempts were made from early 1900 to explain these properties through various theories. We explain here some of the important theories of bonding in metals.

A. Free electron theory: Drude in 1900 considered a metal as a lattice with electrons moving through it as a gas moves in well packed pebbles. This idea was further developed by Lorentz in 1923 who considered a metal crystal as a close packed structure of positive metal ions and the electrons are free to move about in the combined periodic, potential field due to the positve ion. In simple terms the cations of metal atoms are embedded in a sea of free valency electrons which could move in the interstices. This werein results from the electronic attractions between cations and electron cloud. This model is capable of explaining qualitatively various mettalic properties. For example, the high electrical conductivity could be readily explained from the free movement of electrons through the lattice and malleability results from lack of directional property of the mettalic bond. These concepts could not lead to any quantitative methematical approach and this lead to evolution of valence bond theory and molecular orbital theory of metals.

B.Valence Bond theory of metals: According to this theory metals in general have low ionization potential and the valence electrons can be moved in a facile way. In other words,valence electrons are weakly bound to the Kernel, formed by the atomic nucles and its non-valence electrons. Let us consider the case of lithium whose electronic configuration is $1s^2\,2s^1$ with one valence electron, from the X-ray examination it was established that each atom of Li is surrounded by eight other Lithium atoms (cordination number of Li is 8). The single valence electron, therefore, could be shared with one of its neighbours forming a normal two electron bond, some of the different possible arrangements are

Li   Li—Li   Li   Li—Li   Li—Li   Li—Li   Li—Li
| | | |
Li   Li—Li   Li   Li—Li   Li—Li   Li—Li   Li—Li
(a)   (b)   (c)

Li^+ Li—Li   Li   Li—Li   Li—Li
| |
Li—Li   Li—Li^- Li^+ Li—Li^-—Li
(d)   (e)

The real structure is a mixture of all these possible structures, constant interchange of two electron bonds increase the cohesive energy and there by the strength of the metal. The presence of ions in the possible structures explain the high electrical conductions exhibited by metals. This theory is not successful in explaining the other metallic properties like lustre,thermal conduction, etc.,

C. Molecular orbital theory of metals or Band theory: Consider two Li atoms being brought together. The energy levels of each Li atom is peturbed by the presence of the other Li atom and bonding occurs using the 2*s* valence electron. The combination of two atomic orbitals (AO) yield two molecular obritals (MO) one bonding and one antibonding. MO, fig. 11-13(a). Similary, when three Li atoms joined to form Li_3, the three 2*s* AOs, would combine to form three MOs, one bonding one antibonding and one nonbonding, Fig.11-13(b). Extending this argument to n Li atoms joining to form Li_n we get n number of MOs, fig.11-13(c). Since n is very large, all the n energy levels generated are closely spaced and constitute the 2*s* band. There is only one valence electron per atom and each MO formed can hold two electrons. The presence of n energy levels and n electrons make it clear that only half the MOs (bonding) are occupied, therefore, making it clear that the **2*s* band** is only half filled. The partially occupied or empty MOs is a prerequisite for metallic properties.

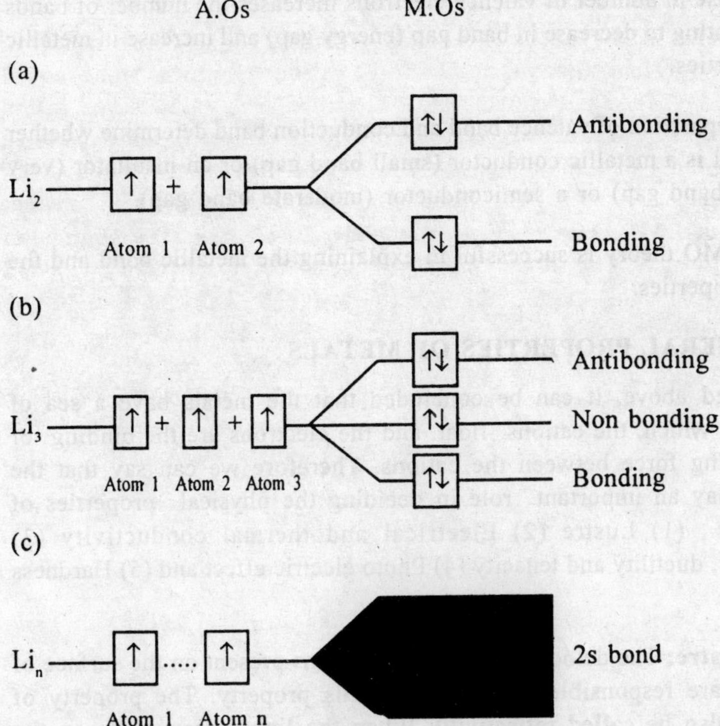

Fig.11-13 (a,b,c) Atomic orbital and molecular orbital diagram of lithium.

Similarly in case of other metal atoms we notice an interactionof 2*p*, 3*s*, 3*p*, etc., energy levels forming the 2*p* band, etc. The existance of partially filled half filled or empty bands explain all the metallic properties.

1. Conduction occur because the half filled MOs,extend over the whole crystal.

2. The closely packed energy levels in any metallic bond allow excitation of electrons within the band, return back to the ground state by the emission of photons which account for the properties like matallic lustre.

3. The partially filled bands are called valence bands (VB) since they contain the valence electrons. The empty bands are called conduction bands (CB).

4. Increase in number of valence electrons increases the number of bands amounting to decrease in band gap (energy gap) and increase in metallic properties.

5. The separation of valence band and conduction band determine whether a solid is a metallic conductor (small band gap) or an insultator (very large band gap) or a semiconductor (moderate band gap).

Thus MO theory is successful in explaining the metallic bond and the metallic properties.

11-10 GENERAL PROPERTIES OF METALS

As discussed above, it can be concluded that the metals have a sea of electrons in which, the cations float, and the electrons are the binding or the cementing force between the cations. Therefore we can say that the electrons play an important role in deciding the physical properties of metals. viz., (1) Lustre (2) Electrical and thermal conductivity (3) Malleability, ductility and tenacity (4) Photo electric effect and (5) Hardness and rigidity.

(1) Lustre: The delocalized mobile electrons present on the surface of the metals are responsible for exhibiting this property. The property of lustre can also be called reflectivity. When the light energy falls on the surface of the metals the electrons get excited and jump into higher energy levels but are immediately deenergized to the ground level. In this process energy in the form of light is absorbed by the electrons and the same is reemitted in the form of electromagnetic radiation. This constant emission of light imparts the property of metallic lustre. However, some of the metals like copper, gold, etc., can absorb selectively some wave lenghts of the light and give corresponding colours to the metal.

(2) Conductivity: Metals are good conductors of both electricity and heat. When an electric current (nothing but a flow of electrons) is passed, immediately the electrons become a part of the lattice and start moving throughout the metal from one end to the other; Fig.11-14.

Similarly when a metal is heated, the mobile electrons absorb heat and get excited with an increase in their vibrational action. During this vibrational motion they start travelling between the interstitial spaces of the metal, colliding with the adjacent electrons and transferring the heat as they move and thus carrying it from hotter end ot colder end, as shown in Fig. 11-15. The best conductors are those which have low ionization potential.

e^- e^- e^- e^-

e^- M^{2+} e^- M^{2+} e^- M^{2+} e^- M^{2+} e^-

e^- e^- e^- e^-

e^- M^{2+} e^- M^{2+} e^- M^{2+} e^- M^{2+} e^-

Electricity enters $\rightsquigarrow$ e^- $\rightsquigarrow$ e^- $\rightsquigarrow$ e^- $\rightsquigarrow$ e^- $\rightsquigarrow$ Electricity exits

e^- M^{2+} e^- M^{2+} e^- M^{2+} e^- M^{2+} e^-

e^- e^- e^- e^-

e^- M^{2+} e^- M^{2+} e^- M^{2+} e^- M^{2+} e^-

e^- e^- e^- e^-

Fig. 11-14 Movement of electrons through metals on passing electricity

Silver has the highest electrical conductance of 6.28×10^5 ohm cm^{-1} (Ω cm^{-1}) followed by copper 0.598×10^5 Ω cm^{-1}, gold, 4.5×10^5 Ω cm^{-1} and aluminium, 3.77×10^5 Ω cm^{-1}. Because of the cost factors silver is not

e^- e^- e^- e^-

e^- M^{2+} e^- M^{2+} e^- M^{2+} e^- M^{2+} e^-

e^- e^- e^- e^-

e^- M^{2+} e^- M^{2+} e^- M^{2+} e^- M^{2+} e^-

e^- e^- e^- e^-

e^- M^{2+} e^- M^{2+} e^- M^{2+} e^- M^{2+} e^-

heat (γ) $\rightsquigarrow$ e^- $\rightsquigarrow$ e^- $\rightsquigarrow$ e^- $\rightsquigarrow$ e^- $\rightsquigarrow$

e^- M^{2+} e^- M^{2+} e^- M^{2+} e^- M^{2+} e^-

e^- e^- e^- e^-

Fig. 11-15 Movement of electrons on heating the metals.

used as electrical conductor but copper and aluminium find a wide application in electrical industry. Electrical and thermal conductivity has a loose relationship between them.

Elecrical conductance is defined as the ability of substance(metals) to transmit electricity under the action of an extrernal electrical field. It is measu ed in ohms per centimeter. The electrical conductance values of a few metals are given in table 11-5.

Table 11-5 Electrical conductance (mΩ) value for few metals

Li 0.117	Be 0.169								
Na 0.238	Mg 0.224								
K 0.163	Ca 0.292	Se --	Ti 0.013	V 0.038	Cr 0.077	Fe 0.102	Co 0.160	Ni 0.146	Cu 0.598
Rb 0.080	Sr 0.043	Y --	Zr 0.024	Nb 0.076	Mo 0.193	Ru 0.131	Rh 0.222	Pd 0.093	Ag 0.628
Cs 0.053	Ba 0.016	La 0.017	Hf 0.031	Ta 0.081	W 0.181	Os 0.11	Ir 0.189	Pt 0.103	Au 0.452

Thermal conductance is defined as the ability of substances (metals) to transmit heat under the action of external heat. It is measured in kcal $cm^{-1}s^{-1}K^{-1}$. The thermal conductance values for a few metals is given in table 11-6.

Table 11-6 Thermal conductance (kcal $cm^{-1}s^{-1}K^{-1}$) value for few metals

Li 0.17	Be 0.38	Cr 0.16	Fe 0.11	Co 0.16	Ni 0.14	Cu 0.99	Zn 0.27
Na 0.32	Mg 0.38	Mo 0.35	Pu --	Rh 0.21	Pd 0.17	Ag 1.01	Cd 0.22
K 0.23	Ca 0.3	W 0.4	Os --	Ir 0.14	Pt 0.17	Au 0.70	Hg 0.07

(3) Malleability, Tenacity and Ductility. In a metal the valency electrons are spread all over the surface more or less uniformly and as mentioned before the positive ions are surrounded by the electrons. If one layer is forced across another layer by using an external force the internal structure of the metal will not change, instead the electrons get themselves adjusted in the interstitial spaces. In such a case the metal can be molded in any form, therefore it can be made into beads(malleability), it can be drawn into sheets(tenacity) and are capable of being drawn into wires (ductility).

(4) Photoelectric effect : Most of the metals specially the IA group (alkali metals) metals possess this property of electron emission when the surface of the metal is exposed to light. Electrons escape from the metals with a resultant decrease in the energy of the incident photon. The best metal which exhibits this property is cesium, therefore it is used in photoelectric cells. Photoelectric effect is explained from the same consideration of excitation of lattice electrons and low ionization potential values of alkali metals.

(5) Hardness and Rigidity : As the metals have closely packed structures, they possess great cohesive forces. The strength of metals is largely due to the electrostatic nature of the binding forces. Smaller the cation the better packed is the metallic lattice and harder is the metal.

Another physical property exhibited by metals is the high melting point with an exception of alkali metals which melt at low temperatures. This is also related to the hardness of metals. Metals possess high densities because of their highly close packed structures.

11-11 CRYSTAL STRUCTURES OF METALS

Of the known nine crystal structures there are three common arrangements for metal atoms in the solid state, face centered cubic (cubic close packed), hexagonal close packed and body centered cubic . The first two correspond to the closest possible packing of similar spheres, just as their names imply. In both of these close packed structures each metal atom has twelve nearest neighbours. In the body centered cubic structures each metal atom has eight nearest neighbours.

Crystal structures of some metals are as follows:

(i) Body centered cubic crystal structure is exhibited by metals like :

Li, Na, K, Rb., Cs,

Fr, Ba, V, Nb, Ta,

Cr, Mo, W, Fe, Pt, Eu.

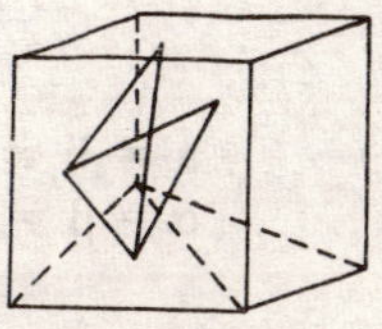

(ii) Face centered cubic structure is represented by the metals like:

Ca, Sr, Rh, Ir, Ni,

Pd, Cu, Ag, Au, Al,

Pb, Ce, Th.

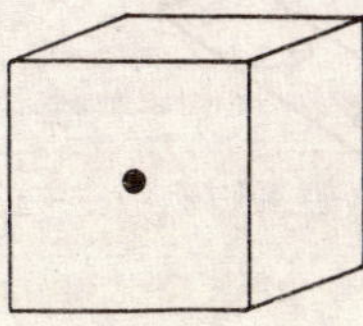

(iii) Close packed hexagonal crystal structure are shown by

Be, Mg, Sc, Y, La, Ti,

Zr, Hg, Re, Ru, Os, Co,

Zn, Cd, Tl, Pr, Nd, Pm, Gd,

Tb, Dy, Ho, Er, Tm, Lu.

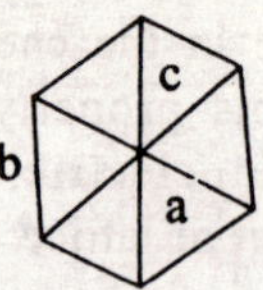

a = b ≠ c
α = β = 90°
α = 120°

(iv) Cubic crystal structure is exhibited amongst the metals by only Mn.

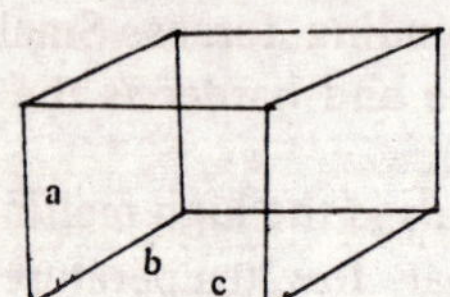

a = b = c
α = β = γ = 90°

(v) Rhombohedral crystal structure is found in

Hg, Sb, Bi, Sm

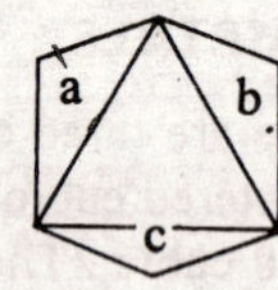

a = b = c
α ≠ β ≠ γ ≠ 90°

(vi) Tetragonal crystal structure is exhibited by In and Sn.

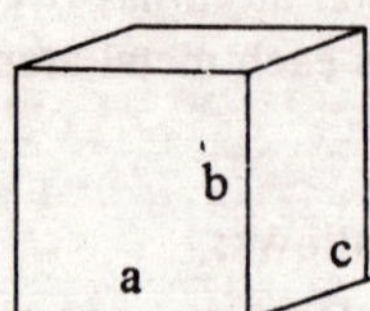

a = b ≠ c
α = β = γ = 90°

(vii) Orthorhombic crystal structure is represented by Ga and U.

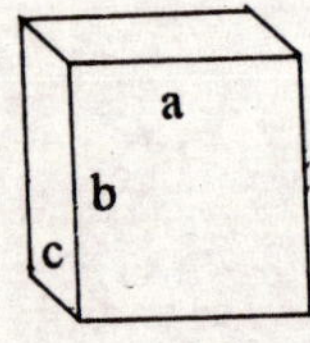

a ≠ b ≠ c
α = β = γ = 90°

(viii) Monoclinic crystal structure amongst the metals is found to be exhibited only by Po.

a ≠ b ≠ c
α = γ = 90°

(ix) Diamond crystal structure amongst the metals is found to be present only in Ge.

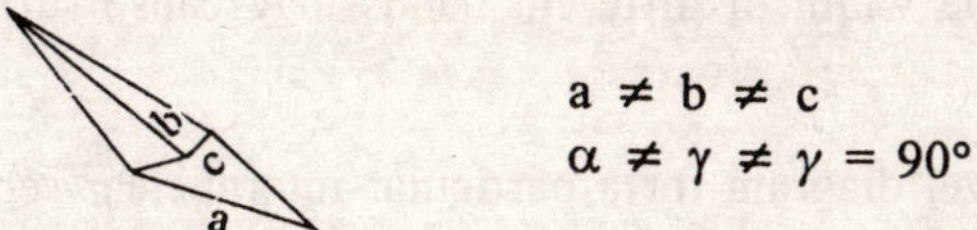

$a \neq b \neq c$
$\alpha \neq \gamma \neq \gamma = 90°$

The difference in the crystal shapes is due to the different types of hybridizations which are present in the metals. The sd^3 and sd^4 hybridization gives a body centred crystal lattice, sd^2 and pd^5 hybridization forms a hexagonal close packed crystal lattice, whereas p^3d^3 and spd^4 hybidization has a face centred cubic crystal lattice. These unusual type of hybridizations of *s*-and *d*-orbitals are possible when the *d*- and *s*- orbitals are of nearly equal energy; for example the $3d{\approx}4s$, $4d{\approx}5s$, etc.

The crystal structures are elucidated by x-ray diffraction. Many inter metallic compunds also have structures that involve the disordered or ordered distribution of metal atoms of 2 or more kinds among the atomic positions for cubic or hexagonal closed packing. Cubic closed packing is present in AuCu, PtCu, $AuCu_3$, $CaSn_3$, $LaPb_3$ etc., body centered cubic crystal lattice is found to be present in CuPd, CuBe, AgCd, SiHg, LaCl etc., The hexagonal crystal lattice is found to be present in $PbAl_3$. the intermetallic compounds also show some unusual crystal systems, eg., $MoAl_{12}$, which shows icosahedral this structure has 8 triangular faces and 6 square faces. In $MoAl_{12}$ the molybdenum atom is present at the corner and it is surrounded by 12 aluminium atoms.

11-12 BAND THEORY OF METALS

The covalent bond approach gives some understanding of the strength of bonding in metals and also to a lesser extent the lattice types found for different metals. However, these theories, so far, have not accounted for the properties like electrical conductivity, metallic luster, etc., that allow us to distinguish metals from nonmetals. These properties may be accounted for by the **band theory of metals**, which is considered below.

The combination of atomic orbitals to form molecular orbitals result in the splitting of the energy levels of the original atomic orbitals. The magnitude of this splitting depends upon the extent of overlap of the atomic orbitals and , accordingly is strongly dependent upon the internuclear separations involved and the orientation of the atomic orbitals with which we are concerned. In a metal, the combination of the **n** atomic orbitals of a given type will give rise to a series of molecular orbitals of closely spaced energies called energy bands. Due to low degree of orbital overlaps the band width for inner shell electrons is very small and the energies of the electrons in these bands are virtually same as the isolated atoms. For the outer or valence electrons, however, the energy range of electrons in a given band may be relatively large. The width of different bands may cause sufficient band overlap.

An energy level diagram for a particular internuclear separation may be constructed by taking the band width and position corresponding to the internuclear separation. The energy level diagram fails to give us information regarding the number of electrons of given energy which the system may contain. Figure 11-16 shows the distribution function of electrons with energy in a given band . As electrons are added to a given metal lattice, the lowest energy levels will be filled first. The spacing of the bands and the filling of the bands determines whether a substance is a **conductor (metal), insulator(nonmetal), or semiconductor (metalloid)** . If the bands are completely filled or completely empty and the energy spacing between bands is large as in Fig. 11-16c the substance will be an **insulator**. Only a substance having a partially filled band will serve as a **conductor.**

Thus Fig. 11-16a & b would show metallic conduction, c would be an insultor, and d would be a semiconductor. Fig. 11-16a is typical of single

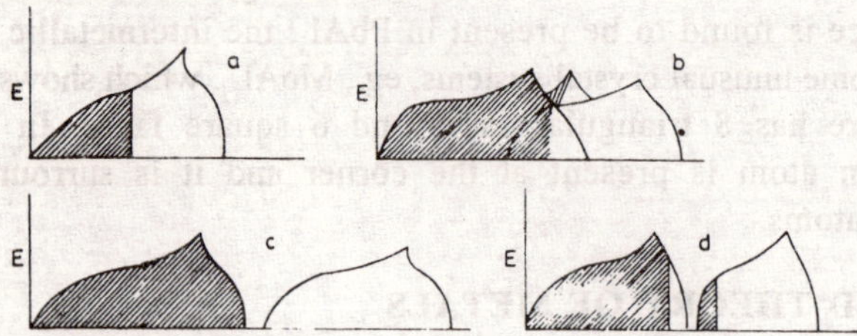

Fig. 11-16 Distribution of electrons in energy states.
(The shaded portion indicate the occupied orbitals)

electron metals which would have exactly half the number of electons the band would hold. The two electrons metals such as Mg and Zn would have enough electrons to completely fill the s-band where it will not overlap with p-band., which causes the p-band to begin to fill before the s-band is completely filled.

For a **semiconductor** with an electron distribution similar to that shown in figure 11-16d , some electrons may be thermally excited from the nearly filled band to the nearly empty band. The number of electrons so excited will increase with temperature and hence the conductivity of a semiconductor, will increase with increasing temperature, unlike that of metal.

Metals are characterized by their high electrical conductivity, for example when sodium metal is considered, its electronic configuration is $1s^2 2s^2 p^6 3s^1$ so each atom has one valence electron in the 3s orbital. In a metal the atoms are packed closely together , so the energy levels of each sodium atom are affected by the immediate neighbours of the atom as a result of orbital overlaps. It is a fact, that, when interaction between two atomic orbitals takes place it leads to the formation of a bonding and an antibonding molecular orbital. Since the number of atoms, in say a gram mole of sodium is 6.023×10^{23}, the corresponding number of molecular orbitals formed is also very large. These molecular orbitals are so closely spaced on the energy scale that they are more appropriately described as forming a band, Fig.11-17, Even though each level can accomudate two electrons there is only one electron to contribute in the case of the sodium atom,. The Fig 11-17a explains, that,

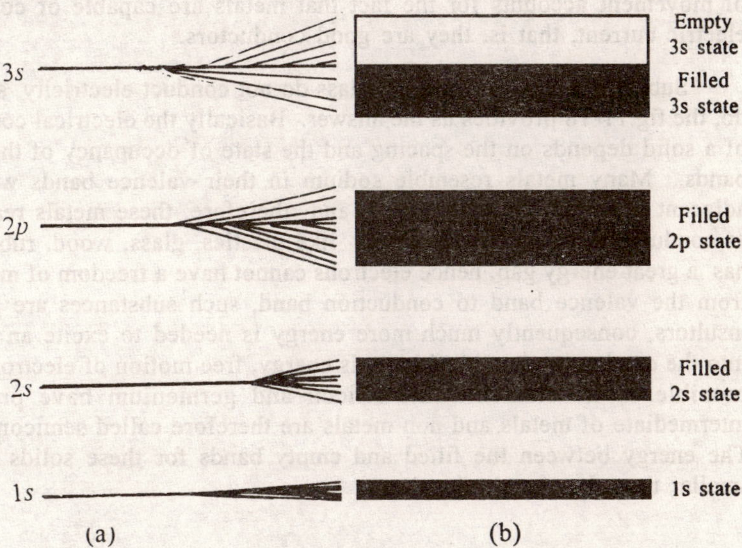

Fig 11-17 Formation of conduction bands in sodium metal.

when lone sodium atoms are far apart, there is no interaction among the coresponding atomic orbitals (in the left portion of the figure the atomic orbitals 1s, 2s , 2p and 3s retain their identity). When they are close to one another, as in a metal, the orbitals overlap to fom bonding and antibonding molecular obitals. Because of a very large orbitals formed are closely spaced giving rise to bands. In the fig 11-17b the band structure in sodium metal is seen. It is formed that the 1s, 2s, 2p bands are filled with electrons, whereas the band formed by the overlap of the 3s orbitals is only half filled, since each sodium atom posses only one 3s electron. Thus, only the lower half of the energy levels (the bonding molecular orbitals) are filled. This set of closely spaced filled levels is called the valence band as seen in fig.11-18. The upper half of the energy levels corresponds to the empty antibonding

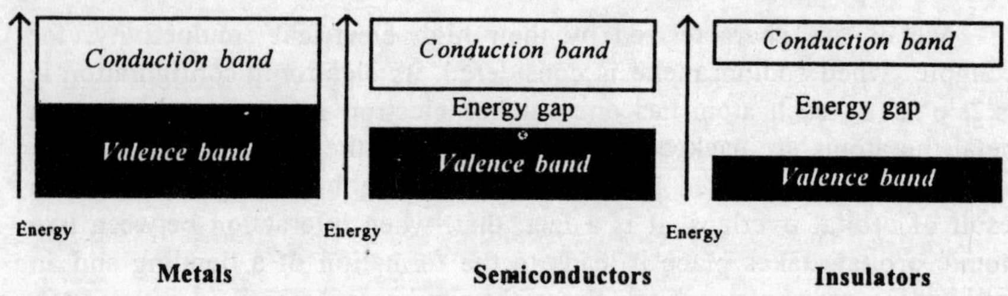

Fig. 11-18 conduction and valence bands

molecular orbitals. This set of closely spaced empty levels is called the conduction band. Becuase the valence band and conduction bands are adjacent to each other only a neglible amount of energy is needed to promote a valence electron to the conduction band, where it is free to move through the entire metal, since the conduction band is void of electrons. This freedom of movement accounts for the fact that metals are capable of conducting electric current, that is, they are good conductors.

Substances like plastics and glass do not conduct electricity, as metals do, the fig.11-18 provides us the answer. Basically the electrical conduction of a solid depends on the spacing and the state of occupancy of the energy bands. Many metals resemble sodium in their valence bands which are adjacent to their conduction bands and therefore, these metals readily act as conductors, whereas, substances like plastics, glass, wood, rubber etc., has a great energy gap, hence electrons cannot have a freedom of movement from the valence band to conduction band, such substances are know as insultors, consequently much more energy is needed to excite an electron into the conduction band lacking this energy, free motion of electrons is not possible. The elements like silicon and germenium have properties intermediate of metals and non metals are therefore called semiconductors. The energy between the filled and empty bands for these solids is much smaller than that for the insultors.

ALLOYS

11-13 INTRODUCTION

Alloys are homogeneous mixtures of two or more metals. Metals are insoluble in the usual solvents like water, benzene, alcohol, ether, etc., but when it is in molten state it can be mixed with another molten metal to make a homogeneous mixture which when on cooling solidifies. The metal can be mixed in any proportion to form various alloys. Alloys can be prepared by mixing a metal with a nonmetal also but, an alloy, should always contain a metal in it. *Hence an alloy can be defined as a metallic, intimately mixed solid mixture of two or more different elements, one of which atleast is essentially a metal. When an alloy contains mercury in it, then the alloy is called as an* ***amalgam****. A copper mercury alloy is thus called a copper amalgam. An alloy in its solid state can also be in a heterogeneous state.*

11-13-1 Purpose of making alloys

Alloys are prepared to improve the general physical properties of a metal. Every metal possesses properties like high malleability, ductility, tenacity, lustre, conductance, etc., as mentioned in the previous sections. To improve all or some of the properties either a suitable metal or a nonmetal is mixed with the metal under observation. Alloy formation is of interest mainly:

(i) to enhance the hardness of the metal;

(ii) to lower the melting point;

(iii) to enhance the tensile strength;

(iv) to decrease the corrosive nature;

(v) to modify the colour of the metal and

(vi) to provide a better castability.

(i) To enhance the hardness of the metal, a suitable metal or a nonmetal is added.

For example: 0.5% Arsenic is added to lead to make very strong lead shots or bullets. To make iron hard and to get steel about 0.15 to 1.5% carbon is added.

(ii) In general the melting point of an alloy is lower than the original metals.

For example : The wood's metal containing bismuth, lead, cadmium and tin has a melting point of only 71°C, which is much lower than the melting points of the individual components Bi (271.3°C), Pb (327.4°C), Cd (320.9°C) and Sn (231.9°C).

(iii) The tensibility of an alloy increases with respect to the metal.

For example: On addition of 1% carbon to pure iron, the tensile strength of the steel formed increase by ten folds.

(iv) The corrosion resistance increases by the formation of an alloy. A pure metal can get tarnished or oxidized or get corroded easily but on making an alloy its corrosion resistance increases.

For example: Pure copper is highly corrosive but when mixed with tin to get bronze, it becomes corrosion resistant.

Likewise, stainless steel an alloy of iron, chromium and nickel is even acid proof.

(v) To modify the colour of a metal, alloys are prepared.

For example: The red copper and silvery zinc when mixed give yellow coloured brass.

(vi) Finally when metals are to be molded in a particular cast, an alloy is preferred. The reason for this is, the alloy can expand on modification and is easily fusible due to its low melting points.

For example: To make jewellery of gold a very small quantity of either silver or copper is added to get better casting and to provide hardness to the gold ornaments.

11-13-2 Classification of alloys

Alloys are classified into two classes

(i) Ferrous alloys and (ii) Non ferrous alloys

Ferrous alloys contain iron as the major component whereas the non-ferrous alloys do not have iron as one of the main components. The ferrous alloys are often known as steels.

11-13-3 Ferrous alloys

In the making of ferrous alloys, carbon, nickel, chromium, cobalt, etc., are added in traces. The main purpose of adding these substances to iron is to:

(a) improve tenacity

(b) improve ductility

(c) improve hardness

(d) improve heat resistance

(e) improve corrosion resistance

(f) improve acid and chemical resistance

(g) improve the ability to retain shape even in hot conditions

(h) minimize the coefficient of expansion

(i) improve smoothness

(j) improve the ability to undergo milling operations

(k) improve magnetic permeability

(l) improve forgebility

(m) improve shock resistance

(n) improve the retention of hardness in hot condition

(o) improve the ability to retain physical properties at high temperature

(p) improve cutting ability

The alloying is done based on the specific effects of the added non-metal or metal.

For example:

Nickel, improves tensile strength, ductility, toughness, elasticity, heat and corrosion resistances.

Chromium improves hardening and toughness.

Manganese increases strength and toughness alongwith brittleness.

Silicon improves the magnetic permeability and corrosion resistance.

Molybdenum is added to strengthen the alloy at elevated temperature and eliminates the brittleness.

Tungsten increases toughness, hardness, shock resistance.

Vanadium increases tensile strength, ductility and shock resistance.

Some of the ferrous steels are discussed below:

Nichrome contains 20% chromium and 10% nickel. This steel can be used at temperatures as high as 1100°C, therefore it is used in making boilers, steam lines, gas turbines, retorts, annealing boxes, etc.

Stainless steels are highly resistant to corrosion and are resistant to strong concentrated acids and bases. They can withstand temperatures up to 800°C. Stainless steel containing 12.22% of carbon is a magnetic type stainless steel while non-magnetic type stainless steel contain 18 to 26% chromium, 8 to 21% nickel with 0.15% carbon in it. The household stainless steel contains 18% of chromium and 8% of nickel. This steel is also used in the making of decorative pieces, surgical instruments, etc. The stainless steel is strong, hard and high resistant towards wear and tear. If it gets soiled during cooking its brightness is restored by washing with any detergent. It has a bright attractive appearance, the only disadvantage of it is, it has no resale or recasting value.

High speed steels are steels which can be used for making, cutting and drilling tools which work at high speeds These are;

(a) 18-4-1 H.S.S. containing 18% tungsten, 4% chromium and 1% vanadium along with 0.75% carbon. It is used in lathe machines and like instruments.

(b) Cobalt H.S.S. contains 5 to 8% of cobalt.

(c) Vanadium H.S.S. contains 1% of vanadium and 0.7% carbon.

(d) Molybdenum H.S.S. contains 6 to 8% molybdenum.

The nickel steels improve tensile strength, ductility, toughness, elasticity, heat and corrosion resistance of the steel. The nickel steels are of six types containing:

(a) 2.5% to 3.5% Ni and 0.1to 0.15% carbon.

(b) 3.5%Ni, 0.2 to 0.4% carbon.

(c) 25-28% Ni, 1.4% Mn, 0.5% Cr and 0.3 to 0.5% carbon.

(d) 30-36% of Ni, 0.3 to 0.5% C and this alloy is called **Invar**. It is the best nickel steel and is used in high precision instruments.

(e) **Platinite** contains 46% Ni and 0.15% carbon and is used in the making of the bulb filaments.

(f) **Parmalloy** has no carbon but contains 80% nickel, this alloy is used in communication engineering.

The **Manganese steels** are of three main types containing either 1 to 2% manganese and 0.25% to 0.55% carbon, or 1 to 8% manganese and 0.9% carbon or 11 to 14% Mn with 1 to 1.3% carbon in them.

The use of **the chromium steels** varies with the composition. Steel with composition 0.5% Cr, 0.25% Mo and 0.35% C is used in making machine gun barrels and gear sets.

Steel with 1% Cr, 0.5% Mo and 0.5% C is used in the making of springs. Steel with 1% Cr and 0.9% carbon is very hard and is used in the making of knives, hammers, etc.

Steel with 0.5% Cr and 1.3% C is used in the making of files, razor blades, etc.

Alnico is another important steel containing 12% Al, 20% Ni and 6% Co. It is highly magnetic and is used in the making of magnets.

11-13-4 Non-ferrous alloys

The non-ferrous alloys usually contain aluminium, copper, lead, zinc, tin and nickel.

Aluminium alloys or Duralumin containing 95% of aluminium, 4% of copper, 0.5% of magnesium and 0.5% of manganese is used in the making of aeroplane bodies, automobile parts, etc.,

The Y-alloy contain 4% Cu, 1.5% Mg, 2% Nickel, 0.5% Si, 0.5% Mn, 0.5% Fe with rest being Aluminium. It is used for making of pistons, cylinder heads and crank shafts.

Magnalumin is a tough strong alloy containing 70% aluminium and 30% magnesium and the proportion can also be varied to 95% aluminium and 5% of magnesium. Magnalumin is used for making cheap balances and scientific instruments.

Electron contains about 9 to 11 per cent of manganese and rest of it is magnesium. It is used in the construction of oil tankers, petrol bunkers, gear boxes etc.

Alpan, an aluminium-silicon alloy containing 10-30 per cent of silicon with the rest being aluminium is a highly corrosion resistant alloy. Therefore it is used in the building materials. Other Copper-Zinc alloys are brasses, bronzes and beryllium copper alloys.

L-5 an aluminium-zinc-copper alloy containing 13.5% of zinc, 12.75% of copper and the rest of it is aluminium, possesses good mechanical properties. It is suitable for sand castings and is used for making of gear boxes, valve covers, crank cases, etc.

Brasses are of eight types depending upon the percentage composition of the copper and zinc present or nickel and tin present along with the main ingredients copper and zinc. These alloys are usually golden in colour in the absence of nickel or tin.

A 90% copper + 10% zinc alloy is popularly known as commercial **bronze** or **French gold** is used in the making of artificial costume jewellery.

The Dutch metal with 80% per cent copper and 20 per cent zinc is used in the making of cheap jewellery, musical instruments, battery caps, name plates, etc.

The cartridge brass contains 70% copper and 30% zinc is easily pressed into any form and is used in cartridge making, sheet fabrications, household articles etc.

The muntz brass with 60% copper and 40% zinc is much stronger than the cartridge brass. The muntz brass is used in the making of screws, springs, radiator cores, chains etc.

The admiral brass with 70% copper, 29% zinc and 1% tin is highly corrosion resistant.

The Naval brass with 62% copper, 37% zinc and 1% tin, is used in the marine works.

German silver with 50% copper, 20% zinc and 30% nickel has good strength. It is resistant to salt water. It is used for making table wares, ornaments, cutlery, coins etc.

The coinage bronze with 89-92% copper and 11 to 8% tin is used in the making of coins, artificial jewellery.

Gun metal with 85% copper, 5% tin, 5% zinc and 5% lead is used in the foundry works, heavy load bearings, water fittings and marine pumps, etc.

Beryllium copper with 97.4% copper and 2.6% beryllium is used in industries, in the making of diaphragms, springs, and tools with high precision.

11-13-5 Substitutional Alloys

Substitutional alloys are obtained by removing the atoms of a metal M gradually and replacing them by the atoms of another metal M' to give an alloy of MM' type. The best known example of this type is a combination of silver with gold, both having a face centered cubic lattice and atoms having almost the same size (Ag = 1.432 Å and Au = 1.442 Å), these metals can be mixed in any proportion. An alloy of this type is possible, when the radius ratio of the constituents does not exceed 1.14 Å and their charge numbers are alike. When gold is mixed with copper, the atoms get segregated giving an alloy having a ratio of Au : Cu as 1:3 even though the metals are initially mixed in the ratio 1:1. This is observed as due to a change in the crystal lattice from cubic to tetragonal with Au occupying the corners of cubic lattice and Cu atoms occupying the face centered position.

11-13-6 Hume Rothery Rules

Composition of an alloy changes with the variation in the phases and vice versa. **These phases are usually designated as β, γ, ε, etc. For example, in the following sets of copper alloys, shown in table 11-7, we notice a change** in composition with the changes in the phase.

Table 11-7 Composition and Phases of some copper alloys

Phase	β	γ	ε
Composition changes	CuZn	Cu_5Zn_8	$CuZn_3$
	Cu_3Al	Cu_9Al_4	Cu_5Al_3
	Cu_5Sn	$Cu_{13}Sn_8$	Cu_3Sn_8
	$CuRh_2CuRh$	Cu_2Rh	

This particular property is usually observed in the alloys of transition metals, with the B group metals of the periodic table. It generally happens when two metals of different valencies are made to combine. For example; copper with a valency one and zinc with valency two form an extensive range of solid solutions (alloys), with different compositions and phases. Replacement of copper atoms by zinc atoms in a solid solution creats a charge imbalance. Crossing a particular limit, the alloy acquires another structure that brings back stability. The succession of phases is observed in the copper-zinc system. Pure copper has face centered cubic structure and on adding increasing amounts of zinc to it, the zinc atoms at first merely replace copper atoms. When a certain limit of this solid solution is reached the face centered cubic lattice gives way to a new phase with a body centered cubic structure, known as β-brass. Further addition in turn gives place to a γ-phase which is a cubic system and finally forms a hexagonal close packed structure. *In all these phases, while the composition of the phases differ, the ratio of the total number of valence electrons to the total number of atoms remain a constant for each phase.* For the β-brass structures this ratio is always 3:2, as illustrated in table 11-8.

Table 11-8 Ratio of the total number of valence electrons to the total number of atoms in β-phase alloys

Alloy	Valence electrons	No. of atoms	Ratio
CuZn	Cu(1) + Zn(2) = 3	2	3 : 2
Cu_5Si	Cu(1x5) + Si(4) = 9	6	3 : 2
Cu_3Al	Cu(1x3) +Al(3) = 6	4	3 : 2
Ag_3Al	Ag(1x3) + Al(3) = 6	4	3 : 2

Similarly in the γ-phase the ratio is 21:13 as observed in the alloys like Cu_5Zn_8, Cu_9Al_4, etc. Distribution of energy states for an electron/atom ratio of 3:2 for the face centred and body centred cubic alloys are exhibited in figure 11-19.

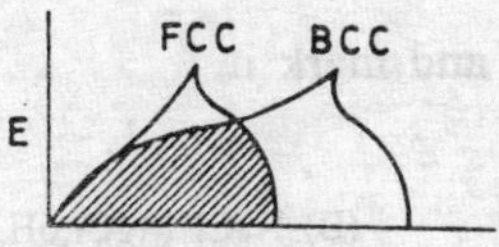

Fig. 11-19 Energy state for a face centred and body centred cubic alloy.

11-13-7 Interstitial alloys

The interstitial alloys are covalent compounds of hydrogen, boron, carbon and nitrogen with transition metals. The hydrides, borides, carbides and nitrides of the transition metals are nonstoichiometric, with refractory characters and possessing true metallic properties. They are obtained by heating the powdered metal to a high temperature with carbon or boron or in a stream of hydrogen, nitrogen or ammonia needed are the temperatures 2200° for the carbides, 1800°-2000°C for the borides, 2000°C for the hydrides and 1100 to 1200°C for the nitrides. The compounds are obtained in the form of powders. They are purified, sintered and compacted in the form of rods in vacuum or in an inert atmosphere. Most of the impurities, being highly volatile get lost in the above treatment. The compressed interstitial compound becomes very hard and its hardness lies between that of topaz (hardness = 8) and diamond (hardness = 10). The hardness of a few interstitial alloys is as follows:

Alloy	TiC	ZrN	WC	ZrB	W_2C
Hardness	8-9	8	9	9	9-10

These compounds show extremely high chemical inertness under usual conditions. The compounds have much in common with the true metals, thus they have a high electrical conductivity. In some cases the metallic character extends so far as to make the compound a super conductor, eg., zirconium nitride.

The name of interstitial alloys or compounds was suggested by Hagg. Hagg was able to determine the position of these small covalent radii nonmentals. He found, that, they get inserted in the interstitial spaces present between the metal atoms. When the radius ratio is less than 0.59, the alloy is normal and the metal interstitial atom arrangement is face cen tered cubic, close packed hexagonal or body centered cubic. The complex interstitial alloys having a radius ratio greater than 0.59 are less stable. Carbon and nitrogen always occupy octahedral holes in the interstitial alloys, hydrogen occupies the smaller tetrahedral interstices.

11-14 SELF TEST QUESTIONS

Choose the correct answer and mark it

11-14-1 Carnallite is

(A) KCl $MgCl_2$ $6H_2O$ (B) NaCl $MgCl_2 6H_2O$

(C) KCl $MnCl_2 6H_2O$ (D) NaCl $MnCl_2O$

11-14-2 Cryolite is

(A) AlF_3NaBr (B) AlF_3 3NaF (C) AlF_3.4NaF (D) AlF_3MgF_2

11-14-3 The metal sulphides are usually referred to as

(A) Heamatites (B) Pyrites (C) Syggests (D) None of these

11-14-4 $Cu(OH)_2 CuCO_3$ is

(A) Cerussite (B) Dolomite (C) Malachite (D) Calamite

11-14-5 Calamine is a carbonate of

(A) Copper (B) Zinc (C) Mercury (D) None of these

11-14-6 Metals like silver, gold, etc., found in free state are called

(A) Cinnaboras (B) Glance (C) Syggests (D) None of these

11-14-7 Tin stone is

(A) SnO_2 (B) SnO (C) $SnCO_3$ (D) $Sn(CO_3)_2$

11-14-8 Fe_2O_3 is referred to as

(A) Ag_2CO_3 (B) Bauxite (C) Haematite (D) Cerussite

11-14-9 Horn silver is

(A) Ag_2CO_3 (B) $AgNO_3$ (C) AgCl (D) AgBr

11-14-10 Forth flotation process is usually applied to

(A) Carbonate ores (B) Sulphide ores

(C) Oxide ores (D) Sulphate ores

11-14-11 Calcination is done in the

(A) Absence of air (B) Presence of air

(C) Vacuum (D) Presence of nitrogen

11-14-12 Roasting is carried out in the

(A) Absence of air (B) Presence of air

(C) Vacuum (D) Presence of nitrogen

11-14-13 To a metallic gangue, when a flux is added it forms

(A) Slag (B) Matte (C) Oxide (D) Nitric

11-14-14 Mond's process is applied for the extraction of

(A) Zr (B) Th (C) Ni (D) Cr

11-14-15 The compound used for reduction to get nickel is

(A) $NiSO_4$ (B) $Ni(CO)_4$ (C) NiS (D) $NiCO_3$

11-14-16 Monel metal contains along with Cu,Fe and Mn

(A) Ni67% (B) Ni23% (C) Ni50% (D) Cr 33%

11-14-17 Invar is an alloy of a

(A) Ni, C,F (B) Mg,C,Fe (C) Mn,Si,Fe (D) Ni,C,Fe

11-14-18 German silver contains

(A) 50% Cu, 20% Ni and 30% Zn (B) 50%Cu,20%Ni and 30%Mg

(C) 50% Cu, 40% Ni and 10% Mg (D) 40%Cu,40%Ni and 20% Zn

11-14-19 Zircon is

(A) $ZnSiO_4$ (B) ZrO_2 (C) $ZrOCl_2$ (D) $Zr(OH)_2$

11-14-20 Zircon is also known as

(A) Baddeleyite (B) Zirconia (C) Hyacinth (D) Zirconate

11-14-21 Zirconia is

(A) $ZrOCl_2$ (B) $Zr(OH)_2$ (C) ZrO_2 (D) $ZrSiO_4$

11-14-22 Uranium can be isolated generally from

(A) Autunite (B) Carnolite (C) Pitch blende (D) Baddeleyite

11-14-23 Uranyl sulphate has a formula

(A) U_2SO_4 (B) $U_2(SO)_4$ (C) $UO_2(SO_4)_2$ (D) $U(SO_4)_2$

11-14-24 Uranium is usually used in the production of

(A) Urea (B) NH_3 (C) Plutonium (D) Magnesium

11-14-25 Thorite is

(A) Th SiO_4 (B) $Th_3(PO_4)_4$ (C) ThO_2 (D) $ThO_2UO_2SiO_2$

11-14-26 Thoria is

(A) Th SiO_4 (B) $Th_3(PO_4)_4$ (C) ThO_2 (D) $ThO_2UO_2SiO_2$

11-14-27 Stainless steel contains iron alongwith

(A) Cr and C (B) Cr and Co (C) Cr and Ni (D) Cr and B

11-14-28 Nichrome contains mainly

(A) Ni 60%, Cr 20% and Fe 25% (B) Ni 60%, Cr 10% and Mg 25%

(C) Ni 60%, Cr 5% and Mn 25% (D) None of the above

11-15 SHORT AND ESSAY TYPE QUESTIONS

11-15-1 Mention two ores of uranium. Starting from sodium uranate, how do you obtain pure uranium ? Give all the reactions in the above process. Mention any two uses of uranium. (O.U.Apr.1991)

11-15-2 What are alloys ? How are they Classified? Describe Hume-Rothery rules. (O.U. Apr.1991)

11-15-3 Mention the ores of nickel? How is it extracted from its ores by Mond's process? Mention the uses of Nickel. (N.U. Mar., 96) (A.U.,Dec.1989)

11-15-4 Write notes on Hume-Rothery rules. (A.U.,Dec.1989)

11-15-5 Mention the industrial uses of thorium and uranium. (A.U.,Dec.1989)

11-15-6 Write an account of bonding in metals.

11-15-7 What are alloys? Give two examples. (A.U., Jun. 1990) (SKU., Apr.1990)

11-15-8 Write the industrial importance of chromium. (A.U., Jun.1990)

11-15-9 What are alloys? How are they formed? How Hume-Rothery rules are applicable in the formation of alloys? (A.U.,Jun.1990)

11-15-10 Mention two important ores of thorium? How is it extracted? Give two uses of thorium. (N.U. Mar., 96) (O.U., Oct.1990)

11-15-11 Write briefly about the metallurgy of Zirconium. (O.U.,Oct.1990)

11-15-12 In what form does nickel occur in nature? Give the details of its extraction by Mond's process. Mention its uses. (N.U. Sep., 95) (O.U.,Apr.1984)

11-15-13 Write about V.B.theory of bonding in metals. (N.U. Mar., 95)(O.U.,Apr.1984)

11-15-14 Mention Hume Rothery rules & Explain. (O.U., Apr,.1984) & (A.U ., Apr.1989)

11-15-15 Write a brief note on Mond's process. (N.U. Sep., 94) (O.U., Apr. 1990)

11-15-16 (a) Describe the extraction of Zirconium from its important minerals.

(b) Write any two industrial applications of zirconium. (O.U., Apr., 1989)

11-15-17 (a) Describe the valence bond theory of metals.

(b) How is malleability explained from VBT of metals? (O.U., Apr. 1989)

11-15-18 Describe any method for the preparation of alloys. (O.U., Apr. 1989)

11-15-19 (a) How does Zr occur in nature? Describe the extraction of pure Zr from its minerals?

(b) What are the uses of Zr. (O.U., Apr. 1989)

11-15-20 How does free electron theory explain the high conductivity of metals?

11-15-21 (a) Mention the minerals of Ni. How is it extracted from its ores, in the pure form?

(b) How is thorium metal useful? (O.U., Oct. 1988)

11-15-22 Explain on the basis of valence bond theory the malleability of metals? (O.U.,Oct.1988)

11-15-23 Name the ores of Zirconium? Give briefly the method of its extraction. (O.U., Apr. 1987)

11-15-24 Discuss briefly valence bond theory of bonding in metals. (O.U., Apr. 1987)

11-15-25 Write a note on inter-metallic compounds. (SKU.,Apr.1990) 7 (K.U., Apr. 1989)

11-15-26 (a) What are the important sources of uranium.

(b) Give a detailed account of its extraction.

(c) What are the various oxidation states of uranium ? Give examples. (A.U.,Apr.1989)

11-15-27 What are the chief ores of uranium? How is uranium extracted from its ores. Mention its uses. (N.U. Mar., 95) (O.U. Mar., 95) (A.U.,Jun.1989)

11-15-28 Mention any two ores of nickel. Write a method of extraction of nickel and its uses. (S.K.U., Apr.1990)

11-15-29 Write electronic configuration of chromium(at.no.24) and copper (at.no.29) (S.K.U.,Apr.1990)

ANSWERS

11-14-1 A, 11-14-2 B, 11-14-4 C, 11-14-5 B, 11-14-6 C, 11-14-7 A,

11-14-8 C, 11-14-9 C, 11-14-10 B, 11-14-11 A, 11-14-12 B, 11-14-13 A,

11-14-14 C, 11-14-15 C, 11-14-16 A, 11-14-17 D, 11-14-18 A, 11-14-19 C,

11-14-20 C, 11-14-21 C, 11-14-22 A, 11-14-23 C, 11-14-24 A, 11-14-25 D.

❋

12
NUCLEAR CHEMISTRY

12-1 INTRODUCTION

Nuclear chemistry, as an important science that has undergone a major and rapid development in the recent years. Three discoveries in the closing years of nineteenth century not only had a profound effect on scientific progress and thought, but also created drastic change in the life pattern of a common man. These discoveries were those of X-rays by Rontgen in 1895, radioactivity by Henry Bacquerel in 1896 and the discovery of the cathode rays by J.J.Thompson in 1897. In 1900, Planck also put forth his fundamental concept of the quantum in order to meet the failure of classical atomic theory, to describe the spectral distribution of the black body radiation, the so called ultraviolet catastrophe. Thus in a brief span of five years, some important foundations for the nuclear era were laid. Henry Becquerel reported that all compounds of uranium, whether fluorescent or not, emitted a type of rays which were sensitive on a photographic plate. These rays were weakened and had little or no effect on the photographic plate when the uranium containing mineral is wrapped in aluminium, copper or other metal foils. The less penetrating effect of these radiations was also noticed when the mineral is wrapped in a black paper. Madam Curie took the lead and cxtended the studies to other sources also, which led to the discovery of the elements such as polonium and radium, although much of this early work devoted to the discovery of new sources of radioactive rays. Rutherford was able to suggest the emmisions to be of two general types. The first type was observed when wapped in a thin aluminium foil while the second type showed a greater penetrating capability. Rutherford called the first type as alpha radiation and the more penetrating type was termed beta radiation. This *self-spontaneous emission of radiation by a nuclei is called* ***radioactive decay.***

A clear understanding of the nature of these radiations came with the application of magnetic deflection techniques. A radio active source was kept in a lead container over which a photographic plate was held. On the application of the magnetic field, it was found that three types of deflections are emitted out of the source, Fig.12-1.

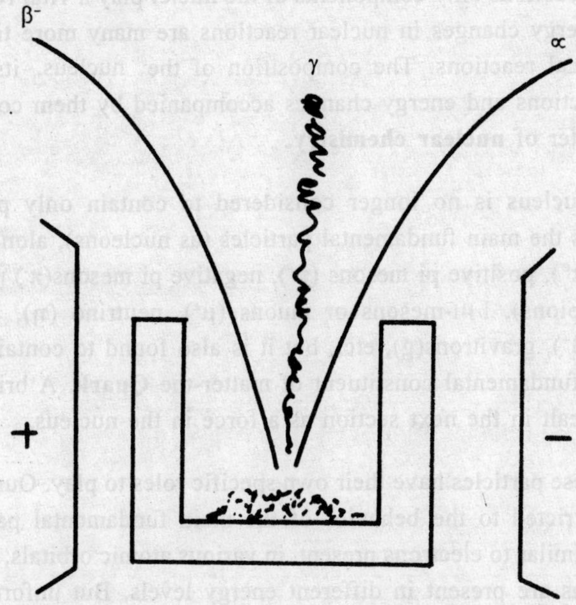

Fig.12-1 Schematic paths of α, β and γ radiation under the influence of magnetic field at right angles to the plane of the drawing.

A large deflection towards positive pole of the magnet having a velocity of 1.6 X 10^{10}cm s^{-1} (less than the velocity of light) and charge to mass ratio (e/m ratio), of 10^7 emu per gram. This is of the same order of magnitude as that for the cathode rays. It was readily concluded that the β- rays were of the same nature as the cathode rays.

A segment of rays were found to deflect towards the negative pole of the magnet and from their deflection in electric and magnetic fields, Rutherford concluded that they are α-rays. These are positively charged particle rays. The velocity is 10^7 times much slower than light and e/m ratios differed from source to source. The atomic weight of the α-particle calculated was 3.84 amu and that of the Helium nucleus was 3.96 amu. So, taking into account of the probable experimental errors in the estimates, it was concluded that the α-particle, is a dipositive Helium nucleus.

A third type of emission was also recorded on the photographic plate. This emissiion had no deflection both in the electric field and magnetic field. These rays were more penetrating and electro magnetic in nature, having wave length similar to those of x-rays. These were named as γ (gamma) rays.

The chemical reactions involve interactions between the valence electrons of different atoms and the atomic nuclei remains unchanged. But in nuclear reactions only components of the nuclei play a vital role. Besides this, the energy changes in nuclear reactions are many more times higher than chemical reactions. The composition of the nucleus,, its stability , nuclear reactions and energy changes accompanied by them constitute the subject matter of **nuclear chemistry.**

The nucleus is no longer considered to contain only protons and neutrons; as the main fundamental particles (as nucleons); alongwith zero-pi mesons(π^o), positive pi mesons (π^+), negative pi mesons(π^-) ((commonly known as pions), l μι-mesons or muons ($\mu^\pm$), neutrino (η), photons(υ), positrons (β^+), gravitrons(g), etc., but it is also found to contain the basic indivisible fundamental constituent of matter-the **Quark.** A brief study of quarks is dealt in the next section as a force in the nucleus.

All these particles have their own specific roles to play. Our discussion will be restricted to the behavior of the main fundamental particles, the nucleons. Similar to electrcns present in various atomic orbitals, the protons and neutrons are present in different energy levels. But unfortunately no rationalization, like in electrons, has yet been made.

18-2 FORCES IN NUCLEUS

According to Yukawa(1935) neutrons and protons are bound by a very rapid exchangeof a π-meson or pion. The three pions are known to exist-positive(π^+), negative(π^-) and neutral (π^o). The mass of a pion is 2 to 3 times that of an electron. According to the exchange theory, a neutron is converted into a proton by the emission of negative pion(π^-). This in turn is accepted by another proton which is converted into another neutron.

$$n^o \longrightarrow p^+ + \pi^-$$

$$\pi^- + p^+ \longrightarrow n^o$$

Similarly the exchange of π^+ and π^0 may be responsible for p-n and p-p and n-n attractions as illustrated below;

$$\beta^+ \longrightarrow \pi^o + \pi^+$$

$$n^o + \pi^+ \longrightarrow p^+$$

$$p^+ \overset{\pi^o}{\longleftrightarrow} p^+$$

$$n^o \overset{\pi^o}{\longleftrightarrow} n^o$$

The exchange of mesons is extremely rapid(nearly 10^{24} transfers per second) and are always busy in keeping the nucleus intact. It is found that these particles(π^-, π°, π^+) do not come out of the nucleus unless they are induced.

The liquid drop model suggests that powerful cohesive forces between nucleons exist and that they effectively overcome the electrostatic forces of repulsions between the protons of the nucleus. Though their nature is far from completely understood, the attractive forces between nucleons are independent of the charge on them, that is, proton-proton, proton-neutron and neutron-neutron attractions are identical. However, these forces are operative only for a very short range.

The Quarks: As mentioned earlier, *quarks are the fundamental indivisible constituents of matter which build the nucleus.* These are charged particles with a wide range of mass. Two type of quarks are found to be present,**u** (up) quarks and **d** (down)quarks. The **u** quark has constant charge of $+2/3e^-$ and the **d** quark has a constant of $-1/3e^-$ (e^- is the electron charge). Hence in the nucleon: proton has a charge equivalence of +1 which is calculated based on the charge of the quarks as u.u.d., that is , +2/3 + 2/3 – 1/3 = + 1 and the charge of the neutron is calculated as d.d.u.or –1/3 – 1/3 +2/3=0. The mass of **u** and **d** quarks are found to be a few MeV/c^2 and a third type of the quarks, that is, t (top) quarks are the heavy ones with a mass around 40 GeV/c^2. These are twice in mass with respect to an electron. The mass defect of a nuclei (explained later in section 18-4) is most probablly balanced by the mass of the quarks. According to the quark thoery not only does the conversion of one **d** quark to an **u** quark can take place but also a neutron can get converted to a proton during such a conversion with the liberation of an electron and an antineutrino.

(d u d) → (u u d) + e + υ

Neutron Proton Electron Antineutrono

A quark has not been isolated till now, but it is being considered to be the ultimate physical particle of an atom.

18-3 PROPERTIES OF NUCLIDES

Nuclide is the term used for an independent nuclear species having its individual identity. For example C^{13} and N^{13} are nuclides, differeing in all respects.

An important characteristic of the nucleus, is its size related to its atom. The radius of an atom is usually of the order of 10^{-8}cm and that of the nucleus is for 10^{-13} cm indicating that the nucleus is 10^5 times smaller than the atom. Generally *for the lighter nuclides the mass number is found to be double that of the nuclear charge,* the difference in atomic mass the charge is attributed to the presence of the neutrons. Thus the atom of Helium(charge2) consists of a tiny nucleus with 9.60572 x 10^{-10} esu of positive charge, equal to the charge of 2 protons and two electrons at a relatively enormous distance from the central nucleus of equally opposite charge. The α-particle or helium nuclei, is merely a helium atom from which two electronsare stripped off. The mass of a helium atom is 6.6456 X 10^{-24}g, that is 4 atomic mass units. The two electrons are responsible for only 1.82166 X 10^{-27}g which is about 1/3600 of the total mass, has to be two neutrons also in the helium nucleus.

An important property of a nuclide is its existence as an isobar, isotone, isotope or as an isomer. These properties of the nuclide are explined below.

18-3-1 Isobars

The term Isobar is used to express the relationship between nuclear species *which share the same mass number but differ in atomic charge.* For example $_{18}Ar^{40}$, $_{19}K^{40}$ and $_{20}Ca^{40}$ share mass number of 40 but differ in atomic charge.

Nuclei	No. of Protons/ atomic charge	No. of Neutrons	Actual mass
$_{18}Ar^{40}$	18	22	39.9641282
$_{19}K^{40}$	19	21	39.9640079
$_{20}Ca^{40}$	20	20	39.9623892

In this regard, the nuclear models too, give only a vague picture regarding their stability. The mass difference seems to be slight, but as we shall see later, it is a very important quantity because it is related to the difference in energies of the atoms.In fact even this small mass difference corresponds to a considerable energy difference.

18-3-2 Isotones

Isotones are nuclei *which have the same number of neutrons but different number of protons.* For example $_4Be^9$, $_5B^{10}$, $_6C^{11}$, $_7N^{12}$ have the same number of neutrons but different number of prtons, hence are isotones of each other.

Nuclei	No. of Protons	No. of Neutrons
$_4Be^9$	4	5
$_5B^{10}$	5	5
$_6C^{11}$	6	5
$_7N^{12}$	7	5

18-3-3 Isotopes

Isotopes are nuclei of the same element possessing different mass numbers. This difference in the mass is due to the different number of neutrons present in them. The most familiar example is the three isotopes of hydrogen,viz., hydrogen - $_1H^1$, deuterium - $_1H^2$, tritium - $_1H^3$.

Nuclei	No. Of Protons	No. of Neutrons
$_1H^1$	1	0
$_1H^2$ (D)	1	1
$_1H^3$(T)	1	2

18-3-4 Isomers

Nuclei possessing same charge and mass but different in their energy content are called isomers. The isomers in an excited state or meta stable state are short lived. Such metastable isomers attain stability by loosing the extra energy in the form of photons or gamma radiations. For example; $_{72}Hf^{180m}$ (10^{-14} sec.) (metastable form) and $_{72}Hf^{180g}$ (ground state) are two isomers of hafnium.

12-4 STABILITY OF THE NUCLEUS

Study of the stability of the nucleus is based on a most important factor, the neutron/proton ratio or n/p ratio. Before we discuss about this ratio, we have to first see how actually the nucleus looks like, its energy levels and the forces within the nuclei.

12-4-1 The Structure of the nucleus

To study the structure of the nucleus we will have to develop a diagram or a model as we usually do for the electronic levels in the atom.

For this purpose, let us consider two large positively charged particles, one of these must be atleast as large as a proton and the other say the size of a deuteron. If these two particles are at a large distance apart, they do not have any effect on each other. Suppose, however,that one of them gets moved towards the other, since like charge particles repel, a repulsive force is created (Fig.12.2a) between them or in simple words, work must be done on one of the particles to move it closer to the other. On bringing these two positively charged particles closer to each other, this colombic repulsion, continues to increase until they are about 1 Fermi, that is 10^{-13} centimetres, which is also the diameter of the nucleus, apart from each other. But very surprisingly at a distance less thant 10^{-13} centimetres the repulsion suddenly decreases, even becoming zero and then becomes a negative repulsion. This however can be represented as an attractive force between the particles. This region in which the attractive force operates can be visualized as the nucleus of the atom. The nuclear radius is given by following equation;

$\mathbf{r = A^{1/3} \times 1.4 \times 10^{-13}}$ where r is the nuclear radius in centimetres A is the atomic mass number.

In the nucleus the coulombic repulsion is not discarded but the attractive nuclear forces are more prominent. The Fig.12-2b and Fig.12-3 shows the continuation of the nuclear diagram. It is intersting from the point of view of elucidating this diagram to imagine of it, as a hill up on which a ball might be pushed. Now work must be done on the ball to push it, to a given point up on the hill, for that some potential energy based on the gravitational forces required (Fig.12-2c) . If the ball is released , then we know, that, its potential energy will get converted to kinetic energy and the ball will roll back, down the hill. A more general description of this type of a fictious diagram is one in which the hill represents a potential energy system. In this case, under our consideration, the potential energy scale not only represents the repulsive and attrative forces working on the particle but it represents the total energy the particle is posessing. The Einstein energy mass equivalence $\mathbf{E = mc^2}$ indicates the total energy of the particle, would include, its own mass energy, as well as the kinetic and potential energies possessed by it. Now, let us suppose a neutron approaching the positively charged nucleus. As the neutron has no electrical charge, repulsion is not observed when it comes closer to the nucleus. But, when it reaches the distance of 10^{-13} centimetres it gets attracted by the forces of the nuclear particles and the neutron gets captured by the nucleus (Fig.12-2d and Fig.12-3). The nuclear forces arise from the fundamental properties of the particles themselves. The property responsible for these nuclear forces among large particles such as protons and neutrons has been called the **"Mesonic charge"** or the **"Nuclearity"** of the particles. As the electrons do not have this property of nuclearity they cannot exist in the nucleus.

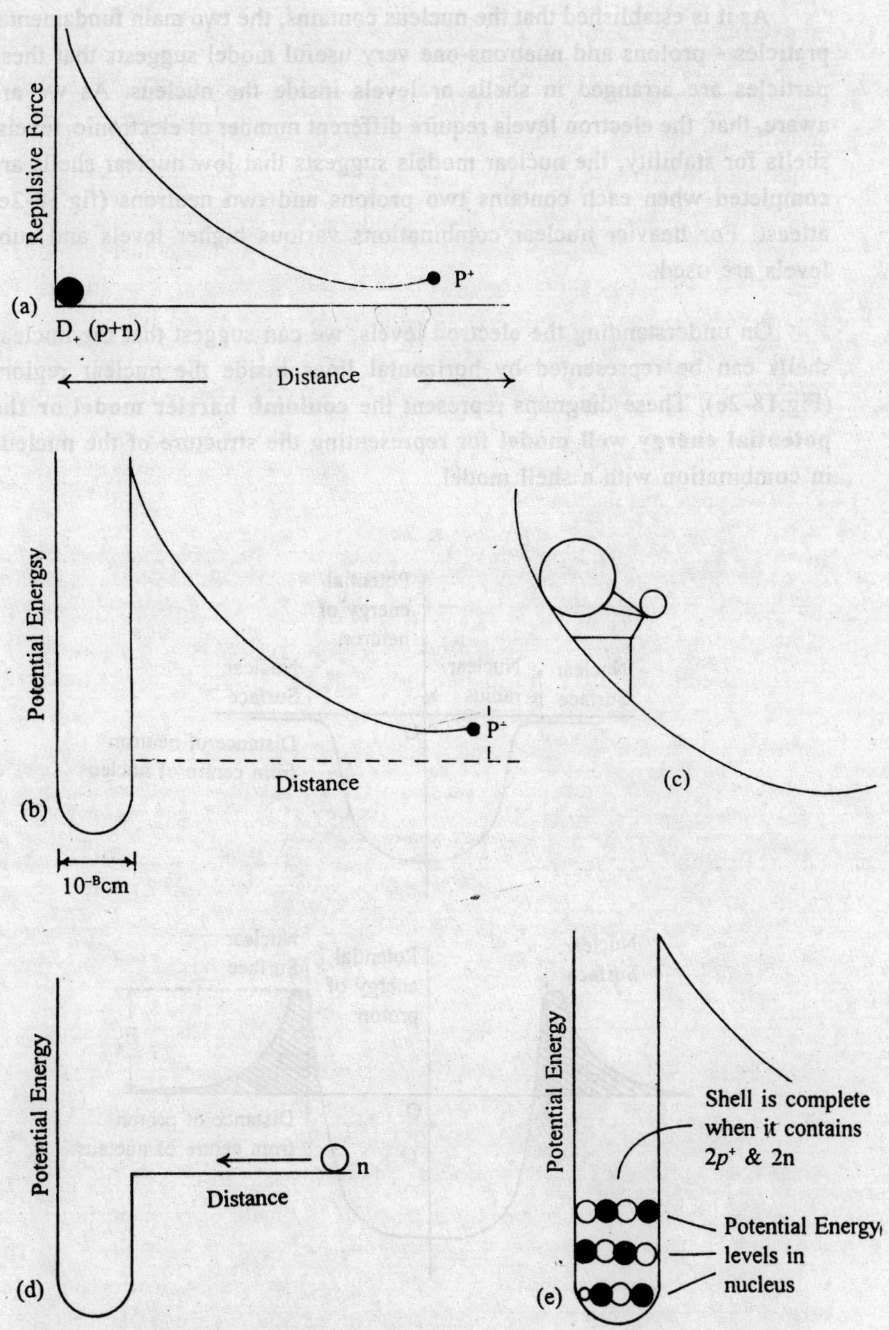

Fig 12-2 (a-e) Potential energy diagram of the nucleus

As it is established that the nucleus contains, the two main fundamental praticles - protons and nuetrons-one very useful model suggests that these particles are arranged in shells or levels inside the nucleus. As we are aware, that, the electron levels require different number of electronic levels/ shells for stability, the nuclear models suggests that low nuclear shells are completed when each contains two protons and two neutrons (fig.18-2e) atleast. For heavier nuclear combinations various higher levels and sub-levels are used.

On understanding the electron levels, we can suggest that the nuclear shells can be represented by horizontal lines inside the nuclear region. (Fig.18-2e). These diagrams represent the **coulomb barrier model or the potential energy well model** for representing the structure of the nucleus in combination with a shell model.

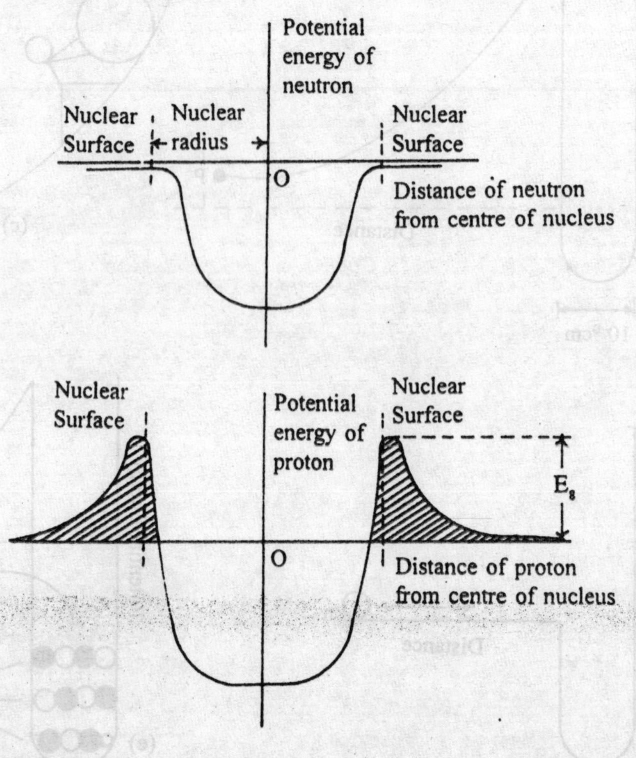

Fig.12-3 The potential energy well, when a neutron or a proton is inthe vicinity of the nucleus.

To establish the credibility of this nuclear diagram let us examine a system consisting of a radon nucleus and a helium nucleus i.e. an alpha particle. The radon nucleus comprises of 86 protons and 136 neutrons,giving a total mass number of 222. The helium nucleus contains two protons and two neutrons, based on these facts we can calculate the energy required to push an alpha particle/helium nucleus to the top edge of the radon nucleus. As per the equation for the radius(r) given earlier, the radius of the radon nucleus is calculated to be 9×10^{-13}cm.The energy of the helium nucleus at any distance from the radon nucleus can be calculated by the equation given below:

$$E = \frac{q_1 \times q_2}{r}$$

Where 'E' is considered as the energy of repulsion 'q_1' & 'q_2' are the charges on the particles and 'r' is the distance separting the particles. From this calculation, it is found that, the energy needed is 20 Mev, hence if this much sufficient energy is given to the alpha particle for it to overcome the nuclear potential barrier and combine with the radon nucleus, it would form a nucleus of radium with mass number 226, after nuclear transformation.

According to traditional mechanics, when an alpha particle needs 20 MeV energy to enter a nucleus, essentially, then it needs almost the same amount of energy to be pushed up, out of the nucleus, climbing down the potential energy barrier hill, out side the rim of the nucleus, but alpha particle cannot achieve this much sufficient energy to come out and cross the potential energy barrier. It has an energy of only 6 MeV. After establishing the structure of the nucleus to be a potential well with horizontal energy levels being occupied by nucleons as such or independent. neutrons in it, it can be concluded that the potential energy of a nucleus is around 20 MeV.

12-4-2 Energy profile in a nulcear interaction

The nucleus is represented as a sort of volcanic cone with a deep crater extending down below the ground level, fig. 12-4. The x-axis represents the changes in potential energy of a positively charged particles approaching the nucleus. As a result of repulsion of like charged particles, i.e.,between the nuclei and any approaching positively charged particle; the approaching particle slows down or in other words its kinetic energy drops. The lowered kinetic energy will contribute in raising its potential energy. To get into the crater, the particle must initially start with suficient kinetic energy so as to reach at least to the rim of the crater. If it has less than this amount, it will simply roll back, after climbing a part of the crater's slope.

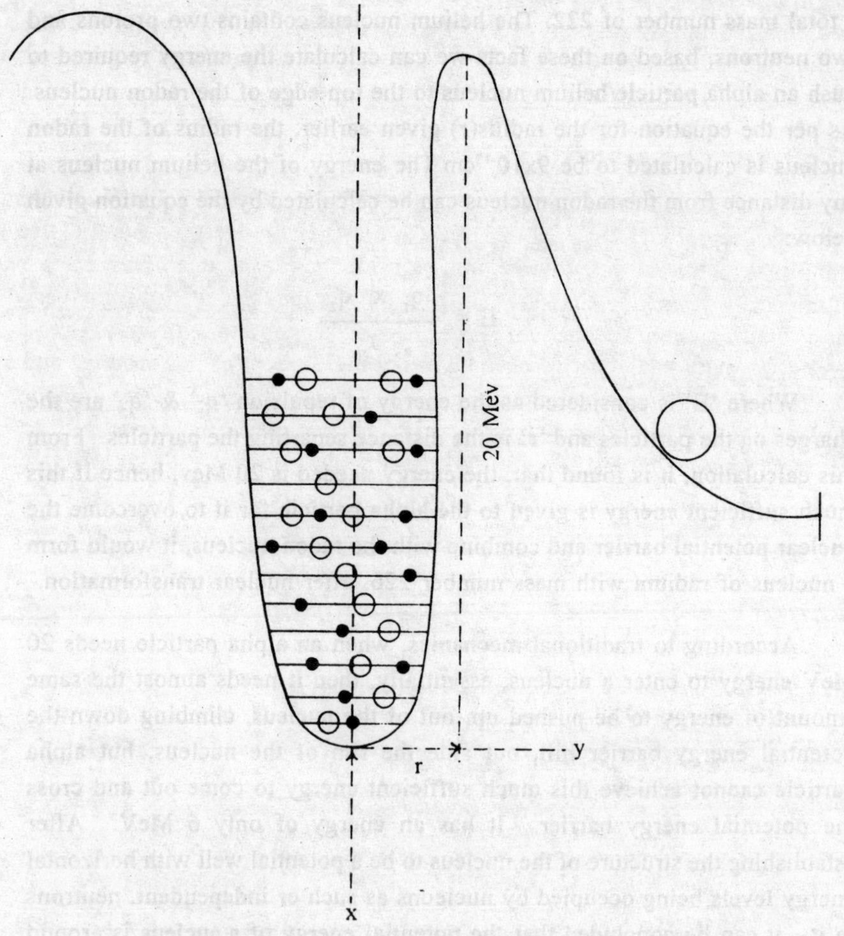

Fig.12-4 Energy profile of the nucleus

However, a particle which succeeds in reaching the top of crater rim will feel the close range forces of attraction, which will draw it towards the nucleus. Under the influence of this force, the particle which was at its potential energy maximum, standing on the top of the rim, will spontaneouly fall down into the nuclear crater. When the particle falls into the crater, its potential energy decreases but its kinetic energy increases by an equal amount (≅20 Mev). With this very high kinetic energy the particle cannot come out again. However, the crater being the real nucleus, containing many protons and neutrons, it often loses some of its kinetic energy to nucleons as result of collisions with each other. Hence in reality it is not possible for the particle to escape from the crater valley, and it gets trapped permanently in it. In real sense the protecting rim around the potential well is callled the coulomb barrier. The height of this barrier depends on the strength of the columbic repulsion, which in turn depends on the charge of the nucleus and of the incoming particles.

12-4-3 Neutron : Proton (n/p) Ratio

Apart from the forces within the nucleus which try to stabilize the nucleus, the neutron/proton ratio plays a very important role in determining the stability of a nuclei. In addition, this ratio also gives an idea of the nature of the decay of the nucleus. We will be seeing in the later sections, the very large binding energy of helium nuclei over hydrogen atoms from an extra strong attractive nuclear force between individual pairs of protons and neutrons. We call this a **pairing force**. For different elements if the number of protons is plotted agaist the number of neutrons, it is found that all the nonradioactive nuclides lie within a region called **the zone of stability.**

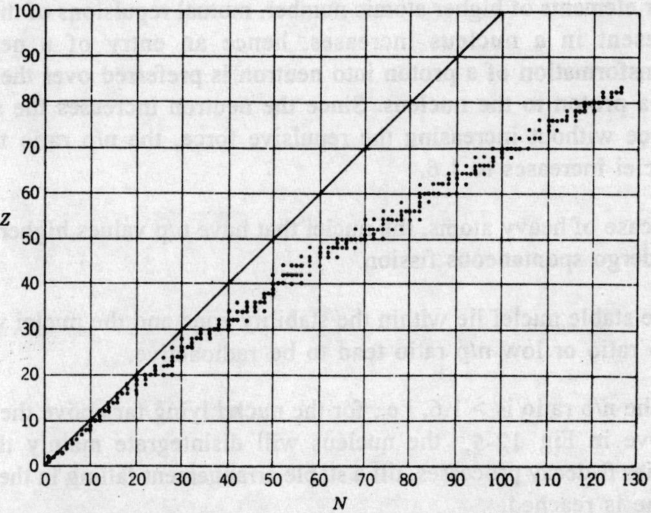

Fig 12-5 The known stable nuclei of a plot of charge (Z) versus no. of Neutrons (N). We can note the gradual increase in the neutron-proton ratio; the 45° line across the graph corresponds to a neutron-proton ratio of unity.

Even though there are only 81 elements from hydrogen to lead, there are all together 267 stable nuclides with in this 81 atomic numbers. Among these, 213 have an even number of protons and 159 have both protons and neutrons in even number, and only 27 elements have a single isotope and all except beryllium have odd number of protons. Only four stable nuclei H^2, Li^6, B^{10} and N^{14} have both odd number of protons and odd number of neutrons in them. Five other nuclides K^{40}, V^{50}, Rb^{88}, La^{138}, and Ta^{180} have both odd number of neutrons and odd number of protons, but are radioactive with long half lives.

	even P	odd P	Total
even N	159	50	209
odd N	54	4	58
	213	54	267

To determine the stability of a nucleus, the facts observed based on the number of neutrons and protons or in other words the neutron/proton ratio, are summarized in a qualitative way and given below.

(i) We find that elements of low atomic number, up to 20, invariably have the most stable nuclei and have n/p ratio equal to unity. In the number of neutrons vs number of protons graph they fall within th^ stability zone.

(ii) For elements of higher atomic number, mutual repulsions of the protons present in a nucleus increases, hence an entry of a neutron or transformation of a proton into neutron is preferred over the addition of a proton to the nucleus. Since the neutron increases the attractive force without increasing the repulsive force, the n/p ratio for stable nuclei increases to 1.6.

(iii) In case of heavy atoms, the nuclei that have n/p values higher than 1.6 undergo spontaneous fission.

(iv) The stable nuclei lie within the stability zone and the nuclei with high n/p ratio or low n/p ratio tend to be radioactive.

(v) If the n/p ratio is > 1.6, i.e., for the nuclei lying far above the stability curve in Fig 12-5, the nucleus will disintegrate mainly through a series β-decay processes, till a stable arrangement falling in the stability zone is reached.

$$_{56}Ba^{141} \xrightarrow[t_{1/2} = 18 \text{ min}]{_{-1}\beta^0} {}_{57}La^{141} \xrightarrow[t_{1/2} = 3.7 \text{ hr}]{_{-1}\beta^0} {}_{58}Ce^{141} \xrightarrow[t_{1/2} = 28 \text{ yrs}]{_{-1}\beta^0} {}_{59}Pr^{141}$$

In the above isobars we find that the n/p ratio decreases till a stable isotope, $_{59}Pr^{141}$, is formed.

(vi) If the isotope is not far away from the stability zone or stable n/p line, just one β-decay may be sufficient to put the isotope in the stability zone.

$$_{6}C^{14} \longrightarrow {}_{7}N^{14} + {}_{-1}\beta^0 + \gamma$$

(vii) In some rare cases, there is an emission of a neutron itself from the nucleus and the resulting final isobar is stable.

$$_{36}Kr^{87} \xrightarrow{-_0n^1} {}_{36}Kr^{86} + {}_0n^1 \xrightarrow{-_1\beta^0} {}_{37}Rb^{87} \xrightarrow{-_1\beta^0} {}_{38}Sr^{87}$$

(viii) Another interesting case is when the n/p ratio is too low and the isotope lies below the curve, then the possible modes of decay is by the capture of a neutron or emitting a positron or an α-particle. This results in an increase of neutrons in the nucleus and thus getting close to n/p - ratio of unity.

(a) Capture of a neutron :

$$_7N^{13} + {}_0n^1 \longrightarrow {}_7N^{14}$$

(b) Emission of a positron :

$$_7N^{13} \longrightarrow {}_6C^{13} + {}_1\beta^+$$

(c) Emission of an α-particle :

$$_{84}Po^{212} \longrightarrow {}_{82}Pb^{208} + {}_2He^4$$

(ix) The nucleus may capture an orbital electron from the K-shell, converting a proton into a neutron thus increasing the n/p ratio, resulting in a stable nuclide formation.

All these decays will be dealt in the following section. Therefore to have a stable nuclide, the nucleus should undergo changes by emission of usually a β-particle. *The phenomenon involved in the changes which take place in a radioactive nuclide by the emission of a particle, resulting in the formation of daughter nuclide/s, either stable or unstable, is called* ***nuclear disintegration.***

12-5 MASS DEFECT, BINDING ENERGY and PACKING FRACTION

If the masses of atoms can be measured accurate enough, then it is possible for us to calculate the total amount of energy with which the nucleus is bound together. The mass of a neutron has been found to be 1.008665 atomic mass units (a.m.u), in comparison to the mass of a C^{12} atom, which is assigned the mass value of 12.00000 amu. Similarly the mass of a proton is found to be 1.007825 amu. According to our usual way of thinking, that the mass of the H^2 (deuterium) nucleus consisting of a proton and a neutron should then be:

$$\begin{aligned} \text{Mass of neutron} &= 1.008665 \text{ amu} \\ \text{Mass of proton} &= 1.007825 \text{ amu} \\ \hline \text{Mass of neutron + proton} &= 2.016490 \text{ amu} \end{aligned}$$

One of the puzzling results of the determination of various isotopic masses with precise mass spectrometers and other instruments was that the ${}_1H^2$ (Deuterium) atomic mass was not found to be the sum indicated above. Instead the isotopic mass of this nucleus was found to be 2.014102 amu.

$$\begin{aligned} \text{Mass of neutron + proton} &= 2.016490 \text{ amu} \\ \text{Mass of deuterium nucleus} &= 2.014102 \text{ amu} \\ \hline \text{Mass difference} &= 0.002388 \text{ amu} \end{aligned}$$

This difference was formerly called the **mass defect** but this has given way to a new term **binding energy**. A different concept is used for the explanation of this discrepancy.

Einstein's mass-energy relationship is expressed by the equation;

$$E = mc^2$$

where c is the velocity of light and m the mass of the body. If c is expressed in cm s^{-1} and m in grams, then E will be obtained in ergs. He showed in his treatment that mass and energy are not to separate phenomena, but are identical, and that we can express quantities of mass and energy in the same or equivalent units. For example, the energy equivalent to 1 amu is obtained by putting $m=1.660 \times 10^{-24}$ g and $c=2.988 \times 10^{10}$cm s^{-1} and we get $E=mc^2=1.492 \times 10^{-3}$ergs. However, energy units in nuclear chemistry which are more popular are electron volts (eV) and the millon electron volts (MeV where 1 MeV $=10^6$ eV). *The electron volt is defined as the energy necessary to raise one electron through a potential difference of 1 volt.*

$$1\text{eV} = 1.602 \times 10^{-12} \text{ ergs}$$
$$1\text{MeV} = 1.602 \times 10^{-6} \text{ ergs}$$

Using these units, we find 1 mass unit = 931.5 Mev and 1 electron mass = 0.5110 MeV. Now to calculate the binding energy of the ${}_1H^2$ (Deuterium) it will be 0.002388 x 931.5 = 2.224422 MeV per nucleon and that of He^4 is therefore approximately 7.07 MeV. Thus, if it were possible to form an atom of H^2 or He^4 from its component parts, these would be energy release of 2.224422 MeV for H^2 and 7.1 MeV for He^4. Alternately it would require these amounts of energy to break their respective atoms into protons and neutrons. Thus **binding energy** is quite analogous to the heat of formation of a chemical compound from its element, but it is very much larger than any known chemical energy. The average binding curve is shown in fig 12-6.

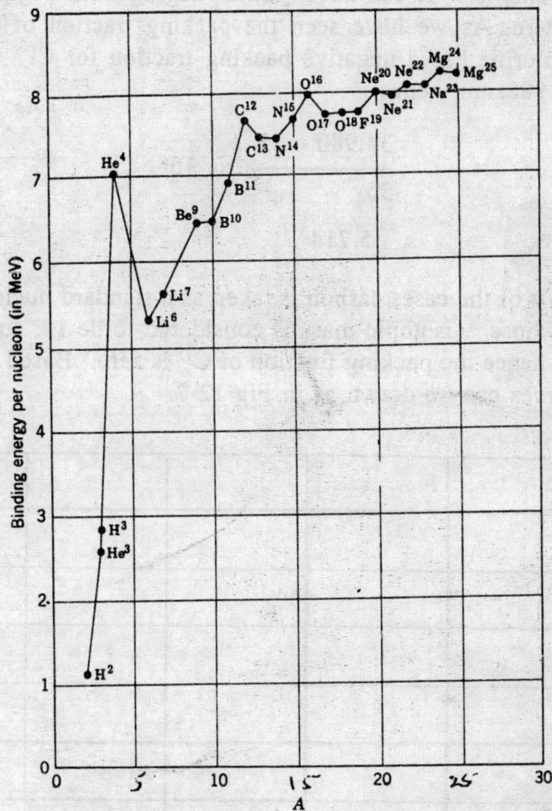

Fig. 12.6 The curves showing average binding energy per nucleon for the lighter nuclides. (Hydrogen to Magnesium including their isotopes)

Packing fraction (f) was expressed by Aston by the following expression

$$f = (\text{Isotopic mass-Mass number})/\text{Mass number}$$

$$= \frac{\Delta m}{A} \times 10^4$$

For example the packing fraction for N^{14} whose isotopic mass is 14.0082 amu will be:

$$f = \frac{14.0082 - 14}{14} \times 10^4 = 5.857$$

It is multiplied by 10^4 to get an easilycomprehensible numerica[illegible]lue and the packing fraction is thus expressed in parts per ten thousand. *Packing fraction is a measure of the stability of the nucleus and varies from one nucleus to another.* It can have values ranging from negative to positive including zero. As we have seen the packing fraction of nitrogen being positive, chlorine has a negative packing fraction for Cl^{35} whose isotopic mass is 34.980amu, hence

$$f = \frac{34.980 - 35}{35} \times 10^4$$

$$= -5.714$$

In most of the cases carbon is taken as a standard nuclei for reference purpose whose isotopic mass is considered to be 12, equal to its mass number12, hence the packing fraction of C^{12} is zero. Based on the packing fration, curves can be drawn as in Fig 12-7.

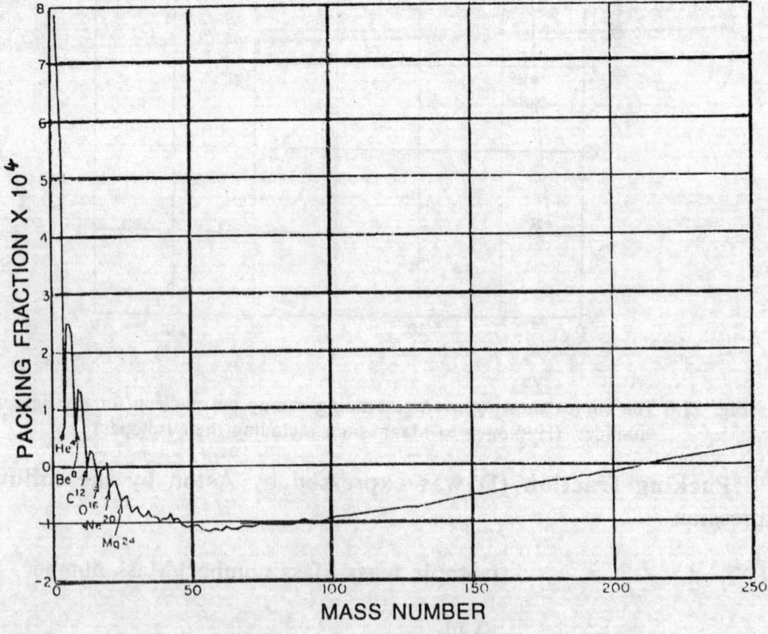

Fig 12.7 Packing fraction curve drawn with C^{12} mass scale

Problem 1: Determine the binding energy and packing fraction of a $_3Li^7$ atom, when the exact mass of Li is 7.01822amu

Mass of neutron = 1.008665 amu

Mass of proton = 1.007825 amu

$_3Li^7$ has 4 neutrons and 3 protons

Mass of 4 x N = 4.03466 amu & Mass of 3 x P = 3.023475 amu

Mass of 4N + 3P	=	7.058135 amu
Mass of $_3Li^7$	=	7.018220 amu
Mass defect	=	0.039915 amu
Binding energy	=	0.039915 x 931
	=	37.160865 MeV
Binding energy per nucleon	=	5.308695 MeV.
Packing fraction	=	26.028

Problem 2 : Determine the amount of mass that is being converted into energy in the following equation.

$$_3Li^7 + {}_1H^1 \rightarrow 2\,{}_2He^4 + ?\ \text{MeV}$$

7.01822 amu 1.00812 amu 4.00391 amu

Given that

$_3Li^7$ = 7.01822 amu; $_1H^1$ = 1.00812 amu; $_2He^4$ = 4.00391 amu

Therefore $_3Li^7$ + $_1H^1$ = 8.02634 amu

$2\,{}_2He^4$ = 2 x 4.00391 = 8.00782 mass units

Mass defect = 8.02634 - 8.00782 = 0.01852 amu

Energy = Mass defect x 931

= 0.01852 x 931.2

= 17.2458 MeV.

Problem 3 : Calculate the binding energy of $_2He^4$ when $_2He^4$ has a mass of 4.0028 amu

Mass of two neutrons = 2 x 1.008665 amu = 2.01733 amu

Mass of two protons = 2 x 1.00782 amu = 2.01565 amu

Mass of 2P + 2N	=	4.03298 amu
Mass of $_2He^4$	=	4.00280 amu
Mass defect	=	0.03018 amu
Binding energy	=	0.03018 x 931.2
	=	28.1036 MeV.
Binding energy per nucleon	=	7.0259 MeV.

12-6 MAGIC NUMBERS

Our knowledge regarding the correlation of chemical properties of an atom, with orbital electronic configaration has led to speculation as to the possibility of a similar concept for the arrangements of nucleons in the nucleus. It has been observed that atoms with an even number of nucleons in their nuclei are more plentiful in nature than those with an odd number. This would indicate that nuclei made of paired up nucleons is more stable than one in which an odd nucleon is present. Additional facts may be correlated if it is assumed that the paired nucleons are also arranged in shells, similar to the electrons in their planetary orbitals.

It is an accepted fact that neutral atoms containing a total of 2, 10, 18, 36, 54 or 86 extranuclear electrons are remarkably stable and are known as noble gases. Atoms with one more or one less electron, the alkali metals and the halogens respectively, are unusually active elements. This may be interpreted as a tendency on the part of the atom to achieve stabilty by completing their octets of outermost electronic shells. For nuclei it so happens that those with 2, 8, 20, 50, 82, 184 or 126 nucleons of the same kind are particularly stable. These numbers are referred to as **magic numbers** and are related to the completion of energy levels (shells) in the nuclei, just as certain numbers of electons are required to fill the orbital level of atoms. The evidence supporting a nuclear shell structure has been largely empirical in nature, although there has been some recent theoretical developments to support this structure. Some of the information which led to the idea of energy levels or shells, in nucleus is;

(i) The binding energies of atomic mass in atoms with the magic number nucleon totals, as calculated by the Einstein mass-energy equation, are higher than their neighbors.

(ii) The abundance of stable atoms containing magic number nucleon totals is greater than that of atoms with nucleon totals above and below these totals.

(iii) Nuclei with nucleon totals just above a magic number are less stable than those with magic number totals. Such nuclei may be neutron emitters and in the process of emitting neutrons they tend to attain the magic number total. Nuclei of this type are extremely short lived and occur only as intermediate products in processes involving artificial radioactivity.

(iv) The alpha emitter radioactive atoms tend to attain the stable configuration associated with the magic number totals.

Similar to building of electron orbits in the atom, one can picturize the building pattern of the nucleus involving a gradual filling up of single particle by neutrons and protons. The orbits can be described by the same lettes as those used to designate quantized orbital angular momentum of the electrons: s=0, p=1, d=2, f=3, g=4, h=5, i=6, etc. Since the neutrons and protons, also obey the Pauli exclusion principle, the s level can accommodate just 2 nucleons only, the p level has an accommodation for 6, d=10, f=14, g=18, h=22, i=26, and so on.

One of the proposed quantum configuration is shown in table below. In this table the bold numbers refer to shells, the lower-case letters designate the orbitals, and the superscript number indicate the number of nucleons in each orbital.

Shell	Configuration
2	$1s^2$
8	$1s^2\ 2p^6$
20	$1s^2\ 2p^6\ 3s^2\ 3d^{10}$
50	$1s^2\ 2p^6\ 3d^{10}\ 4f^{14}\ 5g^{18}$
82	$1s^2\ 2p^6\ 3d^{10}\ 4f^{14}\ 5g^{18}\ 6h^{22}\ 4d^{10}$
126	$1s^2\ 2p^6\ 3d^{10}\ 4f^{14}\ 5g^{18}\ 6h^{22}\ 7i^{26}\ 5f^{14}\ 3p^6\ 4p^6\ 2s^2$

A schematic diagram of a nuclear potential well, showing energy levels of nucleons is shown Fig.12.8. Shaded areas show the energy gaps corresponding to the complete filling of shells, the partial shading denotes the partly filled levels and the non shaded areas are the energy levels which are completely vacant. This nuclear potential well also exhibits the energy required to remove a nucleon from a typical nucleus is only 7-8 MeV. Hence we can conclude that the nucleons within the potential well are not all lying at the bottom of the well. Infact, as seen they occupy a series of discrete levels. Because the coulomb repulsion acts on the protons within the nucleus but not on the neutrons, the depth of the well is about 5MeV less for the protons than for the neutrons. The levels marked E_f (Fermi Energy) for protons/neutrons, represents the energy of the highest level which contains a nucleon. A nucleon in the energy level at E_f will be the easiest nucleon to remove from the nucleons since it already possesses the most energy. The energy required to remove a nucleon from the level E must be equal to the binding energy of a nucleon to the nucleus, and as we have seen already, this amounts to 7-8 MeV. Hence E_f must lie at an energy of about 7-8 MeV below the potential energy zero. In the Fig 12.8 although E_f (proton) is shown below E_f (neutron) in some cases it can be reverseed also, but the E_f difference is always very small. The E_f is function of the atomic mass (A) and the nuclear charge (Z).

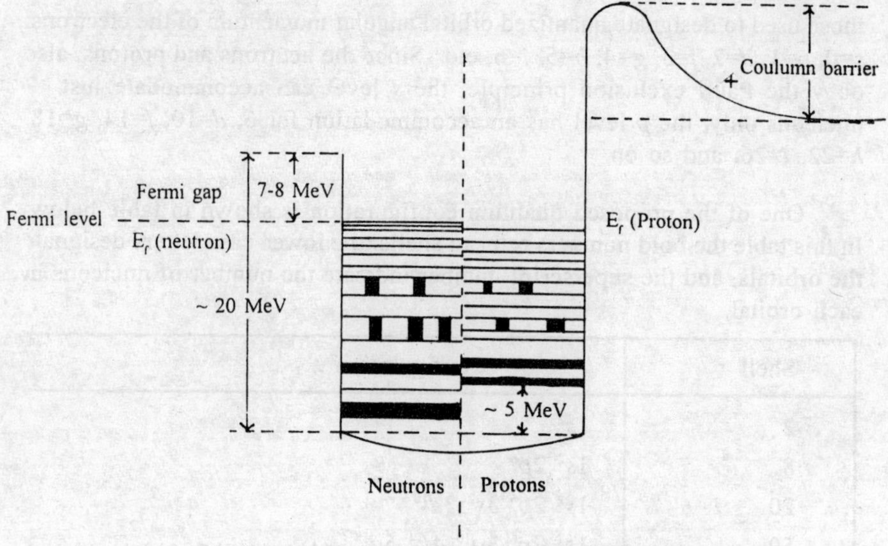

Fig 12.8 Energy levels for neutrons & protons in a nuclear potential well.

12-7 NUCLEAR MODELS

It is obvious that the nucleus is a very complicated system and an exact solution to the problems of nuclear structure is not possible for various reasons. However the behaviour of the nucleus can be explained on the assumption of the two models.

1. Liquid drop model
2. Nuclear shell model

Both these concepts are not mutually, exclusive, but emphasize different aspects of the nucleus.

12-7-1 Liquid drop model

It was proposed by Bohr(1936) that the nucleus resembles a liquid drop with the neutrons and protons enclosed in it in a random and disorderly way similar to molecules in a drop of liquid, the following are some similarities between Nucleus and a liquid drop.

i) Both exhibit surface tension effect

ii) They are capable of interacting only with their immediate close neighbours.

iii) Thermal agitation of molecules in a liquid drop is comparable to kinetic energy of nucleons.

iv) Evoporation of molecules in a drop when heated may be compared to nuclear disintegration when energised.

Thus study of stability of a charged vibrating drop explains nuclear fission.

The strongest support for this model comes from the fact that the density of different nuclei is almost constant, i.e., 10^{13}g cm^3. The repulsive forces of k the protons tend to split the nuclear drop, while the surface tension tends to keep the nucleus together.

Nuclear fission is easily explained by the liquid drop model concept. When neutrons enter the U^{235} nucleus, its mass increases, thereby increasing the corresponding repulsive forces of the proton. The attractive forces, however, incrases much less rapidly. The shape of the nucleus is thus deformed and it then gets split into two or more smaller nuclei, as seen in fig 12-9, Similar to the division of a charged water molecule getting divided by the electrostatic repulsion, in a way much similar to fission of a nucleus. The Fig12.9, shows the last stages in the division of a water drop into three parts. In (a) the first division is just complete and nuclei is formed where the second division (b) will occur and this continues upto final (f) stage.

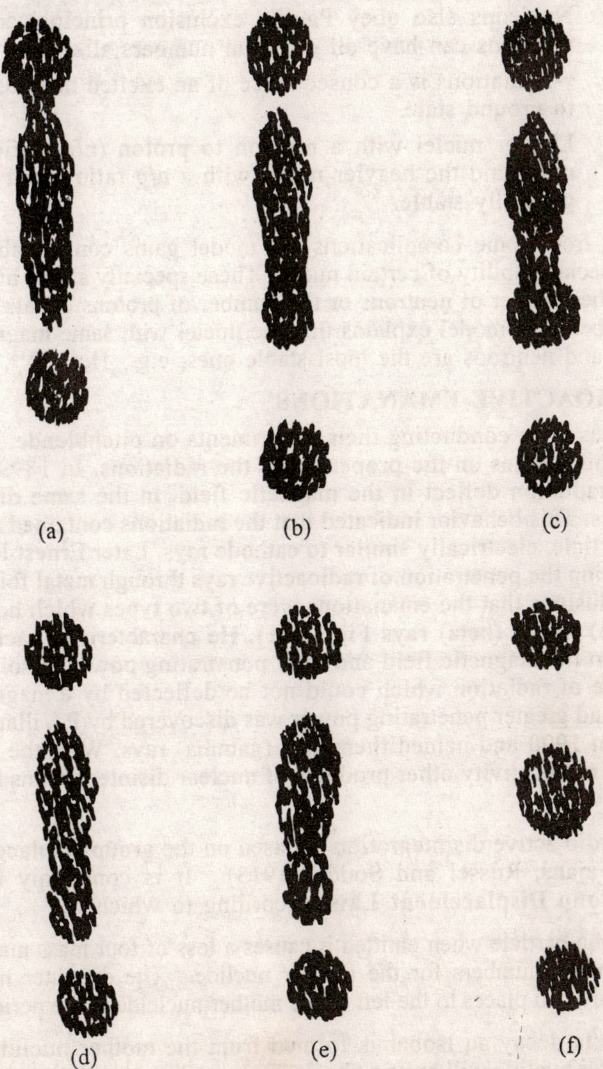

Fig. 12-9 Shapes of a nucleus undergoing fission, according to the liquid drop model. The figures (a to f) show the successive statges in the division of anelectrically charged water drop into three fragments.

12-7-2 Nuclear shell model

This model was suggested by Mayer (1950), according to this model:

i) Nucleons move in definite orbits inside the nucleus and have little mutual interaction.

ii) They have many energies or shells

iii) Nucleons are arranged in a nucleus in a number of energy levels or shells as are the electrons in their valence orbits.

iv) Analogous to the atoms of an inert gas element, the completely filled shell of the nucleus subscribes to its stability.

v) Nucleons also obey Pauli's exclusion principal and no two nucleons can have all quantum numbers alike.

vi) γ-Radiations is a consequence of an excited nucleus returning to ground state.

vii) Lighter nuclei with a neutron to proton (n/p) ratio equal to unity and the heavier nuclii with a n/p ratio equal to 1.6 are generally stable.

Apart from some complications the model gains considerable support from the special stability of certain nuclei. These specially stable nuclei occur whenever the number of neutrons or the number of protons equals one of the magic number. The model explains that the nuclei with same magic numbers of protons and neutrons are the most stable ones; e.g., ${}_2He^4$, ${}_8O^{16}$, ${}_{20}Ca^{40}$.

12-8 RADIOACTIVE EMANATIONS

While Curies were conducting their experiments on pitchblende, Becquerel continued his studies on the properties of the radiations. In 1899 he found that these radiation deflect in the magnetic field, in the same direction as cathode rays. This behavior indicated that the radiations contained negatively charged particle, electrically similar to cathode rays. Later Ernest Rutherford while studying the penetration of radioactive rays through metal foils, arrived at the conclusions that the emanations were of two types which he designed as α (alpha) and β (beta) rays Fig (12.1). He characterized each by their deflection to the magnetic field and their penatrating power in solid bodies. A third type of radiation which could not be deflected by a magnetic field but which had greater penetrating power was discovered by P.Villard (French physicist) in 1900 and named them as γ (gamma) rays. With the discovery of artificial radioactivity other products of nuclear disintegrations have been recognized.

The radio active disintegration is based on the group displacement law stated by Fajans, Russel and Soddy (1913). It is commonly known as **Fajan's Group Displacement Law,** according to which:

(i) An alpha particle when emitted it causes a loss of four mass numbers and two charge numbers for the mother nuclide. The daughter nucleide is therefore two places to the left of the mother nucleide in the periodic table.

(ii) In a beta decay an isobar is formed from the mother nuclide and the daughter nuclide will have a charge increase of one unit but the addition causes a charge decrease of one unit, in the formercase the daughter nuclide moves one place to the right and in the later case one place to the left in the periodic table.

Problem 4 :

Calculate the number of α- and β- particles emitted when $_{94}Pu^{241}$ disintegrates to $_{84}Po^{213}$.

Change in mass number = 241–213 = 28 amu.

∴ α particles emitted = $^{28}/_{4}$ = 7

The charge decrease due to 7 α-emissions = 7 x 2 = 14

∴ The charge of the daughter nuclide = 94 – 14 = 80

Charge difference with respect to the product = 84 – 80 = 4

∴ No. of β-particles emitted = 4

It will not be out of place to mention other types of radioactive emanations such as, mesons, π-mesons, δ-mesons, hyperons, photons, positrons, neutrinos, etc., but they are all short lived and are produced during induced radioactivity. The changes occurring because of these particles are not of significance enough to be dealt with in the scope of this book. Hence the three major rays α, β & γ are being discussed in the following sections.

12-8-1 Alpha Decay

Figures 12-3 and 12-4 represent the nucleus as a potential well and the significance was already presented in section 12-4. Elaborating our discussion, figure 12-11 shows the potential well of a nucleus of heavy element like U^{238}. The coulomb barrier has drawn at 20 MeV height, the radius being 8 x 10^{-13}cm. The horizontal lines present in the valley portion are well represente energy levels of protons and neutrons. Let us suppose that two protons and two neutrons have combined to form an alpha particle. For an alpha particle decay of the nucleus, it must pass the coulombic barrier. If the α-particle formed had sufficient kinetic energy inside the nucleus, it could presumably tear itself away to give-back proton and neutron due to the attractive forces in the nucleus. By chance if an α-particle could reach the rim of the barrier by gaining potential energy at the expense of its initial kinetic enrergy it experiences repulsions due to the electrostatic (coulombic) force and would move farther away from the nucleus Fig (12-2,3). By the time it had gone a distance equal to its nuclear radii, the 20 MeV of potential energy that it had at the top of the barrier would be reconverted into kinetic energy. But no nucleus is observed to emit α-particles having an energy of 20MeV. So we are forced to conclude that α-particaes, cannot escape by passing over the top of the coulombic barrier. Similarly α-particles shot with 6 MeVenergy were found to be incapable of penetrating into the coulombic barrier of a nucleus.

To solve this problem of how an alpha particle is able to come out of the nucleus with less energy than would be required to introduce it cannot be explained except using quantum mechanical property of the particles. For explaining and for the purpose of understanding, we wish to refer to a ficitous character, Mr. Tompkins, invented by the famous physict Professor George Gamov, who will help us to solve this problem. In his research findings and papers Gamov has mentioned about this character at many instances and refering to the different happenipgs with Tompkins, Gamov was able to hypothese and finally establish different theories of physical sciences, specially about the subatomic particles.

Once in his dreams, Mr.Tompkins ascertained that his car always possessed energy (motion) even after switching it off. The following night, he dreamt, that, he drove his car into the garage, and locked the garage door (fig 12-10a) and slept in the night. But to his astonishment the next morning the car was seen outside the garage (Fig 12-10b). This sort of an incident cannot occur, when the garage and the lock were both intact. Tompkins inquisitiveness made him ask Gamov an explanation to this effect. Gamov

Fig. 12-10 (a) and (b) The quantum mechanical tunnelling effect

in the first instance suggested **it can happen.** "In quantum mechanics if an object in motion strikes an obstruction, lying in its way, sufficiently several number of times, then there is a finite probability that the particle may pass through of obstruction and appear on the other side. This is usually impossible for any car to do so, but Gamov explained this is how an alpha praticle emmission takes place.

He suggested that, the alpha particle with only 6 Mev energy cannot pass the potential energy barrier, hence with this small amount of energy in

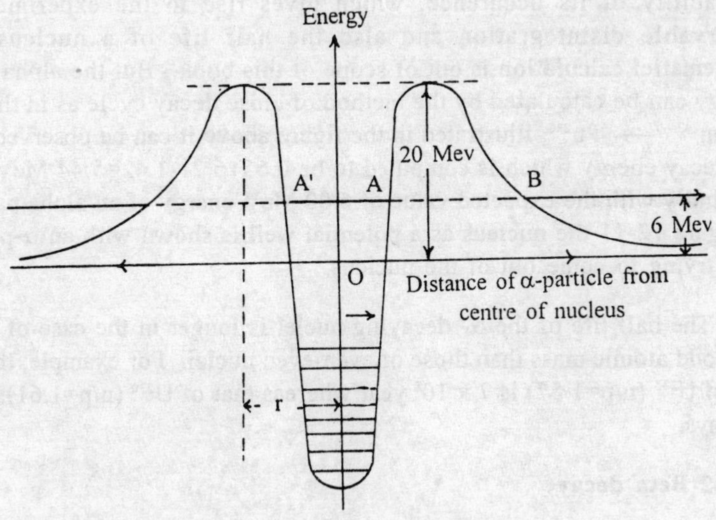

Fig 12-11 Nucleus as a potential well with an α-particle in it.

it, if it strikes the nuclear barrier several number of times then a time is reached when the nuclear wall gets raptured making the way for alpha particle to leave the nucleus forming a tunnel in the electron cloud. This phenomenon is known as Gamov's or quantum mechanical tunneling effect.

Although this explanation seems to be quiet imaginary, but there is no other explanation in classical physical concepts, the only explanation hence is based on the quantum mechanical concepts. We have to accept this concept only because from this we are not only able to calculate mathematically the energy of the observed alpha particle, but also the

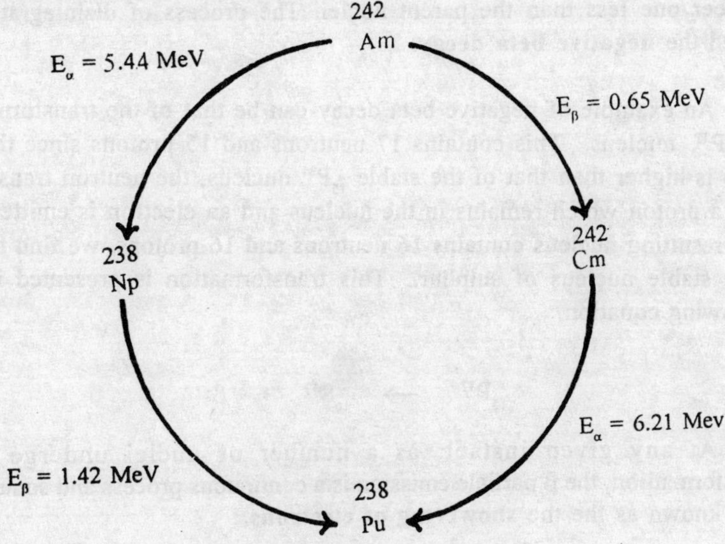

probability of its occurence, which gives rise to the experimentally observable disintegration and also the half life of a nucleus. The mathematicl calculation is out of scope of this book. But the alpha decay energy can be calculated by the method of close decay cycle as in the case of $Am^{242} \longrightarrow Pu^{238}$, illustrated in the figure above it can be observed, that, the decay energy which is computed to be 0.65+6.21-1.42=5.44 MeV, is in proximity with the expected value of 6.00 MeV energy of an alpha particle. In figure 12-11 the nucleus as a potential well is shown with an α-particle in it trying to come out of the nucleus.

The half life of the α- decaying nuclei is longer in the case of nuclei with odd atomic mass than those of even-even nuclei. For example, the half life of U^{235} (n/p=1.57) is 7 x 10^8 year whereas that of U^{230} (n/p=1.61)is only 21 days.

12-8-2 Beta decay

Beta (β^-) particles are high speed electrons, ejected from the nuclei of atoms of certain radioactive elements. The β^- decay usually occurs in those nuclei which have undergone a change, such that the neutron-proton ratios are higher than for the stable isotope of the element. In these cases it seems that the number of neutrons are higher than the number of protons and the most easiest way to bring stability to the nucleus would be by the emission of a neutron. This would result in the formation of adjacent lighter nucleus of the same element, but the extra neutron does not emit out directly but instead undergoes a nuclear transformation into a proton and an electron. The proton gets held-up in the nucleus and the electron being not a component of the nucleus is emitted out of the nucleus and thus leaves the atom. This emission of an electron results in the formation of an isobar with atomic number one less than the parent nuclei. The process of disintegration is called the **negative beta decay.**

An example of negative beta decay can be that of the transformation of ${}_{15}P^{32}$ nucleus. This contains 17 neutrons and 15 protons since the n/p ratio is higher than that of the stable ${}_{15}P^{31}$ nucleus, the neutron transforms into a proton which remains in the nucleus and an electron is emitted out. The resulting nucleus contains 16 neutrons and 16 protons, we find this to be a stable nucleus of sulphur. This transformation is presented in the following equation.

$${}_{15}P^{32} \longrightarrow {}_{16}S^{32} + \beta^-$$

At any given instant, as a number of nuclei undergo such transformatiion, the β^-particle emission is a continuous process and sometimes also known as the the **showering of electrons.**

The emission of a negative beta particle is essentially the only method by which nuclei with high neutron-proton ratios could return to stability. When neutron-proton ratio is low the stability is reached by the trapping of an electron from neighbouring K-orbital by the nucleus. This process is thus referred to as **electron capture** or **K-capture**. or **K-electron capture**

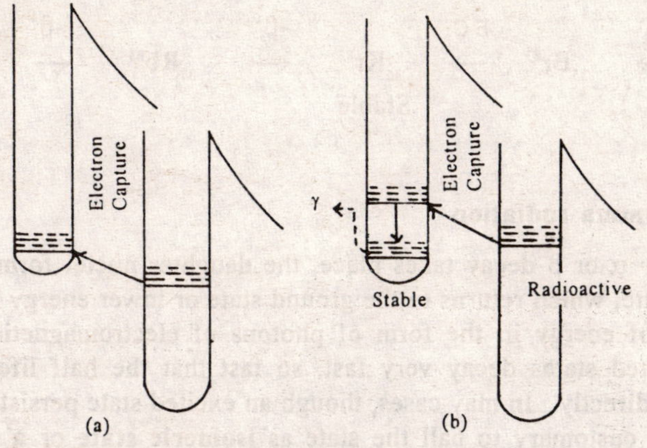

Fig 12-12 : Energy level diagram due to K-capture

(Fig 12-12). The electron trapped by the nucleus, will naturally be attracted by the proton which then on combination transforms into a neutron. The neutrons formed remain in the nucleus increasing the neutron to proton ratio of the nucleus and thus bringing stability to it.

$$p^1 + e^- \longrightarrow n^1 + \gamma$$

$$_{29}Cu^{64} + \text{E.C.} \longrightarrow {}_{30}Zn^{64}$$

$$(n/p = 1.2) \qquad\qquad (n/p = 1.13)$$

It was suggestd by W.Pauli, that during β-decay another particle must always escape simultaneously with the β-particle, carrying off a variable fration of the available energy. The sum of the energies of the two particles would always add up to the energy difference between the decaying nucleus and its product. Since such a particle had never been detected, it was assumed that it had zero mass and zero charge, and it was given the name **neutrino**. As, the nuclei are supposed to contain protons and neutrons, it is natural to wonder from where the electron and neutrino originates. According to Fermi, they are both created actually at the moment of the decay and there is no resons to doubt that this view is incorrect.

It is to be noted, that, some nuclei emit positively charged electrons known as **Positrons** (β^+). These are identical to the electrons (β-particles) in every respect expect that the charge is of the opposite sign. During the positron decay also, neutrinos are emitted as in the case of β-decay.

$$_{29}Cu^{64} \longrightarrow {}_{28}Ni^{64} + \beta^{+} + {}_{1}n^{o}$$
$$(n/p=1.2) \qquad (n/p=1.29)$$

To show all the three types of transformations of β^-,electron capture (E.C) and β^+ we can study the following reaction, in which a stable daughter nuclei Kr^{83} is formed.

$$_{34}Se^{83} \xrightarrow{-\beta^-} {}_{35}Br^{83} \xrightarrow{E.C.} \underset{\text{Stable}}{{}_{36}Kr^{83}} \xleftarrow{-\beta^-} {}_{37}Rb^{83} \xleftarrow{-\beta^+} {}_{38}Sr^{83}$$

12-8-3 Gamma radiation

Whenever α-or β-decay takes place, the daughter nuclei formed is in an excited state, which returns to the ground state or lower energy state by the emission of energy in the form of photons of electromagnetic radiation. Most excited states decay very fast, so fast that the half life cannot be measured directly. In may cases, though an excited state persists for a long time, it is customary to call the state as **isomeric state** or **a metastable state**. Thus Hf^{180m} has an excited energy level 1.142 MeV above the ground

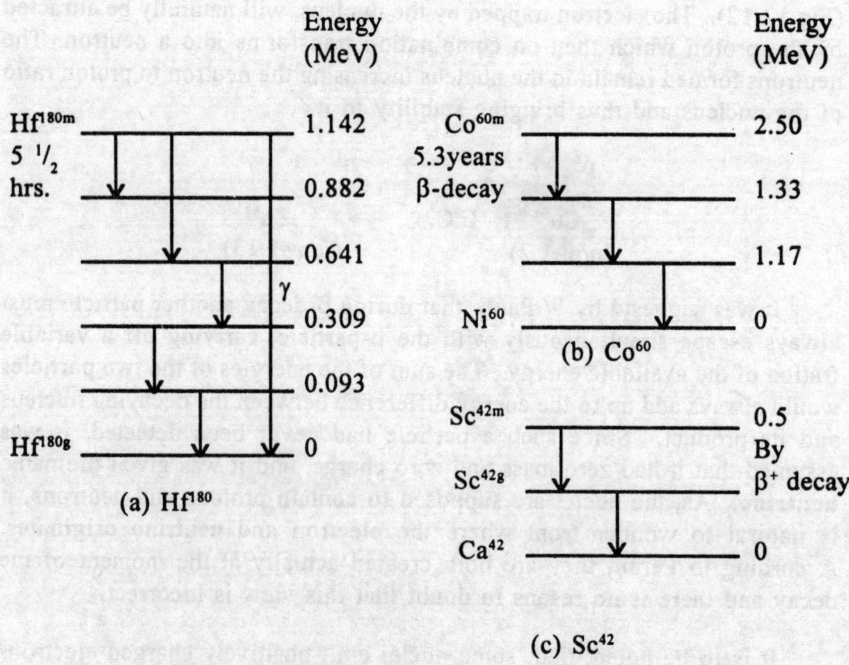

Fig 12-13 Representation of the conversion of Hf^{180m}, Co^{60} and Sc^{42m} to form stable nuclei.

state, which decays with a half life of 5 $^1/_2$ hours. The first excited state of Sc^{42m} decays so slowly to the ground state that in fact the excited nuclei are lost by β^+ decay to Ca^{42} before they have time to undergo γ-decay. The excited state of Hf^{180}, Co^{60} and Sc^{42}, undergoing change to ground state are seen in Fig 12-13 a,b,c.

The superscript m in the nuclear isotope Hf^{180m} & Sc^{42m} indicate a metastable state. Properly speaking any excited state of a nucleus is a metastable state, but usually the term is reserved for those excited states that decay with a half life long enough to be measured $>10^{-9}$ sec.

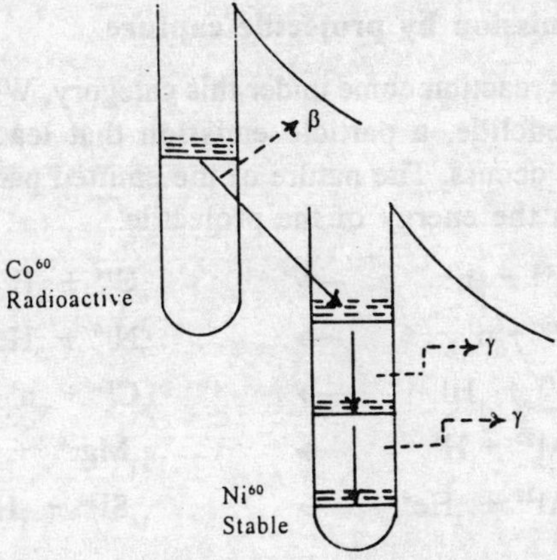

Fig 12-14 : Energy level diagram negative metadecay with gamma radiations

In the Fig 12-14 as seen, there are different energy levels present, the nuclei while cooling down to the lower energy will emit out γ radiation in the form of photons equal to the amount of energy lost on cooling. Hence it is important to realize that gamma radiations are always emitted as photons or discrete energy units.

12-9 NUCLEAR REACTIONS

Whenever two atoms/compounds react with each other, new products are formed with properties quite different from the parent atoms/compounds. In these chemical processes, only the valence electrons play the crucial role without any significant change taking place in the nucleus. In nuclear reactions on the contrary, the nucleus is directly involved undergoing significant changes. It is possible to classify the nuclear changes either in terms of the over all energy transformation that occurs or in terms of the types of projectiles that are used in the bombardments of target nuclei. According to Neil Bohr (1936), the nuclear transformation occur depending more upon the energy of projectile than upon its nature. Except for neutron reactions this theory seems to hold true for projectiles of moderate energies. The nuclear transformations can be classified into five types.

12-9-1 Projectile - capture reations

The absorption of slow neutrons is possible for nearly all nuclides, and proton capture is common only for target nuclei with low mass numbers.

$$_{47}Ag^{107} + {}_{0}n^{1} \rightarrow {}_{47}Ag^{108} + \gamma$$

$$_{6}C^{12} + {}_{1}H^{1} \rightarrow {}_{7}N^{13} + \gamma$$

In both the transformations γ (gamma) radiation is emitted. In neutron capture an isotope of the same element is formed, whereas in proton capture a new nuclei is obtained.

12-9-2 Particle emission by projectile capture

Most of the nuclear reaction come under this category. Whenever a projectile is captured by a nuclide, a particle emission that leads to stability of a compound nucleus occurs. The nature of the emitted particle or particles is largely based upon the energy of the projectile.

$$_{7}N^{14} + {}_{0}n^{1} \rightarrow {}_{6}C^{14} + {}_{1}H^{1}$$

$$_{9}F^{19} + {}_{0}n^{1} \rightarrow {}_{7}N^{16} + {}_{2}He^{4}$$

$$_{5}B^{11} + {}_{1}H^{1} \rightarrow {}_{6}C^{11} + {}_{0}n^{1}$$

$$_{13}Al^{27} + {}_{1}H^{1} \rightarrow {}_{12}Mg^{24} + {}_{2}He^{4}$$

$$_{13}Al^{27} + {}_{2}He^{4} \rightarrow {}_{14}Si^{30} + {}_{1}H^{1}$$

12-9-3 Spallation reactions

When high speed projectiles with energy around 400 MeV are bombarded, they chip fragments from a heavy nucleus, leaving a smaller nucleus with usually an atomic number between 10 and 20 units less than the original nucleus.

$$_{29}Cu^{63} + {}_{2}He^{4} \rightarrow {}_{17}Cl^{37} + 14\,{}_{1}H^{1} + 16\,{}_{0}n^{1}$$

12-9-4 Fission Reactions

When a heavy nucleus breaks into two or more medium heavy fragments, the transformation is said to be fission (A detailed study is presented in section 12.14).

$$_{92}U^{235} + {}_{0}n^{1} \rightarrow {}_{56}Ba^{141} + {}_{36}Kr^{92} + 3\,{}_{0}n^{1} + 200\text{ MeV}$$

12-9-5 Fusion

Lighter Isotopes combine with one another to from heavier and stable nuclei. The potential energy barriers, of the nucleus, for reaction of this type are very high. Typical fusion reaction occurs between the isotopes of hydrogen.

$$_{1}H^{2} + {}_{1}H^{3} \rightarrow {}_{2}He^{4} + {}_{0}n^{1} + 17.5\text{ MeV}$$

Details of this process will be given in section 12.16.

12-10 RATE OF DISINTEGRATION, HALF LIFE PERIOD AND RADIOACTIVE EQUILIBRIUM

As seen earlier, the reation of an alpha or beta particle represents the transformation of one atom of radio element into another radio or stable element. The parent atom is then said to have **disintegrated** into a new atom and an α- or β- particle. This process is similar to a chemical reaction in which one molecule transforms into another molecule. *The actual number of atoms of a parent radioactive isotope which disintegrate in a unit of time is directly proportional to the number of parent nuclei present at that instant.* This statement is known as **law of radioactive decay**. This value is exclusively a characteristic property of the nuclei of the radioactive element under consideration and is known as **rate of disintegration.** The law of radioactive decay may also be stated as the rate of disintegration of a radioactive element at any time is directly proportional to it's concentration of the nuclei at that instant. Unlike a chemical reaction, rate of disintegration of a nuclei is found to be independent of the temperature, pressure and state of chemical combination of the radioactive element. Rate of disintegration of a radioactive atom follows **first order kinetics.**

Mathematical treatment of radioactive decay

Let us consider an atom A disintegrating to give another atom B. Let a is the initial amount of the disintegrating substance and $a-x$ is the amount left after the lapse of a time t. So in a time t the corresponding change in concentration of the decaying substance is $a-(a-x)$, that is equal to x.

$$\text{Rate of disintegration} = \frac{\text{A small change in the concentration of a radioactive element}}{\text{Corresponding change in time}}$$

$$= \frac{dx}{dt} = \frac{-d\,(a-x)}{dt} \qquad (12.1)$$

During radioactive decay the amount of a nuclide decreases. This fact is taken care by putting a - sign before the expression 12.1.

But according to law of radioactive decay

Rate of disintegration α Concentration of radioactive material

$$\text{Hence} = \frac{dx}{dt} \propto (a-x) \qquad (12.2)$$

Putting proportionality constant

$$= \frac{dx}{dt} = k\,(a - x) \qquad (12.3)$$

This proportionality constant *k* is called **disintegration constant or nuclear decay constant** and the equation (12.3) is known as differential from of disintegration process. The disintegration constant is a measure of decay of substance per a unit and is indepentent of initial amount of a radioisotope. Unlike rate constant of a chemical reaction which is temperature dependent, the nuclear decay constant assumes additional importance as it is independent of temperature.

On rearranging equation (12.3)

$$\frac{dx}{a - x} = k\,dt \qquad (12.4)$$

Integrating the above expression we have;

$$kt = 2.303 \log \frac{a}{(a\text{-}x)} \qquad (12.5)$$

The equation (12.5) is the radioactive decay expression. This is highly useful in evaluating the concentration of the radioactiye isotope at any given instant or conversely estimating time required for a particular percentage of radioactive decay.

When $x = a/2$ i.e., half the atoms have got disintegrated the equation (12.5) becomes

$$k = \frac{2.303}{t_{1/2}} \log 2$$

$$k = \frac{2.303}{\tau} \log 2$$

where $\tau = t_{1/2}$

$$k = \frac{0.693}{\tau}$$

Therefore τ, the time required for disintegration of any radioactive element to half its concentration is independent of the initial amount of the substance taken and depends exclusively on the disintegration constant *k*. This is often referred as it's **half life period.**

The reciprocal of the diintegration constant, *k*, gives the **average life** of the disintegrating atom.

$$\lambda = 1/k = 1.44\ \tau$$

or

$$\tau = \frac{1.44}{\lambda}$$

When a radioactive material stands over a period of time, a condition known as **radioactive equilibrium** is established. This equilibrium results from the fact that in a radioactive series after buildup of a particular concentration of a radioactive daughter nuclei concentration, the rate of its decay becomes equal to its formation. Each element in a deacy series is formed at the same rate as that of its decay creating a virtual dynamic equilibrium similar to a chemical equilibrium.

The general principle underlying the formation of a radioactive equilibrium may be illustrated by using radium and its daughter nuclides. It so happens that radium is a fairly long lived element and will disintegrate into radon at a rate in turn, will start disintegrating into Po^{218}, Po^{218} disintegrates into Pb^{214}, at a rate that is proportional to the number of atoms present and their tendency to disintegrate. Eventually the rate at which all the descendants of radium are disintegrating will become equal to the rate at which they are being formed. Under equilibruim conditions the amounts of the raidoactive material will remain constant, because the rate of formation and decay for each particular element will be the same. The concentration of each element present is inversely proportional to the constant of the element. If the decay constants of the members of a series such as radium, radon, polonium and lead are λ_1, λ_2, λ_3, **and** λ_4, respectively the number of atoms of each member of the series present at equilibrium will be $\lambda_1 N_1 = \lambda_2 N_2 = \lambda_3 N_3 = \lambda_4 N_4$.

More over, it is observed that any two members for example A and B in the series, regardless of their position within the series, the following mathematical relationship can be deduced as

$$\frac{N_A}{N_B} = \frac{\lambda_B}{\lambda_A} = \text{constant}$$

Problem 5: Determine the rate constant of C^{14} whose half life is 5600 Yrs.

$$t_{1/2} = 5600 \text{ Yrs}$$

$$= 5600 \times 365 \times 24 \times 60 \text{ min.}$$

$$= 2.95 \times 10^9 \text{ min.}$$

We know

$$t_{1/2} = 0.693 / k$$

substituting $t_{1/2}$ value in the above expression

$$k = \frac{0.693}{2.95 \times 10^9}$$

$$= 2.35 \times 10^{-10} \text{ min}^{-1}$$

Problem 6: Calculate the disintegration constant value of Rn whose life 3.82 days

$$k = \frac{0.693}{t_{1/2}}$$

$$= \frac{0.693}{3.82}$$

$$= 0.1814 \text{ day}^{-1}$$

Problem 7: An old wooden idol was found to show C_{14}/C_{12} ratio only 0.714 of that in the atmosphere today. What is the age of the idol. Given that the half life of C_{14} is 5670 years.

$$k = \frac{0.693}{5670}$$

$$= 1.22 \times 10^{-4} \text{ yr}^{-1}$$

$$-kt = 2.303 \log_{10} N/N_0$$

$$t = \frac{-2.303 \log_{10} (N/N_0)}{k}$$

$$= \frac{-2.303 \times \log_{10} (0.714)}{1.22 \times 10^{-4}}$$

$$= 2760 \text{ years.}$$

12-11 RADIOACTIVE SERIES

All elements found in natural sources with atomic number greater than 83 (bismuth) are radioactive. They belong to chains of successive decays, and all the species in one such chain constitute **a radioactive family or series.** Three such families that include all the naturally occurring radioactive elements mentioned in the periodic chart were identified, along with one artificial series. These radio active series were named after the prominent member of the series, namely thorium series, neptunium (artificial) series, uranium series, and actinium series.

Thorium series : Thorium (Th^{232}) is the parent nuclide of the 4n or thorium series, with Pb^{208} as the stable end product. This series is shown in chart 12-1. It proceeds in 14 transformations (8 of them being α particle emissions and the rest β-emissions). The nuclide masses are given by the expression 4n, where n is an integer.

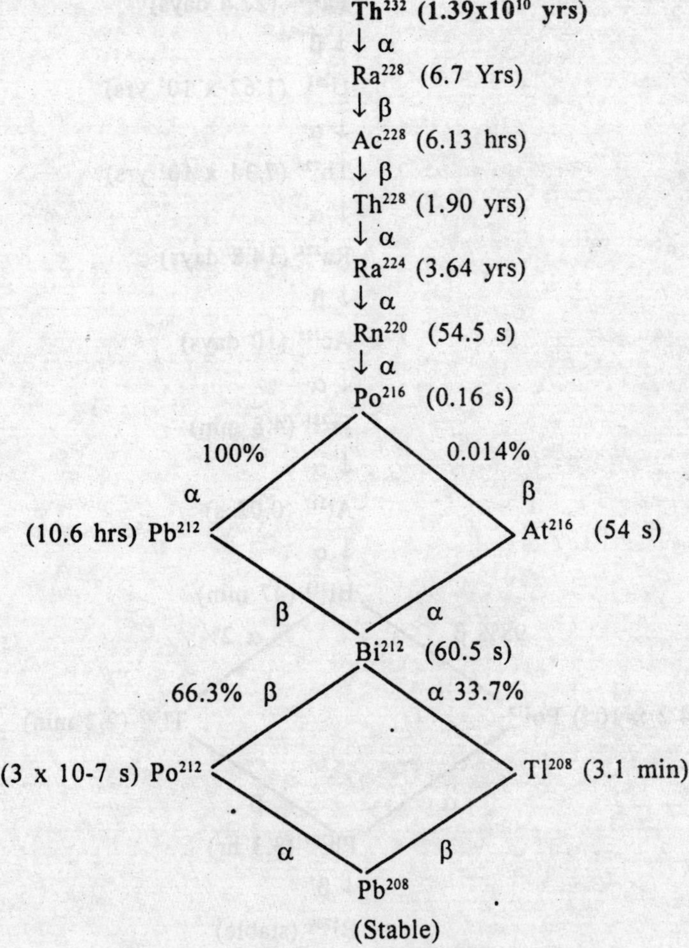

Chart 12-1 Thorium (4n) series of radioactive elements

Neptunium series : The artificial transmutation of the elements present in the periodic table resulted in the **4n+1 or neptunium series**. This series takes the name of the longest lived member of the family Np^{237}. Pu^{241} is the parent of this series and its stable end product is Bi^{209}. The neptunium series is given in chart 12-2. There are 14 transformations of which 10 are α-decays and 5 are β decays.

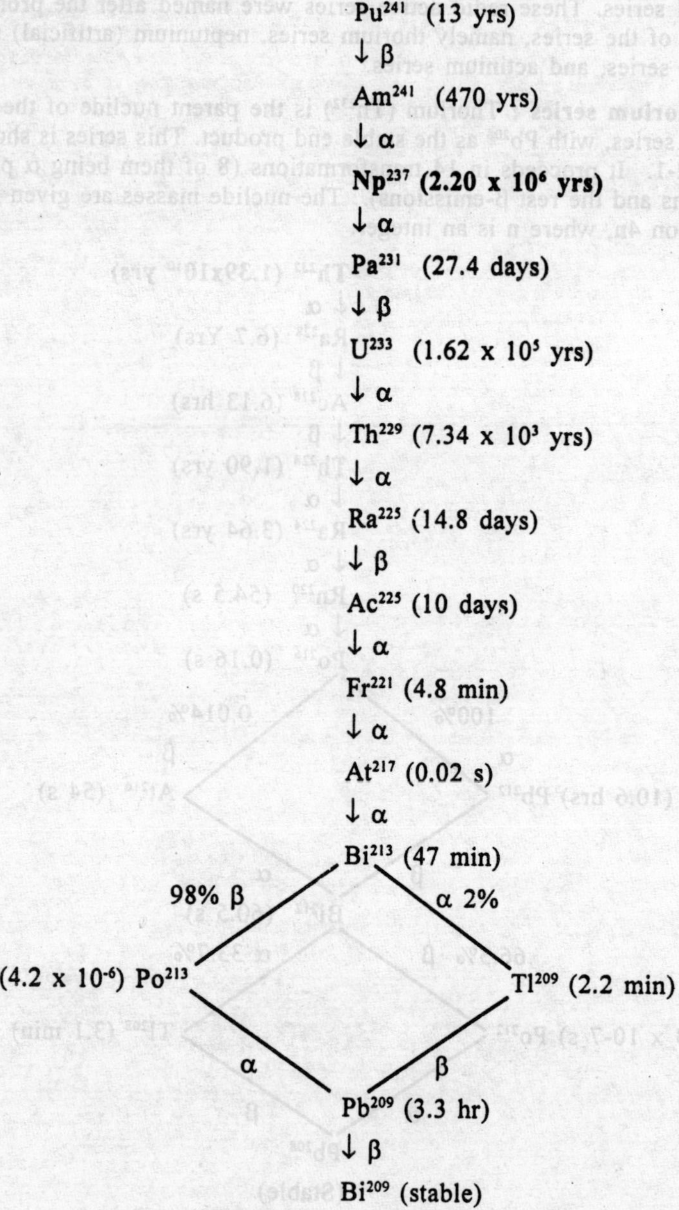

Chart 12-2 Neptunium(4n+1) series of radioactive elements

Uranium series : U^{238} is the parent nucleide of this series and after 14 transformations (8of them by α-particle and 6 by β-particle emissions) the nuclei reaches a stable end product Pb^{206} Since the atomic mass of all the nuclides is given by the mathematic expression 4n+2, where n is an integer, the series is also called **4n+2 series.** The uranium series with all the nuclides involved in the radio active decay are given in chart 12-3.

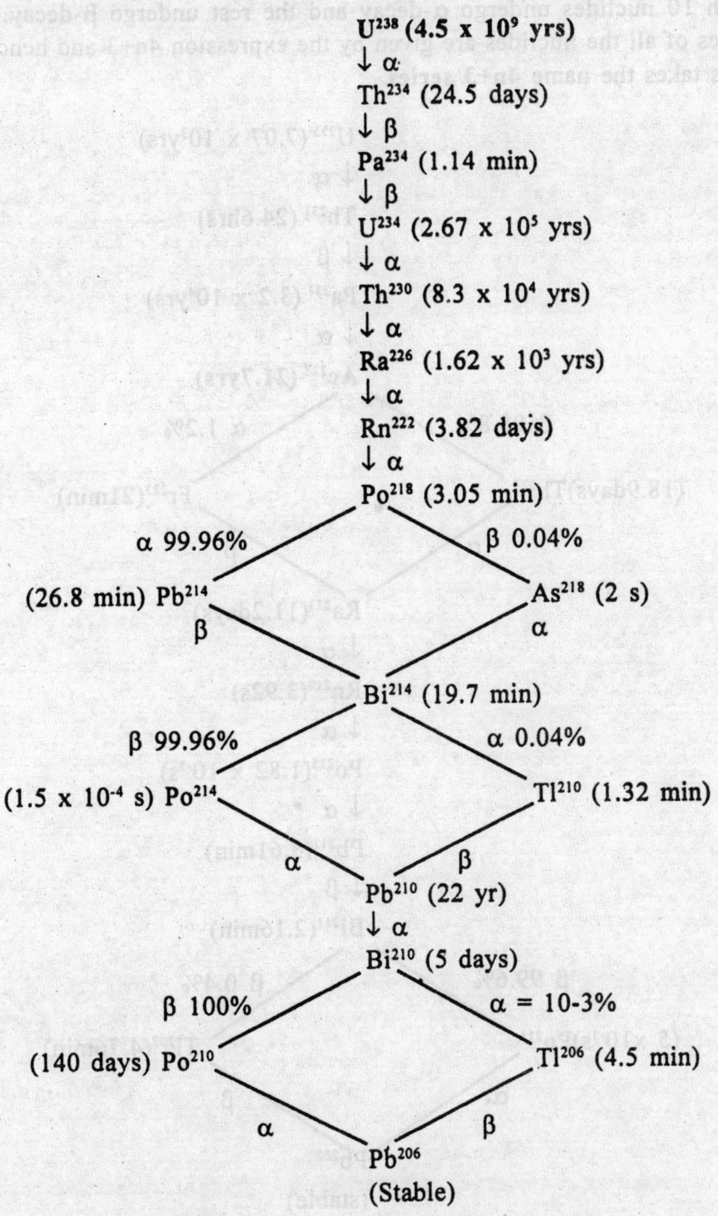

12-3 Uranium (4n + 2) series of radioactive elements

Actinium series: While the parents of uranium and thorium series, U^{238} and Th^{232} respectively are longest-living isotopes in their respective families. The name **actinium series** originated from the thinking of old school that actinium is the parent of this series. Latter discoveries proved irrevocably that the true parent is U^{235} and Pb^{207} is the stable end product. This series is given in the chart 12-4. There are 17 transformations of which 10 nuclides undergo α-decay and the rest undergo β-decay. The masses of all the nuclides are given by the expression 4n+3 and hence this series takes the name **4n+3 series.**

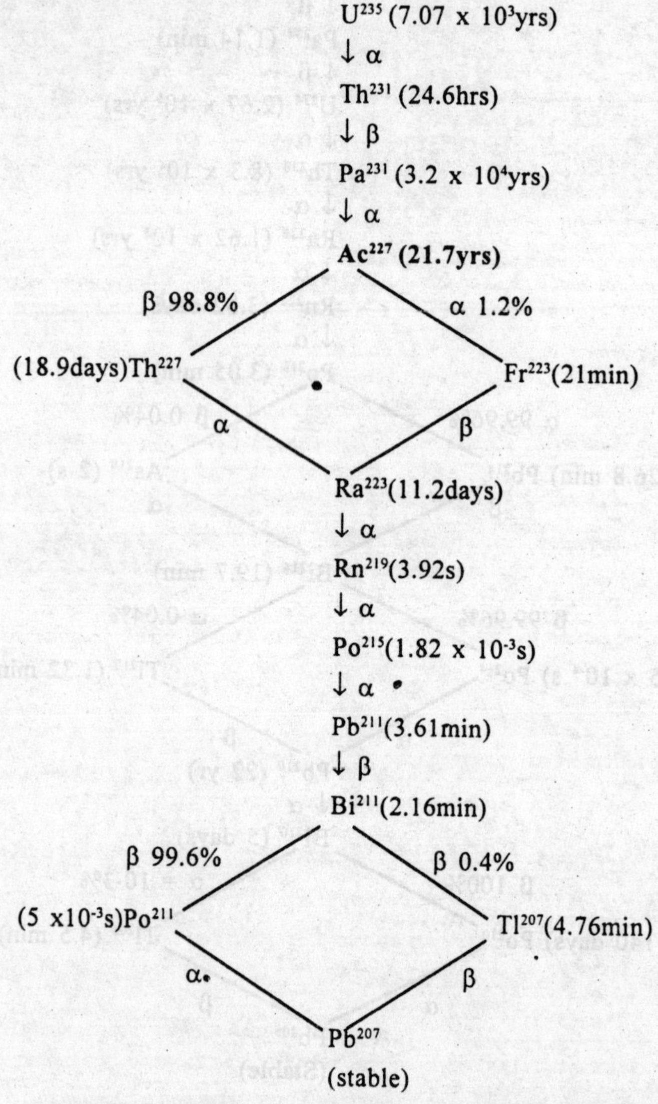

Chart 12-4 Actinium (4n+3) series of radioactive elements

12-12 NUCLEAR FISSION

As seen earlier that heavy nuclei break into two or more medium weight fragments, and this transformation was said to be **fission**. Fission may occur as a spontaneous mode of decay or it may follow as a result of the excitation of a nucleus tend to make heavy nuclei unstable. The fission process initially involves structural deformation of a nucleus, ultimately leading to its breakdown into smaller fragments, Fig 12-15.

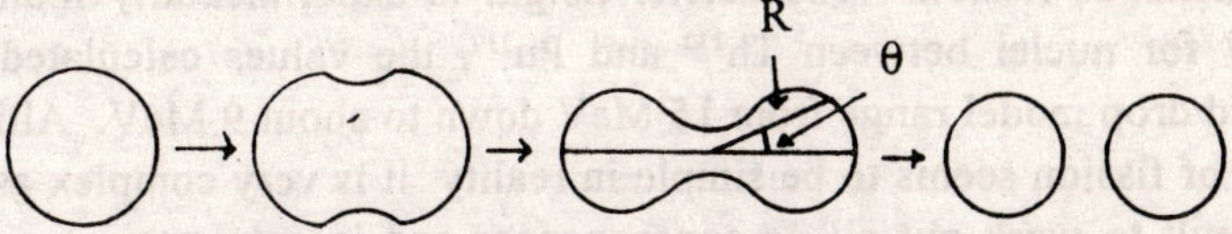

Fig 12-15 Structural deformation of nucleus during fission

Initially the nucleus is spherical, but in order to form two fragments of comparable size, it must first become deformed into some sort of elongated shape. As this amounts to pulling nucleons apart the potential energy steeply rises. Once they reach a saddle point indicated by a maximum potential energy fission barreir, the mutual attractive forces of nucleons get completely weakened allowing a spontaneous separation of daughter nuclei Fig 12-16.

The fig 12-16 shows the variation of the potential energy as a function of α_2 and α_4 which are two coefficients whose values determine the nuclear shape. The sequence of shapes shown in Fig 12-15 can be represented by a series of values of α_2 and α_4. The energy is plotted as a contour map in which the lines join points of equal energy. The spherical nucleus ($\alpha_2 = \alpha_4 = 0$) lies in a hole in the potential energy surface. If the nucleus is given

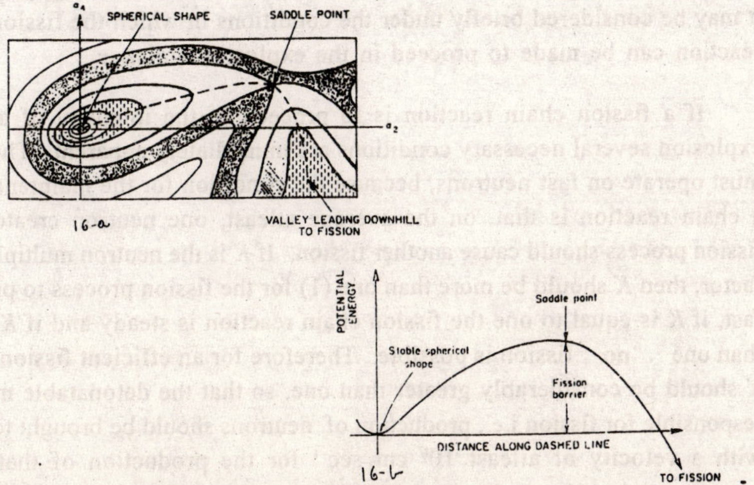

Fig 12-16 Nuclear fission & saddle point

enough energy, however it can escape from this hole and the dashed line in Fig 12-16 a shows a distortion path which leads over a pass or saddle point in the energy surface and into a descending valley leading to fission. The sequence of shapes along this path are those seen in Fig 12-15. The Fig 12-16b shows a cross section of the potential energy surface taken along the dashed lined. The height of the pass corresponds to the fission barrier. It is this barrier which prevents the heavy nuclei from instantaneous decay by sponteneous fission. The barrier height is experimentally found to be 6 MeV for nuclei between Th^{232} and Pu^{239}, the values calculated from the liquid drop model range from 15 MeV down to about 9 MeV. Although this type of fission seems to be simple in reality it is very complex as it is very difficult to work out all its consequences and is not certain too.

In α-decay, a tunnelling through the fission barrier without reaching the saddle point is possible, then it is called **spontaneous fission.** The spontaneous fission is less probable in lower mass particles like the nuclear isotopes of lead. For heavier elements, the greater coulombie repulsions make the fission barrier relatively less important and increase the probability of spontaneous fission, as seen in the case of $_{100}Fm^{256}$. When fission occurs, the two fragments are not always exactly equal to one half of the initial nucleus. The division can occur in many ways. For example the fission of U^{235} yields a nucleus of mass number is equal to 95 and second a fragment of mass number is equal to 140. Symmetric division is almost one thousand times less probable and no convincing explanation has yet been obtained for it.

The Atomic or Fission Bomb is one of the best examples of fission, which is involved in military applications. It is a chain reacting system and it may be considered briefly under the conditions in which the fission chain reaction can be made to proceed in the explosive violence.

If a fission chain reaction is to proceed to the intensity of a major explosion several necessary conditions are immediately apparent. The chain must operate on fast neutrons, because the condition for the maintenance of a chain reaction is that, on the average atleast, one neutron created in a fission process should cause another fission. If K is the neutron multiplication factor, then K should be more than one (1) for the fission process to progress fast, if K is equal to one the fission chain reaction is steady and if K is less than one no fission is possible. Therefore for an efficient fission bomb, K should be considerably greater than one, so that the detonatable material responsible for fission i.e., production of neutrons should be brought together with a velocity of atleast 10^6 cm sec^{-1} for the production of that much energy causing destruction as was done on the cities of Hiroshima and Nagasaki on 6th August 1945, during the World War II.

12-13 NUCLEAR FUSION

As mentioned earlier, lighter isotopes combine with one another to form heavier and more stable nuclei. All such reactions are strongly exothermic. The energy source of the stars, including sun, is believed to be derived from a complex chain of nuclear transformations, which when simplified amounts to union of four hydrogen atoms to form one helium atom.

$${}_6C^{12} + {}_1H^1 \longrightarrow {}_7N^{13} + \gamma$$

$${}_7N^{13} \longrightarrow {}_6C^{13} + e^+ + \gamma$$

$${}_6C^{13} + {}_1H^1 \longrightarrow {}_7N^{14} + \gamma$$

$${}_7N^{14} + {}_1H^1 \longrightarrow {}_8O^{15} + \gamma$$

$${}_8O^{15} \longrightarrow {}_7N^{15} + e^+ + \gamma$$

$${}_7N^{15} + {}_1H^1 \longrightarrow {}_6C^{12} + {}_2He^4 + \gamma$$

on addition $$4\,{}_1H^1 \longrightarrow {}_2He^4 + \text{energy.}$$

In this cycle of reactions the combination of 4 ${}_1H^1$ nuclei result in the formation of one ${}_2He^4$ nucleus. The final energy requirement is of 26.8 MeV cycle. It is estimated that it takes about 6 million years for the completion of one cycle as the temperature of approximately 20,000,000°C is necessary.

Though the actual reaction which is utilized in the hydrogen bomb is not of a public information, however, a number of possibilities are there, as seen in the following equations with the probable amount of energy getting liberated as given in the parenthesis.

$${}_1H^2 + {}_1H^3 \longrightarrow {}_2He^4 + {}_0n^1 \quad (17.8\ \text{MeV.})$$

$${}_1H^1 + {}_1H^3 \longrightarrow {}_2He^4 \quad (20.0\text{MeV.})$$

$${}_1H^2 + {}_1H^2 \longrightarrow {}_2He^4 \quad (24.9\ \text{MeV.})$$

$${}_0n^1 + {}_1H^3 \longrightarrow {}_2He^4 + e \quad (20.7\ \text{MeV.})$$

$${}_3Li^7 + {}_1H^1 \longrightarrow 2\,{}_2He^4 \quad (17.7\ \text{MeV.})$$

$${}_3Li^6 + {}_1H^2 \longrightarrow 2\,{}_2He^4 \quad (22.6\ \text{MeV.})$$

$${}_3Li^6 + {}_1H^3 \longrightarrow 2\,{}_2He^4 + {}_0n^1 \quad (16.4\ \text{MeV.})$$

12-14 USES OF RADIO ISOTOPES

After the discovery of the radioactivity much of its use was not known until its destructive nature came to light in the World War II. Cities of Hiroshima and Nagasaki were charred completely by the bombardment of a nuclear bomb over these cities of Japan. The effects of this incident are still alive and the man is afraid of even hearing about this destruction, and trembles if he even thinks of a nuclear war today. Any nuclear war will wipe away the human race and most of the organism. In the recent past such a situation scraped during the Gulf War.

Radioisotopes produced in various radio active processes posesses numerous uses. They are obtainable from four principal sources.

i) As members of one of the natural radioactive decay series.

ii) Irradiation of stable elements with high speed projectiles from accelerators such as the cyclotron,

iii) Irradiation of stable elements with neutrons from an atomic reactor pile.

iv) The products of atomic fission.

Of the many uses, we will now examine a few of them, which can be within the scope of this book.

12-14-1 Applications of radio isotopes in the analysis

Natural radioactive elements are used to determine the solubilities of sparingly soluble salts. In 1913, for the first time, the solubility of lead chromate was determined by adding a known quantity of a radioisotope of lead, Pb^{212}, to a definite amount of a soluble lead salt.

The solubility of water in benzene or other hydrocarbons is too small to be measured by ordinary physical or chemical methods. If however, a definite quantity of radioactive tritium H^3 in the form of tritium oxide is added to a given amount of organic solvent, the resulting mixture has an activity that can be measured even in trace quantities to give the solubility. Isotopic dilution analysis is becoming increasingly important in the analytical biochemistry. For example; in the determination of the percentage of glycine present in the products of hydrolysis of proteins, isotopic water is used.

12-14-2 Reaction mechanism through tracer chemistry

Radioactive tracers are used for investigating the structure of compounds, the nature of linkages, the mechanism of chemical reactions and the kinetics of chemical reactions. For example they are used in the establishment of structure of thiosulphate $S_2O_3^{--}$ ion.

12-14-3 Radio isotopes in medicine

The application of tracer methods in medicine, although is a new technique, it has made a large progress. The application fall into two categories.

(i) The use of tracers in diagnostic methods for localizing bodily disorders

and (ii) As therapeutic agents in the treatment of certain abnormal conditions in the body.

Certain elements are preferentially absorbed by definite type of tissues and organs. Radioactive phosphorus in the form of a phosphate is widely used in the treatment of certain blood disorders. In the conditions where the bone marrow produces an excess of red corpuscles, an oral dosage of radioactive phosphate containing P^{32} isotope is administered. The radioactive phosphorous is absorbed by the bone marrow and the β-emission of the isotope, restricts the formation of excess red corpuscles. Similar technique is employed in inhibiting formation of excess white corpuscles, there by curing leukemia. The treatment of thyroid is also done by the administration of radioactive iodine which accumulates in thryroid and cures malfunctioning.

12-14-4 Applications in biology

By the use of radioisotopes it was found that the atoms within a living organism undergo a continuous replacement. Earlier it was believed that proteins, carbohydrates and fats get stored in the body and are left undisturbed until they are needed by the body. The isotopic tracers have disproved this contention. It is now believed that a living organism contain a dynamic system of balanced chemical reactions and therefore in a constant state of successive changes in the body. The substances of cells and tissues are constantly balanced between chemically active materials circulating within the system and the noncirculating material which are a part of the system. In other words there is a constant exchange between the elements supplied in the diet and those which make up the body tissues. Transportation of ions across cell membranes and ions present in the cells form one such system. From the study of tracer osmotic pressure measurements, this is correlated to the passage of ions through an animal membrane.

18-14-5 Radioisotopes in agriculture

The transportation of mineral nutrients from roots to leaf and subsequent redistribution within plants can be followed through the use of radioactive isotopes. This information is obtained by means of radioautography. This study makes use of the effect of radioactive radiations upon a photographic plate. Thus a radio autograph is a form of photograph produced by the exposure of a sensitized plate to radioactive emanations coming either from a whole plant or from the sections of plant tissues. This technique is helpful in giving the informations concerning the use and need of trace elements like boron, cobalt, copper, manganese, zinc and molybdenum, in the plant nutrition.

12-14-6 Industrial application of radioisotopes

Radioisotopes are finding acceptance by the industry and day by day finding a greater role. The γ-radiation from radioisotopes are utilized for the destruction of bacteria in foods and drugs, and in the activation of many chemical reaction , in initiating polymerization process involving various organic monomers etc. In the metallurgical industries also radioisotopes play an important role, in the research and development of alloys of different metals.

12-14-7 Determination of the age of the earth

The existence of considerable amounts of the radioactive substances like uranium and thorium in nature and the absence of neptunium provide some basis for the estimation of the age of the earth. It is assumed that geologically all these elements are present in the earth's crust. It is possible by chemical analysis to determine the ratio of uranium or thorium to helium or lead (at one end a radio active element and at the other extreme stable isotopes formed from them) and thereby from a knowledge of decay life times get figures for the estimation of the formation of the parent element and hence earth.

Table 12-1 Half life times of some isotopes

Longest lived isotope	Half life (Million years)
$_{90}Th^{232}$	1300
$_{93}Np^{237}$	2.2
$_{92}U^{238}$	4500
$_{92}U^{235}$	707
Age of earth	2000-3000

From the data given in table 12-1 it is found that except $_{93}Np^{237}$ all the other three isotopes occur in nature. From these values and the chemical analysis of the decay product Th and U, the age of the earth has been estimated to be between 2000-3000 million years.

12-14-8 Radiocarbon dating

The discovery that all carbon in the world's living cycle is kept uniformly radioactive through the production of C^{14} by cosmic rays led Libby to propose the C^{14} **dating method** that has been a very powerful and widely used technique for determining ages of carbon containing specimens. The C^{14}

method is being used fairly routinely in many laboratories for finding the ages of such diverse samples as wood, peat, grains, shells, bones, cloth, charcoal, beeswax, corncobs and other archeological evidences gathered from all over the world. Radiocarbon dating has certainly become a most important tool for archaeologists and geologists. The range of the method is, of course, limited by the lower limit of detectable carbon activity. An age of 50,000 years correspond to a specific activity of 0.03 disintegrations min^{-1} g^{-1} of carbon which is probably about as low as one can hope to measure.

The carbon dating is done by determining the C^{14}/C^{12} ratio present in the specimen. The constancy of the average cosmic ray flux has been checked back to about 5,000 years ago by means of specimens whose ages are known from positive historical records. A problem is soon to arrive in this tecnique is, the C^{14} content that has overwhelmingly increased due to the current bomb explosions and bomb testings, thereby changing the specific activity of atmospheric carbon. These man made effects will surely cause trouble in future of C^{14} dating studies.

12-15 SELF TEST QUESTIONS

Choose the correct answer and mark it

12-15-1 An alpha particle is

(A) Ar nucleus (B) Proton (C) Electron (D) He nucleus

12-15-2 During an alpha decay the mass number changes by

(A) 2 units (B) 3 units (C) 4 units (D) 0 units

12-15-3 During a beta decay the charge changes by

(A) 0 units (B) 1 units (C) 2 units (D) 4 units

12-15-4 An Alpha particle can also be represented as

(A) $_2Be^4$ (B) $_3B^4$ (C) $_3Li^4$ (D) $_2He^4$

12-15-5 The set of nuclides which are isotopic to each other

(A) $_6C^{12}$, $_5B^{12}$, $_4Be^{12}$ (B) $_6C^{12}$, $_7N^{13}$, $_8O^{14}$

(C) $_6C^{12}$, $_6C^{13}$, $_6C^{14}$ (D) $_5B^{12}$, $_7N^{13}$, $_4Be^{12}$

12-15-6 Isotopes of an nuclide have the same number of

(A) Neutrons (B) Protons (C) Electrons (D) Nucleons

12-15-7 The heaviest amongst nucleons is

(A) Neutrons (B) Deuterium (C) Tritium (D) α-Particle

12-15-8 Isotone contains same number of

(A) Electrons (B) Nucleons (C) Neutrons (D) Protons

12-15-9 Emission of an alpha particle followed by a β-particle leads to the formation of an

(A) Isomer (B) Isotope (C) Isobar (D) None of them

12-15-10 Stability of the nucleus is due to

(A) Electron-proton ratio (B) Proton-proton ratio

(C) Proton-neutron ratio (D) Electron-neutron ratio

12-15-11 The isobars amongst the following pais are

(A) $_{18}Ar^{40}$ & $_{20}Ca^{40}$ (B) $_{19}K^{40}$ & $_{19}K^{39}$

(C) $_{20}Ca^{40}$ & $_{20}Ca^{42}$ (D) $_{18}Ar^{40}$ & $_{18}Ca^{39}$

12-15-12 The isotones amongst the following pairs of nuclides are

(A) $_4Be^9$ & $_6C^{11}$ (B) $_4H^2$ & $_1C^3$ (C) $_4Be^9$ & $_6C^{12}$ (D) $_1H^1$ & $_1H^2$

12-15-13 The isotopes amongst the following pairs of nuclides are:

(A) $_1H^2$ & $_1H^3$ (B) $_5B^{10}$ & $_6C^{11}$

(C) $_{90}Th^{230}$ & $_{92}U^{230}$ (D) $_{91}Pa^{234}$ & $_{92}U^{234}$

12-15-14 Isomers amongst the following paris of nuclides are

(A) $_{72}Hf^{180}$ and $_{72}Hf$[illegible] (B) $_{72}Hf^{180m}$ and $_{72}Hf^{180}$

(C) $_{72}Hf^{180m}$ and $_{72}H$[illegible] (D) $_{72}Hf^{180}$ and $_{72}Hf^{182}$

12-15-15 Mass of a neutron is

(A) 1.008885 amu (B) 1.007825 amu

(C) 2.016490 (D) 1.008665 amu

12-15-16 Mass of a proton i

(A) 1.008885 amu (B) 1.007825 amu

(C) 2.016490 amu (D) 1.008665 amu

12-15-17 Mass of a deuteriu nucleus is

(A) 2.014102 amu (B) 1.008885 amu

(C) 1.007825 amu (D) 1.008665 amu

12-15-18 When the neutron raton ratio values are higher than 1.6 the nuclide undergoes

(A) Fusion (B) Repulsion

(C) Fission (D) Nothing happens to the nucleus

12-15-19 Fission occurs in heavy atoms when the neutron to proton ratio values are:

(A) More than 1.6 (B) Less than 1.6

(C) Unity (D) Zero

12-15-20 The phenomenon involved in the changes which take place in a radioactive nuclei by the emissoon of a particle resulting in the formation of a daughter nuclide is called

(A) Nuclear integra n (B) Nuclear fusion

(C) Nuclear fission (D) Nuclear disintegration

12-15-21 The magic numbers in a nuclide are

(A) 2, 8, 20, 50 et (B) 4, 8, 16, 24 etc

(C) 3, 9, 12, 15 et (D) 5, 16, 20, 25 etc.

12-15-22 An α-particle decay takes place by the phenomenon of

(A) Showering eff (B) K-Capture

(C) Tunneling eff (D) Auger effect

12-15-23 An alpha particle will usually have a potential energy of

(A) 2 million elec volts (B) 6 MeV

(C) 20 Mev (D) Will not have any energy

12-15-24 To penetrate into the nucleus an alpha particle has to cross a

(A) Coulombic bar (B) Kinetic barrier

(C) Potential barri (D) Magnetic barrier

12-15-25 A β-decay takes place by the process of

(A) K-capture (B) Tunnelling effect

(C) Isomeric effect (D) Fusion

12-15-26 A Positively charged β-particle (β^+)is referred to as

(A) Electron (B) Proton (C) Positron (D) Meson

12-15-27 The reaction which undergoes change by a positron decay is

(A) ${}_{28}Cu^{64} \longrightarrow {}_{30}Zn^{64}$ (B) ${}_{29}Cu^{64} \longrightarrow {}_{28}Ni^{64}$

(C) ${}_{29}Cu^{64} \longrightarrow {}_{27}Co^{64}$ (D) ${}_{29}Cu^{64} \longrightarrow {}_{27}Co^{60}$

12-15-28 The reaction which undergoes change by a β-decay

(A)${}_{29}Cu^{64} \longrightarrow {}_{30}Zn^{64}$ (B)${}_{29}Cu^{64} \longrightarrow {}_{27}Co^{64}$

(C)${}_{29}Cu^{64} \longrightarrow {}_{28}Ni^{64}$ (D)${}_{29}Cu^{64} \longrightarrow {}_{27}Co^{60}$

12-15-29 Rate of disintegration can be calculated for a radio-active nuclei by the following expression

(A) $k = 2.303 \log(a-x)/a$ (B) $k = 2.303 \log a/(a-x)$

(C) $k = \dfrac{2.303}{t} \log a/(a-x)$ (D) $kt = 2.303 \log(a-x)/a$

12-15-30 The average half life ($t_{1/2}$) of a nuclide can be calculated by the expression

(A) $\lambda/1.44$ (B) λ 1.44 (C) $0.693/\lambda$ (D) $0.693\ \lambda$

12-15-31 The radio isotope useful in chemotherapy is that of

(A) Uranium (B) Thorium (C) Fluorine (D) Cobalt

12-15-32 Determination of the age of a substance is done by

(A) Electrons and protons (B) Protons and neutrons

(C) Electrons and neutrons (D) Electrons and mesons

12-15-33 A nucleus contains

(A) Electrons and protons (B) Electrons and neutrons

(C) Electrons and mesons (D) Protons and neutrons

12-15-35 Nuclei possessing same charge and mass but differing in the energy content are called

(A) Isotopes (B) Isobars (C) Isotones (D) Isomer

FILL IN THE BLANKS:

12-15-36 Nuclei possessing same charge and mass but differing in the energy content are __________________

12-15-37 __________________ of a nuclide have the same mass number.

12-15-38 Isotones are nuclei having the same number of ____________ in them.

12-15-39 ${}_{20}Ca^{40}$and${}_{19}K^{40}$ are ____________ of each other.

12-15-40 A higher nuclide is said to be very stable when its neutron-proton ratio is equal to____________________.

12-15-41 The mass defect is calculated by taking the mass of__________________ with respect to the actual mass.

12-15-42 The proposed quantum configuration of magic number 20 is________________.

12-15-43 The quantum configuration of $1s^22p^62s^23d^{10}$ is a magic number of________.

12-15-44 The nuclear reaction

$$_7N^{14} + _0n^1 \longrightarrow _8C^{14} + _1H^1$$

can be classified as a ________ emission by________capture

12-15-45 An example of spallation reaction is________.

12-15-46 The uranium series is classified as a ________.

12-15-47 4n+3 series is referred to as________series.

12-15-48 The artificial transmutation series is also called as the ________ series.

12-15-49 Through tunneling effect________ particle decay takes place.

12-15-50 Auger effect is the phenomenon by which________decay takes place

12-15-51 Electron capture or________capture is a phenomenon by which β-decay takes place.

12-15-52 Hf^{180m} undergoes transformation to Hf^{180g} during________radiations.

12-15-53 Conversion of $_1H^1$ to $_2He^4$ is a process of nuclear________.

12-15-54 The ratio________can determine the age of a substance which is usually referred to as carbon dating.

12-15-55 An alpha-particle has to cross an energy barrier of________MeV to enter into a nucleus.

12-15-56 A reaction of nuclear fusion can be exemplified as________.

12-16 NUMERICAL PROBLEMS

12-16-1 The amount of ^{14}C isotope in a piece of wood is found to be 1/6 of its amount present in a fresh piece of wood. Calculate the age of wood(half-life of ^{14}C = 5577 years). (O.U.,Nov 1990)

12-15-2 Calculate the binding energy and packing fraction of the following nuclides.

$_4B^9$=9.0134 amu $_5Be^{11}$=10.8271 amu

$_6C^{12}$ 12.0116 amu $_7N^{14}$=14.0082 amu

$_8O^{16}$=15.994710 amu

12-16-3 One gram of radio-active element decays to 0.25g in 30 hr. Calculate the time required for the activity to fall to 10% of the original value. (O.U.,Apr 1991)

12-16-4 Calculate the mean binding energy for the nucleus of carbon (mass of neutron 1.00840 amu and proton 1.00894 amu) (O.U.,Apr1991)

12-16-5 Calculate the mass defect and mean binding energy for helium nucleus.(mass of proton 1.00815 amu, neutron 1.008905 amu) (K.U.,Apr 1990)

12-16-6 Calculate the mean binding energy of helium nucleus given that the mass of a proton is 1.0089 & amu and that of a neutron is 1.00840 amu. (O.U.,Apr 1990).

12-16-7 Calculate the binding energy of the following. (the mass of neutron = 1.008665, proton = 1.007825 amu and e = 0.00054 amu).

a) $_{18}Ar^{40}$ with mass 39.9623 amu

b) $_{3}Li^{7}$ with mass 7.01822 amu

c) $_{13}Al^{27}$ with mass 26.9815 amu

d) $_{14}Si^{30}$ with mass 29.9738 amu

12-17 SHORT AND ESSAY TYPE QUESTIONS

12-17-1 (a) How do you explain that only certain elements are radioactive?

(b) Give a brief account on

i) Mass Defect ii) Binding energy iii) Packing fraction

(K.U.,Dec 1990)

12-17-2 What are the uses of radioactive isotopes (A.U.,Mar 1991)

12-17-3 What is meant by transmutation of elements?Give one example.

(O.U. Mar., 96) (A.U.,Mar 1991)

12-17-4 What are isotopes and isobars? Name methods useful for the separation of isotopes and discuss about any two methods. (A.U.,Mar1990)

12-17-5 Complete the following nuclear reactions

$$_{7}N^{14} + _{2}He^{4} \longrightarrow _{1}H^{1} + \ ?$$

$$_{11}Na^{23} + _{1}H^{1} \longrightarrow _{12}Mg^{23} + \ ?$$

12-17-6 (a) What do you understand by n/p ratio?How does it effect the stability of the nucleus?

(b) Define the term decay constant, half-life period and average life of a radioactive nucleus. How are they inter related? (O.U., Nov 1990)

12-17-7 Explain group displacement law on radio activity. (O.U.,Nov 1990)

12-17-8 How does the rate of decay of a radioactive isotopes change with, and why?

(O.U..Nov 1990)

12-17-9 Define the term isotope, isotone, isobar and isomer and give an example for each. (O.U., Nov 1990, & Apr 1991)

12-17-10 (a) Explain briefly the liquid drop model of the structure of nucleus.

(b) What do you understand by packing fraction.

(c) Define the term decay constant.half-life period and average life. How are they interrelated. (O.U.,Apr 1991)

12-17-11 Mention the different radio active series how are they different from each other?

(O.U.,Apr 1991)

12-17-12 Give the properties of α, β, γ-rays. (O.U.,Apr 1991)

12-17-13 (a) Describe the shell model of nucleus

(b) Define the terms? (i)Isotope (ii)Isotone (iii)Isobar (iv) Isomer. Give on example of each

(c) Calculate the number of neutrons present in $_{92}U^{238}$ atom after emitting an alpha particle.

(d) Write brief note on group Displacement Law. (K.U., Apr 1990)

12-17-14 What is fission and fusion? Explain with suitable examples.

(N.U. Mar., 96) (S.K.U., Sept 1990)

12-17-15 (a) Write notes on group displacement law

(b) What are isotopes? Write their uses in agriculture, industry and medicine

(c)What happens when $_{92}U^{238}$ is bombarded with neutrons? (S.K.U.,Sept 1990)

12-17-16 Explain artificial radioactivities. What, is an isotope? Give two of their uses each in medical, agriculture and industry. (N.U. Mar., 96) (S.K.U.,Apr 1990)

12-17-17 (a) Explain nuclear fission reaction.

(b) What is mass defect and binding energy? Explain with examples.

(c) What is natural radioactivity? Give two examples.

(d) Write the composition of the nucleus. (S.K.U.,Apr. 1990)

12-17-18 (a) What are the differences between nuclear reaction and chemical reaction.

(b) Write about the liquid drop model of the nucleus. (N.U.,Apr 1991)

12-17-19 (a) What is effective atomic number rule? Give two examples where this rule is not obeyed.

(b) Pick up the pair of isotopes and isobars from the following.

$_1H^1$, $_7N^{14}$, $_8O^{16}$, $_1H^2$, $_{20}Ce^{40}$, $_8O^{17}$, $_{78}Pt^{198}$, $_{18}Ar^{40}$, $_6C^{14}$, $_{78}Pt^{192}$,

(c) Define the terms mass defect and packing fraction. (N.U.,Apr 1991)

12-17-20 (a) Complete the following reaction.

(i) $_6C^{12} + ? \longrightarrow {}_7N^{13}$

(ii) $_{92}U^{238} + {}_7N^{14} \longrightarrow {}_{99}Es^{247} + ?$

(iii) $_{17}C^{37} + ? \longrightarrow {}_{18}Ar^{38} + {}_0n^1$

(iv) $_1H^1 + {}_0n^1 \longrightarrow {}_1H^2 + ?$ (O.U.,Apr 1990)

12-17-21 (a) Write a note on radio active equilibrium.

(b) Mention two reactions each for transmutation by α particle and proton. (O.U.,Apr. 1990)

12-17-22 Complete the following nuclear reactions.

a) $_3Li^7 + {}_1H^1 \longrightarrow ? + {}_0n^1$ - 1.621 MeV

b) $_4Be^9 + ? \longrightarrow {}_6C^{12} + {}_0n^1$ 5.5 Mev

c) $? + {}_2He^4 \longrightarrow {}_{14}S^{30} + {}_1H^1$

d) $_7N^{14} + ? \longrightarrow {}_8O^{17} + {}_1H^1$

e) $? + {}_{12}Mg^{26} \longrightarrow {}_{13}A^{128}$

f) $? + {}_2He^4 \longrightarrow {}_{17}C^{137} + 14, H^1 + 16{}_0n^1$

g) $_{94}Pu^{239} + {}_2He^4 \longrightarrow {}_{96}Cm^{242} + ?$

h) $_{92}U^{235} + {}_0n^1 \longrightarrow {}_{56}Ba^{141} + ? + 2{}_0n^1$

12-17-23 Discuss the applications of radio isotopes in different fields of life. (N.U. Sep., 95)

12-17-24 Write a note on induced nuclear reactions. (N.U. Mar., 95)

12-17-25 Calculate the no. of α- and β- particle emissions during the disintegration of $_{90}Th^{234}$ to $_{81}Tl^{206}$.

12-17-26 Calculate the no. of α- and β- particle emissions during the disintegration of $_{95}Am^{241}$ to $_{83}Bi^{209}$.

ANSWERS

12-15-1 **D**	12-15-2 **C**	12-15-3 **B**	12-15-4 **D**
12-15-5 **C**	12-15-6 **B**	12-15-7 **D**	12-15-8 **C**
12-15-9 **C**	12-15-10 **C**	12-15-11 **A**	12-15-12 **A**
12-15-13 **A**	12-15-14 **C**	12-15-15 **D**	12-15-16 **B**
12-15-17 **B**	12-15-18 **C**	12-15-19 **A**	12-15-20 **D**
12-15-21 **A**	12-15-22 **C**	12-15-23 **B**	12-15-24 **A**
12-15-25 **A**	12-15-26 **C**	12-15-27 **B**	12-15-28 **A**
12-15-29 **C**	12-15-30 **C**	12-15-31 **D**	12-15-32 **B**
12-15-33 **D**	12-15-34 **D**	12-15-35 **D**	
12-15-36 Isomers	12-15-37 Isotopes	12-15-38 Neutrons	12-15-39 Isobars
12-15-40 Unit	12-15-41 Neutrons	12-15-42 $1s^2\ 2p^6\ 2s^2\ 3d^{10}$	
12-15-43 20	12-15-44 Particle, projectile		

12-15-45 $_{29}Cu^{63} + _2He^4(400 MeV) \longrightarrow _{17}Cl^{37} + 14\,_1H^1 + 16\,_0n^1$

12-15-46 4n+2	12-15-47 Actinium	12-15-48 Neptunium	
12-15-49 α	12-15-50 β	12-15-51 K-	12-15-52 γ
12-15-53 Fusion	12-15-54 $C^{14}:C^{12}$	12-15-55 20	

12-15-56 $4\,_1H^1 \longrightarrow _2He^4$ + energy.

Part II

Organic Chemistry

6
ALKYNES

6-1 INTRODUCTION

Alkynes are hydrocarbons that contain carbon - carbon triple bond. The general formula of alkynes is C_nH_{2n-2}. An alkyne contains two and four fewer hydrogen atoms than an alkene and an alkane respectively. As the simplest member of this homologous series is acetylene, alkynes are often referred as acetylenes. The functional group of alkynes is $-C{\equiv}C-$. The triple bond present in alkynes determines their chemical reactivity.

6-2 ELECTRONIC STRUCTURE OF ALKYNES

A carbon-carbon triple bond results from the overlap of two *sp* hybridized carbon atoms. The two *sp* hybrid orbitals of carbon adopt a linear geometry. They lie at an angle of 180° to each other along an axis that is perpendicular to the axes of the unhybridized $2p_y$ and $2p_z$ orbitals. When two *sp* hybridized carbons approach each other for bonding, the geometry is perfect for the formation of one *sp-sp* sigma (σ) bond and two *p-p* π-bonds, with the result **a triple bond** is formed. The remaining two *sp* hybrid orbitals form bonds to other atoms at an angle of 180° from the carbon–carbon bond.

Fig 6-1 Formation of triple bond in alkynes

The bond strength for the carbon-carbon triple bond is approximately 196 kcal mole^{-1}, making it the strongest carbon-carbon bond, and the bond length is 1.20Å. Electron diffraction, X-ray diffraction and spectroscopic analysis reveal that acetylene is a linear molecule. It may be noted that C—H bond length is shorter than C≡C bond length because of the greater σ character of former.

6-3 NOMENCLATURE

There are two ways of naming alkynes.

6-3-1 Common system

Commonly, alkenes are named as derivatives of the simplest alkyne i.e. acetylene. Replacement of one or both of the hydrogen atoms by alkyl groups gives a compound that is named by placing the alkyl group names in front of the parent acetylene name.

For example,

Formula	Common name	Trivial name
$H—C{\equiv}C—H$	Acetylene	
$H_3C—C{\equiv}C—H$	Methyl acetylene	Allylene
$H_3C—C{\equiv}C—CH_3$	Dimethyl acetylene	Crotonylene
$H_3C—CH_2—C{\equiv}C—CH_3$	Ethylmethyl acetylene	

6-3-2 IUPAC Nomenclature

In the IUPAC nomenclature, the ending **-ane** of the corresponding alkane is dropped and **-yne** is added as suffix, which also indicates the presence of a triple bond. The parent compound must contain the triple bond in the longest continuous carbon chain, which is numbered so that the triple bond has the lowest number. Substituents are indicated by numbers.e.g.

Formula	Name
$H—C{\equiv}C—H$	Ethyne
$H_3C—C{\equiv}C—H$	Propyne
$H_3C—C{\equiv}C—CH_3$	2-Butyne
$H_3C—CH(CH_3)—C{\equiv}C—H$	2-Methyl-1-butyne

The monovalent radicals obtained by the removal of one hydrogen of alkynes are known as alkynyl radicals and their names are obtained by replacing **-e** of corresponding alkynes by **-yl.**

Formula	Name
$H—C{\equiv}C\cdot$	Ethynyl
$H_3C—C{\equiv}C\cdot$	Propynyl
$C_2H_5—C{\equiv}C\cdot$	Butynyl

If the molecule contains more than one triple bond the suffix **-ane** of the alkane is replaced by **diyne, triyne** etc., depending on number of triple bonds. The positions of triple bonds are indicated by the number of carbon atoms preceding these bonds.

$HC\equiv C—C\equiv C—H$ 1,3-Butadiyne

$H—C\equiv C—CH_2—C\equiv C—H$ 1,4-Pentadiyne

6-4 ISOMERISM

Alkynes exhibit following three types of isomerism:

(i) Chain isomerism : Alkynes with same molecular formula may differ in the structure of their chains.

For example; C_5H_8 can have the structures:

$H—C\equiv C—CH_2—CH_2—CH_3$
1-Pentyne

$H—C\equiv C—CH(CH_3)—CH_3$
3-Methyl-1-butyne

(ii) Position isomerism : Alkynes may possess same molecular formula but differ in the position of triple bond.

For example, C_4H_6 can be given following structures with different positions of triple bonds.

$HC\equiv C—CH_2—CH_3$
1-Butyne

$H_3C—C = C—CH_3$
2-Butyne

(iii) Functional isomerism : Alkynes show functional isomerism with alkadienes.

$H—C\equiv C—CH_3$
Propyne

$H_2C=C = CH_2$
Propadiene

$H_3C—C\equiv C—CH_3$
2 - Butyne

$H_2C = CH—CH = CH_2$
1,3 - Butadiene

6-5 METHODS OF PREPARATION

The general methods of formation of a triple bond in the molecule are essentially similar to those employed for the preparation of alkenes. Only important general methods of formation are discussed below.

6-5-1 Dehydrohalogenation of dihalides

When dihaloalkanes are treated with excess of alcoholic KOH, alkynes are formed by the elimination of two molecules of halogen acids. Dihaloalkanes are of two types (i) 1,2-dihalides or vicinal dihalides (ii) 1,1-dihalides or gem dihalides.

(i) From 1,2-dihalides (vicinal dihalides): 1,2-Dihalides on treatment with alcoholic KOH give alkynes. The reaction occurs in two stages, first an olefinic halide is formed which on subsequent elimination of a molecule of hydrogen halide with $NaNH_2$ yieldsan alkyne.

$$-\underset{H}{\overset{X}{C}}-\underset{H}{\overset{X}{C}}- \xrightarrow[\text{KOH}]{\text{Alcoholic}} \underset{H}{\diagdown}C=C\overset{X}{\diagup} \xrightarrow{NaNH_2} -C\equiv C-$$

1,2 - Dihalide Alkyne

For example,

$$H_3C-\underset{H}{\overset{Br}{C}}-\underset{H}{\overset{Br}{C}}-H \xrightarrow[\text{KOH}]{\text{Alcoholic}} H_3C-\underset{H}{C}=\overset{Br}{C}-H \xrightarrow{NaNH_2} H_3C-C\equiv C-H$$

1,2 Dibromopropane 1-Bromo-1-Propene Propyne

(ii) From 1,1-dihalides: 1,1-Dihalides (gem-dihalides) when treated with alcoholic KOH yield alkynes.

$$-\underset{H}{\overset{H}{C}}-\underset{X}{\overset{X}{C}}- \xrightarrow[\text{KOH}]{\text{Alcoholic}} \underset{H}{\diagdown}C=C\overset{Br}{\diagup} \xrightarrow{NaNH_2} -C\equiv C-$$

1,2 - Dihalide Vinylhalide Alkyne

For example,

$$H_3C-\underset{H}{\overset{H}{C}}-\underset{Br}{\overset{Br}{C}}-H \xrightarrow[\text{KOH}]{\text{Alcoholic}} H_3C-\underset{H}{C}=\overset{Br}{C}-H \xrightarrow{NaNH_2} H_3C-C\equiv C-H$$

Propyne

6-5-2 Dehalogenation of tetrahalides

In this reaction 2 moles of halogens are removed by the action of zinc metal in alcohol on a suitable tetrahalide.

$$-\underset{X}{\overset{X}{C}}-\underset{X}{\overset{X}{C}}- \xrightarrow[\text{Alcohol}\ \Delta]{Zn} -\overset{X}{C}=\overset{X}{C}- + ZnX_2 \xrightarrow[\text{Alcohol}]{Zn} -C\equiv C- + ZnX_2$$

Alkyne

For example,

$$H_3C-\underset{Br}{\overset{Br}{\overset{|}{\underset{|}{C}}}}-\underset{Br}{\overset{Br}{\overset{|}{\underset{|}{C}}}}-CH_3 \xrightarrow{Zn} H_3C-C\equiv C-CH_3$$

2, 2, 3, 3-Tetrabromobutane — 2-Butyne

This method is of limited application because the required tetrahalides are themselves prepared from alkynes. However, it can be employed for the purification of alkynes by converting them into tetrahalides followed by subsequent dehalogenation.

6-5-3 Alkylation of acetylene and 1-alkynes

Acetylene and 1-alkynes can be converted into metal acetylides and metal alkylines respectively by treating them with sodium or sodamide in liquid ammonia.

$$R-C\equiv C-H + Na \xrightarrow{\text{liq. } NH_3} \underset{\text{Acetylide}}{R-C\equiv \overset{-}{C}\,\overset{+}{Na}}$$

The acetylide ion is a good nucleophile and reacts with primary alkyl halide to form higher alkynes. It is an example of S_N2 reaction.

$$R-C\equiv \overset{-}{C}\,\overset{+}{Na} + \underset{1^\circ \text{ halide only}}{R-Br} \longrightarrow R-C\equiv C-R + NaBr$$

This method is very useful in converting a lower alkyne to a higher alkyne.

Alkylation of alkynes can also be accomplished by treating with an appropriate Grignard reagent, followed by the action of an alkyl halide.

$$R-C\equiv C-H + \overset{\delta^-}{CH_3}\overset{\delta^+}{MgI} \longrightarrow R-C\equiv \overset{\delta^-}{C}\,\overset{\delta^+}{MgI} + CH_4$$

$$R-C\equiv C\,MgI + CH_3I \longrightarrow R-C\equiv C-CH_3 + MgI_2$$

6-5-4 Electrolysis of salts of unsaturated dicarboxylic acids

Concentrated solutions of potassium salts of maleic acid and fumaric acid upon electrolysis form acetylene.

$$\begin{array}{l} \text{H—C—}\overset{\overset{\text{O}}{\|}}{\text{C}}\text{—}\overset{-}{\text{O}}\ \overset{+}{\text{K}} \\ \quad \| \\ \text{H—C—}\underset{\underset{\text{O}}{\|}}{\text{C}}\text{—}\overset{-}{\text{O}}\ \overset{+}{\text{K}} \end{array} \xrightarrow{\text{electrolysis}} \begin{array}{l} \text{H—C—}\overset{\overset{\text{O}}{\|}}{\text{C}}\text{—}\overset{-}{\text{O}} \\ \quad \| \\ \text{H—C—}\underset{\underset{\text{O}}{\|}}{\text{C}}\text{—}\overset{-}{\text{O}} \end{array} + 2K^{+}$$

Potassium maleate

Reaction at anode:

$$\begin{array}{l} \text{H—C—}\overset{\overset{\text{O}}{\|}}{\text{C}}\text{—}\overset{-}{\text{O}} \\ \quad \| \\ \text{H—C—}\underset{\underset{\text{O}}{\|}}{\text{C}}\text{—}\overset{-}{\text{O}} \end{array} \xrightarrow{-\,2e^{-}} \begin{array}{l} \text{H—C—}\overset{\overset{\text{O}}{\|}}{\text{C}}\text{—}\dot{\text{O}} \\ \quad \| \\ \text{H—C—}\underset{\underset{\text{O}}{\|}}{\text{C}}\text{—}\dot{\text{O}} \end{array} \longrightarrow H—C\equiv C—H + 2CO_2$$

Reaction at cathode:

$$2K^{+} + 2e^{-} \longrightarrow 2K \xrightarrow{2H_2O} 2KOH + H_2$$

6-5-5 Hydrolysis of calcium carbide

The first member, acetylene is prepared on a large scale by the hydrolysis of a calcium carbide.

$$CaC_2 + 2H_2O \longrightarrow H—C\equiv C—H + Ca^{+2} + 2OH^{-}$$

This is an interesting reaction because the electronic structure of calcium carbide reveals that the carbide ion, C_2^{-2}, is a dicarbanion, which abstracts protons from water. In this case C_2^{-2} is a strong base and water serves as the acid.

$$Ca^{2+} \begin{array}{c} \overline{C} \\ ||| \\ \underline{C} \end{array} + 2H_2O \longrightarrow H—C\equiv C—H + Ca(OH)_2$$

Recently higher homologues of acetylene have been prepared in good yield by passing vapors of methyl alcohol over calcium carbide at about 100°C.

$$Ca^{2+} \begin{array}{c} \overline{C} \\ ||| \\ \underline{C} \end{array} + 2CH_3OH \longrightarrow H_3C—C\equiv C—CH_3 + Ca(OH)_2$$

6-6 PHYSICAL PROPERTIES

The physical properties of alkynes are similar to those of the alkanes and alkenes with the same carbon skeleton.

1. The first four members are gases, next eight are liquids and the higher alkynes are solids.

2. They are all colorless and odorless except acetylene which has characteristic garlic odor. This garlic odor is due to phosphine and arsine present as impurities.

3. Alkynes are compounds of low polarity and hence they are slightly soluble in water but are more soluble in organic solvents such as benzene, acetone etc. They are, however, slightly more soluble in water than the corresponding alkanes and alkenes. This is perhaps due to greater electron density at the triple bond which permits some degree of association with dipolar water molecules.

4. The boiling points and melting points of alkynes are slightly higher than the corresponding alkanes and alkenes. This is presumably due to greater polarity of alkynes than alkenes and alkanes. Further, these properties show a regular increase with raising molecular weight.

5. Lower members are poisonous and when inhaled cause unconsciousness.

6-7 CHEMICAL PROPERTIES

Alkynes undergo two types of reactions. *One is characteristic of terminal hydrogen attached directly to the triple bond.*

This type of alkyne which includes ethyne, H—C ≡ C—H, and more generally, R—C≡C—H, is called a terminal alkyne because the triple bond must be at the end of a carbon chain. The terminal hydrogen on alkyne is moderately acidic and thus can be removed by a strong base.

$$\underset{}{R{-}C \equiv C{-}H} + \underset{\text{base}}{\bar{B}:} \longrightarrow R{-}C \equiv \bar{C}: \;+\; B:H$$

The resulting alkynyl anion, a carbanion, undergoes several important types of reactions.

The second is characteristic of only the triple bond, which undergoes addition in two steps.

$$-C\equiv C- \xrightarrow[\text{A-B}]{\text{1 mole}} -\overset{\text{A}}{\underset{|}{C}}=\overset{\text{B}}{\underset{|}{C}}- \xrightarrow[\text{A-B}]{\text{1 mole}} -\underset{\text{A}}{\overset{\text{A}}{C}}-\underset{\text{B}}{\overset{\text{B}}{C}}-$$

6-7-1 Acidity of acetylene and 1-alkynes

The terminal hydrogen of an alkyne, —C≡C— H is relatively acidic compared with hydrogens in alkanes or alkenes. One consequence of the moderate acidity of the hydrogen atom on a terminal triple bond is that it can be removed by several reagents. A terminal alkyne such as ethyne, for example, reacts with sodium metal to form hydrogen gas and sodium acetylide.

$$\underset{\text{Acetylene}}{H-C\equiv C-H} + Na \longrightarrow \underset{\text{Acetylide ion}}{H-C\equiv \overset{-}{C}\ \overset{+}{Na}} + \tfrac{1}{2}H_2\uparrow$$

The addition of acetylene to sodium amide results in the formation of the acetylide ion and ammonia.

$H-C\equiv C-H$ +	$\overset{-}{N}H_2$	⇄	$H-C\equiv \overset{-}{C}$ +	$H-NH_2$
Acid	Base		Acetylide ion	Ammonia
stronger	stronger		weaker	weaker
acid	base		base	acid
$K_a \approx 10^{-25}$				$K_a \approx 10^{-34}$

This is a typical acid -base reaction in which the amide ion NH_2^- as the base pulls the hydrogen from acetylene as a proton and forms ammonia. Further evidence regarding the reactivity of the acetylide ion comes from its reaction with water. It abstracts a proton from water and becomes a terminal alkyne again.

$H-C\equiv \overset{-}{C}\ \overset{+}{Na}$ +	$H-OH$	⇄	$H-C\equiv C-H$ +	$OH^- + Na^+$
Base	Acid		Acid	Base
stronger	stronger		weaker	weaker
base	acid		acid	base
	$K_a \approx 10^{-15}$			

In this case the acetylide ion serves as a base and water as an acid. The acetylide ion has a greater affinity for protons than does water, or, alternatively water is stronger acid than a terminal alkyne.

We now know the position of a terminal alkyne relative to water and ammonia in terms of acidity. The order of acidity is water > acetylene > ammonia.

To explain the acidity of terminal alkynes, one must look at the electronic structure of the conjugate base, the anion, that is formed. The pair of electrons on carbon in the anion are closer to the carbon nucleus because they occupy a *sp*-hybrid orbitals, which has 50% *s*- character. This means that the electrons in this orbital are close to carbon in the terminal alkyne than they are for carbon atoms of other hydrocarbons. For example, in an alkene, 33.3% *s*- character is in sp^2 hybrid orbitals, and in an alkane, 25% *s*-character in sp^3 hybrid orbitals.

This results in a more stable electronic configuration and therefore more stable anion.

For example,

Hydrocarbon	R — H	R — CH = CH_2	R—C≡C—H
	Alkane	Alkene	Alkyne
≈ K_a	10^{-45}	10^{-35}	10^{-25}
s-character	25%	33.3%	50%

Acetylenic hydrogens can be replaced by alkyl groups by treating acetylene and 1-alkynes with primary and secondary alkyl halides in the presence of sodium and liquid ammonia.

$$R-C\equiv C-H + Na \xrightarrow{\text{liq. } NH_3} R-C\equiv C\,Na$$

$$R-C\equiv \overset{-}{C}\overset{+}{Na} + Cl-CH_2-CH_3 \longrightarrow R-C\equiv C-CH_2CH_3 + NaCl$$

This process does not work with tert alkyl halides which eliminate HX and form the corresponding alkenes.

6-7-2 Addition reactions

1. Addition of hydrogen: In the presence of catalyst like finely divided Raney nickel, palladinum or platinum, alkynes add up two molecules of hydrogen first forming corresponding alkenes and finally the alkanes.

$$H-C\equiv C-H \xrightarrow[\text{catalyst}]{H_2} [H_2C=CH_2] \xrightarrow[\text{catalyst}]{H_2} H_3C-CH_3$$

The reduction can, however, be stopped at the alkene stage by using Pd poisoned with $BaSO_4$ (Lindlar's catalyst) or **nickel boride.**

$$H-C\equiv C-H \xrightarrow[Pd/BaSO_4]{H_2} \begin{matrix} H & & H \\ & C=C & \\ H & & H \end{matrix}$$

The catalytic hydrogenation of disubstituted acetylenes (R—C≡C—R) with Lindlar's catalyst gives the *cis*-alkene exclusively. On the other hand when the reduction is carried chemically with Na or Li, liq. NH_3, predominantly *trans*-alkene is obtained. The *trans* addition of alkenes as above is possible because the intermediate, dianion form, assumes a configuration in which the anions like to be farthest apart. Such reactions which produce predominantly one stereoisomer are generally called **stereoselective reactions.**

2. Addition of halogens: Alkynes add on two molecules of halogens forming first a dihalo alkene and then a tetrahalo alkane.

$$R-C\equiv C-H \xrightarrow{Cl_2} \begin{matrix} R & Cl \\ | & | \\ C & = C \\ | & | \\ Cl & Cl \end{matrix} \xrightarrow{Cl_2} \begin{matrix} & Cl & Cl & \\ & | & | & \\ R- & C- & C & -H \\ & | & | & \\ & Cl & Cl & \end{matrix}$$

1 - alkyne 1,1, 2,2 - tetrachloroalkane

$$H_3C-C\equiv C-H \xrightarrow{Br_2} \begin{matrix} CH_3 & & Br \\ & C=C & \\ Br & & H \end{matrix} \xrightarrow[excess]{Br_2} \begin{matrix} & Br & Br & \\ & | & | & \\ H_3C- & C- & C & -H \\ & | & | & \\ & Br & Br & \end{matrix}$$

1,1, 2,2 - tetrabromopropane

3. Addition hydrogen halides : The addition of hydrogen halides to alkynes can also be made to go in two stages, either stopping at the haloalkene stage or going on to give a dihaloalkane.

$$H-C\equiv C-H + HBr \xrightarrow{HBr} \begin{matrix} H & H \\ | & | \\ C & = C \\ | & | \\ H & Br \end{matrix} \xrightarrow{HBr} \begin{matrix} & Br & \\ & | & \\ H_3C- & C & -H \\ & | & \\ & Br & \end{matrix}$$

1,1, - dibromo ethane

Thus when one molecule of the halide is added to a symmetrical alkyne, the product is an unsymmetrical derivative of alkene. The addition of the second molecule of hydrogen halide, therefore, takes place in accordance with Markovnikov rule. Hence both the halogens become attached to the same carbon atom.

$$H_3C-C\equiv C-H \xrightarrow{HBr} \underset{Br}{\overset{H_3C}{}}C=C\underset{H}{\overset{H}{}} \xrightarrow{HBr} H_3C-\underset{Br}{\overset{Br}{C}}-CH_3$$

2 - bromopropene (vinyl bromide) 1,1, - dibromo ethane

Vinyl halides are less nucleophilic than acetylenes, therefore the addition of the second molecule of hydrogen halide requires more drastic conditions.

4. Addition of hypohalous acids : Addition of two molecules of hypohalous acids, to alkynes is a two step process.

$$H_3C-C\equiv C-H \xrightarrow{HOCl} \underset{HO}{\overset{H}{}}C=C\underset{H}{\overset{Cl}{}} \xrightarrow{HOCl} H-\underset{OH}{\overset{OH}{C}}-\underset{Cl}{\overset{Cl}{C}}-H$$

Unstable

$$\downarrow -H_2O$$

$$H-\overset{O}{\overset{\|}{C}}-\underset{Cl}{\overset{Cl}{C}}-H$$

5. Addition of water (Hydration) : Alkynes can not be hydrated as easily as simple alkenes, because of their lower reactivity toward electrophilic addition. Aqueous H_2SO_4 by itself has no effect on the carbon-carbon triple bond. In the presence of mercuric sulphate catalyst, however dehydration occurs readily.

$$R-C\equiv C-H \xrightarrow[HgSO_4]{H_2O,\ H_2SO_4} \left[R-\overset{OH}{C}=\underset{H}{C}-H\right] \longrightarrow R-\overset{O}{\overset{\|}{C}}-CH_3$$

1 - Alkyne An enol A ketone

Water adds in accordance with Markovnikov's rule to give the first intermediate, a vinyl alcohol, but this alcohol is very unstable and undergoes tautomerisation to form an aldehyde or ketone. It turns out that enol and ketones rapidly achieve an equilibrium, and the process is called tautomerism (For definition of tautomerism and for a more general discussion refer § 6-7-6). With few exceptions the keto-enol tautomeric equilibrium lies heavily on the keto side. Vinylic alcohols are almost never isolated.

$$\underset{\text{(Less favoured)}}{\underset{\text{Enol form}}{\overset{\overset{H}{|}}{\overset{O}{|}}C=C<}} \xrightarrow{\text{rapid}} \underset{\text{(More favoured)}}{\underset{\text{Keto form}}{—\overset{O}{\overset{||}{C}}——\overset{H}{\overset{||}{C}}—}}$$

A mixture of both possible ketones results when an internal alkyne is hydrated, and the reacion is therefore most useful when applied to terminal alkynes (R—C≡CH), since only methylketones are formed.

$$\underset{\text{(an internal alkyne)}}{R—C \equiv C—R} \xrightarrow[HgSO_4]{H_2O,\, H_2SO_4} \underset{\text{a mixture}}{R—\overset{O}{\overset{||}{C}}—CH_2—R + R—CH_2—\overset{O}{\overset{||}{C}}—R}$$

6. *Addition of boron hydride (Hydroboration):* Alkynes, like to add on boron hydride (BH_3) in a cis manner to form vinyl boranes.

$$R—C \equiv C—H + BH_3 \longrightarrow R—\underset{H}{\underset{|}{\overset{H}{\overset{|}{C}}}} \equiv C—\underset{H}{\underset{|}{B}}—H$$

Vinyl boranes can be oxidized with alkaline hydrogen peroxide to form aldehydes or ketones.

7. *Addition of hydrogen cyanide:* Alkynes add hydrogen cyanide across the carbon carbon triple bond under the influence of copper salt catalyst, giving alkenyl cyanides. Thus acetylene yields vinyl cyanide or acrylonitrile.

$$H—C \equiv C—H + HCN \xrightarrow[\Delta]{CuCl} \underset{\text{vinyl cyanide}}{H_2C=CH—CN}$$

6-7-3 Oxidative cleavage of alkynes

The oxidizing agents that are effective aginst C=C bond are usually effective against the C≡C bond also. However, the triple bond is less responsive owing to the π electrons being more firmly held by the carbon nuclei.

(i) Oxidation with potassium permanganate: The oxidation of alkynes with alkaline $KMnO_4$ cleaves the molecule at the site of the triple bond to yield carboxylic acids and carbon dioxide.

$$R—C \equiv C—H \xrightarrow{\text{alk. } KMnO_4} R—COOH + CO_2$$

$$R—C \equiv C—R' \xrightarrow{\text{alk. } KMnO_4} R—COOH + R'—COOH$$

This degradation reaction of alkynes is useful in determining structure of alkynes.

(ii) Ozonolysis: In contrast to alkenes, alkynes react slowly with ozone to produce ozonides. These ozonides on decomposition with water yield carboxylic acids.

$$R—C \equiv C—H + O_3 \longrightarrow R—\underset{\substack{|\\O—}}{\overset{O}{C}}—\underset{\substack{|\\O}}{C}—H \xrightarrow{H_2O} R—\overset{O}{\overset{\|}{C}}—OH + HO—\underset{\underset{O}{\|}}{C}—H$$

6-7-4 Polymerization

Acetylene and propyne on passing through red hot quartz or iron tube form benzene and mesitylene respectively.

$$3\ HC \equiv CH \xrightarrow{500\ ^oC} C_6H_6 \text{ (benzene: ring of six CH)}$$

$$3\ H_3C—C \equiv CH \xrightarrow{500\ ^oC} \text{mesitylene (1,3,5-}(CH_3)_3C_6H_3\text{)}$$

6-7-5 Isomerization

1-Alkynes can be readily isomerized to the more stable 2-alkynes on treatment with potassium hydroxide in ethanol.

$$\underset{\substack{\text{1-Alkyne}\\\text{(less stable)}}}{R{-}CH_2{-}C\equiv C{-}H} \xrightarrow[\text{ethanol}]{\text{KOH}} \underset{\text{Allene}}{R{-}CH=C=CH_2} \longrightarrow \underset{\substack{\text{2-Alkyne}\\\text{(more stable)}}}{R{-}C\equiv C{-}CH_3}$$

As this rearrangement takes place through the formation of allene, it is called as **acetylene-allene rearrangement.**

6-7-6 Tautomerism

Tautomerism is a unique phenomenon among organic compounds. The term tautomerism (Latin: tauto = same; meros = parts) was coined by Laar in 1885 to describe the phenomenon of a substance reacting chemically with two possible structures. It is now applied to the behavior by certain compounds in different reactions, as if, they possessed two or more different constituents. This is due to the different arrangement of the atoms and are expressible by different structural formulae.

Tautomerism may be defined as a phenomenon by which a single compound exists in two or more readily interconvertible structures that differ markedly in the relative position of atleast one atom, generally hydrogen. The different structures are known as **tautomers** of each other.

It is a special case of functional isomerism where the isomers are at equilibrium due to their interconvertible nature. In fact there is no sharp borderline between isomerism and tautomerism. Substances which are isomeric under mild conditions could be tautomers under drastic conditions. Tautomerism, therefore, may be considered as dynamic isomerism. If both the tautomers could be isolated or detected in a mixture, tautomerism is then known as **desmotropism.**

$$\underset{\text{Keto form}}{H_3C{-}\overset{\overset{\displaystyle O}{\|}}{C}{-}CH_2COOEt} \rightleftharpoons \underset{\text{Enolic form}}{H_3C{-}\overset{\overset{\displaystyle OH}{|}}{C}=CH{-}COOEt}$$

Ethylacetoacetate

Inter conversion of tautomers is heterolytic in nature, involving the migration of an atom or group from a carbon to another atom (oxygen or nitrogen) in close proximity, with necessary rerrangement of linkages. If the migrating group is a cation, tautomeric transformations are known as **cationtropic**, where as, if it is an anion the transformations are called **aniontropic**. Most common cationtropic transformations involve the migration of a proton and hence are known as **prototropic transformations.**

Laar's classification of tautomeric compounds, revised by Thorpe and Ingold, based on the type of mobile system present. If the hydrogen atom transfers from one to another polyvalent atom linked together, the system is said to be a dyad, and if hydrogen atom transfers from first to third atom in a chain the system is called a triad.

(i) Dyad system:

e.g. : $\underset{\text{Hydrogen cyanide}}{H\text{—}C \equiv N} \rightleftharpoons \underset{\text{Hydrogen isocyanide}}{C \equiv N\text{—}H}$

(ii) Triad system: This involves the oscillation of hydrogen between first and third of the three polyvalent atoms linked together.

(a) Keto-Enol tautomerism: The most common type of tautomerism is that involving a carbonyl compound having α-hydrogen atom (keto form) and its enol form (a structure haivng –OH group attached to a double bonded carbon).

$$\underset{\text{Keto form}}{R\text{—}\underset{\underset{H}{|}}{CH}\text{—}\underset{\underset{O}{||}}{C}\text{—}R'} \longrightarrow \underset{\text{Enol form}}{R\text{—}CH = \underset{\underset{OH}{|}}{C}\text{—}R'}$$

$$\underset{\text{Keto form}}{H_3C\text{—}\overset{\overset{O}{||}}{C}\text{—}\underset{\underset{H}{|}}{CH}\text{—}COOEt} \rightleftharpoons \underset{\text{Enol form}}{H_3C\text{—}\overset{\overset{OH}{|}}{C} = CH\text{—}COOEt}$$

Ethylacetoacetate

$$H_3C\text{—}\overset{\overset{O}{||}}{C}\text{—}\overset{\overset{H}{|}}{CH}\text{—}\overset{\overset{O}{||}}{C}\text{—}CH_3 \rightleftharpoons H_3C\text{—}\overset{\overset{OH}{|}}{C} = CH\text{—}\overset{\overset{O}{||}}{C}\text{—}CH_3$$

Acetylacetone

For discussion related to alkynes see § 6.7.2 under addition of water.

(b) Other systems : *Nitro - acinitro system:* Aliphatic nitro compounds having hydrogen at α-carbon atom exist in **nitro** and **aci** forms exhibiting tautomerism.

e.g.:

$$\underset{\text{Nitro form}}{R-\underset{\displaystyle H}{\underset{|}{C}}H-\overset{\displaystyle O}{\overset{||}{N^+}}=O \quad (\text{with } N-O^-)} \longrightarrow \underset{\text{Aciform}}{R-CH=\overset{\displaystyle OH}{\overset{|}{\underset{\displaystyle O^-}{\underset{|}{N^+}}}}}$$

Nitro form — Aciform

Characteristics of Tautomerism

(i) Tautomers are discrete chemical entities, capable of isolation under suitable conditions.

(ii) Tautomerism is caused by the migration of a group of atoms (commonly Protons) Between two polyvalent atoms of the same molecule.It is thus a reversible intramolecular transformation.

(iii) Tautomers differ from each other in stability. The less stable form is called **labile form**. The relative proportions of the two forms in a tautomeric mixture varies from compound to compound and also with temperature, solvent, etc. Tautomeric transformations are also catalysed by acids and bases.

(iv) Tautomers exist in dynamic equilibrium. their separation can be achieved only by special methods.

6-8 INDUSTRIAL SIGNIFICANCE OF ACETYLENE

Acetylene the first member of the series is the most widely used alkyne. The important uses of acetylene are discussed below:

(1) Acetylene (5-80%), mixed with air, explodes violently when ignited. The flame of acetylene burning in oxygen is extremely hot, about 4000°C, and therefore oxyacetylene torches are used for this purpose.

(2) Eighty percent of the acetylene produced is used for the production of several chemicals like acetaldehyde, acetic acid, ethylalcohol, acetone, benzene, etc.

(3) It is used for the synthesis of heterocyclic compounds such as thiophene, pyrrole, tetrahydrofuran, etc.

$$HC\equiv CH + HC\equiv CH + NH_3 \longrightarrow C_4H_4NH$$

Pyrrole

$$HC\equiv CH + HC\equiv CH + H_2S \longrightarrow C_4H_4S$$

Thiophene

(4) Acetylene is used for the production of industrial solvents like westron (acetylene tetrachloride) and westrol (trichloro ethylene).

(5) For the preparation of Lewisite gas, used as war **gas**.

(6) It is used for the preparation of vinyl chloride, vinyl acetate and vinylethers used in the production of PVC, polyvinylacetate, polyvinyl ethers.

(7) Acetylene is used as a general anaesthetic under the trade name narcylene.

(8) It is used in the preparation of acrylonitrile ($H_2C{=}CHCN$) used for the production of synthetic rubber (Buna N) and synthetic fibre (orlan).

(9) It is used in the synthesis of benzene.

(10) It is also used in the manufacture of **hexachloroethane** (C_2Cl_6), which is used as artificial camphor.

6-9 SELF TEST QUESTIONS

6-9-1 Acetylene gas is liberated by the hydrolysis of
(A) Be_2C (B) Mg_2C_3 (C) CaC_2 (D) Al_4C_3

6-9-2 When allylene is treated with dil H_2SO_4 in the presence of $HgSO_4$, the major product formed is
(A) Propanal (B) Propanone (C) Propanol (D) Prophydrogensulphate

6-9-3 Which of the following on electrolysis give aceetylene
(A) Potassiunm maleate (B) Potassium fumerate?
(C) Both (D) None.

6-9-4 1-Butyne and 2-butyne can be distinguished by
(A) Ammonical silvernitrate (B) Ammonical cuprous chloride
(C) Both (D) None.

6-9-5 Acetylene when passed through a red hot iron tube, undergoes polymerization to form
(A) Toluene (B) Benzene (C) Mesitylene (D) None

6-9-6 Which among the following is most acidic.
(A) 2-Butyne (B) Ethyne (C) Propyne (D) Propene.

6-9-7 Which of the following molecules is linear in structure
(A) 1-Butyne (B) Ethyne (C) Propyne (D) 2-Butyne

6-9-8 Which of the following on dehydrohalogenation with alcoholic KOH gives acetylene
(A) Ethylene chloride (B) Ethylidene chloride
(C) Both (D) None

6-9-9 Final product of the reaction between propyne and HBr is
(A) 2,2-Dibromopropane (B) 1,1-Dibromopropane
(C) 1,2-Dibromopropane (D) None

6-9-10 Which of the following on oxidation with acidic $KMnO_4$ gives only acetic acid
(A) 1-Butyne (B) 2-Butyne (C) Both (D) Propyne

6-10 SHORT AND ESSAY TYPE QUESTIONS

6-10-1 Outline the synthesis of acetylene from ethylbromide. (O.U., Oct. 1990)

6-10-2 How are the following compounds distinguished by test tube experiments

$H_3C—CH_2—C{\equiv}CH$ and $H_3C—C{\equiv}C—CH_3$ (O.U., Oct. 1990)

6-10-3 How is acetylene prepared? Give its importnat properties. (A.U., Jun. 1990), (N.U., Mar., 1995)

6-10-4 Give any two methods for the synthesis of alkynes. (N.U., Mar., 1996), (K.U., Apr. 1989)

6-10-5 How is 1-alkyne isomerized to 2-alkyne. (K.U., Apr. 1989)

6-10-6 What happens when propyne is treated with dilute H_2SO_4 in presence of $HgSO_4$. (N.U., Mar., 96), (K.U., Apr. 1989)

6-10-7 Write briefly about tautomerism.

6-10-8 How are the following pair of compounds distinguished.

6-10-9 Write two methods of preparation of acetylene. (S.K.U. Apr. 1991)

6-10-10 Starting from acetylene how will you obtain each of the following compounds.

(a) Acetaldehyde (b) Vinyl chloride (c) Ethylene

6-10-11 Why are hydrogen atoms in acetylene more acidic than those in ethylene ? (A.U., Sept., 1994)

6-10-12 What happens when acetylene is passed through H_2SO_4 in the presence of mercuric ions ? (A.U., Mar., 1994)

6-10-13 Draw orbital structure of acetylene. Descibe the structures with reference to bond angles, bond lengths and bond formation.

6-10-14 An alkyne A with molecular formula C_5F_8 reacts neither with sodamide nor with ammonical cuprouschloride, assign structural formula to A.

6-10-15 Complete the following

(a) $H_3C—C \equiv C—H \xrightarrow{HBr}$

(b) $H_3C—CH_2—C \equiv CH$ + ammonical cuprous chloride $\longrightarrow$

(c) $H_3C—C \equiv CH \xrightarrow{\text{Ozonolysis}}$ (N.U., Mar., 1996), (O.U., Mar., 1995)

(d) $H_3C—C \equiv C—CH_3 \xrightarrow{\text{Acidic } KMnO_4}$

(e) $H_3C—C \equiv CH + NaNH_2 \longrightarrow A \xrightarrow{CH_3Cl} B$

6-10-16 How will you distinguish the following ?

(a) Acetylene and Propylene

(b) 1-Butyne and 2-Butyne

6-10-17 What happens when

(a) Acetylene reacts with bromine ?

(b) Acetylene is passed through a red hot tube ? (O.U., Mar., 1994)

(c) Propyne reacts with ammonical cuprous chloride ?

(d) Acetylene reacts with HBr ?

6-10-18 Give two methods for the synthesis of allylene.

(a) How are 1-alkynes isomerized to 2-alkynes ?

6-10-19 Explain the concept of tautomerism and write a short note on industrial significance of acetylene.

ANSWERS

6-9-1 **C** 6-9-2 **B** 6-9-3 **C** 6-9-4 **C** 6-9-5 **B** 6-9-6 **B**

6-9-7 **B** 6-9-8 **C** 6-9-9 **A** 6-9-10 **B**

6-10-10

(a) $HC \equiv CH \xrightarrow[HgSO_4]{H_2SO_4} CH_3CHO$

(b) $HC \equiv CH \xrightarrow{\text{One mole HCl}} H_2C = CHCl$

(c) $HC \equiv CH \xrightarrow[Pd/BaSO_4]{H_2} H_2C = CH_2$

6-10-14 $H_3C - C \equiv C - CH_2 - CH_3$

6-10-15 (a) $CH_3CH(Br_2) CH_3$

(b) $H_3C—CH_2C \equiv CCu$

(c) $H_3C—COOH + CO_2$

(d) H_3CCOOH

(e) A : $H_3C—C \equiv CH$

B : $H_3C—C \equiv C—CH_3$

6-10-16 (a) & (b) Ammonical cuprous chloride or Ammonical silvernitrate

6-10-17 (a) $Br_2CH—CHBr_2$

(b) Benzene

(c) $H_3CC \equiv CCu$

(d) $H_3C—CHBr_2$

7
ARENES

7-1 INTRODUCTION

The word *Arenes* is used to denote a group of compounds which are aromatic hydrocarbons. During the nineteenth century, a large number of organic compounds were isolated from balsams, resins and oils. These organic compounds were classified as **aromatic** because all of them had a characteristic smell. The word **aroma** in Greek means smell. In the earlier days of development of organic chemistry, the division between aromatic and aliphatic compounds was on the basis of smell whereas in later years, the emphasis shifted to chemical reactivity. Subsequently investigations showed that the aromatic compounds differed widely from aliphatic compounds in their chemical behavior. With the developments in chemistry, several compounds were discovered which did not possess any smell but still fell under the category of aromatics, whereas a large number of aliphatic compounds were found to possess characteristic odors. The aromatic compounds exhibit reactions assignable to certain common structural feature in all of them.

Benzene is generally considered to be the parent member of the group of aromatic compounds. In the reactions of aromatic compounds, the most important structural feature, called benzene nucleus, is retained. These are known as benzenoidal aromatic compounds. There is another group of compounds which are aromatic in character, but do not contain the benzene nucleus. These are known as **non-benzenoidal aromatic compounds**, for example; ferrocene, tropolone, azulene etc. There is yet another group of aromatic compounds which are non-carbocyclic in nature, namely, **heterocyclic compounds** like pyrrole, pyridine etc.

7-2 SOURCES OF AROMATIC COMPOUNDS

Coal-tar, obtained as one of the volatile products during the pyrolysis of coal, is the most important source of aromatic compounds. Petroleum is another major source of aromatic compounds. The alkane-fraction of petroleum is used to convert catalytically the open-chain compounds into cyclic aromatic derivatives.

It is believed that several millions of years ago, the organaceous matter, such as plants, animals etc., especially the marine ones, got buried below the surface of the earth and under the combined effect of moisture, air, carbon dioxide, pressure and temperature, was converted into various carbonaceous residues. Today, these carbonaceous residues exist in the form of gases (natural gases), liquids (petroleum) and solids (coal). The carbon content of coal varies with its origin.

7-2-1 Pyrolysis of coal

When coal is subjected to low-temperature carbonization, (i.e., heating in the absence of air to 1000-1400°C), a large number of products (gases, volatile liquids) together with a non-volatile residue, called coke, are obtained. The process is very similar to that of obtaining charcoal, by heating dry wood in the absence of air in large columns known as chimneys. The various products, obtained by the low-temperature carbonization of coal, are shown below in the Chart 7-1. Only the approximate yields of products are given since the actual percentage will depend upon the nature of coal, its origin, temperature of carbonization etc.

Chart 7-1 Pyrolysis of Coal

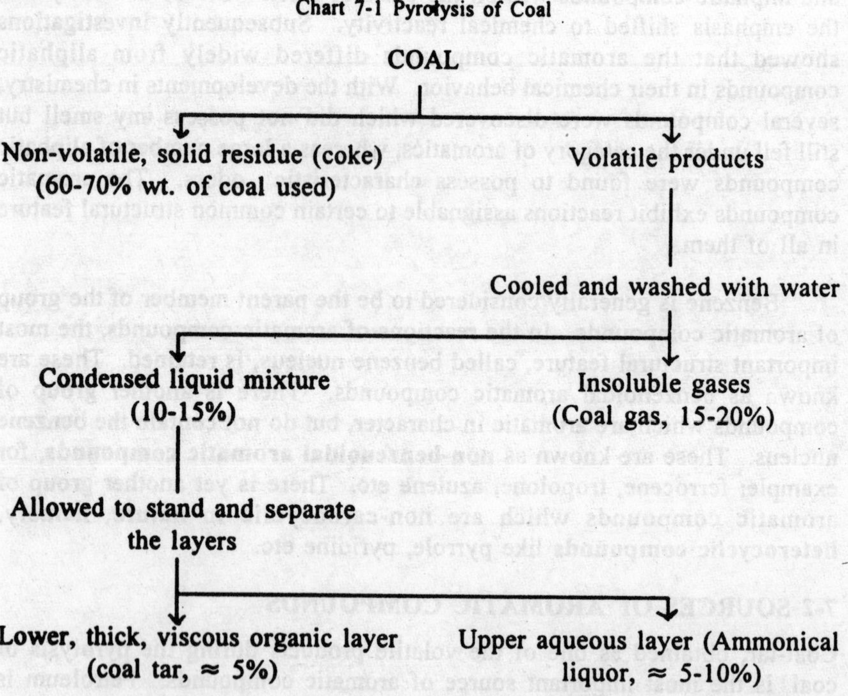

The coal gas, obtained as insoluble gaseous portion, is mostly a mixture of hydrogen (≈ 10%), methane (≈ 30%) and carbon monoxide (≈ 10%). Coal gas finds use as a domestic fuel and illuminant. Coke, which is one of the purest forms of carbon, is used in the manufacture of smoke-less fuel and as a reducing agent in metallurgy, especially iron and steel. The upper, aqueous layer, which consists mostly of ammonia and lower aliphatic amines, is used in the manufacture of ammonium sulphate. The lower, thick, viscous, organic layer, called coal-tar, is subjected to further fractionation to yield a large number of useful aromatic chemicals.

When the pyrolysis of coal is carried out at higher temperatures (i.e., high temperature carbonization 1600-2000°C), larger proportions of volatile constituents are obtained.

7-2-2 Fractionation of coal-tar

Coal-tar is a dark colored, thick, viscous liquid with a strong, unpleasent smell. When preheated coal-tar is subjected to fractionation in cast-iron retorts, different fractions (depending on their boiling ranges) and a non-volatile thick, syrupy residue are obtained. These are shown in Chart-II.

Chart-II: Fractionation of Coal-tar

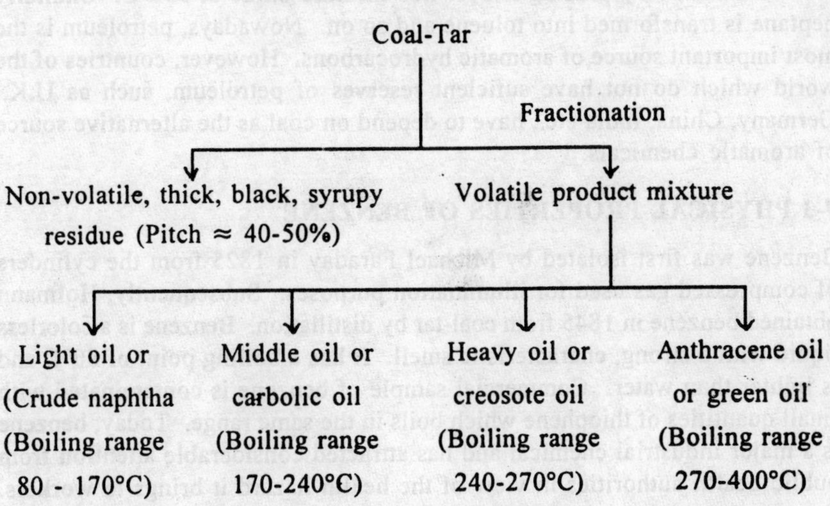

Light oil: Light oil usually consists of benzene, toluene, xylenes, cumene, thiophene etc., in different proportions. It is subjected to further, careful, fractionation to yield various aromatic compounds used as raw material in the manufacture of commercially important dyes, drugs, pharmaceuticals and valuable intermediates for organic syntheses.

Middle oil: Middle oil usually consists of hydroxy derivatives of benzene such as phenol, cresol, xylols etc. Middle oil is subjected to careful fractionation when different phenols are obtained which are used as such as disinfectants, antiseptics etc., or as feed stock for other aromatic chemicals.

Heavy oil : Heavy oil consists of naphthalene derivatives such as naphthalene, naphthols etc. These are used in the manufacture of preservatives, dyes and as starting materials for naphthalene based industrial chemicals.

Anthracene oil : Anthracene oil, also called green oil due to its green Fluorescence, is mostly anthracene with small quantities of phenanthrene and carbazole. It is used primarily to obtain anthracene which finds applications in the manufacture of dyes.

Pitch: Pitch is mostly carbon ($\approx$ 90%). It is used in road-making water-proofing of roofs, protection of wood and in varnish industry.

It may be mentioned here that coal-tar is an important source of aromatic chemicals. However, all the requirements of aromatic compounds cannot be met by coal-tar alone. Alternatively, huge quantities of aromatic hydrocarbons are synthesized from alkanes obtained from petroleum by catalytic processes. For example, hexane is converted to cyclohexane and then to benzene by passing it over hot chromic oxide at 550°C. Similarly heptane is transformed into toluene and so on. Nowadays, petroleum is the most important source of aromatic hydrocarbons. However, countries of the world which do not have suficient reserves of petroleum, such as U.K., Germany, China, India etc., have to depend on coal as the alternative source of aromatic chemicals.

7-3 PHYSICAL PROPERTIES OF BENZENE

Benzene was first isolated by Michael Faraday in 1825 from the cylinders of compressed gas used for illumination purposes. Subsequently, Hofmann obtained benzene in 1845 from coal-tar by distillation. Benzene is a colorless liquid with a strong, characteristic smell. It has a boiling point of 80°C and is lighter than water. Commercial sample of benzene is contaminated with small quantities of thiophene which boils in the same range. Today, benzene is a major industrial chemical and has attracted considerable attention from public health authorities in view of the health hazard it brings to workers. Benzene is a highly toxic chemical. If inhaled or ingested into human body it does not get oxidized or ejected out easily causing very damaging metabolic changes leading to cancerous diseases especially leukemia.

7-4 STRUCTURE OF BENZENE

On the basis of elemental analysis and molecular weight determination, the empirical molecular formula of benzene was found to be C_6H_6. The molecular formula corresponding to a saturated hydrocarbon would be C_nH_{2n+2}, corresponding to an alkene would be C_nH_{2n} and for an alkyne would be C_nH_{2n-2}. The molecular formula of benzene, i.e., C_6H_6, which corresponded to C_nH_{2n-6}, predicted a very high degree of unsaturation in its structure. Therefore, it was expected of benzene to undergo reactions typical of alkenes and alkynes, most notably addition, oxidation and polymerization. On the contrary, the benzene molecule showed the following chemical behavior:

(i) Halogens or hydrogen halides do not add to benzene at room temperature.

(ii) Benzene remained unffected on treatment with potassium permanganate or nitric acid at moderate temperatures.

(iii) Benzene did not show any tendency to undergo polymerization reactions in the presence of free-radical forming catalysts.

(iv) In the presence of certain catalysts, particularly of the Lewis acid types, benzene showed a tendency to undergo substitution reactions involving replacement of its hydrogens.

Some of the important reactions of benzene are nitration, sulphonation, halogenation, Friedel-Crafts reaction etc., all of which involve replacement of benzene hydrogens. Furthermore, it was found that benzene forms only one monosubstitution and three disubstitution products.

The fact that the benzene molecule contains three double bonds and a six-membered carbocyclic ring in its structure was shown by the following reactions:

(a) Benzene adds on three molecules of halogen to form benzene hexahalide under conditions of strong pressure, temperature and a free-radical catalyst.

$$C_6H_6 + 3X_2 \longrightarrow C_6H_6X_6$$

(b) Benzene on reaction with ozone, forms a triozonide.

$$C_6H_6 + 3O_3 \longrightarrow C_6H_6O_9$$

(c) Benzene, on hydrogenation over a catalyst, under very vigorous conditions, yields cyclohexane.

$$\underset{\text{(Benzene)}}{C_6H_6} \xrightarrow[\text{and temperature}]{\text{Catalyst, pressure}} \underset{\text{(Cyclohexane)}}{C_6H_{12}}$$

The above reactions indicate that the double bonds of benzene are of special type, rather unreactive, and are responsible for the unusual stability of this molecule.

7-4-1 Kekule formula and other structures for benzene

Based on the available facts, Kekule in 1865 proposed the following structure-I, for benzene molecule:

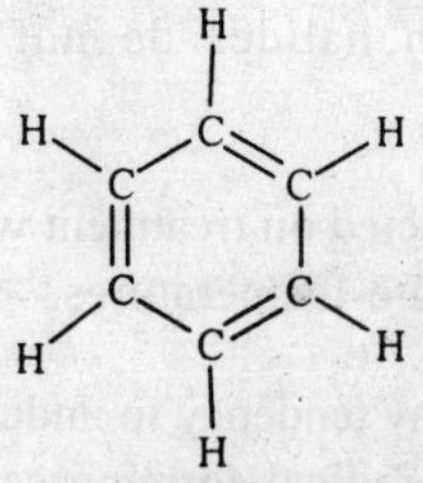

(Structure - I)

Kekule proposed alternate single and double bonds around a six-membered ring. However, it was argued that if this were to be the structure for benzene, then for a disubstituted benzene such as o-xylene there should be two distinguishable isomers, structures III and IV.

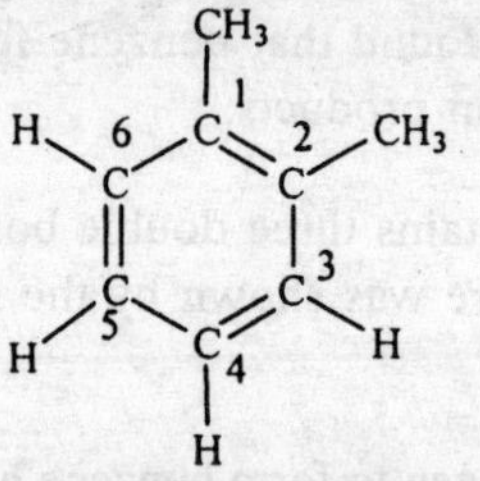

(Structure - III) 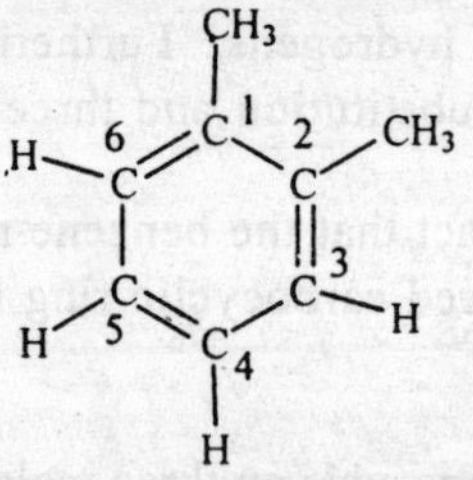 (Structure - IV)

To overcome these objections, Kekule suggested the concept of dynamic structures. He contended that the double bonds in benzene move back and forth, like oscillations, bringing about equilibrium between two structural forms, structures I and II.

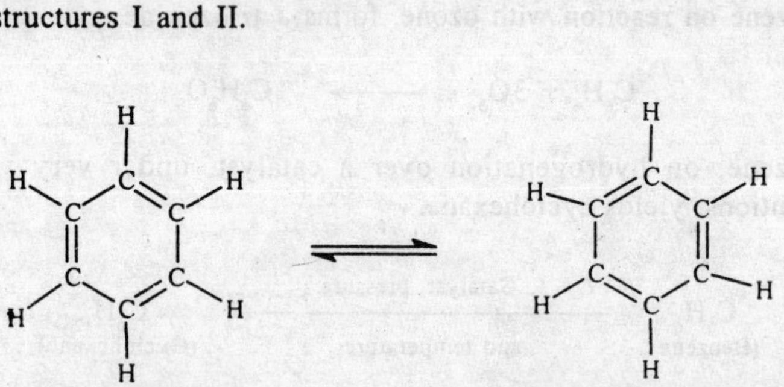

(Structure - I) (Structure - II)

It was very hard for the 19th century chemists to accept the revolutionary suggestions put forward by Kekule and hence over the next 50-60 years, different structures were proposed by several workers which were not accepted for one reason or for another. The following are some of the structures proposed:

$H_3C{-}C{\equiv}C{-}C{\equiv}C{-}CH_3$ (V) $HC{\equiv}C{-}CH{=}CH{-}CH{=}CH_2$ (VI) $H_2C = CH{-}C{\equiv}C{-}CH = CH_2$ (VII)

Structures V to IX were not acceptable for benzene since there was no reason why these should not add on halogens/hydrogen halides at room temperature or why these should not get oxidized with common reagents like $KMnO_4$, HNO_3 etc., at moderate temperatures. Dewar's structure for benzene was also rejected since such a system would respond to reactions

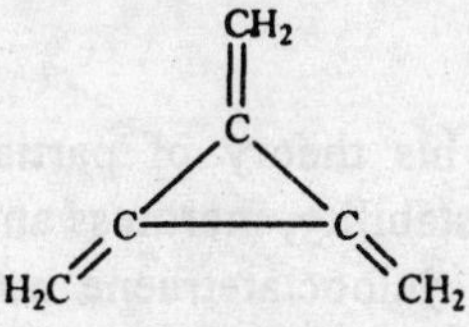

(VIII)

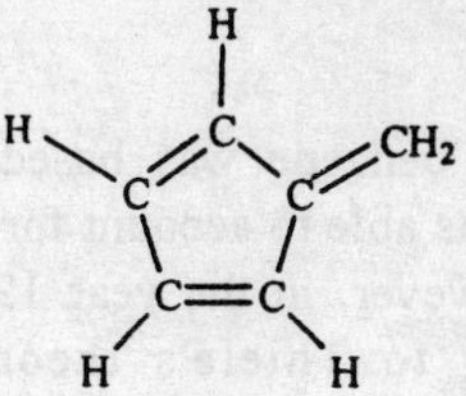

(IX)

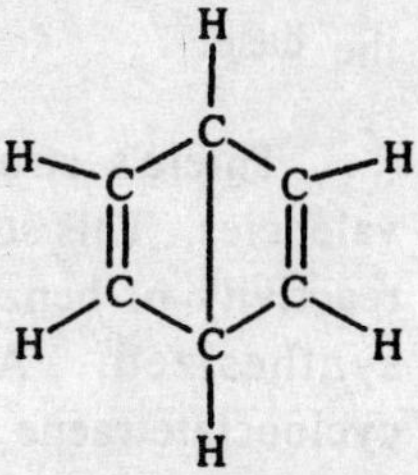

Dewar's structure

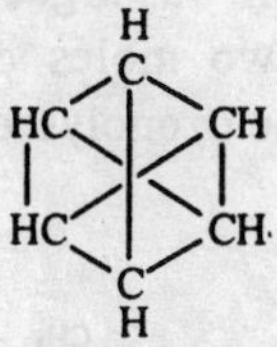

Claus structure (1868)

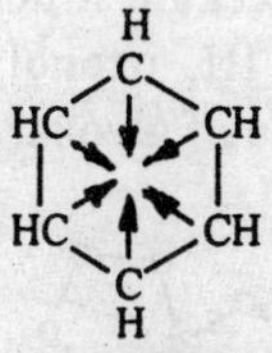

Armstrong-Baeyer's structure (1868)

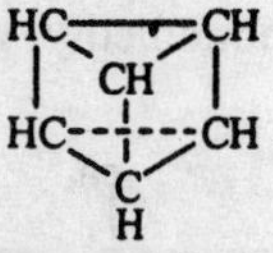

Ladenburg's prism structure (1869)

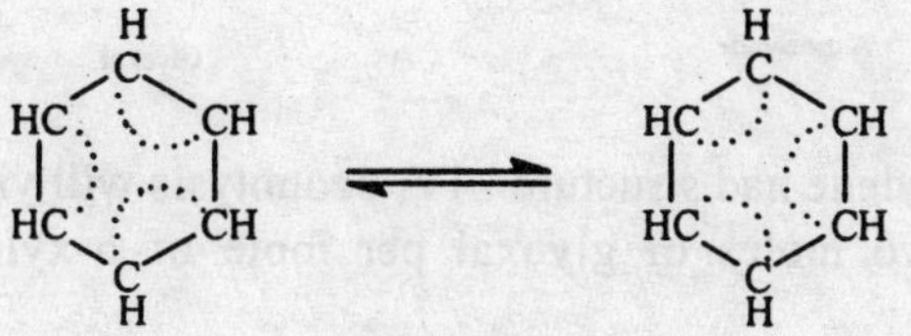

Thiele's structures based on partial valance (1899)

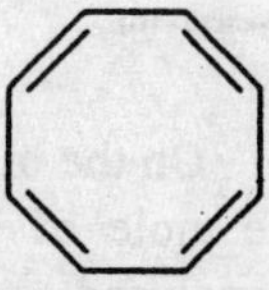

Cyclo Octatetraene

of alkenes. Claus's structure was not accepted on the ground that it did not agree with the number of isomers possible for mono-and di-substituted benzenes.

Armstrong-Baeyer's structure was proposed with the idea that each carbon has its fourth valency directed towards the center of the ring. since the fourth valency of carbon was directed inwards, each carbon was chemically inert, thus accounting for the inertness of benzene. This structure was ruled out on the ground that if the fourth valency was directed inward, it should be independent of other five radial valencies leading benzene to form compounds of the type C_6H_6X, where X is any group.

Ladenburg's prism structure was able to explain very well the inertness of benzene. However, with this structure, a tri-substituted benzene, $C_6H_3X_3$, would be expected to exhibit optical activity which was not found to be the case.

Thiele's structure for benzene was based on his theory of partial valencies. This structure was able to account for the stability, inertness and reactivity of benzene. However, in the year 1909, cyclooctatetraene was synthesized. According to Thiele's theory of partial valencies, cyclooctatetraene should also possess characteristics similar to benzene but in reality it was found to be cyclic, and a highly reactive alkene.

In 1932, Levine and Cole carried out the ozonolysis of o-xylene, which provided strong support for the Kekule structure of benzene. It was argued that if o-xylene had structure III, ozonolysis will yield two moles of monomethyl glyoxal and one mole of glyoxal per mole of xylene ozonolysed.

(Structure - III) A triozonide Glyoxal

On the other hand, if o-xylene had structure - IV, ozonolysis will yield one mole of biacetyl and two moles of glyoxal per mole of o-xylene ozonolysed.

(Structure IV) (A triozonide) Biacetyl Glyoxal

Thus, according to Kekule, if o-xylene was a dynamic equilibrating mixture of equal proportion of two structural forms III and IV, the net result of ozonolysis of o-xylene can be formulated as follows:

$$\text{o-Xylene} \xrightarrow{6\ O_3} H_3C\text{—}\underset{\underset{O}{||}}{C}\text{—}CHO + H_3C\text{—}\underset{\underset{O}{||}}{C}\text{—}\underset{\underset{O}{||}}{C}\text{—}CH_3 + OHC\text{—}CHO$$

Structures III + IV (2 moles)	Methyl glyoxal (2 moles)	Biacetyl (1 mole)	Glyoxal (3) moles

In reality, it was found by Levine and Cole that ozonolysis of o-xylene yields the above products and in the same proportion as seen above.

7-4-2 Objections to Kekule structure

The Kekule structure was able to explain almost all the observed properties of the benzene molecule. The experiment, involving ozonolysis of o-xylene, gave direct evidence in support of the Kekule structure for benzene. However, the concept of oscillating structures or dynamic equilibrating mixtures, which was essential to the Kekule structures could not overcome objections raised by the following two facts regarding behavior of benzene molecule.

(i) Unusual stability of benzene : Although (according to Kekule) the benzene molecule contains three double bonds, it is quite stable and does not normally undergo addition reactions. Consider the hydrogenation of benzene :

The heat of hydrogenation of an alkene is about 28-30 kcal mol^{-1}. When cyclohexene is hydrogenated to cyclohexane, the heat of hydrogenation was found to be 28.6 kcal mol^{-1}. When cyclohexadiene is hydrogenated to cyclohexane, the heat of hydrogenation is found to be 55.4 kcal mol^{-1}, about (28.5 x 2) - 55.4 = 57.2-55.4 = 1.8 Kcal mol^{-1} less than expected. On the other hand, when benzene is hydrogenated to cyclohexane, the heat of hydrogenation is found to be 46.8 kcal mol^{-1}, about (28.6 x 3) - 49.8 = 85.8 - 49.8 = 36 kcal mol^{-1} less than expected. These results are shown in the energy diagram given on the next page:

The results indicate that the benzene molecule is stabler (or its enthalphy/potential energy is less) than hypothetical cyclohexatriene molecule by ≈ 36 kcal $mole^{-1}$. The same inference is arrived at when the heat of combustion data is considered instead of heat of hydrogenation data for the benzene and hypothetical cyclohexatrience molecules respectively.

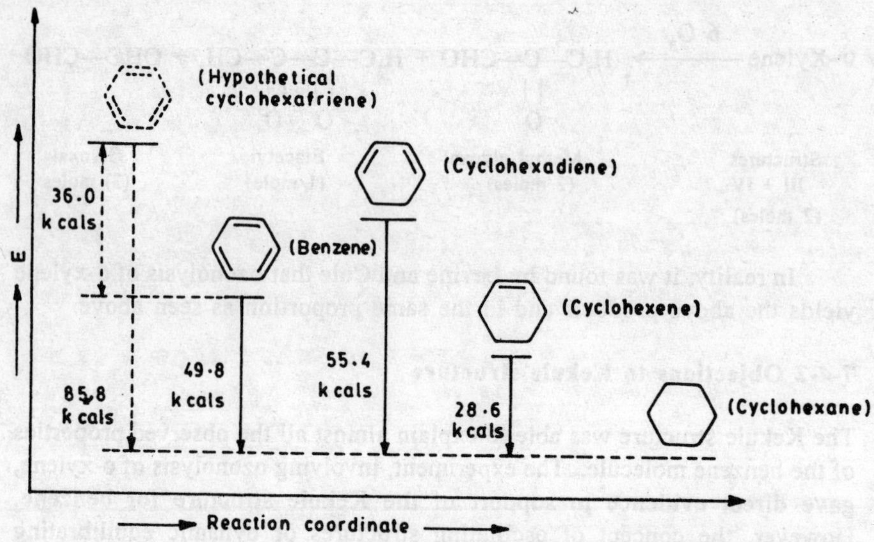

(ii) Equality of all C-C bonds in benzene : According to Kekule, the structure of benzene is a six-membered carbocyclic ring having three C=C double bonds and three C-C single bonds. However, when the bond distances were measured using X-ray diffraction techniques, all the C-C bond lengths in benzene were found to be identical. Moreover, the C-C bond length in benzene (1.40Å) was found to be intermediate between a C=C bond length 1.34Å as in ethylene) and a C-C bond length (1.54Å as in ethane). Further-more, when the force constants were measured using Raman spectroscopy, all the six C-C bonds in benzene had identical force constants and a value intermediate between that for a C=C double bond and that for a C-C single bond. (Force constant is the amount of force required to stretch/bend a given bond in a molecule).

7-4-3 Valence bond/resonance model of benzene

To overcome the objections raised against the Kekule structures, and to account for all the known facts regarding benzene, the valence-bond theory/resonance theory has been suggested. According to this theory, the structure of benzene is considered to be resonance hybrid of the two Kekule structures and several other structures that may be written for it. However, the structures must be interconvertible by single or series of short electron shifts.

(Structure - I) (Structure - II)

Kekule's Structures

The difference in stability or stabilization energy between that of benzene and hypothetical cyclohexatriene is known as resonance energy, 36 kcal mol^{-1}, of the benzene.

It should be noted that whenever it is not possible to write a single structural formula which can explain all the observed properties of the compounds, the concept of ressonance is invoked. *Several, appropriate structures may be written but the molecule will not have any of those structures, rather a weighted composite average of them all. The actual structure of the molecule is a resonance hybrid which cannot be drawn or represented.* However, for writing purposes, any one of those appropriate structures may be used.

The resemblance of the actual structure of benzene to the Kekule forms is known as **Kekule resonance of benzene ring.** This concept of resonance of benzene is in agreement with the ozonolysis experiments. The actual structure of o-xylene is something in between the two Kekule structures having equal chance of reacting in either form. Therefore, half of the time o-xylene reacts as if it had structure - III and the other half as if it has structure IV. The bond distance and force constant data of benzene are in agreement with the resonance concept; i.e., a single structure with all the carbon-carbon bonds being identical. What has been said about o-xylene is applicable to all the derivatives of benzene.

Thus, the benzene molecule is normally represented by a Kekule formula. Either of the two Kekule forms may be used. Usually, the C-H bonds are not shown and only a hexagon is depicted.

(Structure - I) (Structure - II)

7-4-4 Molecular orbital model of benzene

The molecular orbital theory gives a good picture of the structure of benzene. The molecular orbital theory is based on the geometrical description of benzene ring which is derived from X-ray and spectroscopic data. Thus, the benzene molecule is known to be a planar, regular hexagon with all the six carbon atoms lying in the same plane. Each C-C-H or C-C-C bond angle has been found to be 120°, C-C bond length 1.40Å and C-H bond length 1.1Å. Therefore, each carbon of benzene must be in a state of sp^2 hybridization (trigonal hybridization) with three sp^2 hybrid orbitals inclined at an angle of 120° and the fourth p_z orbital in the form of electron-lobe lying perpendicular to sp^2 orbitals. Two of the three sp^2 hybrid orbitals of each carbon overlap with two sp^2 hybrid orbitals of two adjacent carbon atoms forming two carbon-carbon σ-bonds and the third sp^2 hybrid orbital of each carbon overlaps with the 1s orbital of hydrogen to form carbon-hydrogen σ-bonds. Thus, after overlapping of sp^2 orbitals is over, each carbon is left with an electron in the form of a p_z orbital.

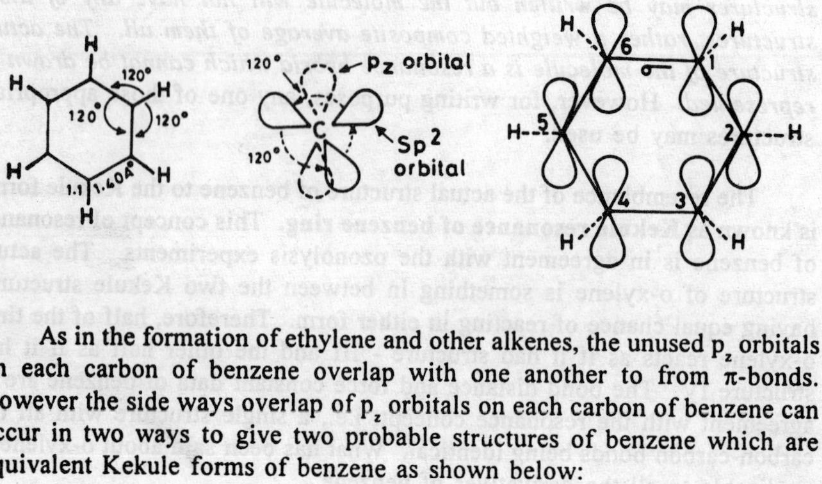

As in the formation of ethylene and other alkenes, the unused p_z orbitals on each carbon of benzene overlap with one another to from π-bonds. However the side ways overlap of p_z orbitals on each carbon of benzene can occur in two ways to give two probable structures of benzene which are equivalent Kekule forms of benzene as shown below:

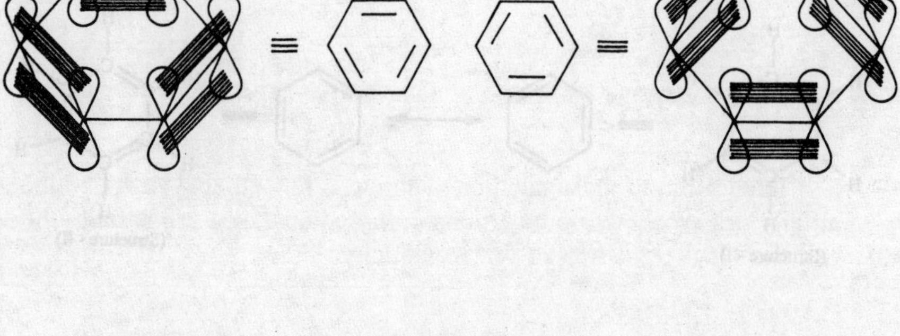

It is obvious from the structures written above that a given p_z orbital can overlap with both adjacent p_z orbitals. Further, since the system is planar, cyclic and equidistant in terms of C-C bond length, all the six p_z orbitals overlap equally and continuously to form a π-molecular orbital. This π molecular orbital is such that it is in the form of cyclic electron clouds, one above and the other below the plane containing all six carbons, they are said to be **delocalized**. Thus, the orbital picture of benzene involves delocalization of six π electrons over a cyclic, planar structure of six carbons.

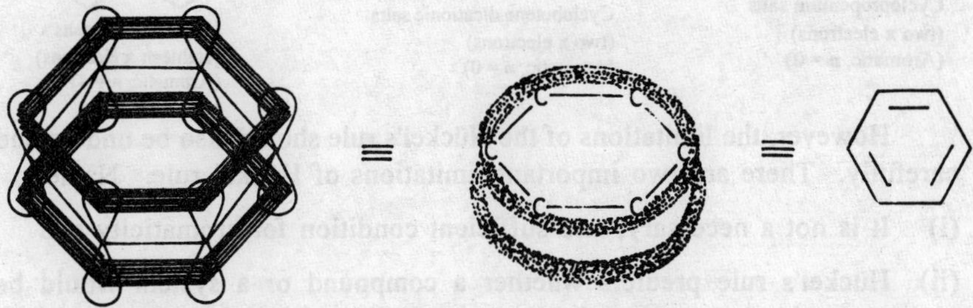

7-5 HUCKEL'S (4n+2)π ELECTRON RULE-CRITERION FOR AROMATICITY

In 1932, the German scientist Hückel made a remarkable contribution to the structural organic chemistry. On the basis of certain molecular orbital calculations, he suggested the criterion for exhibition of aromatic properties by a given ring system which is known as Huckel's (4n+2) π electron rule. According to this rule, a compound or a given system or a cyclic alkene may be regarded as aromatic if and only if it has (4n+2) π electrons, where n can have integral values 1, 2, 3, 4......etc. including zero. Thus according to Hückel's rule benzene is aromatic as it has 4n+2 i.e., 4 X 1 + 2 = 6π electrons. Some other systems which are iso-electronic with benzene, may also be considered aromatic that satisfy the Hückel's rule. Some examples of these systems are given below:

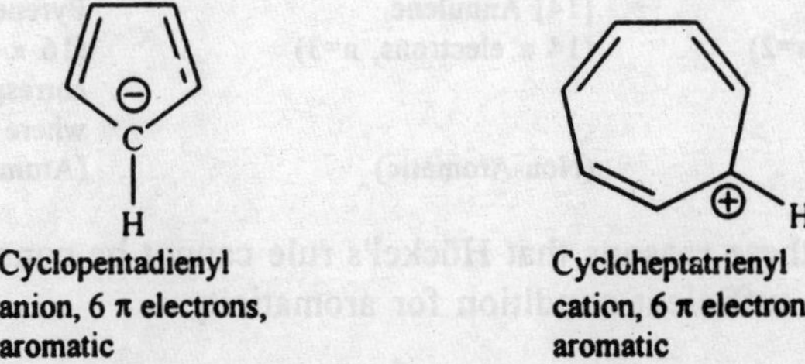

Cyclopentadienyl anion, 6 π electrons, aromatic

Cycloheptatrienyl cation, 6 π electrons, aromatic

In addition to those mentioned above, a few other systems, which were initially not suspected to be aromatic, turned out to be aromatic as they met with the Hückel's criterion for aromaticity.

For example;

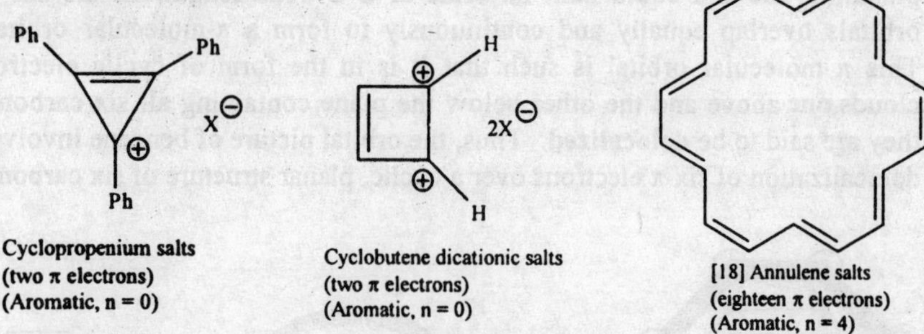

Cyclopropenium salts
(two π electrons)
(Aromatic, n = 0)

Cyclobutene dicationic salts
(two π electrons)
(Aromatic, n = 0)

[18] Annulene salts
(eighteen π electrons)
(Aromatic, n = 4)

However, the limitations of the Hückel's rule should also be understood carefully. There are two important limitations of Hückel rule. Namely;

(i) It is not a necessary, and sufficient condition for aromaticity and

(ii) Hückel's rule predicts whether a compound or a system would be aromatic or not, but it does not indicate the relative degree of aromaticity in isoelectronic systems:

It should be noted that there exist compounds which contain (4n+2)π electrons, yet are not aromatic. For example, [10] annulene, [14] annulene, etc. Further, there exist compounds which do not contain (4n+2) π electrons, yet are aromatic in character, For example, Pyrene.

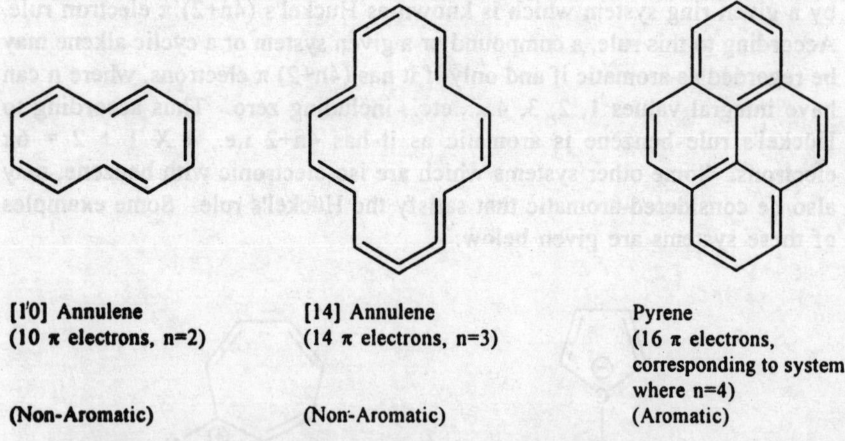

[1'0] Annulene
(10 π electrons, n=2)
(Non-Aromatic)

[14] Annulene
(14 π electrons, n=3)
(Non-Aromatic)

Pyrene
(16 π electrons, corresponding to system where n=4)
(Aromatic)

It is for these reasons that Hückel's rule cannot be considered to be a necessary and sufficient condition for aromaticity.

It should further be noted that Hückel's rule predicts the aromaticity in certain systems which is indeed found to be the case. However, the resonance energy values-which are indicative of the degree of aromaticity in compounds - vary greatly in these systems. For example;

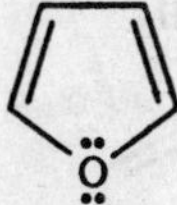

Furan
(Six π-electron system
n=1, aromatic)
(Resonance energy
= 17 kcal mole⁻¹)

Pyrrole
(Six π electrons
n=1, Aromatic)
(Resonance energy
= 21-22 kcal mole⁻¹)

Thiophene
(Six π electrons
n=1, Aromatic)
(Resonance energy
= 26-29 kcal mole)

Thus, Hückel's rule predicts the three isoelectronic heterocyclic systems-furan, pyrrole and thiophene - to be aromatic but does not give any indication about the degree of aromaticity in them.

7-6 NOMENCLATURE OF BENZENE DERIVATIVES

Benzene derivatives are named by adding the name of substituent as prefix to the benzene ring. For example;

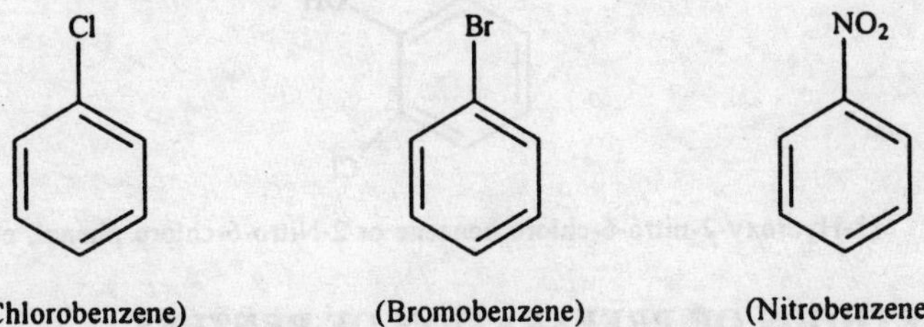

(Chlorobenzene) (Bromobenzene) (Nitrobenzene)

Some derivatives of benzene possess common or trivial names. Because of their usage in chemical literature for a long time, these derivatives are called as shown by their common/trivial names. For example;

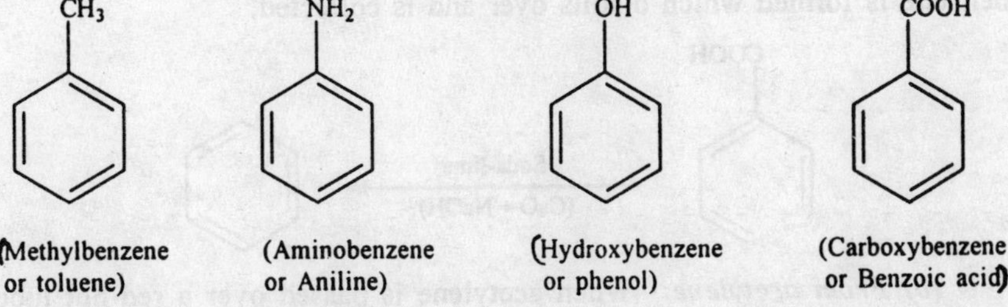

(Methylbenzene or toluene) (Aminobenzene or Aniline) (Hydroxybenzene or phenol) (Carboxybenzene or Benzoic acid)

When there is more than one substituent on the benzene ring, the substituents are numbered with the substituent of highest priority getting the lower number. For disubstituted benzene, only three isomers can exist, and these are called ortho or meta or para depending on whether it is 1,2- or 1, 3- or 1, 4-substitution pattern. For example;

(1, 2-Dimethylbenzene or 1, 2-Xylene ortho-Xylene or o-xylene)

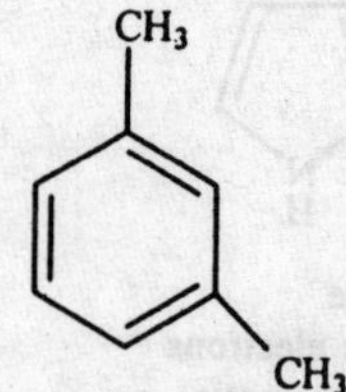

(1, 3-Dimethylbenzene or 1, 3-Xylene, meta-Xylene or m-xylene)

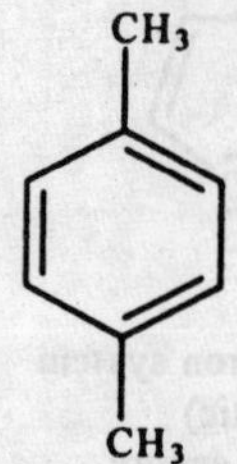

(1, 4-Dimethylbenzene) or 1, 4-Xylene, para - Xylene or p-xylene)

Similarly the substituents in trisubstituted benzenes are numbered and given the names either clubbing with common/trivial name or using a suffix benzene.

For example;

(1-Hydroxy-2-nitro-6-chloro benzene or 2-Nitro-6-chloro phenol, etc.)

7-7 METHODS OF PREPARATION OF BENZENE

The most important sources of benzene are coal-tar and petroleum. However, benzene can be prepared in the laboratory by either of the following two methods:

(a) From benzoic acid: When benzoic acid is fused with soda lime, benzene is formed which distills over and is collected.

COOH —Soda-lime (CaO + NaOH)→ benzene

(b) From acetylene: When acetylene is passed over a red-hot tube, benzene is formed by trimerization, which is collected by cooling the vapors.

HC≡CH, HC≡CH, CH≡CH —Red hot tube→ benzene

7-8 USES OF BENZENE

Benzene is, by far, the most important aromatic chemical used today. Benzene is obtainable in large quantities from natural sources, such as coal-tar, petroleum etc., and the ring hydrogens can be readily substituted by different substituents. Therefore, benzene is the starting material for the largest number of ever-known sequences of chemical reactions producing synthetic intermediates in the laboratory. Other than gasoline which is used as fuel in internal combustion and jet engines, probably benzene has the largest volume use. Huge quantities of benzene are used in the manufacture of phenol, required as raw material for making plastics, rubbers, etc; aniline, required as raw material for making dyes, drugs, pesticides and numerous organic intermediates, etc., nylon, synthetic polymer for fibre-making etc.; detergents, in the form of linear alkyl benzene sulphonates etc., and pesticides as DDT, BHC, etc.,

7-9 CHEMICAL REACTIONS OF BENZENE

7-9-1 Electrophilic substitution

Electrophilic substitution is, by far, the most important chemical reaction undergone by benzene and its derivatives. Despite its high degree of unsaturation, the benzene ring is fairly resistant to hydrogenation, oxidation and addition reactions. It does undergo these reactions, but only under conditions of extreme rigour.

Electrophilic aromatic substitution is a reaction that typifies aromaticity in benzene derivatives and in benzene -like ring systems.Electrophilic aromatic substitution involves successive replacement of benzene ring hydrogens by reactions with an electrophile usually in the presence of a Lewis acid catalyst.

$$C_6H_6 + Y^{\oplus} \longrightarrow C_6H_5Y + H^{\oplus}$$

This reaction has been studied in great detail and the following trend emerges in relation to its mechanism. The benzene ring is an electron-rich system consisting of an electron cloud (due to delocalized six π-electrons) above and below the plane of the ring. Consequently, when an electrophile, say H^+ from a hydrogen halide, approaches the ring, a π-complex is formed which involves partial transfer of electron-density from the ring to the electrophile.

$C_6H_6 + H^{\oplus} \longrightarrow C_6H_6^{\delta\oplus} \cdots \overset{\delta-}{H}$ (π - complex)

The formation of charge-transfer complexes (for example between benzene and iodine) and of picrate derivatives (between picric acid and electron-rich benzene derivatives) is believed to belong to this category. That there is no bond formation (say of σ-type) between benzene and the electrophile (i.e.H^+, picric acid, iodine etc.) is shown by the fact that the two of them can be recovered back quantitatively on quenching the reaction mixture with water. However the existence of a π-complex is proved by spectroscopic data. Further, when a 1:1 mixture of benzene and anhydrous HCl is cooled to -40°C, a solid complex is obtained which decomposes on raising the temperature or on reaction with water.

When the benzene-ring is treated with anhydrous hydrogen halide in the presence of a Lewis acid catalyst like aluminium halide, a σ-complex, also called Wheland intermediate is believed to be formed.

$C_6H_6 + H^{\oplus} \xrightarrow[Cl^{\ominus}]{AlCl_3} C_6H_7^{\oplus}\ AlCl_4^{\ominus}$ (σ - complex)

The σ-complex is a resonance-hybrid of the following contributing structures:

The σ-complex can follow either of the two courses of reaction open to it:

(i) It can add on a chloride ion or

(ii) It can lose a proton to give back benzene.

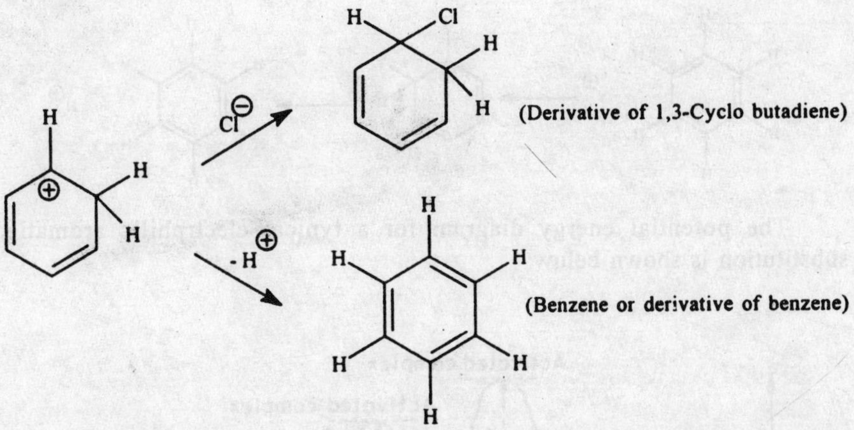

(Derivative of 1,3-Cyclo butadiene)

(Benzene or derivative of benzene)

That, the σ-complex prefers to lose a proton to give back benzene or a derivative of benzene, is proved by the fact that when the reaction is carried out with anhydrous DCl in the presence of aluminium chloride, the benzene ring hydrogens are successively replaced by deuterium.

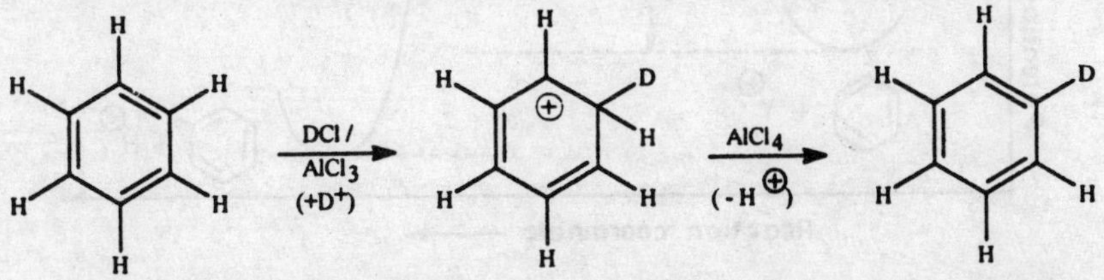

This is due to the fact that a carbon-deuterium bond is much stronger than a carbon-hydrogen bond and so the σ-complex loses hydrogen (as proton) to yield deuterobenzene. Further, the σ-complex prefers to revert back to the benzenoidal structure rather than to the derivative of 1,3-cyclobutadiene, because in the former the aromatic resonance is retained while in the latter a sacrifice of 36kcal mole^{-1} has to be made, which is extremely difficult.

The purpose of the Lewis acid is to help and facilitate the polarization of HCl or DCl. The polarization of HCl or DCl increases the effective electrophilicity of electrphile enabling it to strongly attack the benzene ring leading to π-bond formation.

$$HCl + AlCl_3 \longrightarrow H^+ [AlCl_4]^-$$

Evidence for this type of interaction comes from spectroscopic data and from conductivity measurements on solutions of HCl and $AlCl_3$ in liquid SO_2.

Thus,in general, the reaction of benzene or any of its derivatives with an electrophile (Y^+) can be formulated as follows:

$$C_6H_6 + Y^{\oplus} \longrightarrow [C_6H_6Y]^{\oplus} \longrightarrow C_6H_5Y + H^{\oplus}$$

The potential energy diagram for a typical electrphilic aromatic substitution is shown below:

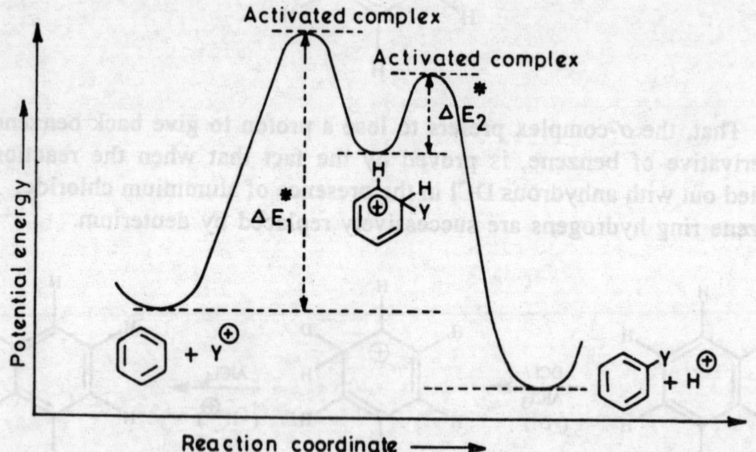

The following are some of the common elcctrphilic susbtitution reactions undergone by benzene and its derivatives. The reactions essentially follow the same mechanism as given above. However the nature of the reagent,nature of the electrophile, its mode of generation etc. vary from reaction to reaction.

(i) Nitration: When benezene is treated with a mixture of concentrated sulphuric and nitric acids at 50-60° for an hour, nitrobenzene is obtained.

$$C_6H_6 \xrightarrow[50 - 60^{0}]{HNO_3 \ + \ H_2SO_4 \ /} C_6H_5NO_2$$

The mechanism of this reaction has been studied in great detail and the main steps in this can be summarized as given below:

(a) The stronger acid, i.e.sulphuric acid, protonates the weaker acid. i.e., nitric acid,to generate a nitronium ion in a fashion similar to any acid-base equilibruium system.

$$H\text{–}O\text{–}SO_2\text{–}OH + H\text{–}O\text{–}NO_2 \longrightarrow HSO_4^- + H_2O^+\text{–}NO_2$$

$$H_2O^+\text{–}NO_2 \longrightarrow H_2O + {}^+NO_2$$

$$H_2SO_4 + H_2O \longrightarrow H_3O^+ + HSO_4^-$$

$$2H_2SO_4 + HNO_3 \longrightarrow H_3O^+ + 2HSO_4^- + NO_2$$

(b) The nitronium ion attacks the benezene ring forming a σ-complex.

$$C_6H_6 + {}^{\oplus}NO_2 \longrightarrow \text{(}\sigma\text{ - complex)}$$

(σ - complex)

(c) The σ-complex loses a proton to form nitrobenzene thereby retaining aromaticity of the ring.

$$\text{(}\sigma\text{ - complex)} \longrightarrow C_6H_5NO_2 + H^{\oplus}$$

(σ - complex)

Evidence for the above mechanism comes from the following facts: Mixtures of solutions of sulphuric and nitric acids indicate the presence of four ions and three species as shown by conductivity and freezing point depression data.That the nitronium ion is the attacking species (electrophic) is shown by the fact that salts such as nitronium acetate (NO_2.OAC).nitronium fluoroborate ($NO_2.BF_4$) etc., act as effcient nitrating agents.Further, the reaction does not show any primary kinetic isotope effect (i.e.if the hydrogens on benzene ring are substituted by deuterium atoms, the rate of nitration does not change). This indicates that the attack by nitronium ion on the benzene ring is the rate-determining step in the reaction. That is,the rate-determing step of the reaction does not involve loss of proton by the σ-complex.

(ii) Sulphonation: In this reaction, a mixture of benzene and sulphuric acid is warmed to 50°C forming benzene sulphonic acid.

$$C_6H_6 + H_2SO_4 \longrightarrow C_6H_5SO_3H$$

(Benzene Sulphonic acid)

The mechanism of this reaction involves the initial decomposition of sulphuric acid to yield sulphur trioxide (SO_3) which attacks the benzene ring

$$H_2SO_4 \rightleftharpoons H_2O + SO_3 \qquad SO_3 \equiv \overset{\ominus}{O}-\overset{++}{S}(=O)-\overset{\ominus}{O}$$

$$H_2O + H_2SO_4 \rightleftharpoons H_3\overset{\oplus}{O} + HS\overset{\ominus}{O_4}$$

$$C_6H_6 + SO_3 \longrightarrow \underset{(\sigma\text{ - complex})}{C_6H_6^{\oplus}SO_3^{\ominus}} \longrightarrow C_6H_5SO_3^{\ominus} + H^{\oplus}$$

$$C_6H_5SO_3^{\ominus} + H^{\oplus} \rightleftharpoons C_6H_5SO_3H$$

in a slow, rate-determing step resulting in the formation of the σ-complex. The latter then undergoes a proton loss in a fast step to give benzene sulphonate, the anion of the benzenesulphonic acid.

Sulphur trioxide is the effective electrophile attacking the benzene ring in sulphonation. This is shown by the fact that reaction can be caried out between benzene and SO_3 under anhydrous conditions leading to a higher yield of benzene sulphonic acid. Further, the use of sulphuric acid enriched with SO_3, known as oleum, in this reaction gives a better yield of the product than the one using conc. sulphuric acid alone. This is due to the fact that the percentage of SO_3 in concentrated sulphuric acid at room temperature is quite low and the decomposition of the acid to SO_3 is also not very high.

(iii) Friedel-Crafts alkylation: Benzene undergoes alkylation with alkylhalides,in the presence of Lewis acid catalyst, to yield alkylbenzenes classically known as arenes.

$$C_6H_6 + R{-}X \xrightarrow{AlX_3} C_6H_5R + HX$$

The First step in this reaction is the polarization of the alkyl halide by the Lewis acid followed by electrophilic attack by the positively polarized carbon of the alkyl halide on the electron-rich benzene molecule leading to σ-complex formation. The later then undergoes a fast proton loss yielding the alkyl benzene.

$$R{-}X + AlCl_3 \longrightarrow \overset{\delta+}{R}{-}{-}{-}X{-}{-}{-}\overset{\delta-}{AlX_3} \quad \text{or} \quad \overset{\oplus}{R}\ \overset{\ominus}{AlX_4}$$

$$C_6H_6 \xrightarrow[\overset{\delta+}{R}---X---\overset{\delta-}{AlX_3}]{+\ R^{\oplus}\ \text{or}} [C_6H_6R]^{\oplus}\ \overset{\ominus}{AlX_4} \longrightarrow C_6H_5R + HX + AlX_3$$

Evidence for the above mechanism comes from the fact that solutions of equimolar mixtures of alkyl halides and Lewis acids are conducting and indicate the presence of two ions on the basis of cryoscopic data. The reaction does not proceed in the absence of the Lewis acid catalyst. The reaction is first order with respect to benzene and first order with respect to the complex. The Lewis acid used in the above reaction is sometimes referred to as *Friedel-Craft's catalyst after its use by these authors in this type of reactions.*

This reaction is disadvantageous from the synthetic point of view since the alkyl benzene first formed reacts further with alkyl halide to yield di-tri- or even poly-substituted benzene. Hence, this reaction should be used in a synthetic sequence only when other methods, such as Friedel-Craft's acylation (given below) do not yield satisfactory results.

(iv) Friedel-Craft's acylation: In this reaction, benzene is treated with an acylhalide in the presence of Friedel-Craft's catalyst (Lewis acid) to yield acyl benzene.

$$C_6H_6 + R-CO-X \xrightarrow{AlX_3} C_6H_5COR + HX$$

$$C_6H_5COR + AlX_3 \rightleftharpoons C_6H_5C(R)=O^{+}-\overset{\ominus}{Al}X_3$$

The course of this reaction is very similar to that in Friedel-Craft's alkylation, except that a large amount of the catalyst (mole per mole of acyl benzene formed) is required. This is due to the fact that the acylbenzene formed binds up the catalyst making it unavailable for any catalysis complexation.

For example,

$$RC(=O)X + AlX_3 \longrightarrow R\overset{\oplus}{C}(X)-O-\overset{\ominus}{Al}X_3$$

$$C_6H_6 + \overset{\oplus}{C}(R)(X)-O\overset{\ominus}{Al}X_3 \longrightarrow [\text{σ-complex}] \longrightarrow C_6H_5COR + \overset{\oplus}{HX}$$

$$C_6H_5C(=O)R \longleftrightarrow C_6H_5\overset{\oplus}{C}(R)-O^{\ominus} \xrightarrow{AlX_3} C_6H_5\overset{\oplus}{C}(R)-O\overset{\ominus}{Al}X_3$$

The advantage of this reaction is that the acylbenzene obtained is less reactive than benzene and hence the reaction apparently stops at monosubstitution stage. The acylbenzene thus obtained may be subjected to *Clemmensen reduction* (Zn –Hg + HCl) yielding alkyl benzene which might otherwise not be accessible through Friedel-Craft's alkylation route.

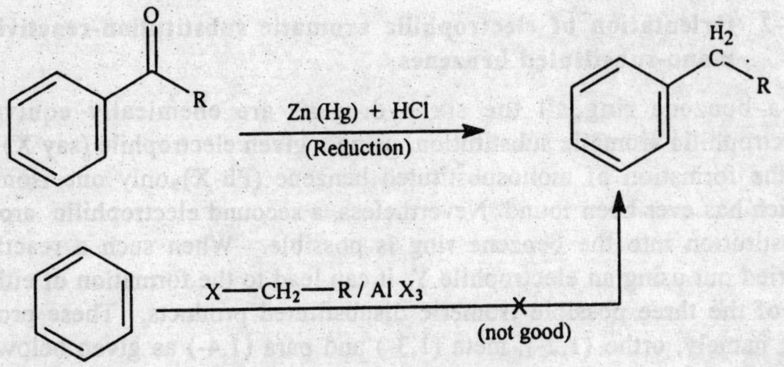

(v) *Halogenation:* Beneze undergoes halogenation on treatment with halogens in the presence of a Lewis-acid catalyst.

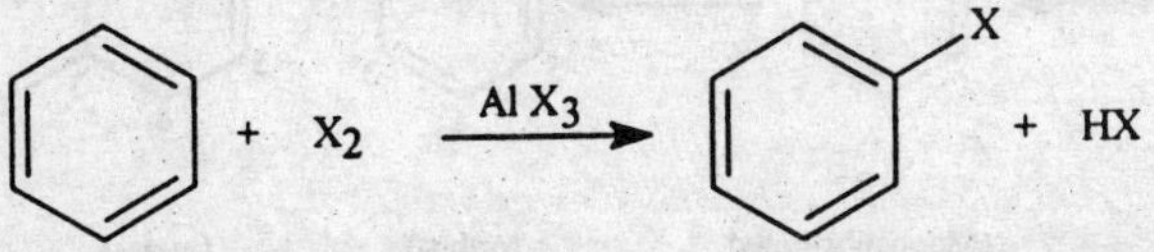

The purpose of the Lewis acid is to facilitate polarization of the halogen-halogen bond. The positively polarized end of the halogen molecule then attacks the benzene ring in a rate determining step to yield the σ-complex. The latter,subsequently, undergoes loss of a proton to yield halobenzene.

$$X{-}X + AlX_3 \longrightarrow \overset{\delta+}{X}{-}{-}{-}X{-}{-}{-}\overset{\delta-}{AlX_3}$$

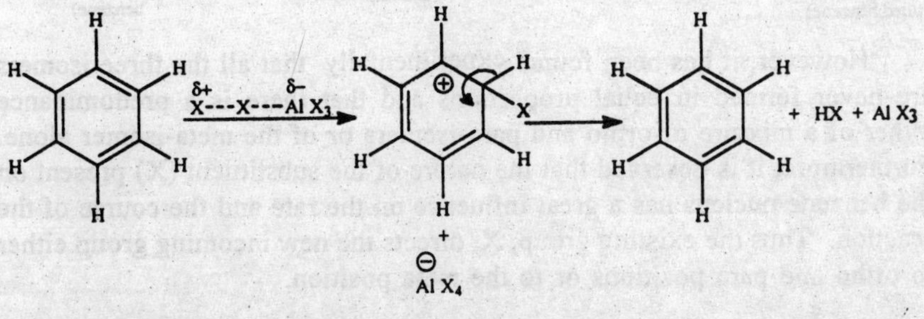

Evidence for the above mechanism is provided by the fact that mixtures of halogens and Lewis acids in anhydrous solvents yield conducting solutions. The reaction is considerably slowed down in the absence of the Lewis acid catalyst. Furthermore, the reaction is first order with respect to benzene and also first order with respect to concentration of the complex. The formation of σ-complex is the rate determining step in this reaction and decomposition of σ-complex to yield halobenzene is a very fast step.

7-9-2 Orientation of electrophilic aromatic substitution-reactivity of mono-substituted benzenes

In a benzene ring, all the six hydrogens are chemically equivalent. Electrophilic aromatic substitution, using a given electrophile (say X) leads to the formation of monosubstituted benzene (Ph-X), only one isomer of which has ever been found. Nevertheless, a secound electrophilic aromatic substitution into the benzene ring is possible. When such a reaction is carried out using an electrophile Y, it can lead to the formation of either or all of the three possible isomeric disubstituted products. These products are, namely, ortho (1,2-), meta (1,3-) and para (1,4-) as given below;

(Benzene) $\xrightarrow{X^{\oplus}}$ (Monosubstituted benzeze) $\xrightarrow{Y^{\oplus}}$ (ortho-) + (meta-) + (para-)

The reaction for disubstitution follows the same mechanism as for the monosubstitution of benzene.

(Monosubstituted benzene) ⟷ $\xrightarrow{Y^{\oplus}}$ (σ-Complex) $\xrightarrow{-H^{\oplus}}$ (Disubstituted-benzene)

However, it has been found experimentally that all the three isomers are never formed in equal proportions and that there is a predominance either of a mixture of ortho and para-isomers or of the meta-isomer alone. Furthermore, it is observed that the nature of the substituent (X) present on the benzene nucleus has a great influence on the rate and the course of the reaction. Thus the existing group, X, directs the new incoming group either to ortho and para positions or to the meta position.

(Toluene) $\xrightarrow{HNO_3 / H_2SO_4 / RT}$ (ortho isomer) (60%) + (para isomer) (35%) + (meta isomer) (≈ 5%)

$C_6H_5NO_2$ $\xrightarrow{HNO_3 / H_2SO_4 / 90^o}$ (meta) + (ortho) + (para) $C_6H_4(NO_2)_2$

(Nitrobenzene) (meta isomer) (≈90%) (ortho isomer) (para isomer)

This influence of the substituent (X) on the benzene ring is known as directive effect of the group or orientation effect. For example, when toluene is subjected to electrophilic substitution,say nitration with $H_2SO_4 + HNO_3$ at room temperature, a predominance of ortho and para products is found whereas nitration of nitrobenzene leads exclusively to m-dintrobenzene with hardly traces of ortho and para isomers.

Based on the above considerations, substituents on the benzene ring have been divided into two categories, namely,

(i)o,p-Directing groups: These are groups which direct an incoming substituent into o- and p-positions. Further, these groups (with the exception of halo substituents) increase the reactivity of benzene towards electrophilic substitution and so they are known as activating groups. Examples of these groups are : —OH, —NH_2, —NR_2, $^-$OR, —SH, —SR, etc.,

(ii) m-directing groups: These are groups which direct an incoming substituent into meta position. Further, these substituents decrease the reactivity of benzene towards electrphilic substitution and so they are also known as deactivating groups. Examples of these groups are : —NO_2, —COOH, —COOR, —CN, —CHO etc.,

7-9-3 Explanation of directing effects on the basis of electronic theory

The orientational effects or the directive influences of the different groups in monosubstituted benzenes towards electrphilic aromatic substitution can be satisfactorily explained on the basis of electronic theory.

It is well-known that the benezene ring is an electron-rich system due to the presence of a delocalized π--electron sextet and so its most important reaction would be with an electrophile. Just like benzene, monosubstituted benzenes may also be expected to undergo substitution of ring hydrogens by electrophiles and the same mechanism. Consequently, σ-complex formation may be considered to be the rate determining step in the reaction of monosubstituted benzenes with electrphiles. There are three alternative positions namely those corresponding to ortho, meta and para substitutions, available for σ-complex formation. The actual point of attack will depend upon the ease of formation (indicated by lower energy of activation for reaching the transition-state) and the stability (indicated by lower potential energy) of the respective σ-complexes which in turn will decide the predominance of that isomer in the reaction. It has been found that substituents present on the benzene ring markedly influence the relative rates and stabilities of the three probable σ-complexes. The effect of the substituents on the formation of σ-complexes is exerted by either electron-release or by electron withdrawal in the following four ways:

(a) Inductive effect (b) Resonance effect

(c) Electromeric effect (d) Hyperconjugation effect.

The influence of these electronic effects, in making a particular group/substituent as o,p-directing type or meta-directing type, can be seen clearly by considering some examples.

(a) Influence of Inductive effect: Substituents of the type $-N^+R_3X^-$, $-CH_2Cl$, $-CHCl_2$,$-CCl_3$, SR_2X^- etc.,have been found to direct an incoming group to meta position on the benzene ring. This is because the substituents exert an **electron-withdrawing inductive effect** on the benzene ring depleting the ortho and para-positions of necessary electron-density. Since electrophilic aromatic substitution involves a preliminary attack by electrophile (Positively charged/electron deficient species) on the benzene ring, attack at ortho or para positions (which are now electron deficient) is resisted leading to meta-substitution as the best alternative.

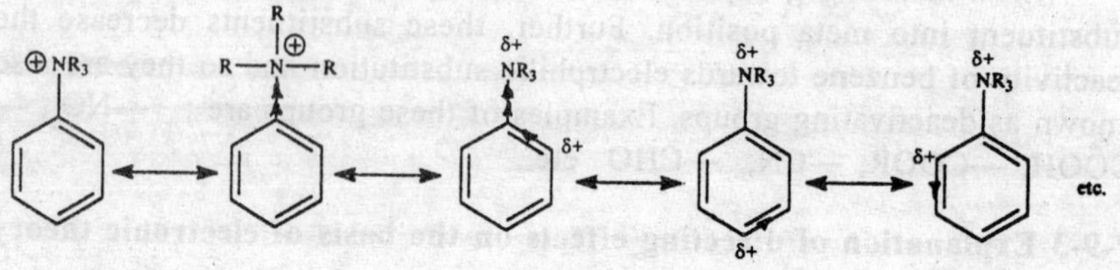

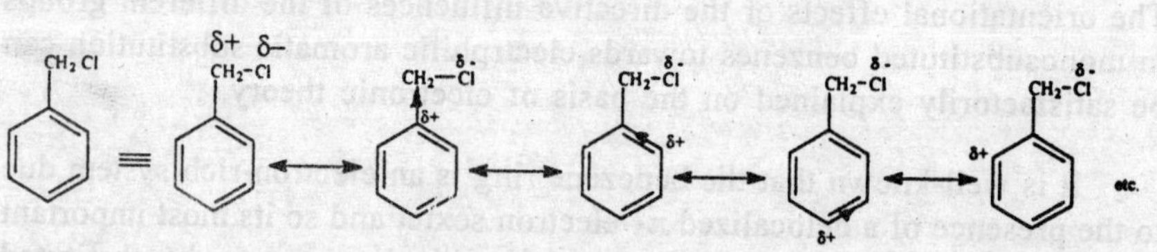

(b) Influence of resonance effect: Substituents like -OH, -OR, $-NH_2$, $-NR_2$, -SR, -SH etc., have been found to be o-,p-directing , whereas those of the type $-NO_2$, -CHO, -COOH, -CN etc., have been found to be meta directing. This is because the former exert an electron-donating resonance effect increasing enormously the electron-density at o- and p- positions facilitating attack by an electrophile at these positions.

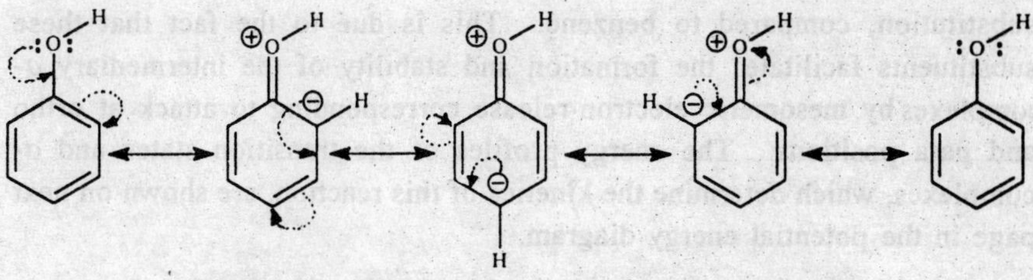

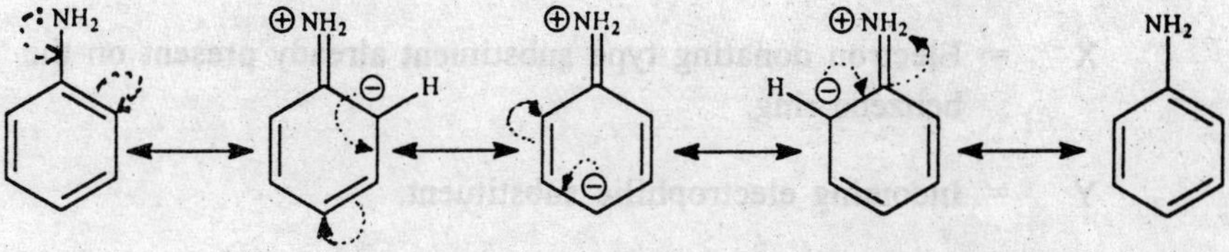

It may be mentioned here that the inductive effect of oxygen or nitrogen,in the above described two cases act in a direction oposite to that of resonance effect.

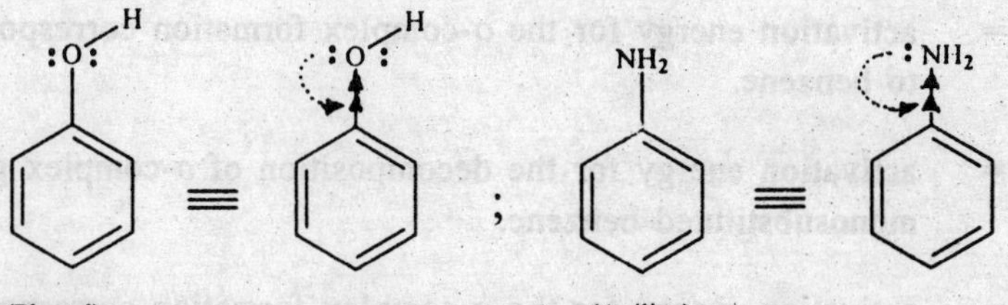

(Phenol) (Aniline)

Nevertheless, the flow of electron density is from the substituent to the benzene ring because the strong,electron-donating resonance effect outweighs the weak, electron withdrawing inductive effect.

Evidence in support of this argument comes from the fact that the dipole moment of phenol is 1.40 D units and is directed such that the positive end of the dipole is towards oxygen, negative end is towards the ring. Similarly, in the case of aniline, the dipole moment is found to be 1.48 D units, the directions of dipole being such that nitrogen is at the positive pole and the rest of the ring is at the negative pole.

The reason why electron-donating resonance effect predominates over electron-withdrawing inductive effect despite the fact that oxygen or nitrogen is a more electronegative element than carbon is because greater stability is attained by the molecule due to generation of several resonance structures.

Furthermore, it has been found that electron-donating substituents increase enormously the reaction rate of the molecule, towards electrophilic substitution, compared to benzene. This is due to the fact that these substituents facilitate the formation and stability of the intermediary σ-complexes by mesomeric electron-release corresponding to attack at ortho and para positions. The energy profiles of the transition states and σ-complexes, which determine the kinetics of this reaction, are shown on next page in the potential energy diagram.

In the potential energy diagram shown on next page ......

X = Electron donating type substituent already present on the benzene ring.

Y = Incoming electrophilic substituent.

It is obvious from the diagram that

$$\Delta E^*_{o_1} \approx \Delta E^*_{op_1} < \Delta E^*_{m_1} < \Delta E^*_{1}$$

where

ΔE^*_1 = activation energy for the σ-complex formation corresponding to benzene.

ΔE^*_2 = activation energy for the decomposition of σ-complex giving monosubstituted benzene.

$\Delta E^*_{o_1}$ = activation energy for the σ-complex formation corresponding to ortho-substitution

$\Delta E^*_{m_1}$ = activation energy for the σ-complex formation corresponding to meta-substitution

$\Delta E^*_{p_1}$ = activation energy for the σ-complex formation corresponding to para-substitution

$\Delta E^*_{o_2}$ = activation energy for the decomposition of σ-complex giving ortho-disubstituted benzene.

$\Delta E^*_{m_2}$ = activation energy for the decomposition of σ-complex giving meta-disubstituted benzene

$\Delta E^*_{p_2}$ = activation energy for the decomposition of σ-complex para-disubstituted benzene.

Since the rates of formation of ortho-, meta-, and para-disubstituted-benzenes and monosubstituted benzene are directly proportional to the rates

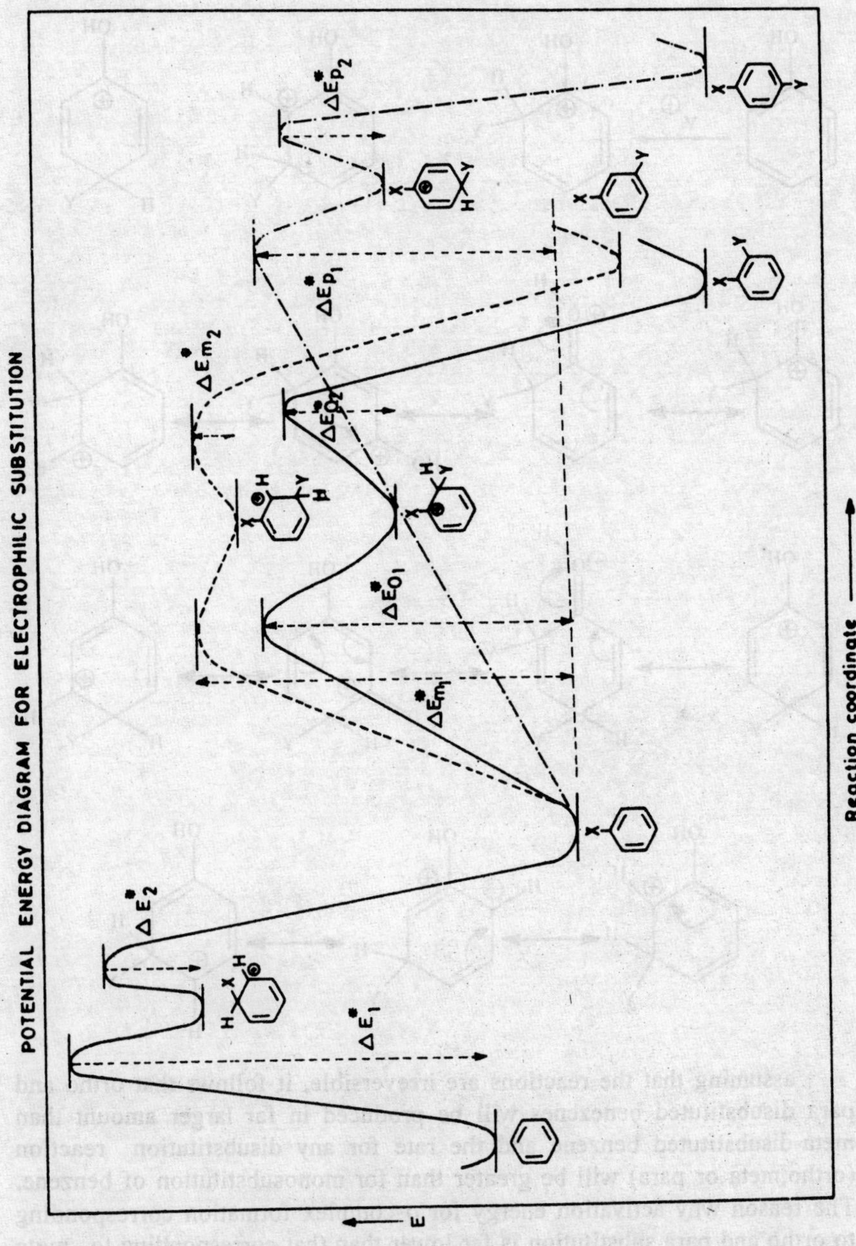

POTENTIAL ENERGY DIAGRAM FOR ELECTROPHILIC SUBSTITUTION
E
Reaction coordinate

of formation of the corresponding σ-complexes which in turn are dependent on the activation energies,

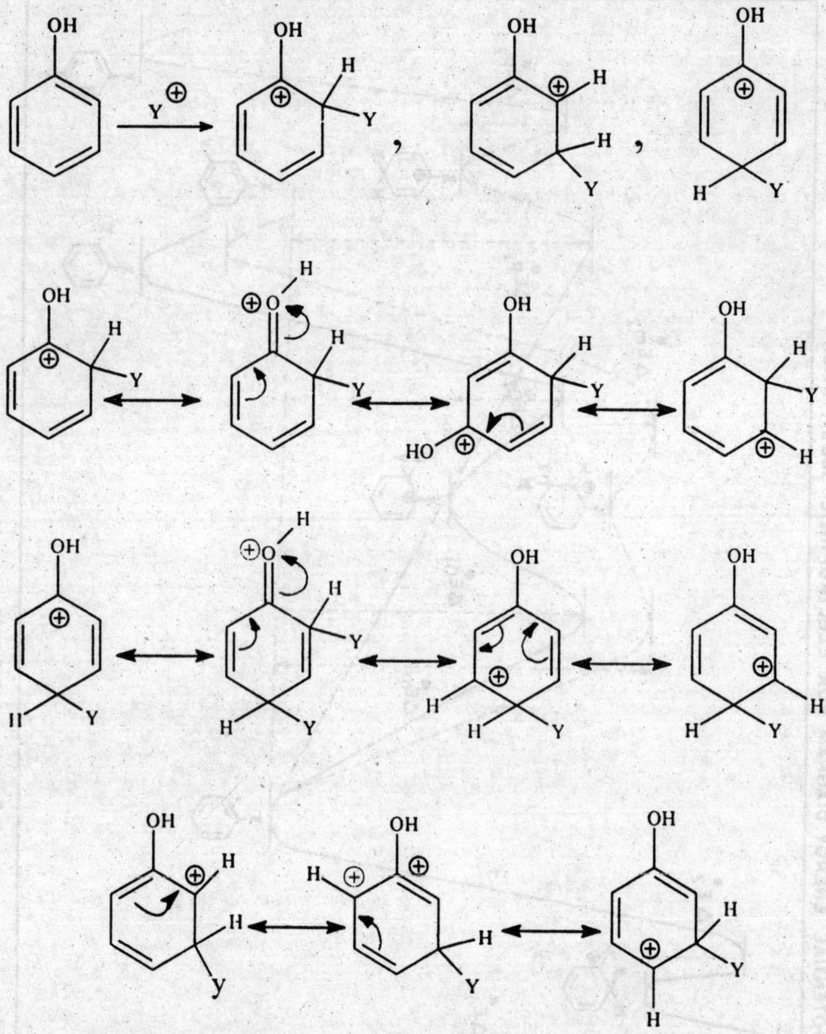

assuming that the reactions are irreversible, it follows that ortho and para disubstituted benezenes will be produced in far larger amount than meta-disubstituted benzene and the rate for any disubstitution reaction (ortho,meta or para) will be greater than for monosubstitution of benzene. The reason why activation energy for σ-complex formation corresponding to ortho and para substitution is far lower than that corresponding to meta substitution is due to **mesomeric electron-release** by the first substituent on benzene. This is shown above with the help of resonance structures taking phenol as an example.

In the case of m-directing groups the substituent exerts a strongly electron-withdrawing resonance effect decreasing electron-density considerably at the o-and p- positions of the benzene ring enabling the electrophile to attack only at m-positions as the available choice. In most cases, the m-directing substituent also exerts, although weakly, an electron-withdrawing inductive effect on the benzene ring. For example;

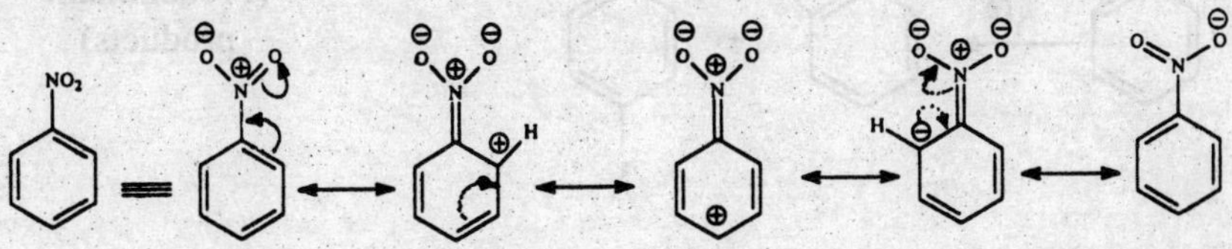

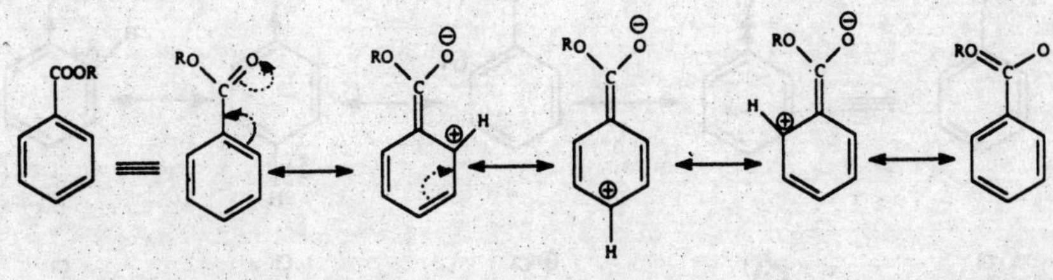

For reasons very similar to those given above, as in the case of o, p-directing groups, a meta-directing group considerably deactivates the benzene ring towards electrphilic substitution. The reactivity of a monosubstituted benzene containing an electron-withdrawing substituent will be far smaller when compared to the parent molecule of benzene. In general, it can be said, that for electrophilic substitution to take place, mild conditions are required for an o-, p-directing monosubstituted benzene and very vigorous conditions for a m-directing monosubstituted benzene. Thus the order of reactivity is:

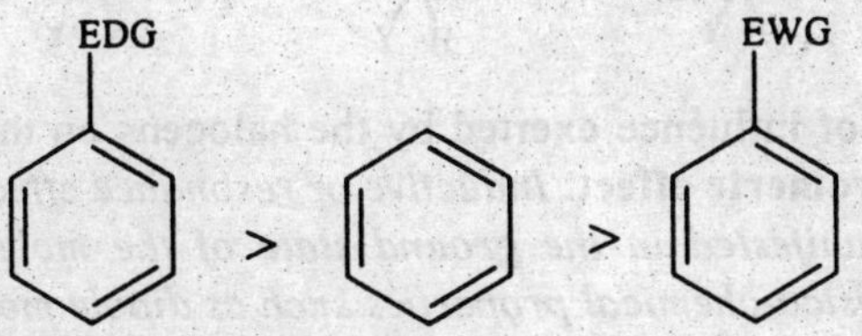

Where EDG and EWG stand for electron-donating and electron-withdrawing groups respectively.

(c) Influence of electromeric effect: Halobenzenes undergo electrophilic substitution predominantly at the ortho and para positions but their reactivity is far less compared to benzene. This is believed to be due to a strongly electron-withdrawing inductive effect exerted by the halogen on the benzene ring resulting in the depletion of electron-density at the ortho- and para-

positions. However, to the extent the halobenzene reacts, it gives o- and p- substitution products due to stabilization of the intermediary σ-complexes by **mesomeric electron-release** involving non-bonding electrons on the halogens. For example:

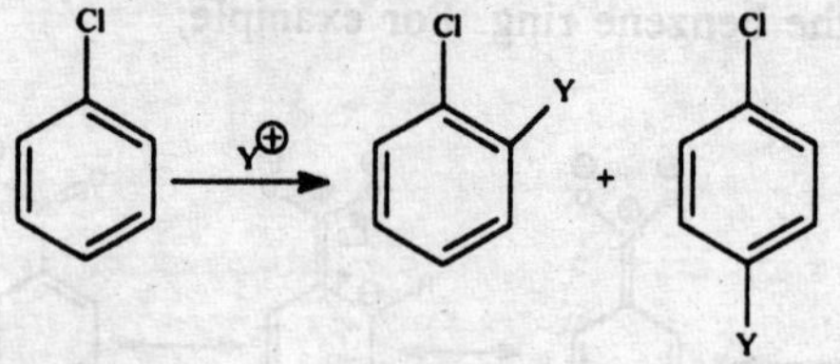

(Predominant products)

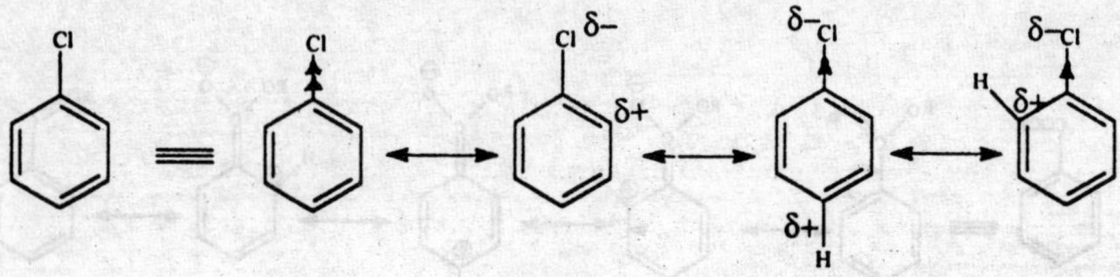

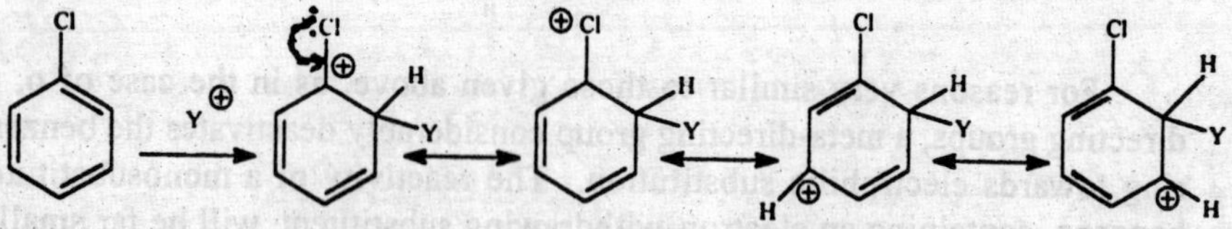

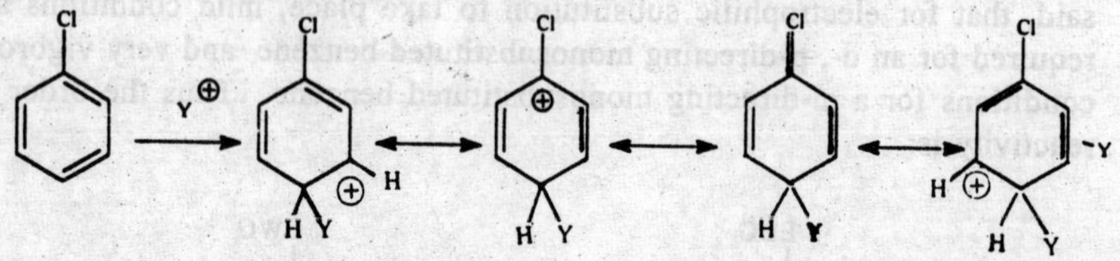

This type of influence exerted by the halogens on the benzene ring is known as **electromeric effect.** *Inductive or resonance effects are permanent polarization manifested in the ground state of the molecule and can be observed in physico-chemical properties such as dipole moment, ultra-violet absorption spectra etc., while electromeric effect (and also hyperconjugation effect) is a dynamic or time variable effect which can be observed only under reaction conditions when the substrate is treated with the reactant.*

(d) Influence of hyperconjugation effect: Alkyl benzenes undergo electrophilic substitution at the ortho and para positions and their reactivity is somewhat higher than benzene. For example, toluene undergoes nitration with conc. HNO_3, plus conc. H_2SO_4 at room temperature to yield a mixture of ortho and para nitrotoluenes at a rate 25 times faster than benzene.

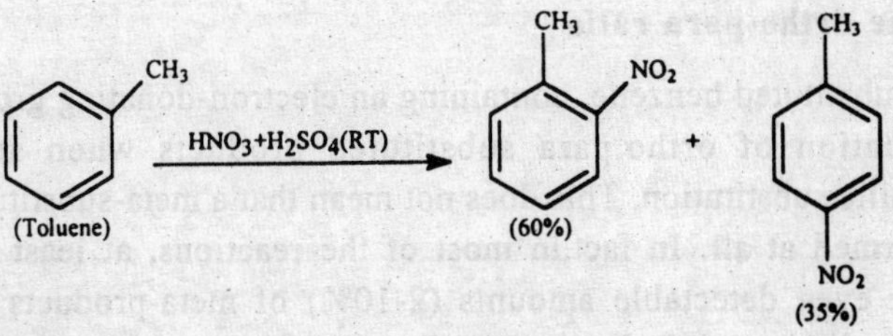

(Toluene) $\xrightarrow{HNO_3+H_2SO_4(RT)}$ (60%) + (35%)

The alkyl groups, being as they are, are not known to exert any electron-donating or electron-withdrawing inductive or resonance effects. However, the activating effect of alkyl group on the benzene ring in directing o, p-substitution and in increasing the rate of electrophilic reaction has been explained by hyperconjugation effect. The latter is a theoretical concept proposed to explain reactivity of alkyl benzenes and there is no experimental evidence in support of it. Like electromeric effect, hyperconjugation is a dynamic effect believed to exist only under conditions of the experiment. Thus, in toluene, the methyl group is presumed to polarize such that hydrogen detaches itself as proton and the resulting carbanion activates the benzene ring especially at ortho and para positions leading to ortho and para substituted products.

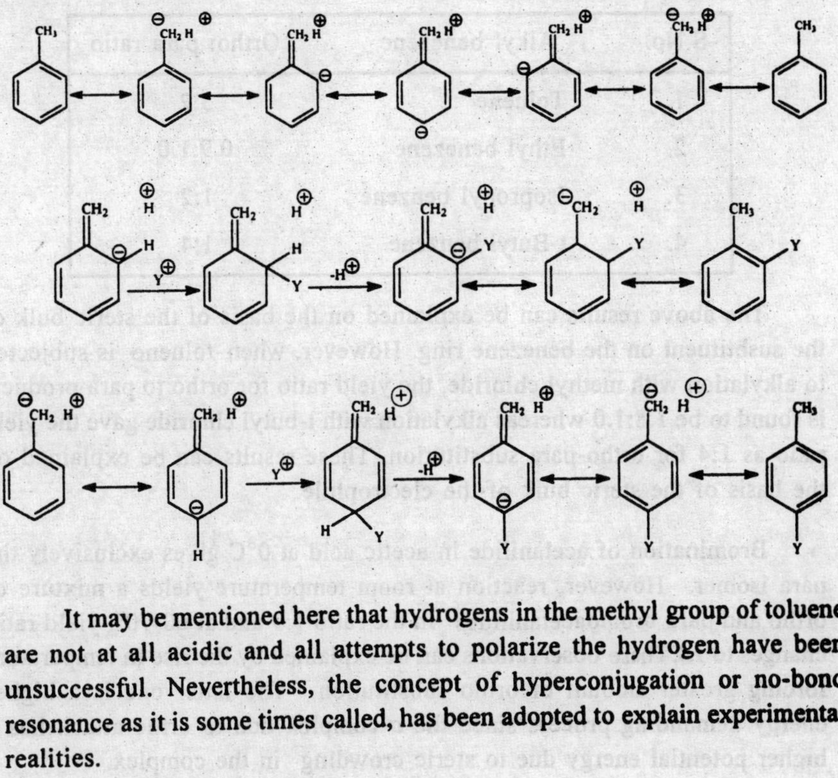

It may be mentioned here that hydrogens in the methyl group of toluene are not at all acidic and all attempts to polarize the hydrogen have been unsuccessful. Nevertheless, the concept of hyperconjugation or no-bond resonance as it is some times called, has been adopted to explain experimental realities.

7-9-4 The ortho-para ratio

A monosubstituted benzene, containing an electron-donating group, leads to the formation of ortho,para substituted products when subjected to electrophilic substitution. This does not mean that a meta-substituted product is not formed at all. In fact,in most of the reactions, at least traces (0.5-2.0%) or even detectable amounts (2-10%) of meta-products are formed which get eliminated during isolation or purification stage. Similarly, when a monosubstituted benzene containing an electron-withdrawing group is subjected to electrophilic substitution, in most cases, at least trace amounts of ortho, para products are formed along with predominating m-isomers.

When a monosubstituted benzene containing o,p-directing group is subjected to electrophilic substitution, the ratio of ortho to para products depends on the nature of the substituents, nature of the electrophile and the reaction conditions. In the absence of strong inductive effects (which generally influence ortho positions), other things being equal, based on probability consideration, the ratio of ortho to para products should be 2:1. However, in most o-,p-directing systems, in the nitration of alkyl benzenes following results were obtained.

S.No.	Alkyl benezene	Ortho: para ratio
1.	Toluene	3:2
2.	Ethyl benezene	0.9:1.0
3.	Isopropyl benzene	1:2
4.	t-Butyl benzene	1:4

The above results can be explained on the basis of the steric bulk of the susbtituent on the benezene ring. However, when toluene is subjected to alkylation with methyl chloride, the yield ratio for ortho to para products is found to be 1.8:1.0 whereas alkylation with t-butyl chloride gave the yield ratio as 1:4 for ortho-para substitution. These results can be explained on the basis of the steric bulk of the electrophile.

Bromination of acetanilide in acetic acid at 0°C gives exclusively the para isomer. However, reaction at room temperature yields a mixture of ortho and para bromoacetanilides in the ratio 1:9 and at 60-70°, yield ratio changes to 1.4. These observations can be explained by the rise in temperature forcing greater amount of ortho substitution. The latter one is a higher energy demanding process since the σ-complex due to ortho-attack has a higher potential energy due to steric crowding in the complex.

7-9-5 Orientation in disubstituted benzenes

When two substituents are present on the benzene ring, orientation of further electrophilic substitution depends on the nature of substituents, position of substituents on the ring, nature of electrophile and reaction conditions. *In general, it can be said that if the two substituents are of similiar nature and if they are appropriately placed, their effects are additive. On the other hand, if the two substituents are of opposite nature, the stronger of the two will direct the course of reaction. The following examples will illustrate these points clearly:*

If the two substituents are not with strong directive effects, mixed products are formed. For example:

p - Chlorotoluene $\xrightarrow{HNO_3/H_2SO_4}$ (60%) + (40%)

(CH_3 : Moderate o,p directing
Cl : Weak o,p - directing)

7-10 ALKYLBENZENES

Arenes is the general term used to denote aromatic hydrocarbons. The most important group of compounds among arenes are the alkyl benzenes, alkenyl benzenes and alkynyl benzenes. Alkyl benzenes are simply, the alkyl

m - Cresol $\xrightarrow{HNO_3/H_2SO_4}$ (Major) + (Minor) + (Almost nil)

(OH : Strong o, p - directing)
CH_3 : Moderate o p - directing)

(Salicylaldehyde) $\xrightarrow{HNO_3}$ (Major) + (Minor)

(OH : Strong o, p - directing)
CHO : Moderate m - directing)

COOH

HNO_3 / H_2SO_4

COOH

NO_2

O_2N NO_2

m - Nitrobenzoic acid

3, 5 - Dinitrobenzoic acid
(sole product)

(COOH : Moderate m - directing
NO_2 : Strong m - directing)

OH Cl

HNO_3 / H_2SO_4

OH Cl NO_2 + OH O_2N Cl (Minor) + OH Cl NO_2 + OH Cl O_2N

(Not detected)

O-Chlorophenol

(OH : Strong o, p - directing)
Cl : Moderate o p - directing)

Cl NO_2

HNO_3 / H_2SO_4

Cl NO_2 NO_2 (Major) + Cl O_2N NO_2 (Minor)

(o - Chloronitro-
benzene)

(Cl: Weak o, p - directing)
(NO_2 : Strong m-directing)

Cl NO_2

HNO_2 / H_2SO_4

Cl O_2N NO_2 (Sole product) + Cl NO_2 NO_2 + Cl NO_2 NO_2 + Cl O_2N NO_2

(Not found)

(m - Chloronitro-
benzene)

(Cl: Weak o, p - directing)
(NO_2 : Strong m-directing)

derivatives of benzene obtained by replacing successively the benzene hydrogens with alkyl groups. Similarly, alkenyl benzenes and alkynyl benzenes are the ethylenic and acetylenic derivatives of benzene. For example,

$CH_2—CH_3$ $CH═CH_2$ $C≡CH$

(Ethylbenzene) (Phenylethylene) (Phenylacetylene)

7-10-1 Nomenclature

The simplest of the alkyl benzenes, that is methylbenzene, is known by its common name toluene. Other alkyl derivatives are named by prefixing the name of the alkyl group to the word benzene. For example:

$CH(CH_3)_2$ CH_2-CH_2-CH_2-CH_3 $C(CH_3)_3$

(i-Propylbenzene) (n-Butylbenzene) (t-Butylbenzene)

Dimethyl benzenes are called by their common name xylene. Three isomers are possible and these are named as ortho,meta or para xylene as the case may be.

CH_3 CH_3 CH_3 CH_3 CH_3 CH_3

(o-Xylene) (m-Xylene) (p-Xylene)

Arenes containing more than one alkyl group at different positions of the benzenes are named by mentioning the individual alkyl groups prefixed by the numbers and finally adding the word benzene as shown below:

CH_3 1 2 3 4 5 6 C_2H_5 C_2H_5 $CH(CH_3)_2$ $CH_2(CH_2)_2CH_3$ CH_3 C_2H_5 CH_2 CH_2 CH_3

(3-Ethyl-methylbenzene or 3-Ethyltoluene) (1-Ethyl-3-isopropyl-4-n-butylbenzene) (1-Methyl-2-ethyl-4-n-propylbenzene)

A compound having a complicated side-chain is named as phenyl derivative of the alkane. For example:

Ph—CH_2—Ph

(Diphenylmethane)

CH_2 - CH - CH_2 - CH -CH_3 with CH_3 CH_3

(1-Phenyl-2,4-dimethylpentane)

7-10-2 Preparation of alkyl benzenes :

Commercially, Coal-tar has been one of the most important sources of alkyl benzenes like toluene, xylenes, cumene (isopropyl benzene) etc., Still large quantities of alkyl benzenes are obtained from petroleum hydrocarbons by a process called catalytic reforming. This process involves heating an aliphatic hydrocarbon (from petroleum fractionation) with a suitable catalyst at high temperature when the aliphatic hydrocarbon undergoes cyclization/ isomerization/aromatization yielding arenes. For example:

(n-Heptane) $\xrightarrow[200-300^{\circ}]{Cr_2O_3/Al_2O_3}$ [ ] $\xrightarrow{-3H_2}$ (Toluene)

(n-Octane) $\xrightarrow[(-H_2)]{Catalyst/\Delta}$ [ ] $\xrightarrow{-3H_2}$ (o-Xylene)

(n-Octane) $\xrightarrow{Catalyst/\Delta}$ [ ] $\xrightarrow{-3H_2}$ (Ethylbenzene)

Synthetically, alkyl benzenes are prepared by the **Friedel-Crafts reaction** and by the **Wurtz-Fittig reaction.** These reactions are described below in brief.

1. Friedel-Crafts reaction: As discussed already this reaction is of two types, namely,

(a) Friedl-Crafts alkylation: In this method, benezene or its derivative is treated with an alkyl halide in the presence of a Lewis acid catalyst to yield the alkylated benzene. For example:

$$C_6H_6 \xrightarrow{R\text{-}X/AlX_3/\Delta} C_6H_5R$$

The main disadvantage of this reaction is that the product, an alkylbenzene, by virtue of its higher reactivity towards electrophilic aromatic substitution (alkyl groups are o, p-directing and ring activating type), undergoes further alkylation giving multiple substituted products and it competes very effectively with benzene for the alkyl halide.

$$C_6H_5R \xrightarrow[AlX_3/\Delta]{R\text{-}X/} C_6H_4R_2 \xrightarrow[AlX_3/\Delta]{R\text{-}X/} C_6H_3R_3 \quad \text{etc.}$$

The result is that a mixture of differently alkylated product is obtained in a given reaction.

$$C_6H_6 \xrightarrow[AlX_3/\Delta]{R\text{-}X/} C_6H_5R + C_6H_4R_2 + C_6H_4R_2 + C_6H_3R_3 \quad \text{etc.}$$

Such a reaction is not considered to be disadvantageous in industry, since there exist efficient fractionating columns and the components of the mixture can be easily fractionated out by distillation.

(b) Friedel-Crafts acylation: In this method,benzene or derivative of benzene is treated with an alkyl halide in the presence of a Lewis acid catalyst to yield an acylated benzene. For example;

$$C_6H_5R \xrightarrow{R^1\text{-}CO\text{-}X/AlX_3/\Delta} R\text{-}C_6H_4\text{-}CO\text{-}R^1$$

The reaction apparently comes to a halt after acylation of benzene derivatives since the acylated compound is at once bound up by the Lewis acid catalyst. The bound derivative has deactivating substituent on the benzene ring which presents further acylation on the ring as Friedel-Crafts reaction is a relatively weak form of electrophilic substitution.

The complex can be decomposed (by quenching the reaction mixture with water) and the acylated compound subsequently reduced to yield alkylated benzene derivative.

This method of preparation of alkyl benzenes is very good and it can be adopted very easily at the laboratory scale. The only disadvantage is that it is a two-stage process when compared to the previous one which is a single stage process.

2. Wurtz-Fittig reaction: In this reaction, an alkyl benzene is obtained by treating a mixture of alkyl halide and aryl halide with sodium metal. For example;

$$C_6H_5X \text{ (Halobenzene)} + CH_3CH_2CH_2CH_2\text{-}X + Na \longrightarrow C_6H_5CH_2CH_2CH_2CH_3 + 2NaX$$

The reaction probably follows a free-radical mechanism as given below:

$$Na + C_6H_5X \longrightarrow \overset{\oplus}{Na} + [C_6H_5X]^{\overline{\cdot}}$$

$$[C_6H_5X]^{\overline{\cdot}} \longrightarrow C_6H_5^{\cdot} + X^{\ominus}$$

$$C_6H_6 + CH_3\text{-}CH_2\text{-}CH_2\text{-}CH_2\text{-}X \longrightarrow C_6H_5CH_2CH_2CH_2CH_3 + \dot{X}$$

$$X + Na \longrightarrow \overset{\oplus}{Na}\overset{\ominus}{X}$$

The advantage in the above reaction is that an alkyl benzene of known structrure is synthesized. However,the disadvantage is that diphenyl and an alkene are the side products formed in this reaction.

7-10-3 Properties of alkyl benzenes :

The alkyl benzenes are colorless liquids with a characteristic odour. Many are less dense than water and highly immiscible with it. Being hydrocarbons, they are essentially non-asociated substances. Therefore, the alkyl benzenes have normal melting and boiling points. Since they are non-polar, the alkyl benzenes show a tendency to dissolve in non-polar solvents. In the chromatographic behavior too, the alkyl benzenes do not get absorbed very strongly on the adsorbents and so get eluted with relatively non-polar elluents.

In their chemical reactions, the alkyl benzenes show the properties of both the aliphatic alkyl part and the aromatic benzene part. Thus, the alkyl benzenes undergo free-radical substitution in the aliphatic side chain and electrophilic substitution in the benzene ring. For example:

$$\underset{\text{(Toluene)}}{C_6H_5CH_3} \xrightarrow[h\nu]{Cl_2/\Delta} \underset{\text{(Benzyl chloride)}}{C_6H_5CH_2Cl} \xrightarrow[h\nu]{Cl_2/\Delta} \underset{\text{(Benzal chloride)}}{C_6H_5CHCl_2} \xrightarrow[h\nu]{Cl_2/\Delta} \underset{\text{(Benzotri chloride)}}{C_6H_5CCl_3}$$

In the halogenation of aliphatic side chain by the free radical process,alkyl benzenes react faster than aliphatic hydrocarbons and require relatively milder conditions. This is due to the stabilization of the intermediary benzyl radicals. Thus, the mechanism can be formulated as given below:

$$Cl_2 \xrightarrow{\Delta / h\nu} \dot{Cl} + \dot{Cl} \text{ (Initiation step)}$$

$$C_6H_5CH_3 + \dot{Cl} \longrightarrow C_6H_5\dot{C}H_2 + HCl$$

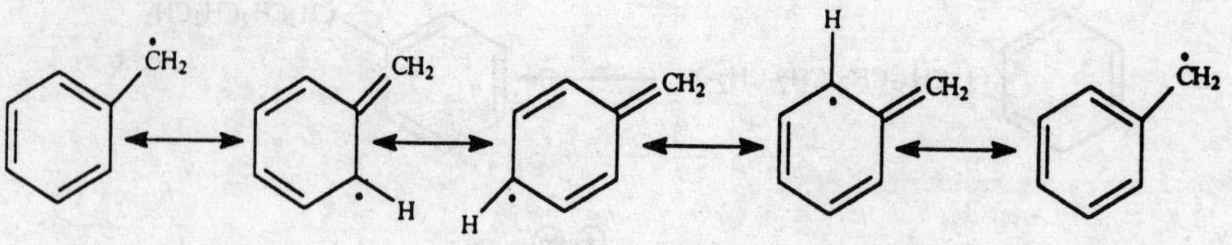

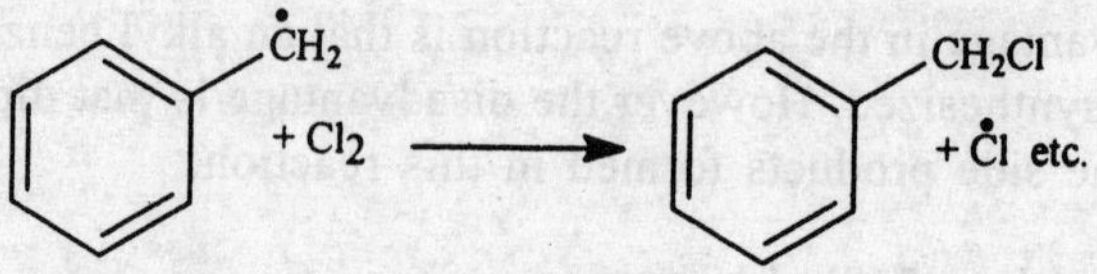

Since the alkyl group is an o-, p-directing or mild ring activating type, alkyl benzenes react faster than benzene towards electrophilic substitutions.

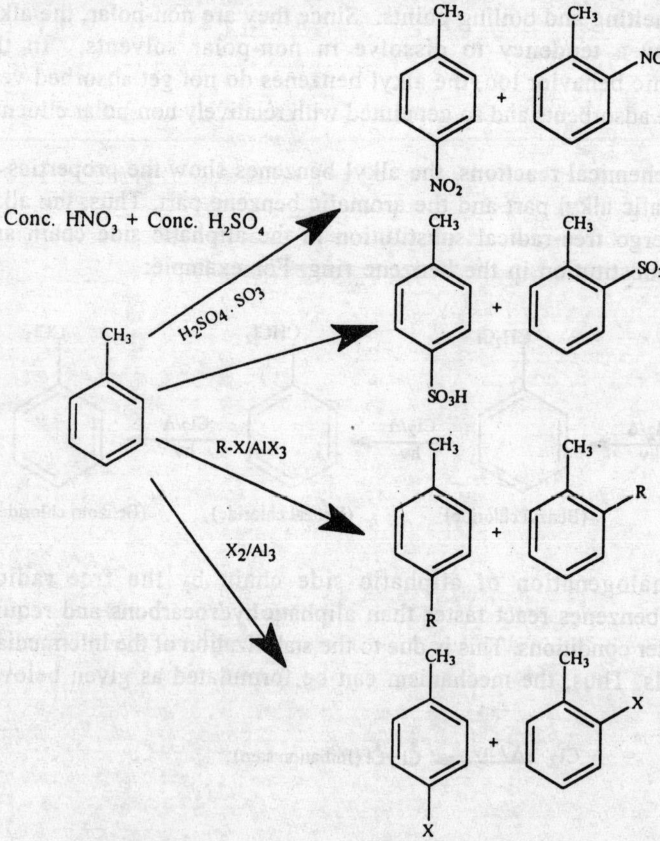

Alkyl benzenes undergo oxidation in the aliphaic side chain, α-carbon to the benzene ring getting preferentially oxidized giving a derivative of benzoic acid. For example;

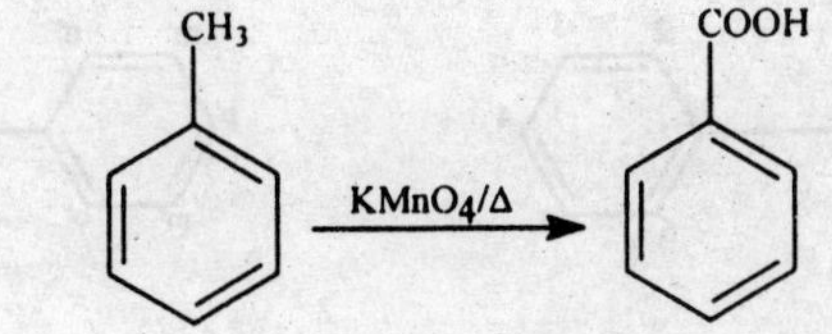

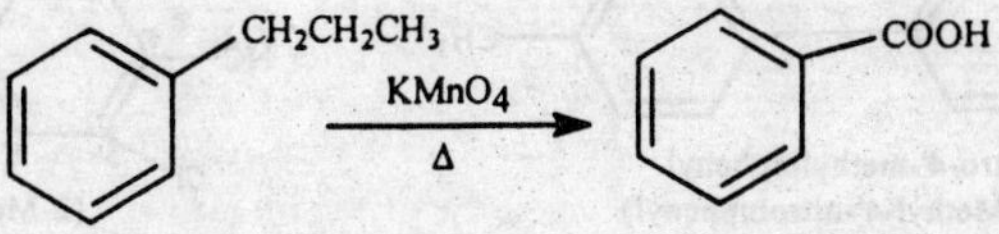

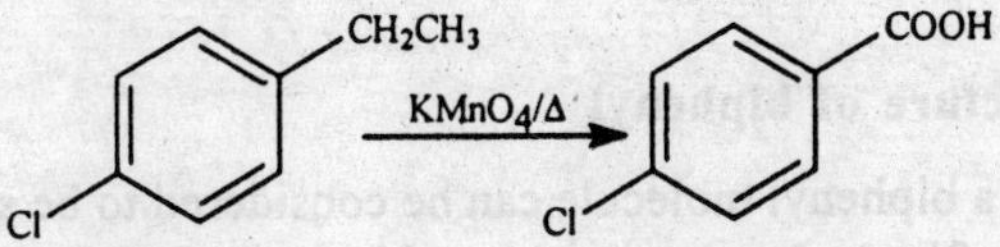

(p-Chloroethyl-benzene) (p-Chlorobenzoic acid)

7-10-3 Uses of alkyl benzenes

The alkyl benzenes serve as important starting materials for the manufacture of several crucial raw materials, which in turn are used in the industry for making useful commerical products. For example, o-xylene is used for making phthalic anhydride required in the manufacture of polymers, plastics and resins. Similarly, p-xylene is used for the preparation of terephthalic acid required in the manufacture of synthetic fibers like viscose rayon. Linear Alkyl Benzenes (LAB) are used in making soaps, detergents, shampoos and other consumer products. Toluene is used in the manufacture of trinitrotoluence (TNT) which is the standard military explosive. Several alkyl benzenes find application as starting materials for making other synthetic intermediates used in the manufacture of dyes, drugs, pesticides, pharmaceuticals, etc., For example, isobutyl benzene is used in the manufacture of well-known pain-relieving drug ibuprofen.

7-11 BIPHENYLS

Biphenyls can be considered to be formally derived from benzene or derivatives of benzene by replacing one of the hydrogens with an alkyl group. For example :

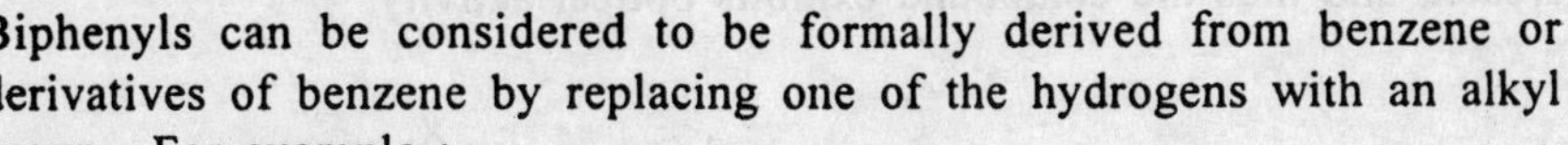

(Biphenyl) (4-Methylbiphenyl) (Dinitrobiphenyl)

For nomenclature purposes,substituents in both the rings are numbered and the compound is named as derivative of biphenyl after mentioning the substituents. For example:

(4-Nitro-4'-methylbiphenyl
or 4-Methyl-4'-nitrobiphenyl)

(2-Methyl-5-nitro-3'-
chloro-4'-cyanobiphenyl

7-11-1 Structure of biphenyl

Structurally, a biphenyl molecule can be considered to be substituted benzene wherein one of its hydrogens is replaced by an aryl ring. The parent biphenyl molecule is essentially planar. It might be expected that the two phenyl rings exist independent of each other and there may be free-rotaion across C_1-C'_1 (Pivot) single bond. However the latter is not bound to be the case with biphenyle since there is considereable amount of double bond character across C_1-C'_1 due to resonance.

etc.

However, when 2 bulky ortho substituents are present in each of the rings, the substituents force the phenyl rings to take up perpendicular orientation. This perpendicular orientation of the rings gives rise to disymmety in the molecule wherein a nonsuperimposable mirror image is created and thus the compound exhibits optical activity.

(Biphenyl molecule-
completely planar)

(Tetrasubstituted biphenyl-
perpendicular orientation of the
two phenyl rings)

(Mirror)

7-11-2 Properties of biphenyls

The parent biphenyl is a neat, colourless crystalline solid, m.p. 71°C. Other biphenyls are also colourless, crystalline solids. The biphenyls are nonassociated substances and so their physical constants, like melting points, boiling points etc., are as expected on the basis of their molecular weights. The biphenyls are highly insoluble in water but readily dissolve in ether, Chloroform, acetone, benzene etc.

In electrophilic substitution, a biphenyl is slightly more reactive when compared to benzene and directs the incoming substituents into ortho or para positions of the other phenyl ring.

(Biphenyl) $\xrightarrow[H_2SO_4]{HNO_3}$ (2-Nitrobiphenyl) + (4-Nitrobiphenyl)

The orientation of substitution is ortho or para in the other phenyl ring because a phenyl ring stabilizes the σ-complex (corresponding to ortho or para attack in the other ring by mesomeric electron-release.

(etc.)

However,in the case of substitued biphenyl, same rules of orientation are followed for mono-or di-substituted benzenes. For example, electrophilic substitution, of 4-methyl-4-nitrobiphenyl with HNO_3/H_2SO_4 at 30-40°C leads to a product having electrophile in the methyl group containing phenyl ring.

H_3C–C_6H_4–C_6H_4–NO_2 $\xrightarrow[H_2SO_4]{HNO_3}$ H_3C–$C_6H_3(NO_2)$–C_6H_4–NO_2

7-11-3 Preparation of biphenyls

There are a number of methods for the prepartion of biphenyls Following are some of them:

(a) Fittig reaction: This reaction is similar to Wurtz reaction except that an aryl halide is used in place of alkyl halide. For example;

$$R\text{-}C_6H_4\text{-}X + 2Na + X\text{-}C_6H_4\text{-}R \longrightarrow R\text{-}C_6H_4\text{-}C_6H_4\text{-}R + 2NaX$$

This method is useful only for preparation of symmetrical biphenyls.

(b) Ullmann reaction : In this method, an aryl iodide is treated with powdered copper to yield a biphenyl.

$$R\text{-}C_6H_4\text{-}I + Cu + I\text{-}C_6H_4\text{-}R \longrightarrow R\text{-}C_6H_4\text{-}C_6H_4\text{-}R + CuI_2$$

The reaction has limited scope because of non-ready availability of aryl iodides.

(c) Coupling with diazonium salts:

$$R\text{-}C_6H_4\text{-}NH_2 \longrightarrow R\text{-}C_6H_4\text{-}N_2^{\oplus} \xrightarrow[(-N_2)\ (-H^+)]{C_6H_3R^1_3} R\text{-}C_6H_4\text{-}C_6H_2R^1_3$$

7-11-4 Uses of biphenyl

The parent biphenyl when mixed with diphenyl ether is known as dowtherm oil.The oil is used extensively in industry as a heat-transfer agent.Dowtherm oil is also used in organic synthesis as an inert, high boiling liquid medium especially for pyrolytic rections, cyclization reactions etc., Chloro derivatives of biphenyl are used as transformer oil. Benzidine (4,4-diamonobiphenyl) is used as an intermediate in dye manufacture.

7-12 FUSED RING SYSTEMS OF BENZENE, NAPTHALENE

Naphthalene is obtained from coal-tar by fractionation. It is also obtained by heating alkylbenzenes from petroleum fractionation over a metal-oxide catalyst.

(2-n-Propyltoluene) → Metal oxide/Δ (-H_2) → [(Tetrahydronaphthalene)] → -2H_2 → (Naphthalene)

7-12-1 Structure

The empirical molecular formula of naphthalene was found to be $C_{10}H_8$. In its chemical reactions, naphthalene resembled benzene. Thus, naphthalene did not undergo the usual chemical reactions of unsaturated compounds and instead showed a tendency to undergo characteristic electrophilic aromatic substitution. However, complete hydrogention of naphthalene under vigorous conditions yielded decahydronaphthalene ($C_{10}H_{18}$), or decalin, a compound of proven structure.

(Naphthalene) ($C_{10}H_8$) → 5H_2 (vigorous) → Decahydronaphthalene or decalin ($C_{10}H_{18}$)

Further, oxidation of naphthalene with CrO_3 yielded phthalic acid, which was also obtained by oxidation of o-xylene.

(Naphthalene) ($C_{10}H_8$) → CrO_3 → (Phthalic acid) ($C_8H_6O_4$) ← o-Xylene (C_8H_{10})

Based on the above considerations, Erlenmeyer proposed the structure of naphthalene as shown below;

The above structure for naphthalene was confirmed by Graebe on the basis of oxidation experiments as follows:

(i) Naphthalene on nitration with conc. HNO_3/H_2SO_4 gave 3-nitrophthalic acid.

(ii) Reduction of the nitronaphthalene with H_2 catalyst yielded aminophthalene and the latter on oxidation with CrO_3 gave phthalic acid.

It is well-known that a nitro group, being an electron-withdrawing group, stabilizes the benzene ring towards oxidation and an amino group, being an electron-donating group, destabilises the benzene ring towards oxidation. Therefore, it was inferred that naphthalene molecule must consist of two benzene rings and the reactions formulated as given below:

NO_2 NH_2 HNO_3/H_2SO_4 H_2/cat [O] COOH COOH (Nitronaphthalene) (Phthalic acid) NO_2 HOOC CrO_3 HOOC (3-Nitrophthalic acid)

7-12-2 Resonance and bond-fixation in naphthalene

The resonance energy of naphthalene, as determined by heats of hydrogenation data was found to be 61 kcals mol^{-1}. Since the naphthalene molecule is formally derived by fusion of two benzene rings in the ortho positions, the resonance energy or stability of naphthalene system is less than expected by $(2 \times 36 - 61) = 11$ kcal mol^{-1}. In other words, if the resonance energy of one of the benzene rings is approximately taken to be 36 kcal mol^{-1}, that of the other is $61-36 = 25$ kcal mol^{-1}. This lower resonance energy value should make the naphthalene ring system more reactive and less stable than the usual benzene ring system. That is, lower value of resonance energy is indicative of less effective delocalisation, rather higher alkenyl character. The resonance structures of naphthalene molecule in its ground state can be written down as follows:

An examination of the resonance structures shows that if one of the benzene rings (symbolized by full kekule form) is assigned the full value of resonance energy (36 kcal mol^{-1}), the other benzene ring is left with large amount of 1,3-dienyl struture. The consequence of this higher dienyl character in naphthalene when compared to that in benzene is that a considerable amount of fixation of the double bonds in naphthalene takes place. *That is, the C_1- C_2 bond in naphthalene has more double bond character and less of single bond character.* X-ray analytical data on naphthalene shows the C_1-C_2 bond lengths as 1.37Å and C_2-C_3 bond length as 1.41Å when compared to 1.40Åas uniform bond length value for benzene.

7-12-3 Nomenclature

The naphthalene skeleton is numbered as shown below. In the older system, positions on the naphthalene ring system were designed as *α, β, β, α* etc. Some representative examples are given below:

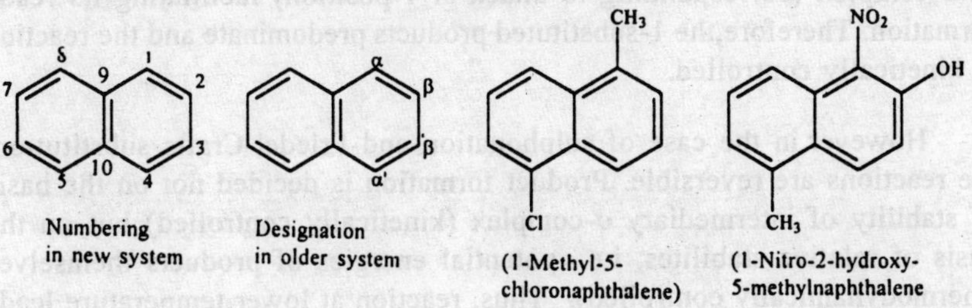

Numbering in new system

Designation in older system

(1-Methyl-5-chloronaphthalene)

(1-Nitro-2-hydroxy-5-methylnaphthalene

7-12-4 Properties of naphthalene

Naphthalene is a white, crystalline solid m.p. 80°C with a charateristic smell. It is a fairly non-associated substance, largely insoluble in water but highly soluble in organic solvents like ether, chloroform, benzene etc.,

In its chemical reactions, naphthalene shows the typical aromatic property of benezenoidal hydrocarbons electrophilic substitution. Naphalene undergoes electrophilic aromatic substitution predominantly at the 1-position except in the case of sulphonation and Friedel-Craft's reaction, when substitution occurs at 2-position. In reactions such as nitration, halogenation etc., 1-substitution occurs which is understandable in view of the following contributing resonance strutures for naphthalene.

For attack at 1-position: 7-Charge separated contributing structures:

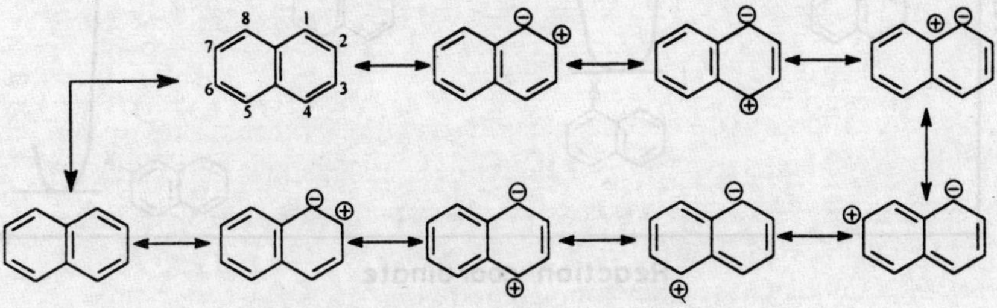

For attack at 2-poisition : 5-Charge separated contributing strutures:

Attack at 1-position leads to the formation of σ-complex to which larger number of resonance structures contribute. This enchances the stability of σ-complex (corresponding to attack at 1-position) facilitating its ready formation. Therefore,the 1-substituted products predominate and the reaction is kinetically controlled.

However,in the case of sulphonation and Friedel-Crafts substitution, the reactions are reversible. Product formation is decided not on the basis of stability of intermediary σ-complex (kinetically controlled) but on the basis of relative stabilites, i.e., potential energies of products themselves (thermodynamically controlled). Thus, reaction at lower temperature leads to 1-substituted products (kinetic control) but at higher temperatures, i.e., under equilibruium conditions, leads to 2-substituted products (themodynamic control).

The potential energy diagram for a reversible electrophilic substitution reaction (electrophile represented by X) is shown below:

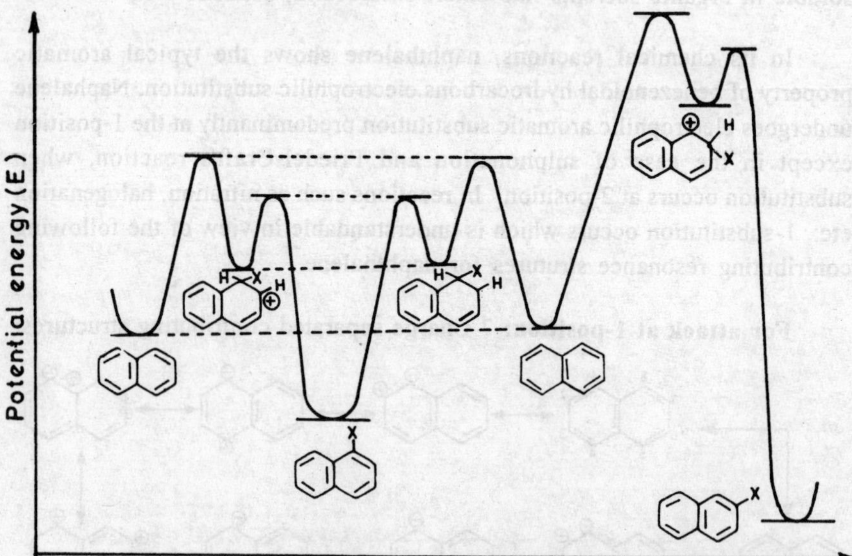

Nitration: The nitration of naphthalene can be carried out with nitrating mixture at 50-60°C to yield 1-nitronaphthalene.

HNO_3/H_2SO_4, 50 - 60°

NO$_2$

(1-Nitronaphthalene)
(90 - 95%)

Halogenation: Naphthalene undergoes halogenation on heating it with the halogen in a suitable solvent. For example;

$Br_2/CCl_4/\Delta$ (No Lewis acid needed)

Br

(1-Bromonaphthalene) (~75%)

Sulphonation: Naphthalene reacts with conc. H_2SO_4 at 80°C to give predominantly the 1-sulphonic acid products and at 160°C to yield the 2-sulphonic acid product. The 1-isomer can be converted to 2-isomer by heating it with conc.H_2SO_4 at 160°.

H_2SO_4

80° → SO$_3$H (Naphthalene-1-sulphonic acid)

160° → SO$_3$H (Naphthalene-2-sulphonic acid)

Friedel-Craft's reaction: Acylation of naphthalene with $CH_3COCl/AlCl_3$ in 1,2-dichloroethane yields the 1-isomer wheras in nitrobenzene the 2-isomer is formed.

7-12-5 Synthesis of naphthalene and its derivatives

In the laboratory, naphthalene and its derivatives are prepared by *Haworth's method.* In this method, benzene (or its derivative) is treated with succinic

$COCH_3$

$C_2H_4Cl_2$

(1-Acetylnaphthalene)

CH_3COCl

$AlCl_3$

$CO—CH_3$

$PbNO_2$

(2-Acetylnaphthalene)

Friedel-Craft's reaction

anhydride/$AlCl_3$ to yield β-benzoylpropionic acid. The latter on Clemmensen reduction followed by cyclisation, reduction and dehydrogenation yield naphthalene.

$AlCl_3$ Redn. Cyclization $-H_2O$ HOOC HOOC

Redn. $-2H_2$ Pd/C

(Naphthalene/s)

7-12-6 Uses of naphthalene

Naphthalene is used as a moth repellant for preservation of old cloths, paper etc., Large quantities of naphthalene are converted to phthalic anhydride via phthalic acid. Phthalic anhydride is used in the manufacture of polymers, plastics, resins etc.. Naphthalene is used in the preparation of a number of naphthalene based synthetic intermediates, which in turn are used in the manufacture of dyes, drugs, pesticides etc. One such example is β-naphthyl amine.

7-13 ANTHRACENE AND PHENANTHRENE

Both anthracene and phenanthrene are obtained from the anthracene oil fraction on coal-tar distillation. The structures of these two compounds were established as given below on the basis of molecular weights, analytical data and oxidation reactions.

(Anthracene) (Phenanthrene)

Thus, the structure of antharacene may formally be derived by the linear fusion of three benzene rings whereas that of phenanthrene may be obtained by angular fusion of a benzene ring to a naphthalene ring.

For nomenclature purposes, the susbtituents on the two rings are numbered and the compounds are named by indicating the positions, name of the substituent followed by adding the name of the system. In the older system of nomenclature,position of substituents was indicated by Greek alphabets, α, β, γ, etc., as shown below:

(Anthracene) (Anthracene)

(phenanthrene)

(9,10-Anthraquinone) (1-Methyl-5-chloroanthracene)

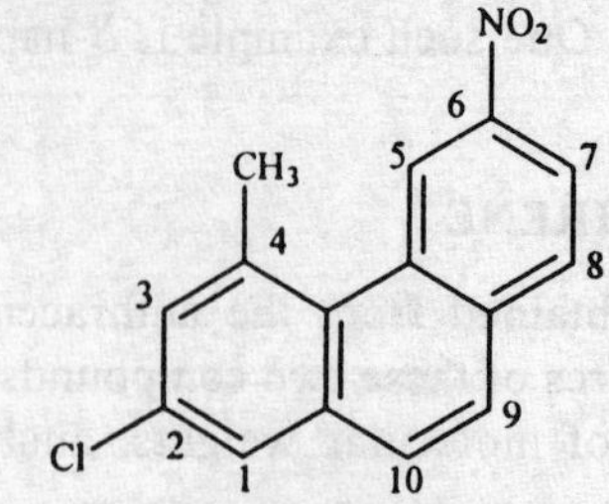

(2-chloro-4-methyl-6-nitrophenanthrene)

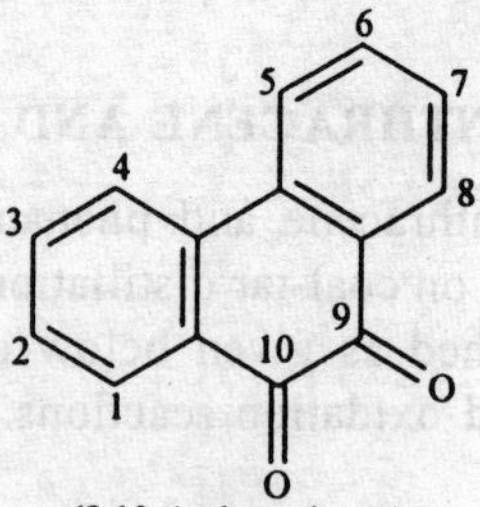

(9,10-Antha-quinone)

Heats of hydrogenation data indicated the resonance energies of anthracene and phenanthrene to be 84 kcal mol^{-1} and 92 kcal mol^{-1} respectively. This indicates that the anthracene and phenanthrene molecules are less stable than expected by (3 x 36-84) = (108-84) = 24 kcal mol^{-1} and (3 x 36-92) = (108-92) = 16 kcal mol^{-1} respectively. Without considering the charge separated strutures, the resonance structures (of the Kekule type) for anthracene and phenanthrene may be written down as shown below:

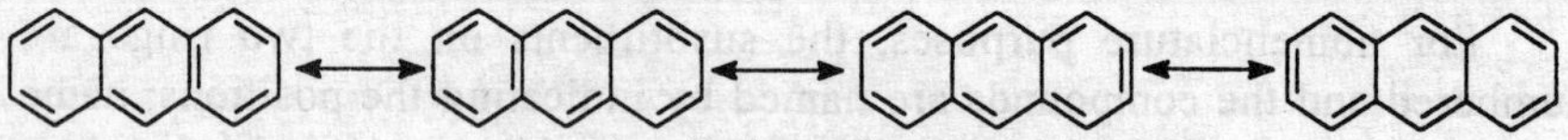

(Anthracene)

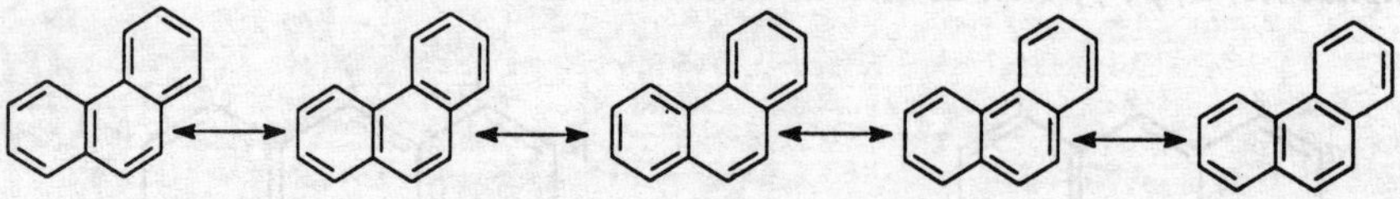

7-13-1 Properties

Anthracene is a colorless, crystalline solid with a m.p.50°C. It is a non-associated benzenoidal hydrocarbon insoluble in water but soluble in hot benzene, ether, chloroform etc. Phenanthrene is a colorless, crystalline solid of m.p. 99°C. It dissolves in organic solvents giving a bright blue fluorescence.

Towards oxidizing agents, both anthracene and phenanthrene are even less stable than naphthalene. Thus, oxidation with dichromate leads to the respective quinones. This can be readily explained since in case of anthracene, oxidation to anthraquinone involves only a sacrifice of (84-2x36) = (84-72) = 12 kcal mol^{-1} of resonance energy (two of the benzene rings are intact). In the case of phenanthrene, it involves a sacrifice of (92-2x36) = 92 - 72 = 20 kcal mol^{-1} of resonance energy.

$K_2Cr_2O_7$

(Anthracene) (9,10-Anthraquinone)

$K_2Cr_2O_7$

(Phenantherene) (9,10-Phenanthraquinone)

For the same reason,both anthracene and phenanthrene undergo ready reduction of the 9,10-double bond to yield the respective dihydro derivatives.

Na/EtOH/Δ

(Anthracene) (9,10-Dihydroanthracene)

Na/EtOH/Δ

(Phenantherene) (9,10-dihydrophenanthrene)

Both the ring systems undergo elecrtrophilic substitutions at 9-positions. These electrophilic substitution are not of much significance in organic sysnthesis. Furthermore, due to considerable amount of double bond character in the 9, 10 bond in both the systems, there is usually a tendency for addition reaction to take place. For example:

Br_2/ Fe Br_3

(Anthracene) (9-Bromoanthracene) + (9,10-Dibromophenanthrene)

(Phenanthrene) —Br/ Fe Br_3→ (9-Bromophenanthrene) + (9,10-Dibromophenanthrene)

For this reason,electrophilic substitution in both the systems are carried out on the corresponding quinones followed by reduction of the quinone groups to yield back anthracene and phenanthrene respectively.

7-13-2 Synthesis

Both the ring systems can be suitably synthesised by appropriate modification of Haworth's method given below:

(Phthalic anhydride) + benzene —$AlCl_3$→ o-benzoylbenzoic acid (COOH) —$-H_2O$, PPA→ (Anthraquinone)

Reduction → o-benzylbenzoic acid (COOH) —i) $-H_2O$ ii) Redn→ 9,10-dihydroanthracene —$(-H_2)$→ (Anthracene)

(Naphthalene) + (Succinic anhydride) —$AlCl_3$→ keto acid (COOH) —Redn.→ butyric acid (COOH) —$-H_2O$→ ketone —Redn.→ (Tetrahydrophenanthrene)

keto acid —$-H_2O$, PPA etc.→ diketone —Redn.→ (Tetrahydrophenanthrene)

(Tetrahydrophenanthrene) —Pd - C $(-2H_2)$→ (Phenarthrene)

7.13.3 Uses

Anthracene as such finds very little use in organic syntheses. However, anthraquinone is a valuable starting material for several anthraquinone dyes. Similarly phenanthrene is not having much use in organic chemistry for synthetic work. Furthermore, various substituted anthracenes and phenanthrenes can be readily prepared from the appropriately substituted materials. Phenanthraquinones also find extensive use in the dye-manufacture.

7-14 SELF TEST QUESTIONS

Choose the correct answer for the following questions.

7-14-1 Which one of the following is not a natural source of benzene?
(a) Coal-tar (b) Petroleum
(c) Natural Gas (d) Trimerisation of acetylene.

7-14-2 Benzene shows tendency to undergo
(a) Addition reactions (b) Elimination reactions
(c) Substitution reactions (d) Polymerization reactions

7-14-3 All the six carbon-carbon bond lengths in benzene have been found to be
(a) Equal (b) Unequal
(c) Identical with C=C double bond length (d) None of the first three.

7-14-4 The carbon-carbon bond length in benzene has been found to be
(a) 1.548Å (b) 1.37Å (c) 1.21Å (d) 1.40Å

7-14-5 In the nitration of benzene with nitrating mixture, the effective electrophilic species is
(a) Nitrate ion (b) Nitronium ion
(c) Nitrite ion (d) Nitrate + Sulphate ion.

7-14-6 Sulphonation of benzene is very effective in
(a) Dil.H_2SO_4 (b) Conc.H_2SO_4
(c) Oleum (d) Sulphuric acid enriched with SO_2

7-14-7 The best laboratory approach to alkyl benzenes is
(a) Friedel-Craft's alkylation
(b) Friedel-Craft's acylation
(c) Friedel-Craft's acylation followed by reduction of the keto group
(d) None of the above three

7-14-8 Aromaticity is due to a
(a) Cyclic structure
(b) Planar structure
(c) Presence of $(4n + 2)$ π electrons
(d) Cyclic, planar structure with delocalization of $(4n+2)$ π electrons.

7-14-9 From the points of view of directive influence, substitutents can be divided into
(a) o,p-directing type and meta-directing type
(b) o,m-directing type and para-directing type
(c) m,p-directing type and ortho-directing type
(d) all the three.

7-14-10 In an electrophilic substitution reaction such as chlorination, which one of the following will be most reactive?

(a) Benzene (b) Chlorobenzene (c) Phenol (d) Nitrobenzene

7-14-11 Which one among the following is a m-directing group when attached to benzene ring in electrophilic substitution?

(a) Methyl (b) Bromo (c) Methoxy (d) Nitro

7-14-12 Which one of the following is least reactive towards electrophilic substitution such as nitration?

(a) Nitrobenzene (b) Acetanilide

(c) Chlorobenzene (d) Benzaldehyde

7-14-13 Nitration of cyanobenzene yields predominantly

(a) 2-Nitrocyanobenzene (b) 3-Nitrocyanobenzene

(c) 4-Nitrocyanobenzene (d) All of the three products.

7-14-14 Chlorination of toluene in the presence of anhydrouss $FeCl_3$ yields largely

(a) Benzoyl chloride (b) Benzyl chloride

(c) m-Chlorotoluene (d) Mixture of o- and p-chlorotoluenes

7-14-15 Oxidation of ethylbenzene with potassium permanganate yields

(a) Phenylacetic acid (b) Benzoic acid

(c) Benzaldehyde (d) Phenylacetaldehyde

7-14-16 Biphenyl can be readily prepared by

(a) Heating chlorobenzene with sodium metal

(b) Heating benzene with sodium chloride

(c) Heating chlorobenzene with sodium chloride

(d) Heating chlorobenzene with anhydrous aluminium chloride.

7-14-17 Which one among the following has the highest resonance energy?

(a) Benzene (b) Naphthalene (c) Anthracene (d) Phenanthrene

7-14-18 Nitration of naphthalene gives very largely

(a) 2-Nitronaphthalene (b)1-Nitronaphthalene

(c) 1,2-Dinitronaphthalene (d) Mixture of 1- and 2-nitronaphthalene

7-15 SHORT AND ESSAY TYPE QUESTIONS

7-15-1 What were the main reasons that led to Kekule structures for benzene?

7-15-2 Describe the different strutures proposed for benzene and explain why they were rejected, other than Kekule structures?

7-15-3 What is aromaticity? Explain the reasons for classifying a compound as aromatic in nature?

7-15-4 What factor determines the degree of aromaticity in a compound ? What are non-benzenoidal aromatic compounds? Give two examples?

7-15-5 What is the mechanism of electrophilic aromatic substitution in benzene ?

7-15-6 How is nitration and sulphonation of benzene carried out? What are the mechanisms of these two reactions?

7-15-7 How are the Friedel-Craft's alkylation and acylation of benzene carried out? What are the advantages and disadvantages of these reactions?

7-15-8 What is meant by orientation of substitution in monosubstituted benzenes? Explain the effects involved and classification of groups according to their directive influence?

7-15-9 How is alkyl benzene prepared? How do the alkyl benzenes react in halogenation and oxidation reactions?

7-15-10 What are biphenyls? Give two examples. How are biphenyls prepared?

7-15-11 How does naphthalene behave towards electrophilic substitution reactions?

7-15-12 Describe a method each for the preparation of naphthalene, anthracene and phenanthrene?

7-15-13 Describe by simple reaction sequences, a method for the preparation of the following compounds from benzene and other readily available reagent chemicals.

(a) Ethyl benzene
(b) 2,4-Dinitrochlorobenzene
(c) p-Nitrobenzoic acid
(d) Phenylacetic acid
(e) m-Nitrochlorobenzene
(f) Phenylacetylene
(g) Styrene
(h) Ethyl benzoate
(i) m-Bromonitrobenzene
(j) p-Bromobenzoic acid

7-15-14 Explain what product/s are formed in the following reactions;

i) $C_6H_5CH_3 \xrightarrow[AlCl_3]{Cl_2} A \xrightarrow{KMnO_4} B$

ii) $C_6H_5OCH_3 \xrightarrow{HNO_3} C \xrightarrow[FeBr_3]{Br_2} D$

iii) $C_6H_5CHO \xrightarrow{HNO_3} E \xrightarrow{KMnO_4} F$

iv) Toluene (CH_3) $\xrightarrow{Cl_2/h\nu}$ G $\xrightarrow{KOH}$ H $\xrightarrow{Me_2SO_4}$ I $\xrightarrow{HNO_3}$ J

v) Benzene $\xrightarrow[H_2SO_4]{HNO_3}$ K $\xrightarrow[HSO_4]{HNO_3}$ L

vi) Naphthalene $\xrightarrow[160^0]{H_2SO_4}$ M $\xrightarrow{KOH}$ N $\xrightarrow[AcOH]{Br_2}$ O

vii) 2,6-Dichlorotoluene (H_3C, Cl, Cl) + succinic anhydride $\xrightarrow{AlCl_3}$ P $\xrightarrow[Zn(Hg)]{Redn}$ Q $\xrightarrow{PPA}$ R $\xrightarrow[Zn(Hg)]{Redn}$ S $\xrightarrow[(-2H_2)]{Pd-C}$ T

7-15-15 Give explanation for the following observations.

(a) Towards nitration, toluene reacts faster than benzene ?

(b) Nitration of nitrobenzene requires more vigorous conditions than benzene itself?

(c) Friedel-Crafts alkylation of nitrobenzene with benzyl chloride/$AlCl_3$ does not occur at all?

(d) Halogenation of aniline occurs with a great speed to give, 2,4,6-Trihaloaniline?

(e) Phenol undergoes nitration with explosive violence to give polynitro phenols?

7-15-16 Discuss with examples how the presence of an electron withdrawing and electron-releasing groups in benzene nucleus influences substitution reactions. (O.U., Oct.1991)

7-15-17 What is aromaticity? Describe the aromatic character of benzene based on molecular orbital theory and resonance. (N.U.,Apr.1990)

7-15-18 Discuss the structures of benzene. (SKU.Apr.,1990)

7-15-19 Explain why benzene undergoes electrophilic substitution and not nucleophilic substitution. Give mechanism for the nitration of benzene. (O.U.,Apr.1991)

7-15-20 With the aid of resonance structures explain why toluene gives a mixture of o-nitrotoluene and p-nitrotoluene as manjor products? (O.U.,Apr.1991)

7-15-21 What are the sources of aromatic compounds? How is benzene obtained? (A.U.,Dec.1989)

7-15-22 Discuss electrophilic substitution in benzene taking nitration as an example? What is the reason for the formation of 93% para product when toluene is subjected to alkylation using t-butyl bromide? (K.U.,Apr.1990)

ANSWER TO PROBLEMS & EXERCISE

7-14-1 **D**	7-14-2 **C**	7-14-3 **A**	7-14-4 **D**	7-14-5 **B**	7-14-6 **D**
7-14-7 **D**	7-14-8 **D**	7-14-9 **D**	7-14-10 **D**	7-14-11 **D**	7-14-12 **D**
7-14-13 **B**	7-14-14 **D**	7-14-15 **B**	7-14-16 **A**	7-14-17 **D**	7-14-18 **B**

7-15-13

a) Benzene $\xrightarrow[AlCl_3]{(CH_3CO)_2O}$ $C_6H_5COCH_3$ $\xrightarrow{Zn(Hg)}$ $C_6H_5CH_2–CH_3$

b) Benzene $\xrightarrow{Cl_2/AlCl_3}$ (Chlorobenzene) $\xrightarrow[H_2SO_4]{HNO_3}$ (o-Nitrochloro-benzene) + (p-Nitrochloro-benzene) $\xrightarrow{HNO_3/H_2SO_4}$ 1-Cl-2,4-$(NO_2)_2$ benzene

c) Benzene $\xrightarrow[AlCl_3]{CH_3I}$ Toluene $\xrightarrow[H_2SO_4]{HNO_3}$ p-$CH_3C_6H_4NO_2$ + o-$CH_3C_6H_4NO_2$ $\xrightarrow{\text{Separation}}$ (p-Nitrotoluene) $\xrightarrow{KMnO_4}$ (p-Nitrobenzoicacid)

d) (Benzene) —$Cl_2/h\nu$, $AlCl_3$→ (Toluene, CH_3) —$Cl/h\nu$→ (Benzyl chloride, CH_2Cl) —KCN→ (Benzylcyanide, CH_2CN)

$CH_2–CN$ —H_3O^+→ $CH_2–COOH$ (Phenylacetic acid)

e) Benzene → CH_3 —Cl_2, $AlCl_3$→ CH_3, Cl (para) + CH_3, Cl (ortho)

→ CH_3, Cl (p-Chlorotoluene) —HNO_3, H_2SO_4→ CH_3, NO_2, Cl —$KMnO_4$→ COOH, NO_2, Cl —$-CO_2$→ NO_2, Cl (m-Nitrochlorobenzene)

(f) & (g)

CH_2-CH_3 $CH(Cl)-CH_3$

i) $CH_3COCl/AlCl_3$ ii) Zn (Hg) → (Ethylbenzene) $Cl_2/h\nu$ →

$CH=CH_2$ $CH(Cl)-CH_2Cl$ $C\equiv CH$

-HCl → (Styrene) Cl_2/CCl_4 → -2HCl → (Phenylacetylene)

(h)

CH_3 COOH $COOC_2H_5$

→ → (Benzoic acid) → (Ethylbenzoate)

(i)

CH_3 CH_3 CH_5 NO_2

CH_3 I / $AlCl_3$ → Br_2 / AlBr3 → Br HNO_3 / H_2SO_4 → Br

COOH NO_2 NO_2

K Mn O4 → Br KOH/Δ (-CO_2) → Br

(4-bromo-2-nitro-benzoic acid) (m-Bromo-nitrobenzene)

(j)

CH_3 CH_3 COOH

→ Br_2 / $AlBr_3$ → $KMnO_4$ →

Br (p-Bromotoluene) Br (p-Bromobenzoic)

7-15-14 Several structural schemes

i) CH_3 (toluene) $\xrightarrow{Cl_2,\ AlCl_3}$ (p-chlorotoluene + o-chlorotoluene) (A) $\xrightarrow{KMnO_4}$ (p-chlorobenzoic acid, COOH, Cl + o-chlorobenzoic acid, COOH, Cl) (B)

ii) OCH_3 $\xrightarrow{HNO_3}$ OCH_3, Cl, NO_2 (C) $\xrightarrow{Br_2,\ FeBr_3}$ OCH_3, Br, NO_2 (D)

iii) CHO $\xrightarrow{HNO_3}$ CHO, NO_2 (E) $\xrightarrow{KMnO_4}$ COOH, NO_2 (F)

iv) CH_3 $\xrightarrow{Cl_2,\ h\nu}$ CH_2Cl, NO_2 (G) $\xrightarrow{KOH}$ CH_2OH, NO_2 (H) $\xrightarrow{Me_2SO_4,\ \ominus OH}$

CH_2OMe (I) $\xrightarrow{HNO_3}$ CH_2OMe, NO_2 (J)

v) (benzene) $\xrightarrow{HNO_3,\ H_2SO_4}$ NO_2 (K) $\xrightarrow{HNO_3,\ H_2SO_4}$ NO_2, NO_2 (L)

vi) $\xrightarrow[160^o]{H_2SO_4/\Delta}$ (M) SO_3H $\xrightarrow{KOH/\Delta}$ (N) OH $\xrightarrow[AcOH]{Br_2}$ (O) Br, OH

vii) H_3C, Cl, Cl $\xrightarrow{AlCl_3}$ (P) $\xrightarrow[Zn(Hg)]{Redn.}$ (Q) COOH $\xrightarrow{PPA}$

(R) $\xrightarrow{Zn(Hg)}$ (S) $\xrightarrow[(-2H_2)Cl]{Pd\text{-}d}$ (T)

7-15-15

(a) The methyl group activates the toluene molecule by increasing electron-density at o-and p-positions. Thus, it facilitates attack by an electrophile whereas there is no such effect in the case of benzene molecule. Therefore, toluence reacts faster than benzene.

(b) The nitro group is an electron-withdrawing group. It deactivates the benzene ring by withdrawing electron density very strongly at o-and p-positions and slightly weakly at m-position. Therefore, nitrobenzene reacts very slowly when compared to benzene. Further, when nitrobenzene reacts, requiring very vigorous conditions, it gives a meta product, m-dinitro benzene.

(c) Friedel-Craft's reaction is a very weak reaction in which the electrophile is the R–Cl—$AlCl_3$ complex. In the case of nitrobenzene, the nitro group is a strong deactivator and electrophilic substitution requires very vigorous conditions. The result is that a weak electrophile (R-Cl...$AlCl_3$) is unable to attack a highly deactivated ring system. In fact, nitrobenzene, due to its inertness and high polarity, is used as a solvent in Friedel-Craft's reactions.

(d) Aniline is a highly activated molecule towards electrophilic substitution. It reacts with halogens without a catalyst at great speed undergoing electrophilic substitution giving 2,4,6-trihalo products.

(e) Phenol is a highly activated system towards electrophilic substitution. Many times, the reaction is so vigorous that an explosion may occur. Furthermore, the products of nitration of phenol such as DNT, TNT etc. are themselves explosive in character.

8

HALOGEN COMPOUNDS

8-1 INTRODUCTION

Organic halogen compounds can be considered to be those obtained by formally replacing one or more hydrogens of hydrocarbons (aliphatic, alicyclic or aromatic) with halogens. The halogen derivatives of alkanes, alkenes and alkynes are known as alkyl halides, haloalkenes and haloalkynes respectively. The halogen derivatives of alicyclic and aromatic hydrocarbons are known as halocyclohydrocarbons and haloaromatic hydrocarbons respectively.

Not very many organic halogen compounds have been known to occur naturally. Those like chloromycetin (chloramphenicol) and aureomycin (chlorotetracycline) obtained from natural sources, have been found to be useful as antibiotics.

(Chloromycetin)

(Aureomycin)

Due to their ease of formation and highly reactive nature, the organic halogen compounds hold a pre-eminent position in synthetic organic chemistry. Apart from their extensive use as reagents and intermediates in organic synthesis, the halogen compounds find application as solvents, insecticides, fungicides, plant protecting agents, explosives, refrigerants, drugs and pharmaceuticals.

8-2 NOMENCLATURE

The organic halogen compounds are named by mentioning the name of the hydrocarbon followed by the ending *"halide"*. In certain compounds, the halogen may be considered to be a substituent and the compound is merely called as the hydrocarbon with the halogen's name preceding its name with the position indicated by a number. The following examples will illustrate the nomenclature more clearly;

$CH_3—CH_2—CH_2—Cl$
(n-Propylchloride or 1-Chloropropane)

$(H_3C)_2CH—CH_2Cl$
(Isobutyl chloride or 1-Chloro-2-methyl propane)

$CH_2 = CH—Br$
(Vinyl bromide 1-Bromoethene)

$CH_3—CH(NO_2)—CH_2—CH_2—CH_2—F$
(4-Nitro-n-pentyl fluoride or 1-fluoro-2-nitropentane)

$C_6H_5—CH_2—I$
(Phenylmethyl iodide or Iodo-phenylmethane or Benzyl Iodide)

(4-Methyl-2-nitrochlorobenzene or 4-chloro-3-nitrotoluene)

8-3 CLASSIFICATION

The organic halogen compounds may broadly be classified, as shown in chart-1, into two types; namely, aliphatic halohydrocarbons and aromatic halohydrocarbons. The aliphatic hydrocarbons may further be divided into alkyl halides, haloalkenes, haloalkynes and halocycloalkanes. The alkyl halides, can be further subdivided into three categories; namely primary alkyl halides, secondary alky1 halides and tertiary alkyl halides. This subdivision is based on whether the halogen/s is/are attached to primary, secondary or tertiary carbon atoms. The above classification of organic halogen compounds, as will be seen later, is largely done on the basis of the variation in the properties and reactivities of these classes of compounds.

CHART - 1

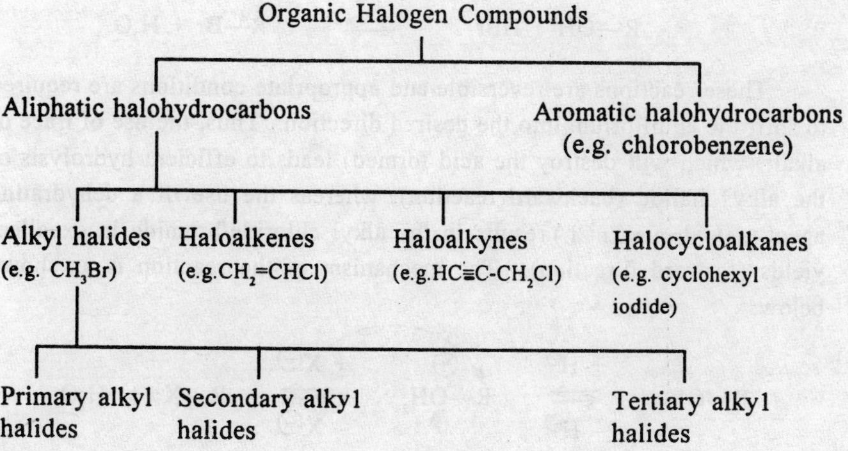

8-4 ALKYL HALIDES

8-4-1 Methods of preparation

There are several methods available for the preparation of alkyl halides. However, the following four are most widely used.

(a) From alkanes : When alkanes are heated with chlorine or bromine in the presence of ultra-violet light/free-radical forming agents, alkyl chlorides/bromides are formed. For example;

$$R\text{—}CH_3 \xrightarrow{Cl_2\ /\ \Delta} R\text{—}CH_2\text{—}Cl$$

$$R\text{—}CH_3 \xrightarrow{Br_2\ /\ \Delta} R\text{—}CH_2\text{—}Br$$

The reaction proceeds by a free-radical process and usually yields a mixture of polyhalogenated products. This method is not suitable for laboratory work and is mostly employed in industry since the polyhalogenated mixture of products can be readily separated using efficient fractionating columns.

(b) From alcohols : The alcohols can be converted into the alkyl halides by using either of the following three methods :

(i)By the action of HX on alcohols : When alcohols are treated with anhydrous hydrogen chloride or hydrogen bromide, the former are converted into alkyl chlorides/bromides.

$$R\text{—}OH + HCl \rightleftharpoons R\text{—}Cl + H_2O$$

$$R\text{—}OH + HBr \rightleftharpoons R\text{—}Br + H_2O$$

These reactions are reversible and appropriate conditions are required to shift the equilibrium into the desired direction. Thus, the use of trace of alkali (which will destroy the acid formed) leads to efficient hydrolysis of the alkyl halide (backward reaction), whereas the use of a dehydrating agent (anhydrous $ZnCl_2$) results in the alkyl chloride/bromide in excellent yields (forward direction). The mechanism of the reaction is as shown below:

$$R\text{—}OH \underset{-\,H^{\oplus}}{\overset{+\,H^{\oplus}}{\rightleftharpoons}} R\text{—}\overset{\oplus}{O}H_2 \underset{-\,X^{\ominus}}{\overset{+\,X^{\ominus}}{\rightleftharpoons}} R\text{—}X + H_2O$$

(ii) By the action of $SOCl_2$ on alcohols : When refluxed with thionyl chloride ($SOCl_2$), alcohols are converted to alkyl chlorides. This method is one of the best for the preparation of alkyl chlorides in the laboratory, since the byproducts (HCl and SO_2), being gaseous, can be easily removed.

$$R—OH + SOCl_2 \longrightarrow R—Cl + SO_2\uparrow + HCl\uparrow$$

The reaction proceeds by the following mechanism :

$$SOCl_2 = Cl—\overset{O}{\overset{||}{S}}—Cl \longleftrightarrow Cl—\underset{\oplus}{\overset{O^{\ominus}}{\overset{|}{S}}}—Cl \xrightarrow{R—\ddot{O}—H} \underset{H}{\overset{R}{}}\overset{\oplus}{O}—\overset{O^{\ominus}}{\overset{|}{S}}—Cl \quad (—Cl)$$

$$\longrightarrow R\overset{O}{\diagup\diagdown}S = O \; (Cl) + HCl \longleftrightarrow R\overset{O}{\diagup\diagdown}\overset{\oplus}{S}—\overset{\ominus}{O} \; (Cl) \longrightarrow R—Cl + SO_2$$

(iii) By the action of phosphorous pentachoride on alcohols : Treatment of alcohols with PCl_5 leads to the formation of alkyl chlorides. The reaction is very similar to that of alcohols with $SOCl_2$.

$$R—OH + PCl_5 \longrightarrow R—Cl + POCl_3 + HCl$$

The reaction proceeds by the following mechanism :

$$R—\ddot{\underset{\cdot\cdot}{O}}—H + PCl_5 \longrightarrow R—\underset{H}{\overset{\oplus}{O}}—PCl_4 \;(\overset{\ominus}{Cl}) \longrightarrow R—O—PCl_4$$

$$\longrightarrow R—Cl + \overset{\ominus}{O}—\overset{\oplus}{P}Cl_3 \;(\equiv POCl_3)$$

In the above reaction sometimes instead of PCl_5, PCl_3 is used. In the latter case, the reaction is less efficient and yields are poor. This is understandable in the light of the mechanism given above which involves an initial nuclephilic attack by weakly basic oxygen of the alcohols on the positively polarized phosphorous atom of the reagent.

If alkyl bromides are desired PBr_5 is required in place of PCl_5 in the above reaction. Since PBr_5 is unstable and decomposes in the free state, it is prepared *in situ* by reaction between red phosphorous and liquid bromine.

(c) From alkenes : When alkenes are treated with hydrogen chloride, alkyl chlorides are formed. The reaction follows Markovnikov rule as far as orientation of addition is concerned.

$$R\text{—}CH = CH_2 + HCl \longrightarrow R\text{—}\underset{\displaystyle Cl}{\underset{|}{CH}}\text{—}CH_3$$

If anti-Markovnikov addition is desired, HBr is allowed to react in the presence of a peroxide catalyst.

$$R\text{—}CH_2 = CH_2 + HBr \xrightarrow[R^1\text{—}CO\text{—}O\text{—}O\text{—}CO\text{—}R^1]{R^1\text{—}O\text{—}O\text{—}R^1 \text{ or}} R\text{—}CH_2\text{—}CH_2\text{—}Br$$

Dihalo derivatives are formed when direct halogens are used in place of HX in the above reactions. For example;

$$R\text{—}CH = CH_2 + Cl_2 \longrightarrow R\text{—}\underset{\displaystyle Cl}{\underset{|}{CH}}\text{—}\underset{\displaystyle Cl}{\underset{|}{CH_2}}$$

$$R\text{—}CH = CH_2 + Br_2 \longrightarrow R\text{—}\underset{\displaystyle Br}{\underset{|}{CH}}\text{—}\underset{\displaystyle Br}{\underset{|}{CH_2}}$$

(d) From alkyl halides by exchange: The above methods, (a) to (c), are suitable for the preparation of alkyl chlorides and bromides only. It is very difficult to prepare alkyl iodides using the above methods. However, the latter can be prepared conveniently by halide exchange reactions involving treatment of alkyl chlorides/bromides with NaI or KI.

$$Ph\ CH_2Cl + NaI \xrightarrow{\text{acetone}/\Delta} PhCH_2\text{–}I + NaCl$$

$$CH_3\text{-}CH_2\text{-}CH_2\text{-}CH_2\text{-}Br + KI \xrightarrow{\text{acetone}/\Delta} CH_3\text{-}CH_2\text{-}CH_2\text{-}CH_2\text{-}I + KBr$$

8-4-2 Physical properties

The lower members of the alkyl halides,such as CH_3Cl, C_2H_5Cl, CH_3Br, CH_3-CH_2-Br, CH_3I, etc., are either gases or low boiling liquids. The higher members, above C-4 are liquids and those above C-12 are solids. The boiling points of alkyl halides increase in the order,

Iodide > Bromide > Chloride > Fluoride

Although, the alkyl halides are polar compounds they are incapable of forming H-bonds. Consequently. , except for lower molecular weight compounds,most alkyl halides are insoluble in water, but dissolve freely in ether,benzene, $CHCl_3$, etc. The alkyl fluoride and chlorides are usually less dense than water whereas the bromides and iodides are heavier than water.

8-4-3 Chemical properties

The alkyl halides are highly reactive compounds. Because of their easy availability and very reactive nature, the alkyl halides can be trasformed into a large variety of compounds containing different functional groups. They are extremely important in chemical industry and hold a strategic position in organic syntheses. The following are some of their important chemical reactions.

(i) Reduction : The alkyl halides undergo reduction to yield hydrocarbons. A variety of redμcing agents, such as metal + acid, H_2/catalyst, phosphorous/HI, $LiAlH_4$, etc., can be employed.

$$\text{R—X} \xrightarrow{\text{Reduction}} \text{R—H}$$

Almost all reductions appear to proceed through the intermediacy of radical anions involving electron-transfer processes.

(ii) Wurtz reaction : The alkyl halides, on heating with sodium metal, undergo reductive dimerization to yield hydrocarbons.

$$\text{R}\!-\!\boxed{\text{X + 2Na + X}}\!-\!\text{R} \longrightarrow \text{R—R + 2NaX}$$

Here again, the reaction appears to proceed through an electron-transfer process.

(iii) Grignard formation : Synthetically, this is one of the most important reactions of alkyl halides. In this reaction an alkyl halide is treated with Mg metal in an ethereal solvent to yield, what is called, a Grignard reagent. It is so named in honour of its discoverer, Victor Grignard. This reagent has approximately the composition, R-MgX and is an extremely reactive compound. It undergoes decomposition in air, moisture, CO_2,etc. So it is usually generated *in situ*, in an inert atmosphere, preferably of nitrogen and used immediately as such for reactions.

Preparation of Grignard reagent is a fairly simple reaction. The alkyl halide is dissolved in an ethereal solvent and Mg metal (in the form of its turnings) is added to it.

$$R-X + Mg \xrightarrow[\text{THF etc.}]{\text{Ether or}} R-MgX$$

The mixture is warmed with stirring in the presence of a trace of iodine. in an inert atmosphere. The Mg metal goes into solution forming the Grignard reagent. Only ether or ethereal solvents can be used since reactive solvents decompose the Grignard reagent and highly inactive ones such as

R X R X

Mg Or Mg

Et—O O—Et R'—O O–R'

Et Et R' R'

hydrocarbons are unsuitable for its formation. Ethereal solvents are known to stabilize the Grignard reagent through solvation.

Despite its high sensitivity, the Grignard reagent can be made to react suitably with any compound which has an acidic hydrogen. For example:

$$\overset{\delta-}{R}-\overset{\delta+}{Mg}X + \overset{\delta+}{H}-\overset{\delta-}{OH} \longrightarrow R-H + Mg\begin{matrix}X\\OH\end{matrix}$$

etc.

$$\overset{\delta-}{R}-\overset{\delta+}{Mg}X + \overset{\delta+}{H}-\overset{\delta-}{OR'} \longrightarrow R-H + Mg\begin{matrix}X\\OR'\end{matrix}$$

The use of Grignard reagents in organic syntheses is probably the first example of *syntheses umpolung* meaning syntheses involving reversal of polarities. The alkyl halide possesses a carbon-halogen bond in which the partial positive charge is on the carbon and partial negative charge on halogen. In its reaction with nucleophiles, particularly the carbanions, the alkyl halide attacks through its partially positive carbon leading to C-C bond formation. Malonic ester syntheses and acetoacetic ester syntheses are the two famous examples belonging to this class of reactions.

$$\overset{\delta+\ \delta-}{R{-}X} + \overset{\ominus}{-\underset{|}{C}-} \longrightarrow -\overset{R}{\underset{|}{\overset{|}{C}}}- + \overset{\ominus}{X}$$

However, when an alkyl halide reacts with Mg metal leading to formation of Grignard reagent, in the latter, the partial positive charge rests on the metal and partial negative charge on the alkyl carbon. Consequently, the Grignard reagent reacts with an electrophile such as a carbonium ion, (as distinct from the reaction of alkyl halide with a nucleophile such as carbanion) leading to C-C bond formation. Syntheses of alcohols and amines by reaction of Grignard reagents with carbonyl compounds and Schiff's bases respectively are the two famous examples belonging to this class of reactions.

$$\overset{\delta-\ \delta+}{R{-}MgX} + (R^1)(R^2)C{=}O \xrightarrow{Et_2O} (R^1)(R^2)(R)C{-}O{-}MgX \xrightarrow{H_2O} (R^1)(R^2)(R)C{-}OH + Mg(X)(OH)$$

(an alcohol)

$$\overset{\delta-\ \delta+}{R{-}MgX} + (R^1)(R^2)C{=}N{-}R^3 \longrightarrow (R^1)(R^2)(R)C{-}N(R^3){-}MgX \xrightarrow{H_2O} (R^1)(R^2)(R)C{-}N(R^3){-}H + Mg(X)(OH)$$

(iv) Nucleophilic substitution : This is one of the most important reactions undergone by alkyl halides. In this reaction, an alkyl halide is treated with a nucleophile resulting in displacement of the halide ion. For example;

$$R{-}X + Y^- \longrightarrow R{-}Y + X^-$$

The most common nucleophiles are $\bar{O}R$, $\bar{O}H$, $\bar{C}N$, $\bar{S}R$, etc., including carbanions like $-\overline{\underset{|}{C}}-$. Even neutral molecules containing a non-bonding electron pair such as NH_3, amines, water, alcohol, thiols, etc., can act as nucleophiles bringing about substitution of the halogen.

The nucleophilic substitution reaction of an alkyl halide is due to the presence of a carbon-halogen bond which is highly polarized. Now, although the carbon-halogen bond is supposedly covalent in character, the pair of electrons constituting this bond is never shared equally between carbon and halogen. Due to the high electronegativity of halogens compared to carbon, the pair of electrons in the bond is drawn more towards halogen resulting in a permanent partial positive charge on carbon and a permanent partial negative charge on halogen. This is known as inductive effect.

$$R—X \equiv R \longrightarrow X \equiv \overset{\delta+}{R}—\overset{\delta-}{X}$$

Consequently, when an alkyl halide is treated with a nucleophile, the latter attacks the alkyl group displacing the halogen as halide ion.

(v) Dehydrohalogenation : Alkyl halides possessing an α-hydrogen in their structures undergo dehydrohalogenation on treatment with a base. For example;

$$R\text{-}CH_2—CH_2—X \xrightarrow{\text{Base}} R—CH = CH_2 + \text{Base.HX}$$

The orientation of elimination is determined by **Saytzeff's rule**. Since in most cases, base happens to be alkali, nucleophilic substitution is a competing reaction. Whether, elimination is a predominant reaction or nucleophilic substitution, depends on the nature of the alkyl halide, nature of the base, reaction conditions, etc.

The Saytzeff's rule, given empirically, states that *"in the dehydrohalogenation of an alkyl halide, hydrogen atom is preferentially eliminated from that carbon which has the least number of hydrogen atoms".* For example, in the dehydrohalogenation of sec-butyl chloride, two products are possible:

$$CH_3—CH_2—CH(Cl)—CH_3 \xrightarrow{-HCl} H_3C—CH_2—CH = CH_2 \text{ (1–Butene)}$$

$$CH_3—CH_2—CH(Cl)—CH_3 \xrightarrow[-HCl]{} H_3C—CH = CH—CH_3 \text{ (2–Butene)}$$

According to Saytzeff, the predominating product in the above reaction would be 2-butene which is indeed found to be the case. The alkene which is obtained in accordance with the Saytzeff's rule is commonly referred to as Saytzeff olefine and the process as Saytzeff elimination.

In general, elimination reactions (like substitution reactions, please see the next section) are divided into two types depending on the mechanisms they follow. These are known as E_2 and E_1. In the E_2 reactions, the rate is second order overall, first order with respect to substrate (i.e. alkyl halide) and first order with respect to reagent (i.e. base). Thus,

Rate α [Alkyl halide] [Base]

Mechanistically, the reaction involves abstraction of a hydrogen as proton from the α-carbon by the base (which is usually the alkoxide ion) with the simultaneous loss of halogen as the halide ion

$$R{-}\underset{|}{\overset{H}{\overset{|}{C}}}H{-}\underset{\underset{X}{|}}{C}H_2 \xrightarrow{\overset{\ominus}{O}Et} R{-}CH = CH_2 + HOEt + \overset{\ominus}{X}$$

The concentrations of base as well as that of the substrate, both, are involved in the rate-determining step of the reaction.

In the E_1 type eliminations, reaction rate depends only on the concentration of the substrate, ie., alkyl halide, and is independent of the concentration of the base. Thus,

Rate α [Alkyl halide]

Mechanismwise, the reaction involves loss of halogen as the halide ion leading to the formation of a carbonium ion in a rate determining step followed by the loss of hydrogen from the α-carbon as proton is a non-rate-determining step as shown below:-

$$R{-}CH_2{-}\underset{\underset{X}{|}}{C}H_2 \xrightarrow[\text{(Slow)}]{-\overset{\ominus}{X}} R{-}\overset{\overset{H}{|}}{C}H{-}\overset{\oplus}{C}H_2 \xrightarrow[\text{(Fast)}]{-\overset{\oplus}{H}} R{-}CH = CH_2$$

There is yet another mechanism possible for first order elimination reactions. This mechanism involves abstraction of an α-carbon as proton from the alkyl halide by the base in a fast - step followed by the loss of halogen as halide ion in a slow (ie rate - determining) step.

$$R{-}\overset{\overset{H}{|}}{C}H_2{-}\underset{\underset{X}{|}}{C}H_2 \xrightarrow[\text{(Fast)}]{\overset{\ominus}{O}Et\ (-H^+)} R{-}\overset{\ominus}{C}H{-}\underset{\underset{X}{|}}{C}H_2 \xrightarrow[\text{(Slow)}]{-\overset{\ominus}{X}} R{-}CH = CH_2$$

The above reaction is commonly referred to as E_1 cb elimination. It involves first-order kinetics but proceeds through the intermediacy of the conjugate base (ie carbanion) of the acid (ie. alkyl halide). It should be noted that E_1 reactions proceed througn carbonium ion intermediates and E_1

cb reactions through carbanion intermediates. In the former case, formation of carbonium ion is the rate determining step whereas in the latter case dissociation of carbanion is the rate-determining step.

Evidence for the above mechanisms (E_2, E_1 and E_1 cb) of elimination reactions of alkyl halides comes from spectral studies, isotope labelling data and kinetic studies. Just which mechanism is being followed in a given reaction depends on the structure of alkyl halide, nature of base, reaction conditions, etc.

Hydrolysis of alkyl halides: This reaction is a particular case of nucleophilic substitution of an alkylhalideusing hydroxyl ion as a nucleophile. This reaction has been extensively studied and its mechanism established in great detail. Sir Christopher Ingold and his group were one of the earliest researchers and have done pioneering work in this field. Essentially, the reaction involves treatment of an alkyl halide, under appropriate conditions, with an alkali to yield an alcohol.

$$\underset{\text{(Alkyl halide)}}{R\text{—}X} + \overline{O}H \longrightarrow \underset{\text{(An alcohol)}}{R\text{—}OH} + X^-$$

Mechanistically, this reaction is believed to be of two types depending on the nature of alkyl halide used.

(i) Substitution Nucleophilic Bimolecular (S_N2) : When a primary alkyl halide, such as methyl bromide, is treated with $\overline{O}H$, a primary alcohol, methyl alcohol, is obtained.

$$CH_3Br + \overline{O}H \longrightarrow CH_3 - OH + Br^-$$

Kinetically the rate of this reaction has been found to depend upon the concentration of both the methyl bromide and $\overline{O}H$ ions.

$$k_1 \; \alpha \; [CH_3Br] \;\; [OH^-]$$

This means that the rate-determining step in this reaction involves the concentration of both the reactants. In other words, the rate determining step should involve two molecular species whose concentration changes during reaction. This reaction has been referred to as **Substitution Nucleophilic Biomolecular**, designated as S_N2.

The reaction involves attack by HO^- on the alkyl halide from the rear side of the carbon-halogen bond leading to the formation of a 5-membered intermediate which has the partially formed C-OH bond and partially broken (rather polarized) C-X bond. Finally, the carbon halogen bond is completely broken, the C-OH bond is completely formed leading to inversion of tetrahedron in a fast, non-rate determining step of the reaction.

$$\bar{O}H + H\text{---}\underset{H}{\overset{H}{C}}\text{—}Br \longrightarrow HO\text{---}\underset{H}{\overset{H\quad H}{C}}\text{---}Br \longrightarrow HO\text{—}\underset{H}{\overset{H}{C}}\text{---}H + \bar{B}r$$

The inversion of the tetrahedron, in going from alkyl halide to the alcohol, is known as **Walden inversion** and has been likened to the turning of an umbrella inside out in a strong wind. The 5-membered intermediate is sterically considered to be such that the carbon and the three hydrogens are oriented like the spokes of a wheel with the carbon atom at its center and the HO –C– Br bond placed in plane perpendicular to the plane of the wheel like an axle. Evidence for this mechanism comes from the stereochemical data. When an optically pure secondary alkyl bromide was used in this reaction, an optically pure secondary alcohol was obtained with completely inverted configuration.

$$\overset{\ominus}{O}H + (C_6H_{11})(H)(H_3C)C\text{—}Br \longrightarrow HO\text{---}C(H)(C_6H_{11})(CH_3)\text{---}Br \longrightarrow HO\text{—}C(C_6H_{11})(CH_3)\text{---}H$$

(+) Alkyl bromide

[(-) Alcohol with inveried configuration]

$$\overset{\ominus}{O}H + (H_3C)(H)(H_{11}C_6)C\text{—}Br \longrightarrow HO\text{---}C(H)(CH_3)(C_6H_{11})\text{---}Br \longrightarrow HO\text{—}C(CH_3)(H)(C_6H_{11})$$

(–) Alkyl bromide

[(+) Alcohol with inverted configuration)]

(ii) Substitution Nucleophilic Unimolecular (S_N1): When a tertiary alkyl halide, such as t-butyl chloride, is treated with $H\bar{O}$, a tertiary alcohol, t-butyl alcohol, is obtained.

$$C(CH_3)_3 - Cl + H\bar{O} \longrightarrow C(CH_3)_3\text{— }OH + Cl^-$$

The rate of this nucleophilic substitution has been found to depend upon the concentration of only the alkyl halide and has been found to be independent of the concentration of hydroxyl ions.

$$k_2 \propto [C(CH_3)_3 - Cl]$$

This indicates that the rate-determining step of this reaction involves the concentration of only one of the reactants. That is to say, the rate-determining step involves only one molecular species, viz., the alkyl halide, whose concentration changes during reaction. This reaction is precisely called the *Substitution Nucleophilic Unimolecular*, and designated as S_N1.

The mechanism of this reaction involves a first, slow (rate-determining) decomposition of the tertiary alkyl halide to the tertiary carbonium ion and the halide ion followed by a fast (non-rate determining) attack by the hydroxyl ion on the carbonium ion yielding the tertiary alcohol.

$$(H_3C)_3C{-}Cl \longrightarrow (H_3C)_2\overset{\oplus}{C}{-}CH_3 + Cl^{\ominus} \xrightarrow{OH^{\ominus}} (H_3C)_3C{-}OH$$

(t-Butyl Chloride) (t- Butyl alcohol)

Since the tertiary carbonium ions are believed to be flat, trigonal, planar structures, attack by HO can occur on it either from top or from bottom. Therefore, there should be a 50:50 chance of either of the tetrahedron-shaped alcohols being formed. This has, indeed, been found to be the case when an optically pure tertiary halide was used in above reaction. For example;

$$(H_{11}C_6)(H_3C)(H_5C_2)C{-}Cl \longrightarrow (H_3C)(C_6H_{11})\overset{\oplus}{C}{-}C_2H_5 + Cl^{\ominus} \xrightarrow{HO^{\ominus}} (H_{11}C_6)(H_3C)(H_5C_2)C{-}OH + OH{-}C(C_6H_{11})(CH_3)(C_2H_3)$$

(-) Alkyl halide (optically pure) (±) Alcohol (Racemic Mixture)

$$(H_5C_2)(H_3C)(H_{11}C_6)C{-}Cl \longrightarrow (H_3C)(C_2H_5)\overset{\oplus}{C}{-}C_6H_{11} + Cl^{\ominus} \xrightarrow{OH^{\ominus}} (H_5C_2)(H_3C)(H_{11}C_6)C{-}OH + OH{-}C(C_2H_5)(CH_3)(C_6H_{11})$$

(+) Alkyl halide (optically pure) (±) Alcohol (Racemic Mixture)

It may appear in the first instance as to, why do the S_N1 and S_N2 reactions proceed by different mechanisms depending on the nature of the alkyl halide? In answer to this question it can be said that S_N1 pathway cannot be followed with primary alkyl halides due to the poor stability of the primary carbonium ions leaving no other choice than S_N2. Similarly, S_N2 mechanism cannot be observed with tertiary alkyl halides due to steric hindrance or crowding in the 5-membered transition state proposed in S_N2 reactions. The relative energy demands for both the mechanisms in respect of the two alkyl halides are shown in the energy-diagram given below.

It is obvious from the above discussion that, in general, the primary alkyl halides react by S_N2 pathway and tertiary alkyl halides by S_N1 mechanism. *In the case of secondary alkyl halides, depending on the steric nature of substituent groups and on the reaction conditions, it may react exclusively by S_N2 or by S_N1 mechanism or by a combination of both the mechanisms.*

Nucleophilic substitution - Effect of different factors: **It has been found that the course of reaction and reactivity of an alkyl halide in a nucleophilic substitution with hydroxyl ions depends on a number of factors. These factors are discussed below.**

(i) Effect of solvent: **In an S_N2 reaction, the intermediate is a 5-membered species containing the hydroxyl ion and the chloride ion bonded to the same carbon atom. This 5- membered intermediate is strongly polar, but nonionic**

Hydrolysis of alkyl halides : Energy Diagrams

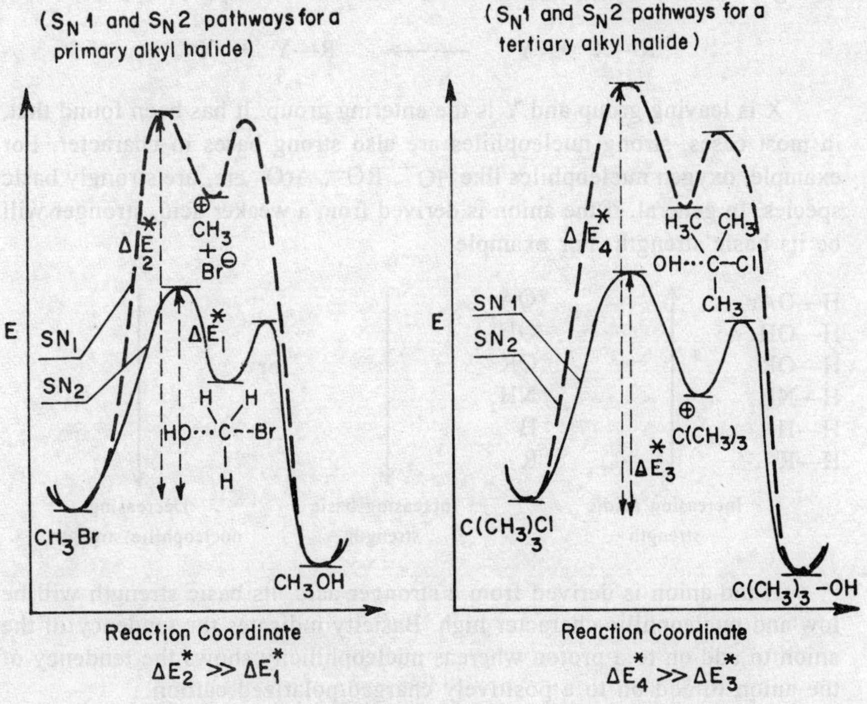

in nature. Since the trasition-state leading to this intermediate in an S_N2 reaction may be expected to resemble the same intermediate, strongly polar but non-ionising solvents can be expected to promote this reaction which is indeed found to be the case. On the other hand, in an S_N1 reaction, the intermediate is the carbonium ion, a discrete ionic species formed by the ionization of the alkyl halide, and the reagent, the hydroxyl ion and the product, a halide ion both being distinctly ionic species, will be in the solvated form. Therefore,*strongly ionizing solvents and those capable of solvation will promote S_N1 reaction.* Furthermore, the concentration of alkali does not figure in the rate equation of an S_N1 reaction. Therefore, *a low concentration of alkali will favour the S_N1 pathway of a nucleophilic substitution reaction.*

In general, it can be said that strongly polar but non-ionizing, non-solvating solvents like DMSO, DMF, etc., and a high concentration of alkali will favour S_N2 reaction. On the other hand, strongly ionizing, solvating solvents like water, methanol, ethanol, etc., and lower concentration of alkali will promote S_N1 pathway for the reaction.

(ii) Effect of nucleophiles: In the hydrolysis of alkyl halides, the hydroxyl group enters and the halide ion leaves the structure of the molecule. That is, hydroxyl ion is the entering group and halide ion the leaving group. In a general nucleophilic substitution reaction of the type;

$$R—X + Y^- \longrightarrow R—Y + X^-$$

X is leaving group and Y is the entering group. It has been found that, in most cases, strong nucleophiles are also strong bases in character. For example, oxygen nucleophiles like HO^-, RO^-, ArO^-,etc., are strongly basic species. In general, if the anion is derived from a weaker acid, stronger will be its basic strength. For example;

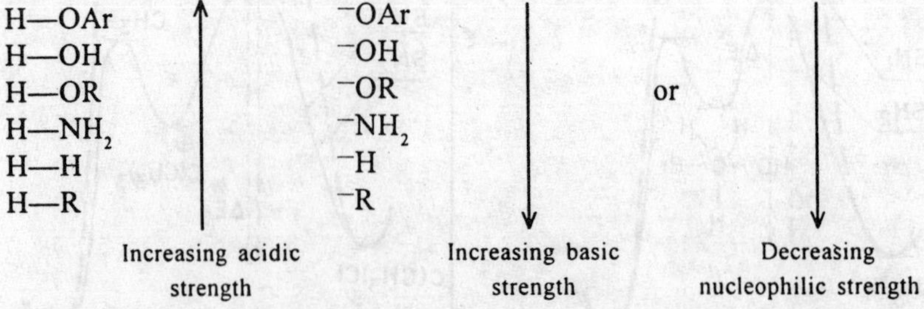

If the anion is derived from a stronger acid, its basic strength will be low and nucleophilic character high. Basicity indicates the tendency of the anion to add on to a proton whereas nucleophilicity shows the tendency of the anion to add on to a positively charged/polarized carbon.

For example, R—SH (thiol) is a stronger acid than R-OH (alcohol) and hence a stronger nucleophile. If ever,there is a competition between RS^- and RO^-, the former will react first and at a rate far greater than the latter. Furthermore,strongly basic anions have a tendency to bring about eliminations which are sometimes faster and competitive reactions compared to substitutions.

A somewhat similar situation holds in the case of leaving groups too. The stronger is the base, the more difficult it is to be good leaving group. Thus, RO^-, HO^-, ArO^-, etc., are poor leaving groups. Reaction of alcohol with halide ions does not yield even detectable amounts of alkyl halides.

$$R—OH + X^- \longrightarrow R—X + H\bar{O}$$

If alkyl halides are to be prepared from alcohols, the reagent should be HX,in which case the reaction proceeds not by substitution, but by protonation followed by ionization - addition process as shown below;

$$R—OH + HX \longrightarrow R—X + H_2O$$

$$R—OH \xrightarrow{H^+} R—OH_2^+ \xrightarrow{-H_2O} R^+ \xrightarrow{X^-} R—X$$

Due to the reversible nature of the reaction, $ZnCl_2$ is used as a catalyst in the above reaction to tie up the water formed.

Ambient Nucleophiles: Some nucleophiles possess more than one nucleophilic site in their structures. Such nucleophiles are known as **ambient nucleophiles**. In their reaction with alkyl halides, the ambient nucleophiles can give rise to more than one product on nucleophilic substitution.

For example :

Soft-Hard interaction:

$$R—X + Ph—\overset{\overset{O}{||}}{S}—O^{\ominus} Na^{\oplus} \longrightarrow R—O—\overset{\overset{O}{||}}{S}—Ph + Nax$$

(Sod. benzene sulfinate) (Alkyl benzene-sulphinate-sulphinic acid ester)

Soft-Soft interaction:

$$R—X + Ph—\overset{\overset{O}{||}}{\underset{\cdot\cdot}{S}}—O Na \longrightarrow R—\overset{\overset{O}{||}}{\underset{\underset{O}{||}}{S}}—Ph + NaX$$

(Alkyl phenyl sulphone a Sulphone)

In the above reactions, it has been found that alkyl halides react with sodium benzene sulphinate to yield predominantly the sulphone derivatives. This can be explained by the fact that the positively polarized carbon of alkyl halide constitutes a soft acidic center and in the ambient nucleophile, sulphur is the soft base, the Oxy-anion,a hard base. Since soft-soft interaction

Hard-hard interaction;

$$R{-}SO_2{-}Cl + Ph{-}\overset{O}{\overset{||}{S}}{-}\overset{\ominus}{O}\,\overset{\oplus}{Na} \longrightarrow R{-}\underset{O}{\underset{||}{\overset{O}{\overset{||}{S}}}}{-}O{-}\overset{O}{\overset{||}{S}}{-}Ph + NaCl$$

(Sulphonic-sulphinic an hydride)

Hard-soft interaction;

$$R{-}SO_2{-}Cl + Ph{-}\overset{O}{\overset{||}{S}}{-}O\,Na \longrightarrow R{-}\underset{O}{\underset{||}{\overset{O}{\overset{||}{S}}}}{-}\underset{O}{\underset{||}{\overset{O}{\overset{||}{S}}}}{-}Ph + NaCl$$

(a disulphone)

is more favoured, the ambient nucleophile attacks through its stronger nucleophilic center-sulphur - on the alkyl halide yielding a derivative of sulphone. Evidence for this argument is provided by the fact that the reaction of sodium benzene sulphinate with sulphonyl halides yields the sulphonic-sulphinic anhydrides as the product rather than the disulphone derivatives.

In the above case,the sulphonyl sulphur of sulphonyl chloride constitutes a hard acid and therefore prefers to react with the harder base the oxy-anion of the sulphinate salt(first reaction).

(iii) Effect of structure: If alkyl halides are arranged in the order of methyl, primary, secondary and tertiary, etc., the pathway followed by these in their nucleophilic substitutions obey the order given below;

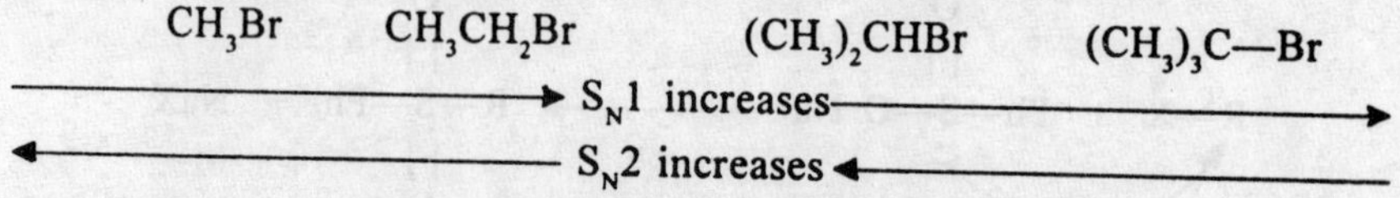

If allylic halides, vinylic halides and tert alkyl halides are considered for any simple substitution reaction, the following has been found to be the order of reactivity:

Allyl > t-Alkyl > Vinylic

This order can readily be explained by considering the structures of the halide and their intermediates in the substitution reactions. When allylic halides react, the intermediates are the allylic cations, which are highly stabilized by resonance and so are readily formed.

$$CH_2{=}CH{-}CH_2{-}Cl \longrightarrow CH_2{=}CH{-}\overset{\oplus}{C}H_2 \xrightarrow{HO^-} CH_2{=}CH{-}CH_2OH$$

$$\updownarrow$$

$$\overset{\oplus}{C}H_2 - CH = CH_2$$

On the other hand, in the case of vinyl halides, reactions are highly sluggish due to back-donation resonance commonly found in them, reducing highly the polarizability of the carbon-halogen bond thereby preventing attack by the nucleophile.

$$\underset{\text{(Vinyl chloride)}}{H_2C = CH{-}Cl} \xrightarrow[\text{Very difficult}]{\overset{\ominus}{O}H} \underset{\text{(Vinyl alcohol)}}{H_2C = CH{-}OH} \rightleftharpoons \underset{\text{(Acetaldehyde)}}{H_3C{-}\overset{\overset{O}{||}}{C}{-}H}$$

$$H_2C = CH{-}Cl \longleftrightarrow H_2\overset{\ominus}{C}{-}CH = \overset{\oplus}{Cl}$$

Benzyl halides possess a reactivity comparable to that of allyl halides towards nucleophiles. This is due to the high stability of the intermediate benzyl cations facilitating their formation. For example;

$$Ph{-}CH_2{-}Cl \xrightarrow{\text{solvent}} \underset{Cl^-}{PhCH_2^+} \xrightarrow{HO^-} Ph{-}CH_2{-}OH$$

The great stability of benzyl cations is due to resonance.

$$C_6H_5{-}\overset{\oplus}{C}H_2 \longleftrightarrow \text{(ortho } \oplus\text{)}{=}CH_2 \longleftrightarrow \text{(para } \oplus\text{)}{=}CH_2 \longleftrightarrow \text{(ortho } \oplus\text{)}{=}CH_2 \longleftrightarrow C_6H_5{-}\overset{\oplus}{C}H_2$$

If more than one phenylgroup is present on the carbon, the reaction proceeds by increasingly S_N1 pathway and at a considerably faster rate. Thus,

$$PhCH_2{-}Cl < (Ph)_2CH{-}.Cl < (Ph)_3C{-}Cl$$

$\longrightarrow$ increasingly S_N1 reaction $\longrightarrow$

The reactivity of triphenyl methyl chloride is highest due to the fact that three phenyl groups stabilize far greatly the intermediary carbonium ion by resonance.

8-5 ARYL HALIDES

8-5-1 Methods of preparation

The methods employed for the syntheses of alkyl halide cannot be used for the preparation of arylhalides and vice-versa. The following are some of the methods used for the preparation of aryl halides.

(a) By direct halogenation: The aryl halides are usually prepared by electrophilic substitution of aromatic hydrocarbons using the halogenation reaction. For example;

$Cl_2/AlCl_3/\Delta$

(Chlorobenzene)

CH_3 CH_3 Br

$Br_2/FeBr_3/\Delta$ +

(Toluene)

Br

(p-Bromotoluene) (major product) (o-Bromotoluene) (Minor product)

$NHCOCH_3$ $NHCOCH_3$ $NHCOCH_3$ Br

$Br_2/AcOH$ 0 - 5° +

(Acetanilide)

Br

(p-Bromoacetanilide) (Very major) (o-Bromoacetanilide) (Minor product)

OH OH Br Br

$Br_2/AcOH$

(Phenol)

Br

(2,4,6-Tribromophenol)

NO_2 NO_2 NO_2 Cl

$Cl_2/AlCl_3/\Delta$ +

(Nitrobenzene)

Cl

(o-Chloronitrobenzene) (p-chloronitrobenzene)

The last reaction (Friedel-Crafts halogenation of nitrobenzene) does not take place to give products because halogenation is a weak reaction and nitro group, besides being a meta-directing group, is a strong deactivator of the benzene ring. In fact, the aryl halides, p-chloronitrobenzene and o-chloronitrobenzene, are prepared by the following sequence of reactions.

(Benzene) $\xrightarrow{Cl_2/AlCl_3/\Delta}$ (Chlorobenzene) $\xrightarrow[H_2SO_4]{HNO_3}$ (p-Chloronitrobenzene) + (o-Chloronitrobenzene)

(b) From diazonium salts: Aryl diazonium salts, when treated with cuprous halides, undergo loss of nitrogen yielding aryl halides. For example;

(Benzenediazonium Chloride) $\xrightarrow{CuCl/\Delta}$ (Chlorobenzene) + $N_2 \uparrow$

The aromatic diazonoium salts are obtained from the corresponding nitro compounds by reduction followed by reaction with nitrous acid.

(Benzene) $\xrightarrow[H_2SO_4]{HNO_3}$ (Nitrobenzene) $\xrightarrow{Redn.}$ (Aniline) $\xrightarrow[HCl]{HONO}$ (Benzenediazonium Chloride)

The diazonium salt method complements the direct halogenation method in the preparation of aryl halides. For example, synthesis of fluorobenzene from benzene by fluorination is an explosively violent and hazardous reaction but fluorobenzene can, however, be conveniently prepared via the diazonium salt route. Synthesis involves treatment of the diazonium chloride with sodium fluoroborate to yield benzene diazonium fluoroborate as a nice precipitate followed by pyrolysis of the latter to give fluorobenzene.

(Aniline) $\xrightarrow{HONO/\ HCl}$ (Benzenediazonium Chloride) $\xrightarrow{NaBF_4}$ (Benzenediazonium fluroborate) $\downarrow$ + NaCl

(Benzenediazonium fluroborate) $\xrightarrow{\Delta}$ (Flurobenzene) + $N_2 \uparrow$ + $BF_3 \uparrow$

Similarly, *m*-dibromobenzene cannot be synthesized from benzene or from bromobenzene by Friedel-Crafts bromination. It can,however, be indirectly prepared from *m*-dinitrobenzene via the diazonium salt route.

$$C_6H_6 \xrightarrow[H_2SO_4,\ 50-60^\circ]{HNO_3} C_6H_5NO_2 \xrightarrow[H_2SO_4,\ 90-100^\circ]{HNO_3} m\text{-}C_6H_4(NO_2)_2 \xrightarrow{\text{Redn.}} m\text{-}C_6H_4(NH_2)_2$$

(m-Dinitrobenzene) (m-Diaminobenzene)

$$m\text{-}C_6H_4(NH_2)_2 \xrightarrow[HBr]{HONO} m\text{-}C_6H_4(\overset{\oplus}{N_2}\overset{\ominus}{Br})_2 \xrightarrow{CuBr/\Delta} m\text{-}C_6H_4Br_2$$

(m-Diaminobenzene) (m-Tetrazonium dibromide) (m-Dibromobenzene)

(c) By Hunsdiecker reaction: The method involves heating of a silver salt of carboxylic acid with halogens to yield aryl halides. For example;

$$C_6H_5COOAg + Br_2 \longrightarrow C_6H_5Br + CO_2 + AgBr\downarrow$$

The method is not a very important one since it gives lower yields of products and silver salts of carboxylic acids are very expensive.

8-5-2 Physical properties

The aryl halides are non-associated substances and possess normal melting and boiling points. In spite of their polarity, due to lack of hydrogen-bonding characteristics and due to a larger hydrocarbon portion, aryl halides are usually insoluble in water. The aryl halides are denser than water and many are pleasant smelling substances.

8-5-3 Chemical properties

The aryl halides are much less reactive substances when compared to alkyl halides. The two types of halides are constitutionally different from each other and exhibit wide variation in physical and chemical behavior. The differences in chemical reactivity are due to the nature of carbon-halogen bonds in the two cases. Following are some of the important reactions undergone by aryl halides.

(i) Wurtz-Fittig reaction: Aryl halides react with alkyl halides in the presence of sodium metal to yield arenes. For example;

$$C_6H_5Br + 2Na + Br-C_2H_5 \longrightarrow C_6H_5C_2H_5 + 2\,NaBr$$

When two moles of an aryl halides react with sodium metal to produce a diaryl hydrocarbon, then it is known as **Fittig reaction.**

R–C_6H_4–Br + 2 Na + Br–C_6H_4–R ⟶ R–C_6H_4–C_6H_4–R + 2 NaBr

A diaryl compd.

Treatment of an aryl iodide with copper metal powder to yield a derivative of biphenyl is known as **Ullmann reaction.**

R–C_6H_4–I + Cu + I–C_6H_4–R ⟶ R–C_6H_4–C_6H_4–R + CuI_2

(ii) Grignard formation: Aryl halides react with Mg metal in an ethereal solvent to form *Grignard reagents.* The reaction is very slow when compared to that of alkyl halides, sometimes requiring a higher boiling solvent(e.g.THF) and longer reaction times. The Grignard reagent, obtained, may be treated with water to yield the aromatic hydrocarbon. This reaction is sometimes used in a multi-step synthesis to dehalogenate halohydrocarbons.

R–C_6H_4–Br $\xrightarrow{Mg}$ R–C_6H_4–MgBr $\xrightarrow{H_2O}$ R–C_6H_4–H + Mg(Br)(OH)

(iii) Electrophilic substitution : The aryl halides, by virtue of the presence of an aromatic ring in them, undergo electrophilic substitution to yield ring-substituted products. The halogen substituent deactivates the aromatic ring but directs an incoming substitutent to *o*-and *p*-positions. For example;

C_6H_5Cl (Chlorobenzene) $\xrightarrow{HNO_3/H_2SO_4}$ (4-Nitrochlorobenzene) + (2-Nitrochlorobenzene)

C_6H_5Br (Bromobenzene) $\xrightarrow{H_2SO_4/SO_3}$ (o- Bromobenzene sulphonic acid) + (o-Bromobenzene sulphonic acid)

(iv) Nucleophilic substitution: The halogen substituent of an aryl halide is involved in resonance with aryl ring. Furthermore, the polarizability of carbon-halogen bond in aryl halides being quite small, the aryl halides do not undergo substitution of halogen by nucleophiles, a reaction very common with alkyl halides. The aryl halides undergo nucleophilic substitution with great reluctance and even if they do so,only under forcing conditions. If an activating group, such as nitro, is present on the ring,substitution is greatly facilitated. For example;

Cl — Aq.NaOH/100°, 4hrs. ✗→ OH

Cl, NO_2 — Aq.NaOH/100°, 4hrs. → OH, NO_2 (≃ 60%)

Cl, NO_2, NO_2 — Aq.NaOH/100°, 1 hr. → OH, NO_2, NO_2 (> 90%)

Cl, O_2N, NO_2, NO_2 — HOH/RT/4hrs. → OH, O_2N, NO_2, NO_2 (≃ quantitative)

Since the nucleophiles in many cases happen to be strong bases, in the reaction of simple aryl halides with such species the substitution proceeds by complicated mechanism. Thus, in the reactions of *p*-chlorotoluene with

Cl, CH_3 — $\overset{\ominus}{N}H_2$ (-HCl) → CH_3 — NH_3 → NH_2, CH_3 + NH_2, CH_3

(p-Chlorotoluene) (A Benzyne) (p-Aminotoluene) (m-Aminotoluene)

$NaNH_2$, in addition to the desired substitution product, *p*-aminotoluene, a considerable amount of *m*-aminotoluene is formed. The reaction is believed to proceed through the intermediacy of a benzyne intermediate.

8-5-4 Uses of halogen compounds

Besides their intensive use in organic syntheses as crucial alkylating reagents or as crucial intermediates, the organic halogen compounds find commercial use as insecticides, plant protecting agents, solvents, antiseptics, anaesthetics refrigerants, drug intermediates etc. Thus, compounds like chloroform ($CHCl_3$), carbon tetrachloride (CCl_4), methylene dichloride (CH_2Cl_2), symmetrical tetrachlorethane (CH_2Cl_2—CH_2Cl_2) etc., find extensive use as solvents. Iodoform is a powerful antiseptic used in hospitals and in surgical dressings. In Surgery, chloroform, vinyl chloride and ethylenedichloride are used as general anaesthetics. DDT and BHC have been used for ages as general insecticides and plant protecting agents. However,in most of the advanced countries of the world,the use of DDT and BHC has been discontinued, in view of the ecological disturbances these insecticides produce and the public outcry against them by the Health Agencies.Notable in this regard is the publication in 1962 by Rachel Carson's *Silent Spring*. These insecticides have now been replaced by more safer and less stable ones like chlordane, methoxychlor, dieldrin etc. The organic fluoro compounds like the chlorofluoroalkanes are used as refrigerants. Several aryl halides are used as high volume intermediates in drug manufacture. For example, *p*-nitro-chlorobenzene is an important intermediate used in paracetamol manufacture.

8-5-5 Fluoro compounds

Alkyl fluorides are generally prepared by the **Swart's reaction** which involves heating an alkyl halide with a metal fluoride like SbF_3, HgF_2, etc.

$$R—X + M—F \longrightarrow R—F + M—X$$

Yields are found to be quite low when KF is used as the metal fluoride. The alkyl fluorides may,alternatively, be prepared from alkenes by reaction with HF.

The fluoro compounds are highly inert, non associated substances. They are usually low boiling liquids, easily vapourizable, approaching at times ideal gas behavior. Therefore the fluorocarbons are used very frequently as refrigerants. Polyfluorcarbons, due to their very wide range as liquids, inertness and viscous nature, are used as excellent lubricants.

8-6 SELF-TEST QUESTIONS

Choose the correct answer from the following questions:

8-6-1 Which one of the following compounds is a secondary alkyl halide?

(A) *n*-propyl chloride (B) Benzyl chloride

(C) 2-Chlorobutane (D) 2,2–Dimethylethyl chloride

8-6-2 Which one of the following compounds is an aryl halide?

(A) Benzyl chloride (B) Benzoyl chloride

(C) Chlorobenzene (D) Phenylacetyl chloride

8-6-3 Alkyl halides react with nucleophiles to give

(A) Substitution products (B) Addition products

(C) Rearrangement products (D) None of the three

8-6-4 Reaction of benzyl chloride with aq.NaOH yields

(A) Benzoic acid (B) Benzyl alcohol

(C) Benzaldehyde (D) A mixture of all the three compounds

8-6-5 Reaction of benzyl alcohol with $POCl_3$ results in

(A) Chlorobenzene (B) Benzyl chloride

(C) Benzoyl chloride (D)A mixture of all the three compounds

8-6-6 Wurtz reaction involves

(A) Heating two moles of an alkyl halide with one mole of sodium.

(B) Heating two moles of an acid halide with one mole of sodium.

(C) Heating two moles of an alkyl halide with one mole of zinc.

(D) All the three reactions.

8-6-7 Benzyl chloride can be prepared from toluene

(A) By treatment with chlorine in the presence of anh.$AlCl_3$.

(B) By treatment with chlorine in the presence of anh.$FeCl_3$.

(C) By treatment with chlorine in the presence of UV light.

(D) By none of the three methods.

8-6-8 Markovnikov addition of HCl to 1-hexene yields predominantly

(A) 1-Chlorohexane (B) 1, 2-Dichlorohexane

(C) Mixture of A and B (D) 2-Chlorohexane.

8-6-9 Chlorobenzene can be prepared from

(A) Phenol by refluxing with $POCl_3$

(B) Aniline by refluxing with HCl.

(C)Benzoic acid by decarboxylation.

(D)Benzene by treatment with Cl_2 gas in presence of anh.$AlCl_3$.

8-6-10 Chloronitrobenzene can be synthesized by

(A) Chlorinating nitrobenzene in the presence of $AlCl_3$.

(B) Nitrating chlorobenzene in the absence of $AlCl_3$.

(C) Treating benzene with nitrosyl chloride.

(D) Treating chlorobenzene with nitrous acid.

8-7 SHORT AND ESSAY TYPE QUESTIONS.

8-7-1 Describe any two methods for the preparation of alkyl halides taking at least one example of each method.

8-7-2 What are Grignard reagents? How are they prepared? What is the reaction of Grignard reagents with water? (N.U., Sept., 1994)

8-7-3 When 2-chlorobutane reacts with alc. KOH, dehydro chlorination occurs along with substitution. What are the products formed?

8-7-4 How are aryl halides prepared? Explain any two methods giving one example in each method. (O.U., Mar., 1994)

8-7-5 What are S_N1 and S_N2 mechanisms for the hydrolysis of alkyl halides?

8-7-6 What is the effect of solvent and nucleophile in S_N1 and S_N2 substitution reaction of alkyl halides?

8-7-7 How is the hydrolysis of alkyl halides proceeding by S_N1 or S_N2 pathways dependent upon the structure of the alkyl halide? (N.U., Mar., 1995)

8-7-8 Give the products for the following reactions:

(a) $CH_3—CH_2—CH_2—CH_2—Br + NaOMe \longrightarrow$

(b) $Ph—CH_2—Cl + KCN \longrightarrow$

(c) $Ph—CO—CH_2Br + EtS^- \longrightarrow$

(d) Bromobenzene $\xrightarrow{Mg}$ Product $\xrightarrow{H_2O}$ Product

(e) $Ph—CO—CH_2—Cl \xrightarrow{CN^-}$ Product

(f) Benzyl alcohol $\xrightarrow{POCl_3}$ Product $\xrightarrow{NaI}$ Product

8-7-9 How can one distinguish by qualitative test-tube experiments, the following pairs of compounds?

(i) Ethyl alcohol and ethyl chloride

(ii) Benzyl alcohol and benzyl bromide

(iii) Benzyl chloride and *o*-chlorotoluence

(iv) Chlorobenzene and benzyl chloride

(v) Ethyl chloride and vinyl chloride

(vi) Alkyl chloride and vinyl chloride

8-7-10 Give an account of nucleophillic substitution reactions undergone by alkyl halides. (SKU., Apr.1990)

8-7-11 How is ethyl bromide prepared from

(i) An alkane (ii) an alkene (iii) an alcohol? (O.U., Apr.1991)

8-7-12 Formulate the mechanism of chlorination of methane? What are the products obtained on reaction of phenyl magnesium bromide with water? (O.U., Apr. 1990)

8-7-13 How can one distinguish by qualitative test-tube experiments between *p*-chlorotoluene and benzyl chloride?

Describe a method to convert 2-bromopropane to acetone. (O.U.,Apr.1989)

SOLUCTIONS TO PROBLEMS & EXCERCISES

8-6-1 C	**8-6-2 C**	**8-6-3 A**	**8-6-4 B**	**8-6-5 B**	**8-6-6 A**
8-6-7 C	**8-6-8 D**	**8-6-9 D**	**8-6-10 B**		

8-7-8 (a) $CH_3CH_2CH_2CH_2 \rightarrow OMe$ (Major) + $CH_3—CH_2—CH = CH_2$ (Minor)

(b) $PhCH_2CN$

(c) $PhCOCH_2—S—Et$

(d) Bromobenzene $\xrightarrow{Mg}$ Ph—MgBr $\xrightarrow{HOH}$ Ph—H + Mg(Br)(OH)

(e) $Ph—CO—CH_2CN$

(f) $Ph—CH_2—OH \xrightarrow{POCl_3} Ph–CH_2–Cl \xrightarrow{NaI} PhCH_2—I$

8-7-9 (i) Ethyl alcohol on treatment with $AgNO_3$ will not give any reaction. On the other hand, with aq.$AgNO_3$ ethyl chloride will give a precipitate of AgCl.

(ii) Benzyl alcohol will not react with alc.$AgNO_3$ whereas benzylbromide will at once give a precipitate of AgBr.

(iii) Benzyl chloride with alc.$AgNO_3$ will give a precipitate of AgCl whereas *o*-chlorotoluene will not react with it.

(iv) Chlorobenzene will not react with alc. $AgNO_3$ but benzyl bromide will give a precipitate of AgBr.

(v) Ethyl chloride will react with $AgNO_3$ (alc) to give a precipitate of AgCl, whereas vinyl chloride will not respond to this test.

(vi) Allyl chloride will precipitate AgCl from alc.$AgNO_3$ solution whereas vinyl chloride will not react with it.

9

HYDROXY COMPOUNDS

9-1 INTRODUCTION

The Hydroxy compounds may formally be derived by replacing one or more hydrogens of hydrocarbons by Hydroxy groups. The compound obtained by replacing the hydrogen of an aliphatic or alicyclic hydrocarbon by a hydroxy group is known as an **alcohol**. The compound obtained by formally replacing the hydrogen of an aromatic hydrocarbon by a hydroxy group is known as a **phenol**.

The Hydroxy compounds, apart from being extremely important raw materials in several synthetic processes, find extensive commercial uses. For example, ethyl alcohol is a crucial starting material for manufacture of several industrial chemicals like acetic acid, ethyl acetate, diethyl ether, etc.

CH_2OH
|
$CHOH$
|
CH_2OH

(Glycerol from fats & oils)

(Menthol from peppermint oil)

(α - Terpineol from Neroil & Camphor oils)

(Cholesterol from animal tissue)

(D(+)–Glucose from fruits)

Similarly, Phenol is used in the manufacture of several polymers, resins, etc. Many hydroxy compounds occur naturally and are of biological importance too. Some of these compounds together with their structures are shown on the previous page.

9-2 CLASSIFICATION

Hydroxy compounds may be classified into three types depending upon the number of hydroxyl groups present in the compound. These are: monohydroxy, di-hydroxy and poly-hydroxy compounds. The monohydroxy compounds may, again, be divided into two types - alcohols and phenols - depending upon the attachment of hydroxy group to an aliphatic/alicyclic hydrocarbon or to an aromatic ring. The alcohols may further be divided into three types, primary, secondary or tertiary, on the basis of attachment of the hydroxyl group to a primary, secondary or tertiary carbon atom. These classifications are summarised in chart-1 given below with examples.

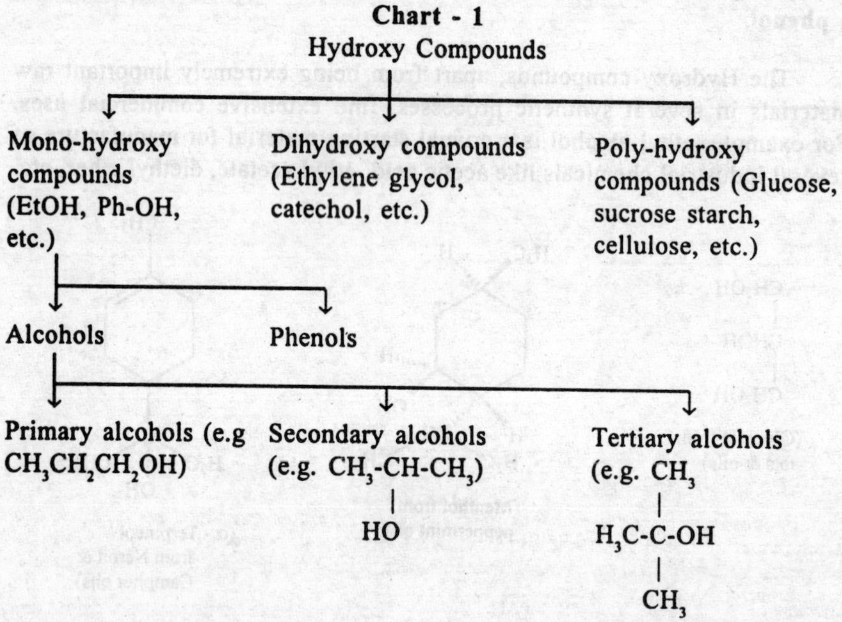

9-3 NOMENCLATURE

In the older system, the hydroxy compounds were called by their classical names. For example: CH_3OH as methyl alcohol, $CH_3—CH_2—OH$ as ethyl alcohol, CH_3-CH(OH)-CH_3 as i-propyl alcohol, Ph-OH as phenol, etc. In the IUPAC system, the class ending for alcohol is *ol*. Compounds are named as derivatives of hydrocarbons and then indicating the position of hydroxyl group by a number followed by the ending *ol*. For example; ethyl

$$CH_3—CH_2—\underset{\displaystyle OH}{\underset{|}{CH}}—CH_3$$

(2-Butanol)

$$(H_3C)_2CH—CH_2—\underset{\displaystyle OH}{\underset{|}{CH_2}}—CH_2OH$$

(4-Methyl-1-pentanol)

Ph-CH_2-OH
Phenyl methanol

(2-Chloro-4-nitrophenol or 2-chloro-1-hydroxy-4-nitrobenzene)

alcohol would be called ethanol, i-propyl alcohol as 2-propanol, etc. Phenols are named as derivatives of benzene, the hydroxyl group being treated as a substituent. The examples given above illustrate these points.

9-4 PREPARATION OF ALCOHOLS

There are several methods for the preparation of alcohols. However, the following are considered to be fairly general and important among them, based on the nature of the starting material.

9-4-1 From allkyl halides

Alkyl halides undergo hydrolysis with alkali to yield alcohols. Primary, secondary and tertiary alkyl halides respectively, yield primary, secondary and tertiary alcohols. Thus,

$$R—X + HO^- \longrightarrow R—OH + X^-$$

The only disadvantage of this method is that, sometimes alkene is an undesirable side-product in this reaction.

9-4-2 From alkenes

Alkenes react with cold, conc. H_2SO_4, to form alkyl hydrogen sulphates following the Markovnikov's rule. The alkyl hydrogen sulphates on boiling with water yield alcohol. The overall reaction involves addition of elements of water in the Markovnikov fashion, sulphuric acid acting as an acid catalyst.

$$R—CH = CH_2 \xrightarrow{H_2O/H_2SO_4} R—\underset{\displaystyle OH}{\underset{|}{CH}}—CH_3$$

The mechanism of this reaction is as follows:

$$R—CH=CH_2 \xrightarrow[(+H^{\oplus})]{H_2SO_4} R—\overset{\oplus}{C}H—CH_3 \xrightarrow{HSO_4^{\ominus}} R—CH(—O—SO_2—OH)—CH_3$$

$$R—CH(—O—SO_2—OH)—CH_3 \longleftrightarrow R—CH(—O—\overset{++}{S}(O^{\ominus})_2—OH)—CH_3 \xrightarrow{H_2\ddot{O}} R—CH(O^{\ominus})—CH_3 + H—\overset{\oplus}{O}(H)—S(O^{\ominus})_2—OH$$

$$\longrightarrow R—CH(OH)—CH_3 + HO—SO_2—OH$$

If the isomeric alcohol is desired, involving anti-Markovnikov addition of elements of water to the alkene, the reaction is carried out in the presence of diborane followed by treatment with alkaline hydrogen peroxide.

$$R—CH=CH_2 \xrightarrow[\text{(Diborane)}]{B_2H_6} (R—CH_2—CH_2)_3B \xrightarrow[HO^-]{H_2O_2\,/} 3R—CH_2—CH_2OH \text{ (Normal chain alcohol)} + B(OH)_3$$

The reactions appear to proceed by the following mechanism;

$$R—CH=CH_2 \xrightarrow[(BH_3 \times 2)]{B_2H_6} \left[R—\overset{\oplus}{C}H—CH_2—\overset{\ominus}{B}H_3\right] \equiv R—\overset{\oplus}{C}H—CH_2—H—\overset{\ominus}{B}H_2$$

$$\longrightarrow [R—CH_2—CH_2—BH_2] \xrightarrow{R—CH=CH_2} [(R—CH_2—CH_2)_2BH]$$

$$\xrightarrow{R—CH=CH_2} [(R—CH_2—CH_2)_3B] \xrightarrow[(+\,O—O—H)]{^-OH+H_2O_2}$$

$$(R{-}CH_2{-}CH_2)_3\overset{\ominus}{B}{-}O{-}O{-}H \equiv R{-}CH_2{-}CH_2{-}\overset{CH_2-CH_2-R}{\underset{CH_2-CH_2-R}{\overset{\ominus}{B}}}{-}O{-}O{-}H$$

$$\longrightarrow \left[R{-}CH_2{-}CH_2{-}\overset{CH_2-CH_2-R}{B}{-}O{-}CH_2{-}CH_2{-}R + \overset{\ominus}{O}H \right] \longrightarrow$$

$$[\, R{-}CH_2{-}CH_2{-}\overset{OH}{B}{-}CH_2{-}CH_2{-}R + \overset{\ominus}{O}{-}CH_2{-}CH_2{-}R \,] \longrightarrow$$

$$R{-}CH_2{-}CH_2{-}\overset{OH}{B}{-}CH_2{-}CH_2{-}R + HO{-}CH_2{-}CH_2{-}R$$

$$R{-}CH_2{-}CH_2{-}\underset{OH}{B}{-}CH_2{-}CH_2{-}R \longrightarrow\longrightarrow\longrightarrow 2RCH_2CH_2OH + B(OH)_3$$

The main advantage of the diborane method over the sulphuric acid method lies in the fact that normal chain alcohols can readily be prepared which are not otherwise easily preparable.

9-4-3 From carbonyl compounds

Alcohols can be prepared from carbonyl compounds by two methods;

(a) Nucleophilic addition of Grignard reagents to aldehydes and ketones.

(b) Reductionof c arbonylderivativessuch as aldehydes, ketones, carboxylic acids, esters, amides, acid chlorides etc., by suitable reducing agents.

These methods are discussed below.

(a) By reaction with Grignard reagents: This is one of the most versatile methods for the laboratory synthesis of all tyes of alcohols viz., primary, secondary and tertiary. The method merely involves addition of Grignard reagent to the carbonyl compound followed by addition of water after sometime to the reaction mixture in order to bring about hydrolysis of the Grignard adduct.

$$R{-}X + Mg \xrightarrow{Et_2O/I_2/\Delta} R{-}MgX$$

$$R^1R^2C{=}O + R{-}MgX \longrightarrow R^1R^2C(R)(OMgX) \xrightarrow{H_2O} R^1R^2C(R)(OH) + Mg(X)X$$

The mechanism of Grignard addition is as follows;

$$R^1R^2C{=}O \longleftrightarrow R^1R^2\overset{\oplus}{C}{-}\overset{\ominus}{O}\ ; \qquad R{-}MgX \equiv \overset{\delta-}{R}{-}\overset{\delta+}{MgX}$$

$$R^1R^2\overset{\oplus}{C}{-}\overset{\ominus}{O} + \overset{\delta-}{R}{-}\overset{\delta+}{MgX} \longrightarrow R^1R^2C(R)(O^{-}MgX) \xrightarrow{H_2\ddot{O}} R^1R^2C(R)O^{\ominus} + Mg(X)(\overset{\oplus}{O}H_2) \xrightarrow{H_2O} R^1R^2C(R)(OH) + Mg(X)(OH)$$

When formaldehyde is used as the carbonyl reagent, the product is a primary alcohol. When any other aldehyde is used as the carbonyl reagent, the product is a secondary alcohol and ketone yields a tertiary alcohol.

$$H_2C{=}O + R{-}MgX \longrightarrow H_2C(R)(OMgX) \xrightarrow{H_2O} H_2C(R)(OH) + Mg(H)(OH)$$

(Formaldehyde)

$$HR^1C{=}O + R{-}MgX \longrightarrow HR^1C(R)(OMgX) \xrightarrow{H_2O} HR^1C(R)(OH) + Mg(X)(OH)$$

(An aldehyde)

$$R^1R^2C{=}O + R{-}MgX \longrightarrow R^1R^2C(R)(OMgX) \xrightarrow{H_2O} R^1R^2C(R)(OH) + Mg(X)(OH)$$

(A Ketone)

It is obvious from the above equations that in the preparation of secondary and tertiary alcohols, appropriate choice of alkyl groups in the substrate (a carbonyl compound) and the reagent (a Grignard derivative), can be made depending on the availability of suitable raw materials. In the synthesis of primary alcohols, instead of formaldehyde even ethylene oxide can be used.

$$\underset{\text{(Ethylene oxide)}}{CH_2\!-\!CH_2 \text{ (via O)}} + \overset{\delta-}{R}\!-\!\overset{\delta+}{MgX} \longrightarrow R\!-\!CH_2\!-\!CH_2\!-\!OMgX \xrightarrow{H_2O} R\!-\!CH_2\!-\!CH_2\!-\!OH$$

(Ethylene oxide)

(b) By reduction: In this method, a derivative of carbonyl compound is reduced to yield an alcohol. Only primary and secondary alcohols can be prepared by this method. A variety of reducing agents can be used in this reaction.

$$\underset{\text{(Aldehyde)}}{R(H)C=O} \xrightarrow[\text{agent}]{\text{Reducing}} \underset{\text{(Primary alochol)}}{R\!-\!CH_2\!-\!OH}$$

$$\underset{\text{(Ketone)}}{R(R^1)C=O} \xrightarrow[\text{agent}]{\text{Reducing}} \underset{\text{(Secondary alcohol)}}{R_2CH\!-\!OH}$$

Reducing agent = $LiAlH_4$, H_2/Catalyst, $Na(Hg)/H_2O$, Fe/H^+ etc.

In the reduction involving $LiAlH_4$, the reaction proceeds by simple nucleophilic addition of complexed hydride ion to the carbonyl carbon, followed by proton abstraction during quenching by water.

$$R(R^1)C=O + \overset{\ominus}{AlH_4} \xrightarrow[Li^{\oplus}]{(+H^{\ominus})} R(R^1)C(H)\!-\!\overset{\ominus}{O}Li^{\oplus} + AlH_3$$

$$R(R^1)C(H)\!-\!\overset{\ominus}{O}Li^{\oplus} \xrightarrow{H_2O} R(R^1)C(H)\!-\!OH$$

When hydrogen over a catalyst is used as the reducing agent, the reaction involves simple *cis*-addition of hydrogen to the carbonyl group.

$$\underset{\text{(a Ketone)}}{RR^1C=O} + H_2 \longrightarrow \underset{\text{(a secondary alcohol)}}{RR^1C(H)OH}$$

Metal-in-acid reductions usually involve electron transfer processes, radicals/radical-anions being intermedicates in the reaction. For example, consider reduction with Na(Hg).

$$RR^1C=O \xrightarrow{Na} \underset{\text{(Radical anion)}}{[RR^1C=O]^{\overset{\bullet}{-}}} + Na^+$$

$$[RR^1C=O]^{\overset{\bullet}{-}} \longleftrightarrow RR^1\dot{C}-O^{\ominus} \xrightarrow{H^{\oplus}} RR^1\dot{C}-OH$$

$$RR^1\dot{C}-O^{\ominus} \xrightarrow{HOH} RR^1\dot{C}-OH + \overset{\ominus}{OH}$$

$$RR^1\dot{C}-OH + Na \longrightarrow RR^1\overset{\ominus}{C}-OH + \overset{\oplus}{Na}$$

$$RR^1\overset{\ominus}{C}-OH \xrightarrow[HOH]{H^{\oplus}\text{ or}} RR^1C(H)OH$$

When esters are used as the source of carbonyl compounds, reduction can be carried out using $LiAlH_4$ or H_2/catalyst to yield primary alcohols. Metal/acid reagents are generally not found suitable for complete reduction to the alcohol stage.

$$\underset{\text{(Ester)}}{R'O-\overset{R}{C}=O} \xrightarrow[\text{H}_2\text{/Catalyst}]{\text{LiAlH}_4 \text{ or}} \underset{\text{(A primary alcohol)}}{R-CH_2-OH}$$

These reductions appear to proceed through the following mechanisms;

With Lithium aluminium hydride;

$$R(R'O)C=O \xrightarrow[(+H^{\ominus})]{\overset{\ominus}{AlH_4}} R(R'O)C(H)-O^{\ominus}Li^{\oplus} + AlH_3$$

$$R(R'O)C(H)-O^{\ominus}Li^{\oplus} \xrightarrow[(+H^{\ominus})]{\overset{\ominus}{AlH_4}} \left[R(H)C(H)-O^{\ominus}Li^{\oplus} + AlH_3 + \overset{\ominus}{OR^1} \right]$$

$$\left[R(H)C(H)-OLi + AlH_3 + \overset{\ominus}{OR^1} \right] \xrightarrow{HOH} \underset{\text{(Primary alcohol)}}{R(H)C(H)-OH} + Al(OH)_3 + R^1-OH$$

With hydrogen/catalyst

$$R(R'O)C=O \xrightarrow{H_2\text{/Catalyst}} R(R'O)C(H)-O-H \xrightarrow{-R_1OH} \left[R(H)C=O \right]$$

$$\left[R(H)C=O \right] \xrightarrow{H_2\text{/Cat.}} \underset{\text{(Primary alcohol)}}{R(H)C(H)-OH}$$

When amides are used as starting materials, the only suitable reducing agent is H_2/catalyst. Thus,

The reaction goes through the following mechanism:

$$\underset{H_2N}{\overset{R}{>}}C=O \xrightarrow{H_2/Catalyst} \underset{H_2N}{\overset{R}{>}}C\underset{OH}{\overset{H}{<}} \equiv \underset{H_2N}{\overset{R}{>}}C\underset{O-H}{\overset{H}{<}} \longrightarrow \left[\overset{R\ \ H}{\underset{O}{\underset{||}{C}}} + \right]$$

$$\left[\overset{R\ \ H}{\underset{O}{\underset{||}{C}}} + NH_3\right] \xrightarrow{H_2/Catalyst} \left[\underset{H}{\overset{R}{>}}C\underset{OH}{\overset{H}{<}} + NH_3\right] \equiv R-CH_2OH + NH_3$$

9-4-4 From esters

Esters undergo hydrolysis in acidic or alkaline medium to yield a carboxylic acid and an alcohol.

$$\underset{\text{(An Ester)}}{R-\underset{O}{\underset{||}{C}}-O-R^1} \xrightarrow[H_2O/\ominus OH]{H_2O/H^{\oplus}\text{ or}} \underset{\text{(A carboxylic acid)}}{R-\underset{O}{\underset{||}{C}}-OH} + \underset{\text{(An alcohol)}}{R^1-OH}$$

This method is not very much useful for the synthesis of alcohols since esters themselves are obtained by reaction of carboxylic acids with alcohol.

$$\underset{\text{(Carboxylic acid)}}{R-COOH} + \underset{\text{(Alcohol)}}{R'-OH} \xrightarrow{-H_2O} \underset{\text{O (Ester)}}{R-\underset{||}{C}-OR'}$$

Complicated alcohols or even simple alcohols cannot be synthesized by this method. However, polyhydric alcohols can be prepared by this method. For example, glycerol is obtained during soap manufacture by hydrolysis of oils and fats which are natural esters of fatty acids and glycerol.

$$\begin{array}{l} R-(CH_2)_n-\overset{O}{\overset{||}{C}}-O-CH_2 \\ R-(CH_2)_n-\overset{O}{\overset{||}{C}}-O-CH \\ R-(CH_2)_n-\underset{O}{\underset{||}{C}}-O-CH_2 \end{array} \xrightarrow{H_2O/\overset{\ominus}{O}H} 3\ R-(CH_2)_n-\underset{O}{\underset{||}{C}}-\overset{\ominus}{O}\ \overset{\oplus}{Na}\ \text{(Soap)} + \underset{\text{OH}\quad\text{OH}\quad\text{OH}}{CH_2-CH-CH_2}$$

(Oil or fat where R=CH_3, n=10-14) (Glycerol)

9-4-5 Preparation of arylcarbinols

In the older system, alcohols have been referred to as carbinols. In this system, all alcohols are considered to be derived from methyl alcohol and the simplest member, methyl alcohol itself named as **carbinol**. some of the other alcohols are named as follows;

$H_3C-[CH_2-OH]$ (Methyl carbinol)

$H_3C-CH_2-[CH_2-OH]$ (Ethyl carbinol)

$H_3C-[C(CH_3)(CH_3)H_2-OH]$ (1,1,1–Trimethyl carbinol etc.)

Aryl carbinols are simply the aryl derivatives of methanol obtained by replacing successively the hydrogen of the methyl group of carbinol. The aryl carbinols are the substituted benzyl alcohols. Following examples illustrate the nomenclature system used earlier for some of these derivatives of hydroxy compounds;

$Ph-CH_2-OH$	$p-Cl-C_6H_4-CH_2-OH$	Ph_2CH-OH	Ph_3C-OH
(Phenyl carbinol or phenyl methanol or Benzyl alcohol	(p-chlorophenyl carbinol ie. p-chlorobenzyl alcohol) or p-chlorophenyl methanol)	(Diphenyl carbinol ie. α-phenyl benzyl alcohol or Dephenyl methanol)	(Triphenyl carbinol ie. α α-Diphenyl-benzyl alcohol or Triphenyl methanol)

The aryl carbinols have properties very similar to those of alcohols. Even the general methods of preparation of aryl carbinols greatly resemble those used for alcohols. Following are some of the methods used for the syntheses of aryl carbinols.

(a) By Hydrolysis: Arylmethyl halides (i.e. benzylic halides) undergo hydrolysis with alkali to yield carbinols. For example;

$$Ph-CH_2-Cl + HO \longrightarrow Ph-CH_2-OH + Cl^-$$

The aryl methyl halides, themselves, are obtained from substituted benzenes by chloromethylation. The latter reaction involves heating a derivative of benzene with HCHO and anh. HCl.

$$C_6H_6 + HCHO + HCl \longrightarrow C_6H_5CH_2Cl$$

The reaction proceeds by the following mechanism;

$$H_2C{=}O \longleftrightarrow H_2\overset{\oplus}{C}-\overset{\ominus}{O} \xrightarrow[(+H^{\oplus})]{HCl} H_2\overset{\oplus}{C}-OH \longleftrightarrow H_2C{=}\overset{\oplus}{O}H$$

$$C_6H_6 \longleftrightarrow \overset{\oplus}{C_6H_6}{}^{\ominus} \xrightarrow{H_2\overset{\oplus}{C}-OH} [C_6H_6(H)CH_2OH]^{\oplus} \xrightarrow{-H^+} [C_6H_5CH_2OH]$$

$$[C_6H_5CH_2OH] \xrightarrow[(+H^{\oplus})]{HCl} C_6H_5CH_2-\overset{\oplus}{O}H_2 \xrightarrow{Cl^{\ominus}} C_6H_5CH_2Cl$$

(b) By Grignard method : Aryl halides react with Mg metal in anh. ether to form arylmagnesium halides which condense with aldehydes and ketones yielding aryl carbinols. For example;

$$Ph{-}Cl + Mg \xrightarrow{Et_2O} Ph{-}MgCl$$

$$\underset{\text{(Phenyl magnesium chloride)}}{Ph{-}MgCl} + \underset{\text{(Formaldehyde)}}{H_2C{=}O} \longrightarrow Ph{-}CH_2{-}OMgCl \xrightarrow{H_2O} \underset{\text{(Phenyl carbinol)}}{Ph{-}CH_2OH} + Mg(Cl)(OH)$$

$$Ph{-}MgCl + \underset{\text{(Benzophenone)}}{Ph_2C{=}O} \longrightarrow Ph_3C{-}OMgCl \xrightarrow{H_2O} \underset{\text{(Triphenyl carbinol)}}{Ph_3C{-}OH} + Mg(Cl)(OH)$$

(c) By Reduction method: Aromatic aldehydes, ketones, carboxylic acids and esters undergo reduction with different reagents to yield arylcarbinols. For example;

$$\underset{\text{(Aromatic aldehyde)}}{Ar{-}CHO} \xrightarrow{Na\,(Hg)} \underset{\text{(Aryl Carbinol)}}{Ar{-}CH_2OH}$$

$$\underset{\text{(Diaryl ketone)}}{Ar{-}CO{-}Ar} \xrightarrow{Zn(Hg)/HCl} \underset{\text{(Diaryl carbinol)}}{Ar{-}CH(OH){-}Ar}$$

$$\text{Ar-C(=O)-R} \xrightarrow{\text{Zn(Hg)/HCl}} \text{Ar-CH(OH)-R}$$

(Aryl alkyl ketone) (a substituted aryl carbinol)

$$\text{Ar-C(=O)-OH} \xrightarrow{H_2\text{/catalyst}} \text{Ar-}CH_2\text{-OH}$$

(Aromatic carboxylic acid)

$$\text{Ar-C(=O)-OR} \xrightarrow[H_2\text{/catalyst}]{LiAlH_4 \text{ or}} \text{Ar-}CH_2\text{-OH}$$

(Aromatic acid ester)

9-4-6 Preparation of phenols

The methods used for the preparation of phenols may be divided into two types.

(a) Laboratory method

(b) Industrial method

These methods are briefly described below:

(a) Laboratory method: In the laboratory, phenols are generally prepared from aromatic diazonium salts. This is one of the simple and elegant methods used for the preparation of phenols. The procedure involves treatment of an aromatic amine at 0-5°C, in dil. H_2SO_4 with $NaNO_2$ followed by warming up of the reaction mixure resulting in phenol.

$$\text{Ar-}NH_2 \xrightarrow[+HONO]{NaNO_2/H_2SO_4} \text{AR-}N_2^+ \; HSO_4^- \xrightarrow[(+HOH)]{\Delta} \text{Ar-OH} + N_2\uparrow + H_2SO_4$$

Alternatively, phenols may also be prepared by the fusion of aromatic sulphonic acids with alkali. For example;

$$CH_3C_6H_4SO_3H \xrightarrow[(250\text{-}300^o)]{\text{NaOH/}} CH_3C_6H_4O^{\ominus}Na^{\oplus} \xrightarrow{H_3O^{\oplus}} CH_3C_6H_4OH$$

(p-Toluene-sulphonic acid) (Sodium salt of p-methyl phenol) (p-Methylphenol or p-cresol)

The latter method is somewhat limited in scope due to non-avalability of aryl sulphonic acids in a pure state. Generally, these sulphonic acids are prepared by sulphonation of aromatic hydrocarbons but are difficult to isolate due to the high water - solubility of sulphonic acids/the corresponding sodium salts. The only known case of this method, which can be used with equal ease, in the laboratory as well as in industry is the preparation of β-naphthol from naphthalene by sulphonation followed by fusion with NaOH at 300°C.

(Naphthalene) $\xrightarrow[160^{0}/\Delta]{\text{Conc. } H_2SO_4}$ (Naphthalene-2-sulphonic acid)) $\xrightarrow[\text{then } H_3O^+]{NaOH/\Delta,\ 300^0\ \&}$ (2-Naphthol or β-Naphthol)

The mechanism of this reaction probably involves a simple nucleophilic displacement of the sulphonic acid group by the hydroxyl group.

$\overset{\ominus}{O}H$ + $NaHSO_3$

(≡ $NaHSO_3$' sodium bisulphite)

$\xrightarrow{\overset{\ominus}{O}H}$ (Sodium salt of 2-naphthol) + Na_2SO_3 (Sodium sulphite)

(b) Industrial method: A few phenols (the parent phenol and cresols) are generally obtained during fractionation of coal-tar. The phenols occur in the middle-oil fraction of coal-tar and are isolated by treatment with alkali followed by separation and acid hydrolysis.

Synthetically, phenols are obtained on a commercial scale by either of the following two processes:

The first, known as **cumene hydroperoxide process**, involves oxidation of cumene (isopropyl benzene) with oxygen/air to yield cumene hydroperoxide followed by acid hydrolysis resulting in phenol. Acetone is obtained as a valuable by-product in this process.

H₃C CH₃ CH

+ $H_2C{=}CH{-}CH_3$ Acid Catalyst

(Benzene) (propene) (cumene or isopopyl benzene)

(Both raw-materials from petroleum)

O₂/or air; $H_3O^{\oplus}$

(Cumene) (Cumene hydroperoxide) (Phenol) (Acetone)

The mechanism of this reaction involves a simple free-radical mediated oxidation of cumene to yield cumene hydroperoxide. The latter on acid hydrolysis undergoes ionic transformation to phenol and acetone as shown below:

$O_2 \equiv \dot{O}{-}\dot{O}$ (Triplet oxygen)

(Aresonance-stabilised tertiary free radical)

Chain continuation steps

(Cumene hydroperoxide)

$H^{\oplus}$

$-H_2O$

$H_3C-\overset{O}{\overset{||}{C}}-CH_3$ + (Acetone)

$\ddot{O}H_2$

$-H^{\oplus}$

(Phenol)

In the second process, known as **Dow process**, aryl halides are hydrolysed to yield phenols. As is well-known, the aryl halides are not so readily susceptible to nucleophilic substitution by hydroxyl ions, unlike in the case of alkyl halides, very forceful conditions are employed to bring about this difficult transformation. For example, chlorobenzene is hydrolysed with sodium hydroxide at 300-350° and 200 atm. pressure to yield phenol.

Cl

+ NaOH $\xrightarrow[\text{200 atoms}]{300\text{-}350^{o}}$ + NaCl + H_2O

(Chlorobenze) (Sodium phenolate)

$\overset{\ominus}{O}\overset{\oplus}{Na}$ $\xrightarrow{H_3\overset{\oplus}{O}/RT}$ OH

(Sodium phenolate) (Free phenol)

However, if one or more electron-withdrawing groups are present on the benzene nucleus in the ortho and para position/s to chlorine, hydrolysis is greatly facilitated. For example;

Cl … Aq. NaOH/100°/4hrs. → OH

NO_2 (p-Nitrochlorobenzene) → NO_2 (p- Nitrophenol $\simeq$ 60%)

Cl, NO_2 … Aq. NaOH/RT/1Hr → OH, NO_2

NO_2 (2,4-Dinitrochlorobenzene) → NO_2 (2,4-Dinitro phenol) (> 90%)

Cl, O_2N, NO_2 … HOH/RT/4hrs. → OH, O_2N, NO_2

NO_2 → NO_2

9-5 PHYSICAL AND CHEMICAL PROPERTIES

Physical and chemical properties of alcohols and phenols will be discussed in separate heads.

9-5-1 Physical properties of alcohols and carbinols

The alcohols are colorless, neutral compounds with a characteristic smell. The lower alcohols are liquids while higher ones are usually white, waxy solids. The alcohols exhibit the phenomena of H-bonding. As is well-known, the phenomenon of H-bonding is due to the binding of hydrogen atom to a highly electronegative element such as fluorine, oxygen or nitrogen. Due to the high electronegativity of these elements, the covalent bond between this element and hydrogen is unequally shared resulting in a net dipole. These dipoles give rise to electrostatic forces of attraction between molecules of the same substance.

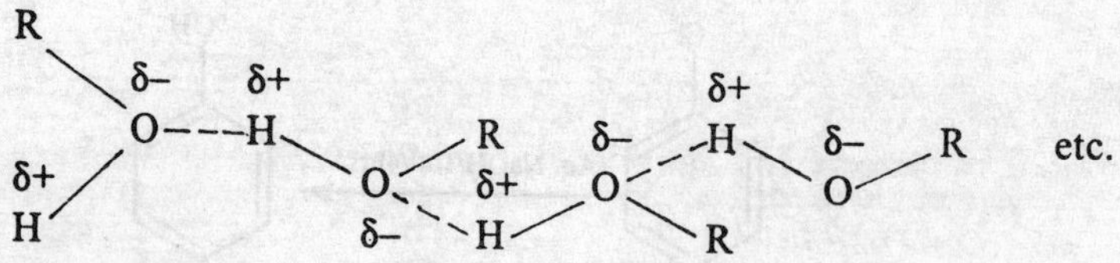

```
 R
  \   δ-    δ+                                δ+
   O---H\            R      δ-             -H\      δ-    R     etc.
δ+ /      \O\      /  δ+     O\  - -  - /      \O/
 H          δ-  \ -H/            \R
```

Consequently, alcohols are strongly associated substances and thus exhibit high melting and boiling points. Again, due to the phenomenon of H-bonding, the alcohols form strong intermolecular, electrostatic bonds with water molecules, resulting in high/considerable solubilities of these compounds in water. Table 9.1 shows the physical constants of somé of the alcohols in common use.

Table 9.1 Physical constants of alcohols

S.No. No.	Name of the alcohol	Structure of the alcohol	M.P. °C	B.P. °C	Solubility (g/100 ml H_2O at RT)
1.	Methanol	CH_3-OH	-98	64.5	complete
2.	Ethanol	CH_3CH_2OH	-115	78.1	complete
3.	n-Propanol	CH_3-CH_2CH_2OH	-126	97	complete
4.	n-Butanol	$CH_3CH_2CH_2CH_2OH$	-90	118	8.0
5.	n-Pentanol	$CH_3(CH_2)_3CH_2OH$	-78	138	2.3
6.	n-Hexanol	$CH_3(CH_2)_4CH_2OH$	-52	156	0.6
7.	n-Octanol	$CH_3(CH_2)_6CH_2OH$	-15	195	0.05
8.	n-Decyl alcohol	$CH_3(CH_2)_8CH_2OH$	+ 6	228	0.001
9.	i-Propanol	$CH_3(CH)(OH)CH_3$	-86	83	complete
10.	i-Butanol	$(CH_3)_2CHCH_2OH$	-108	+108	10.0
11.	i-Pentanol	$(CH_3)_2CHCH_2CH_2OH$	-117	132	2.0
12.	t-Butanol	$(CH_3)_3C$-OH	25.5	83	complete
13.	Benzyl alcohol	Ph-CH_2OH	-15	205	5.0
14.	β-Phenyl ethanol	$PhCH_2CH_2OH$	-27	220	1.5
15.	Diphenyl carbinol	$(Ph)_2CH$-OH	70	297	0.06
16.	Triphenyl carbinol	$(Ph)_3C$-OH	162	—	—

It should be noted that, structurally, alcohols, consist of a hydrocarbon portion and a -OH group. The -OH group participates in H-bonding, whereas the alkyl group/hydrocarbon group does not. Therefore, for higher alcohols where the hydrocarbon group is quite large compared to the hydroxyl group, the melting point/boiling point are not quite high and so is the solubility in water. Similar is the situation with carbinols especially of the benzylic types, which although being strongly associated (due to H-bonding) do not have significant solubilities in water.

9-5-2 Chemical properties of alcohols and carbinols

The chemical reactions of alcohols and carbinols are due to the presence of the -OH group. In spite of presence of a hydrocarbon portion, the properties are mainly due to hydroxyl group and hence are significantly similar in reactivity unless the hydrocarbon portion has other reactive functional groups in it. Since the chemical properties of alcohols and carbinols are due to -OH group, their reactions may be divided into two types: (i) Those in which -H of -OH is replaced and (ii) Those in which the -OH group itself is replaced. These reactions are discussed below in detail.

(A) Those in which the -H of -OH is replaced

(a) Formation of alkoxides : The -OH group of an alcohol consists of a covalent bond between the highly electronegative oxygen atom and the poorly electronegative hydrogen atom. Consequently, the pair of electrons constituting the covalent bond, between O and H, is drawn more towards oxygen than towards hydrogen rendering a partial negative charge on oxygen and partial positive charge on hydrogen. Thus, the hydrogen atom of an alcohol is very weakly acidic and, therefore, it can be replaced by a strongly basic metal. Hence, alcohols react with sodium metal/potassium metal or strong bases such as sodium amide, sodium hydride, etc., yielding alkoxides.

$$R—\overset{\delta^-}{O}—\overset{\delta^+}{H} + Na \longrightarrow R—\overset{-}{O}\overset{+}{Na} + ½ H_2$$

Since alcohols are very weakly acidic in nature and sodium metal strongly basic, the alkoxides thus obtained are strong bases. They undergo hydrolysis on treatment with water to give hydroxides.

$$R—\overset{-}{O}\overset{+}{Na} + HOH \longrightarrow R—OH + NaOH$$

The alkoxides being strong bases, react with weak acids such as CO_2, diethyl malonate, ethyl acetoacetate, etc., generating, addition products in case of CO_2 and carbanions - valuable organic intermediates in the latter cases. For example;

$$R—\overset{\ominus}{O}\overset{\oplus}{Na} + O=C=O \longrightarrow R—O—\underset{\underset{O}{||}}{C}—\overset{\ominus}{O}\,\overset{\oplus}{Na}$$

$$(H)_2C(COOEt)_2 + \overset{\oplus}{Na}\,\overset{\ominus}{OEt} \longrightarrow EtOH + \left[H\overset{\ominus}{C}(COOEt)_2\right]\overset{\oplus}{Na}$$

$$H_3C–\underset{\underset{O}{||}}{C}–CH_2–COOEt + \overset{\oplus}{Na}\,\overset{\ominus}{OMe} \longrightarrow MeOH + [H_3C–\underset{\underset{O}{||}}{C}–\overset{\ominus}{C}H–COOEt]\,\overset{\oplus}{Na}$$

Since alkoxide anions are formed from alcohols or carbinols by replacement of the hydrogen of the –OH group of a metal, the stability of alkoxide ions will depend on the ease with which the negative charge on oxygen is accommodated. In general, *electron-withdrawing groups on alkoxy carbon will facilitate alkoxide formation and electron donating groups on alkoxy carbon will retard alkoxide formation.* Since alkyl groups are electron-donating types, the stability of alkoxide anions follows the order given below;

$$(CH_3)_3C\text{—}O^{\ominus} < (CH_3)_2CH\text{—}O^{\ominus} < CH_3CH_2\text{—}O^{\ominus} < H_3C\text{—}O^{\ominus}$$

If this is the order of stability then the order of reactivity towards acidic reagents or in other words, order of basicity would be

$$(CH_3)_3C\text{—}O^{\ominus} > (CH_3)_2CH\text{—}O^{\ominus} > CH_3CH_2\text{—}O^{\ominus} > H_3C\text{—}O^{\ominus}$$

That is to say, t-butoxide ion will be the strongest base in the series. In fact, potassium t-butoxide is one of the strongest bases in organic chemistry, particularly in polar solvents like DMSO (dimethyl sulphoxide), which stabilize the cation through complexation increasing very greatly the basic power of anion.

(b) Formation of esters: The alcohols, which contain a basic oxygen atom, react with electrophiles like the positively polarized carbon of a carbonyl group, say in a carboxylic acid, to form esters. The reaction is reversible and is general acid-catalysed. For example;

$$\underset{\text{(Alcohol)}}{R\text{—}O\text{—}H} + \underset{\text{(Carboxylic acid)}}{R'\text{—}C(=O)\text{—}OH} \overset{H^+}{\rightleftharpoons} \underset{\text{(Ester)}}{R'\text{—}C(=O)\text{—}O\text{—}R} + H_2O$$

The mechanism of this reaction has been studied in great detail and is reported to be as shown in the next page;

$$R'-\underset{\parallel}{\overset{}{C}}(=O)-OH \xrightarrow{H^{\oplus}} R'-\overset{\oplus}{C}(-O-H)-OH \xrightarrow{R\ddot{O}H} R'-C(OH)(\overset{\oplus}{O}HR)-OH \xrightarrow{-H^{+}}$$

$$R'-C(OR)(OH)-OH \equiv R'-C(OR)(OH)-OH \longrightarrow R'-\underset{\underset{O}{\parallel}}{C}-OR$$

It is worth noting that in the ester formation, the acylato oxygen could be either from acid or from alcohol;

$$R'-\underset{\underset{O}{\parallel}}{C}-\overset{*}{O}-[H + H-O]-R \overset{H^{\oplus}}{\rightleftharpoons} R'-\underset{\underset{O}{\parallel}}{C}-\overset{*}{O}-R + H_2O$$

$$R'-\underset{\underset{O}{\parallel}}{C}-[O-H + H]-\overset{\odot}{O}-R \overset{H^{\oplus}}{\rightleftharpoons} R'-\underset{\underset{O}{\parallel}}{C}-\overset{\odot}{O}-R + H_2O$$

However, it has been conclusively established by isotopic tracer studies that R-O part of ester comes from alcohol and $R'-\underset{\underset{O}{\parallel}}{C}$ part from acid in accordance with the mechanism given above.

It may be mentioned here that the ester formation reaction given above is usually slow and reversible. Reaction in the forward direction can be speeded up by using a dehydrating agent or by using more reactive derivatives of acids like anhydrides or acid halides.

$$\underset{\text{(Carboxylic acid)}}{R'-CO-OH} + R-OH \xrightarrow[(-H_2O)]{H^{+}} R'-\underset{\underset{O}{\parallel}}{C}-OR$$

$$\underset{\text{(Carboxylic acid anhydride)}}{R'-CO-O-CO-R'} + R-OH \longrightarrow R'-CO-O-R + R'-\underset{\underset{O}{\parallel}}{C}-OH$$

$$\underset{\text{(Carboxylic acid chloride)}}{R'-CO-Cl} + R-OH \longrightarrow R'-\underset{\underset{O}{\parallel}}{C}-OR + HCl$$

Inorganic acids such as nitric acid, too, react with alcohols to form esters.

$$\underset{}{R—OH} + \underset{\text{(Nitric acid)}}{H—O—NO_2} \longrightarrow \underset{\text{(Alkyl nitrate)}}{R—O—NO_2} + H_2O$$

With sulphuric acid, different products are formed depending upon the conditions of the reaction. At room temperature with conc. H_2SO_4, an alkyl ester, alkylhydrogen sulphate, is produced. For example;

$$Et—O—H + \underset{\text{(Sulphuric acid)}}{H—O—\overset{\overset{O}{||}}{\underset{\underset{O}{||}}{S}}—OH} \xrightarrow[(-H_2O)]{RT} \underset{\text{(Ethyl hydrogen sulphate)}}{Et—O—\overset{\overset{O}{||}}{\underset{\underset{O}{||}}{S}}—OH} (\equiv RHSO_4)$$

The ethyl hydrogen sulphate, produced, reacts differently under different conditions to give diffrent products. For example;

(α) When heated alone at 110°C, it forms diethyl sulphate.

$$EtO—\overset{\overset{O}{||}}{\underset{\underset{O}{||}}{S}}—\ddot{O}—H + Et—O—\overset{\overset{O}{||}}{\underset{\underset{O}{||}}{S}}—OH \xrightarrow[\Delta]{110°/} \underset{\text{(Diethyl sulphate)}}{EtO—\overset{\overset{O}{||}}{\underset{\underset{O}{||}}{S}}—OEt} + \underset{\text{(Sulphuric acid)}}{HO—\overset{\overset{O}{||}}{\underset{\underset{O}{||}}{S}}—OH}$$

(β) When heated with excess ethanol at 140°C (under pressure), it gives diethyl ether.

$$Et—\ddot{O}—\overset{\overset{O}{||}}{\underset{\underset{O}{||}}{S}}—OH + H—\ddot{O}—Et \xrightarrow[\Delta]{140°/} \underset{\text{(Diethyl ether)}}{Et—O—Et} + \underset{\text{(Sulphuric acid)}}{HO—\overset{\overset{O}{||}}{\underset{\underset{O}{||}}{S}}—OH}$$

(γ) When heated with excess sulphuric acid at 160°C, it forms ethylene.

$$\overset{H}{\overset{|}{CH_2}}—\underset{|}{CH}—O—\overset{\overset{O}{||}}{\underset{\underset{O}{||}}{S}}—OH \xrightarrow[\Delta]{H_2SO_4/160°} \underset{\text{(Ethylene)}}{CH_2 = CH_2} + \underset{\text{(Sulphuric acid)}}{HO—\overset{\overset{O}{||}}{\underset{\underset{O}{||}}{S}}—OH}$$

(B) Those in which the -OH group is replaced:

(a) Formation of alkyl halides: Alcohols react with phosphorous chlorides and thionyl chloride to yield alkyl chlorides.

$$R\text{-}OH + PCl_5 \longrightarrow R\text{-}Cl + POCl_3 + HCl$$

$$3\ R\text{-}OH + PCl_3 \longrightarrow 3\ RCl + P(OH)_3$$

$$R\text{-}OH + SOCl_2 \longrightarrow R\text{-}Cl + SO_2 + HCl$$

Alcohols also react with hydrogen halides, although reversibly, to yield alkyl halides. Usually, a dehydrating agent and anhydrous acid will shift the equilibrium to the alkyl halide side.

$$R\text{-}OH + HCl \xrightarrow{ZnCl_2} R\text{-}Cl + H_2O$$

In the above reaction, where the -OH group is being replaced by a halogen, the order of reactivity is as follows;

Tertiary alcohol > Secondary alcohol > Primary alcohol > Methanol

The above order of reactivity sometimes forms the basis of a qualitative test, known as **Lucas test**, to distinguish between the three types of alcohols. The Lucas reagent consists of a solution of anh. $ZnCl_2$ in very conc. HCl. Treatment of this reagent with tertiary alcohol yields the corresponding tertiary alkyl halide in about 2-5 minutes, which separates as an immiscible oily layer. Secondary alcohols yield the corresponding secondary alkyl halide as oil layer in 10-15 minutes and primary alcohols usually require half hour to react completely.

(b) Oxidation of alcohol: Alcohols undergo oxidation with several oxidizing agents to yield different oxidation products depending upon the nature of alcohol used and the reaction conditions. For example, primary alcohols under mild conditions yield the corresponding aldehyde which on further oxidation leads to carboxylic acid. Secondary alcohols give rise to ketones and tertiary alcohols are difficult to oxidize. However, under very vigorous conditions tertiary alcohols yield degradation products.

$$\underset{\text{(Alcohol)}}{R\text{—}CH_2\text{—}OH} \xrightarrow[\text{(mild)}]{[O]} R\text{—}CHO \xrightarrow{[O]} \underset{\text{(Carboxylic acid)}}{R\text{—}COOH}$$

$$R\text{—}COOH \xrightarrow[\text{(Vigorous)}]{[O]} \text{Degradation products}$$

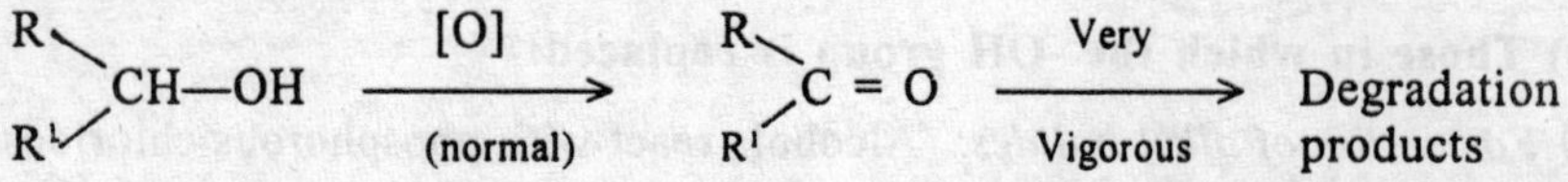

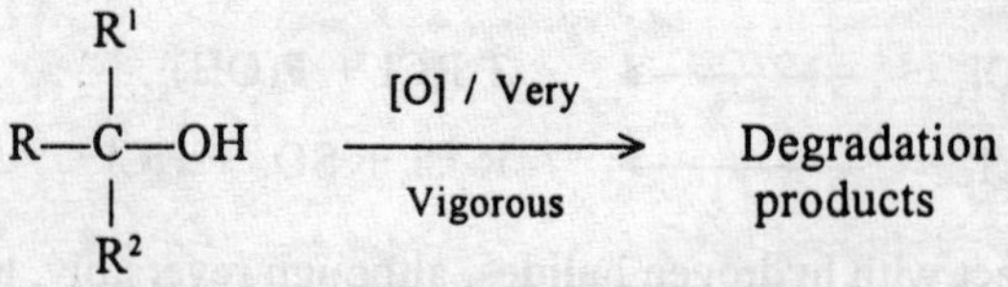

$$\begin{matrix} R \\ R^1 \end{matrix}\!\!>\!CH\!-\!OH \xrightarrow[\text{(normal)}]{[O]} \begin{matrix} R \\ R^1 \end{matrix}\!\!>\!C=O \xrightarrow[\text{Vigorous}]{\text{Very}} \text{Degradation products}$$

$$R\!-\!\overset{\overset{\Large R^1}{|}}{\underset{\underset{\Large R^2}{|}}{C}}\!-\!OH \xrightarrow[\text{Vigorous}]{[O]\ /\ \text{Very}} \text{Degradation products}$$

The dehydrogenation of secondary alcohols to ketones may also be carried out by heating with dehydrogenating agents such as sulphur, nitrobenzene, cyclohexene, etc. When $KMnO_4$ is used as oxidizing agent, primary alcohols are directly converted to carboxylic acids and secondary alcohols to ketones.

9-5-4 Physical properties of phenols

Most phenols, unless chromophoric groups are present in the aromatic nucleus, are colorless substances when freshly distilled. However, on standing exposed to air or oxygen, phenols undergo oxidation giving colored products and hence appear dark colored. Simple phenols are high boiling liquids or low melting solids. Higher molecular weight phenols are neat crystalline solids. The phenomenon of hydrogen bonding, present in alcohols, which is due to the –OH group, is also found in phenols. Hydrogen bonding makes phenols strongly associated substances.

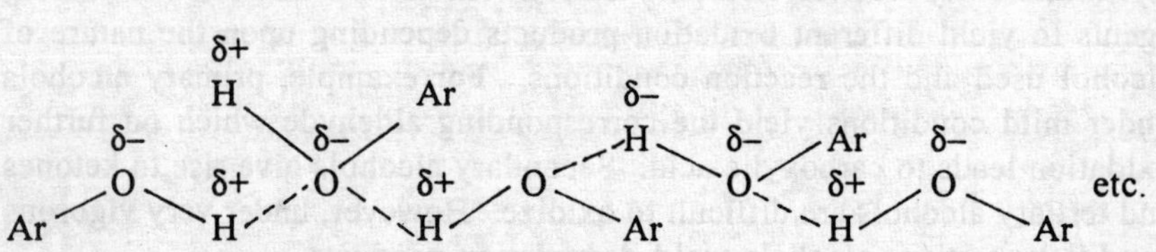

δ+ H Ar δ– δ– δ– H δ– Ar δ– O δ+ O δ+ O O δ+ O etc. Ar H H Ar H Ar

Phenol itself is somewhat soluble in water (≈9%) due to H-bonding. However, higher, molecular weight phenols are almost insoluble in water because the hydrocarbon portion (which is water insolubilizing part) is quite large in comparison to -OH group (which is water-solubilizing part). The physical constants of some important phenols are shown in Table 9.2.

Table 9.2 Physical constants of Phenols

NAME	m.p(°C)	b.p (°C)	Density (g/cm^3)	nD
Phenol	43	182	1.0722	1.5509
4-Bromophenol	66.4	238	1.840^{15}	---
4-Chlorophenol	43	219.8	1.2651^{40}	1.5579
4-Methoxyphenol	53	243	---	—
2-Nitrophenol	44.9	216	1.485^{14}-	---
3-Nitrophenol	97	194^{70}	1.2797^{100}	---
4-Nitrophenol	114	279d	1.479	---
2,4,6-Tribromophenol	95-6	282-90	2.55	---
2,4,6-Tinitrophenol	122-3	Above 300 (exp.)	1.763	---

An examination of Table 9.2 shows that *o*-nitrophenol has a much lower melting and boiling point than its isomers, particularly the *p*-isomer. Furthermore, the solubilities of these compounds have been measured in water at room temperature. As the temperature increases, it is observed, that the solubility of *o*-nitrophenol remains almost same whereas that of *p*-nitrophenol goes on increasing. These differences in the solubility behaviours of the two compounds can be explained by considering the structures of *o*-nitro and *p*-nitro phenols. The *o*-isomer shows the phenomena of H-bonding, of the *intramolecular type*, whereas the *p*-isomer shows the phenomenon of H-bonding of the *intermolecular type*.

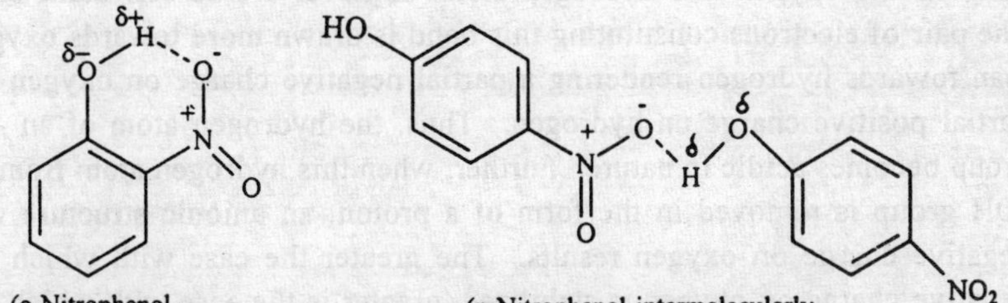

(o-Nitrophenol-chelated structure)

(p-Nitrophenol-intermolecularly H-bonded)

The consequence of intra-molecular H-bonding is that it reduces association between the different molecules of the same substance and also it reduces interaction between this molecule and the molecules of the solvent, i.e., water. Furthermore, intra-molecular H-bonding is almost independent of temperature. Therefore, *o*-nitrophenol has lower melting and boiling points and its solubility in water is quite poor. Converse is the case for *p*-nitrophenol which is inter-molecularly associated. The result is that *o*-nitrophenol is steam-distillable and vacuum sublimable whereas *p*-nitrophenol is not. Thus, a mixture of *o*-nitro and *p*-nitro phenols can be readily separated from each other either by steam distillation or by vacuum sublimation. Same will be the case for salicylaldehyde and *p*-hydroxybenzaldehyde.

(Salicylaldehyde or
(o-Hydroxybenzaldehyde):

(p-Hydroxybenzaldehye)
(Intermolecularly H-bonded)

9-5-5 Chemical properties of phenols

The chemical properties of phenols may be studied from three points of view. These are : (A) Acidity of phenols, (B) Reactions of the -OH group. (C) Nuclear reactions of phenols. These properties are discussed below in detail.

(A) Acidity of Phenols: As was discussed in the case of alcohols, the -OH group of any hydroxy compound consists of a covalent bond between the highly electronegative oxygen atom and the weakly electronegative hydrogen atom. Due to electronegativity differences, the covalent bond between the oxygen and hydrogen atoms is never a true convalent bond. The pair of electrons constituting this bond is drawn more towards oxygen, than towards hydrogen rendering a partial negative charge on oxygen and partial positive charge on hydrogen. Thus, the hydrogen atom of an -OH group becomes acidic in nature. Further, when this hydrogen atom from the -OH group is removed in the form of a proton, an anionic structure with negative charge on oxygen results. The greater the ease with which this negative charge on oxygen is stabilized, greater is the ease with which this oxyanion is produced. *In alcohols, there is no particular structural feature of the molecule which can stabilize the charge on the oxygen except for the electronegative nature of oxygen itself.* However, in the case of phenols, this is not found to be the case. Since the -OH group of phenols is directly linked to the aromatic ring, when the hydrogen of -OH departs, *the resulting negative charge on the oxyanion is stabilized due to resonance with the aromatic ring.* Consequently, phenols are acidic in nature and are more strongly acidic than alcohols. For example, the pK_a of phenol is 10 ($K_a \sim 10^{-10}$) whereas that of ethanol is 16 ($K_a \sim 10^{-16}$). The resonance stabilization of phenoxide ion is shown below;

$$Ph—OH \rightleftharpoons H^{+} PhO^{\ominus}$$

The potential energy diagram for the formation of anion from the neutral molecule in respect of alcohols and phenols is shown in Fig. 9.1.

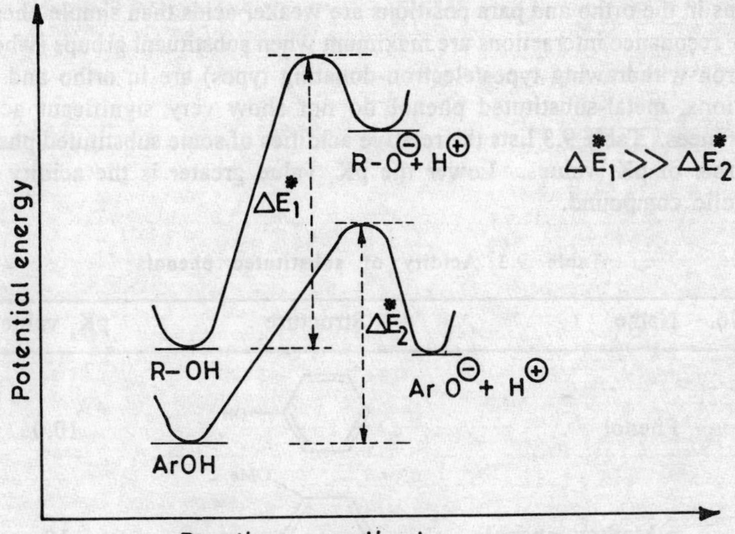

Fig. 9.1 Potential energy diagram of alcohols and phenol and their anions.

It may be mentioned here that the resonance involving non-bonding electron pairs on oxygen and the π-electron cloud of the aromatic ring is present even in the neutral molecule of phenol as given below;

However, what makes resonance more significant in the phenoxide ion than in phenol is the fact that *resonance in the anion does not involve any charge separation* unlike in the case of the neutral molecule itself. This makes the stability of the contributing structures and the degree of resonance quite high in the case of phenoxide ion, when compared to that in case of phenol molecule.

Since stabilization of negative charge on oxygen of the phenoxide by resonance involving the aromatic ring is the criterion for acidity of phenols, greater the resonance interaction, greater will be the stability of phenoxide ion, and greater will be the acidity of phenol. In general, electron withdrawing groups in the ortho and para positions of a benzene ring stabilize the negative charge on oxygen by a resonance interaction (or by inductive effect or by both). Therefore, phenols of this type are stronger acids than simple phenols. By the same reasoning, phenols containing electron-donating groups in the ortho and para positions are weaker acids than simple phenols. Since resonance interactions are maximum when substituent groups (whether electron-withdrawing types/electron-donating types) are in ortho and para positions, metal-substituted phenol do not show very significant acidity differences. Table 9.3 lists the relative acidities of some substituted phenols in terms of pK_a values. Lower the pK_a value greater is the acidity of a phenolic compound.

Table 9.3 Acidity of substituted phenols

Sl. No.	Name	Structure	pK_a value
1.	Phenol	OH	10.0
2.	*o*-Methoxyphenol	OMe, OH	10
3.	*p*-Methoxyphenol	MeO, OH	10.2
4.	*o*-Methylphenol	OH, Me	10.2
5.	*p*-Methylphenol	Me, OH	10.1
6.	*o*-Chlorophenol	OH, Cl	8.5

7.	*p*-Chlorophenol	Cl— —OH	9.4
8.	p-Aminophenol	H_2N— —OH	10.3
9.	*o*-Nitrophenol	—OH, NO_2	7.2
10.	*p*-Nitrophenol	O_2N— —OH	7.1
11.	2,4-Dinitrophenol	2,4-$(O_2N)_2C_6H_3$-OH	4.0
12.	2,4,6-Trinitrophenol (Picric acid)	2,4,6-$(O_2N)_3C_6H_2$-OH	0.7

Due to their acidity, most phenols dissolve is aq.NaOH forming the corresponding sodium phenoxides. Though stronger than alcohols (pK_a ~ 16-18 or K_a ~ 10^{-16} to 10^{-18}), most phenols are weaker acids (pK_a ~ 10 or K_a ~ 10^{-10}) than carbonic acid (pK_a = 7.0 or K_a = 10^{-7}). Therefore, phenols do not react with aq. $NaHCO_3$. In fact, phenols, can be conveniently precipitated from their aq. NaOH solutions on bubbling carbon dioxide. This property is made use of very often to qualitatively distinguish between phenols and benzoic acids and even to separate quantitatively phenols from benzoic acids in a mixture of the two. The benzoic acids are stronger acids than phenols having pK_a ~ 5.0 (K_a ~ 10^{-5}).

$$\text{Ar—O—H} + \text{NaOH} \longrightarrow \text{Ar—}\overset{-}{\text{O}}\ \overset{+}{\text{Na}} + H_2O$$

$$\text{Ar—}\overset{-}{\text{O}}\ \overset{+}{\text{Na}} + CO_2 + H_2O \longrightarrow \text{Ar—OH} + NaHCO_3$$

$$\text{Ar—COOH} + \text{NaOH} \longrightarrow \text{Ar—CO}\overset{-}{\text{O}}\ \overset{+}{\text{Na}} + H_2O$$

$$\text{Ar—CO}\overset{-}{\text{O}}\ \overset{+}{\text{Na}} + CO_2 + H_2O \not\longrightarrow \text{Ar—COOH}\downarrow + NaHCO_3$$

(B) Reactions of the –OH Group

(i) Formation of esters : Phenols react with derivatives of carboxylic acids such as anhydrides or acid chlorides to form esters. The reaction is both acidcatalysed and also base-catalysed. For example;

$$\underset{\text{(Substituted phenol)}}{Ar{-}OH} + \underset{(X = Cl \text{ or } OCOR)}{R{-}\overset{O}{\overset{\|}{C}}{-}X} \xrightarrow[\text{Aq. NaOH or pyridine}]{\text{Conc. } H_2SO_4 \text{ or}} \underset{\text{(Substituted phenolic ester)}}{Ar{-}O{-}\overset{O}{\overset{\|}{C}}{-}R}$$

In the acid catalysed process, reaction involves attack by the activated carbonyl group of carboxylic acid derivative on the phenolic oxygen.

$$R{-}\overset{O}{\overset{\|}{C}}{-}X \xrightarrow[(+H^{\oplus})]{H_2SO_4} R{-}\overset{\oplus}{C}(OH){-}X \xrightarrow{Ar{-}\ddot{O}{-}H} Ar{-}\overset{H}{\overset{|}{\underset{\oplus}{O}}}{-}C(OH)(X){-}R \longrightarrow Ar{-}O{-}\overset{O}{\overset{\|}{C}}{-}R$$

An example of this method is the preparation of *aspirin* in the laboratory from salicylic acid and acetic anhydride in the presence of a trace of sulphuric acid as catalyst.

$$\underset{\text{(o-Hydroxybenzoic acid or salicylic acid)}}{C_6H_4(OH)COOH} + \underset{\text{(Acetic anhydride)}}{H_3C{-}\overset{O}{\overset{\|}{C}}{-}O{-}CO{-}CH_3} \xrightarrow{H_2SO_4} \underset{\text{(o-Acetoxybenzoic acid or acetyl salicylic acid or "Aspirin")}}{C_6H_4(OCOCH_3)COOH} + H_3C{-}\overset{O}{\overset{\|}{C}}{-}O{-}H$$

In the base-catalysed process, reaction involves attack by the electrophilic carbon of the acid derivative on the phenoxide anion.

$$Ar{-}OH \xrightarrow{\text{Aq. NaOH}} Ar{-}\overset{\ominus}{O}\ \overset{\oplus}{Na}$$

$$Ar{-}\overset{\ominus}{O}\ \overset{\oplus}{Na} + R{-}\overset{X}{\overset{|}{C}}{-}\overset{\ominus}{O} \longrightarrow Ar{-}O{-}\overset{R}{\overset{|}{\underset{\overset{\ominus}{O}\ \overset{\oplus}{Na}}{C}}}{-}X \longrightarrow Ar{-}O{-}\overset{O}{\overset{\|}{C}}{-}R + NaX$$

An example of this method is the familiar **Schotten-Baumann procedure** used in the laboratory for the derivatization of phenols using benzoyl chloride in the presence of aq. NaOH.

$$\underset{\text{(Substitued phenol)}}{Ar{-}O{-}H} + \underset{\text{(Benzoyl chloride)}}{Ph{-}\overset{O}{\overset{||}{C}}{-}Cl} \xrightarrow[RT]{\overset{\ominus}{O}H/} \underset{\text{(Substituted phenolic ester of benzoic acid)}}{Ar{-}O{-}\overset{O}{\overset{||}{C}}{-}Ph} + NaCl + H_2O$$

The phenolic ester on heating with anhydrous aluminium chloride undergo isomerization to yield a mixture of ortho and para hydroxy ketones. For example;

OCOR — $AlCl_3/\Delta$ → OH, C(=O)R (ortho) + OH, C(=O)R (para)

(a phenolic ester) (o-Hydroxy-ketone) (p-Hydroxy ketone)

The above reaction is known as **Fries migration** and proceeds by the following mechanism;

OCOR ≡ O–C(=O)–R — $AlCl_3$ → $Cl_3\overset{\ominus}{Al}{-}\overset{\oplus}{O}{-}C(=O){-}R$ → $O{-}\overset{\ominus}{Al}Cl_3$ + $R{-}\overset{\oplus}{C}{=}O \leftrightarrow R{-}C{\equiv}\overset{\oplus}{O}$

$O{-}\overset{\ominus}{Al}Cl_3$ — $R{-}\overset{\oplus}{C}{=}O$ → (arenium ion with H and C(=O)R) → $O{-}\overset{\ominus}{Al}Cl_3$, C(=O)R + $H^{\oplus}$

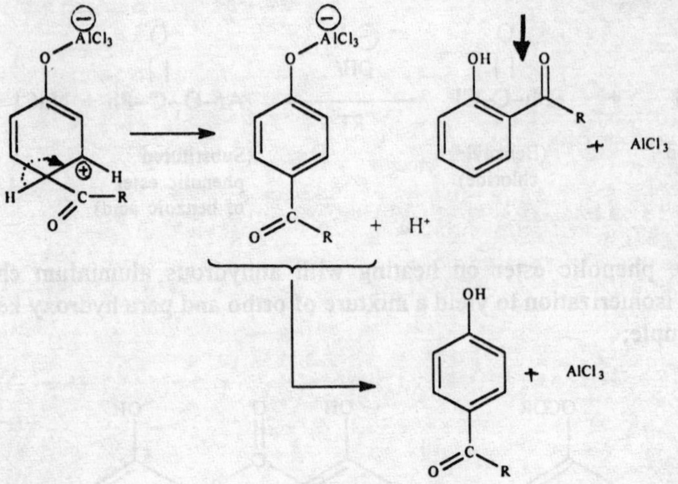

(ii) Reaction with PCl_5: The familiar reaction of alcohols with PCl_5 to yield alkyl halides does not proceed with phenols to yield halobenzenes. However, if electron withdrawing groups such as nitro are present in the benzene ring in ortho or para positions, reaction can be forced to yield chloronitrobenzene. For example;

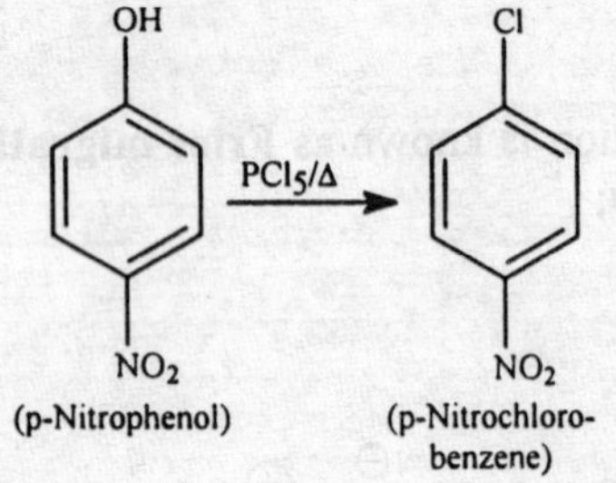

(p-Nitrophenol) $\xrightarrow{PCl_5/\Delta}$ (p-Nitrochloro-benzene)

However, the above reaction is not of much synthetic importance.

(iii) Coloration with neutral $FeCl_3$: Phenols (which are mostly colourless) react with neutral $FeCl_3$ (which is yellowish-orange in colour)to form dark-coloured solutions (which are greenish to blue-violet in colour). The colour is believed to be due to the formation of a complex coordination compound in solution in which iron is hexavalent. The complex is unstable in acid solution and original color of $FeCl_3$ is restored on addition of a few drops of HCl.

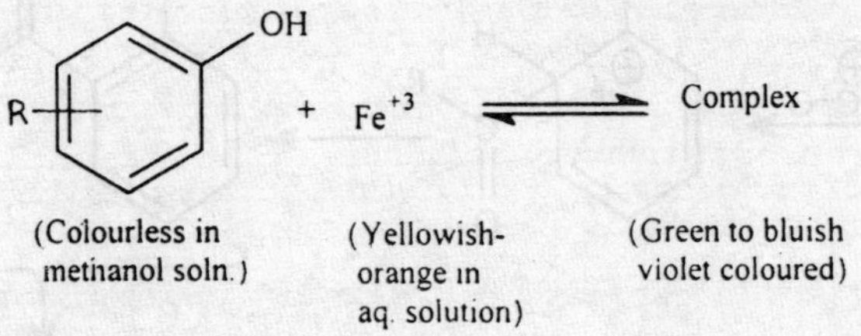

$$R\text{-}C_6H_4OH + Fe^{+3} \rightleftharpoons \text{Complex}$$

(Colourless in methanol soln.) (Yellowish-orange in aq. solution) (Green to bluish violet coloured)

Alcohols do not answer this test. Enols, however, interfere with phenols in this test. This reaction, development of dark coloured solution with neutral $FeCl_3$, is used in organic qualitative analysis for the identification of phenols or enols.

(iv) Dehydration of phenols: It was mentioned earlier that alcohols undergo dehydration on heating with acids to yield alkenes. However, such a reaction does not proceed with phenols. This may be explained in two ways: Firstly, when the proton from the acid attacks the -OH group of the phenol molecule to yield initially an oxonium ion followed by (hypothetically) the loss of a molecule of water, a highly unstable vinyl cation is the resulting intermediate.

(Phenol) $\xrightarrow{H^{\oplus}}$ (an oxonium ion) $\xrightarrow{-H_2O}$ (Vinyl cation) (not stable)

Secondly, if dehydration of phenol is to be successful, the vinyl cation should lose a proton from an ortho-position, (i.e an *sp*[2] carbon atom) to form the so-called dehydrated product, which is in fact the highly unstable benzyne *intermediate.*

(Vinyl cation) $\xrightarrow{-H^{\oplus}}$ (Benzyne)

Both these processes (formation of vinyl cation and of benzyne intrmediate) are highly unlikely to occur and so dehydration of phenol normally does not occur.

(C) Nuclear Reactions of Phenols: These may, once again, be divided into two types;

(i) Normal electrophilic substitutions

(ii) Special reactions of phenolic nucleus

These two types of reactions are discrussed below in detail:

(i) Normal electrophilic substitutions: The benzene ring of a phenolic compound undergoes the normal electronphilic substitution reactions such as nitration, sulphonation, halogenation, Friedel-Crafts alkylation and acylation to yield ring-substituted products. However, the hydroxyl group of a phenolic system strongly activates the aromatic ring and directs an incoming substituent to ortho and para positions.

(a) Nitration: Nitration of phenol with nitrating mixture at RT usually gives a mixture of products, picric acid being one of them. The yield of picric acid is quite low due to formation of various side-products and oxidation of phenolic nucleus with nitric acid. Furthermore, reaction is very vigorous and if not controlled properly, may become explosively violent.

OH (Phenol) —HNO_3/H_2SO_4/RT→ OH, O_2N (6), (1), (2) NO_2, (4) NO_2 (2 4 6 - Trinitrophenol or picric)

It should be noted that picric acid itself is a dangerous chemical having been the standard military explosive until it was replaced by the more safer TNT in the 1920s. However, its present use is only in the preparaion of derivatives for the hydrocarbons/halohydrocarbons, called *picrates*, which are crystalline adducts known as charge-transfer complexes. In the laboratory picric acid may be prepared as follows;

Cl (Chlorobenzene) —HNO_3/H_2SO_4→ Cl, NO_2, NO_2 (2,4-Dinitrochlorobenzene) —Aq. Na_2CO_3→ OH, NO_2, NO_2 (2,4-Dinitrophenol) —HNO_3/H_2SO_4→ OH, O_2N, NO_2, NO_2 (Picric acid)

Picric acid is a very strong acid, stronger than even carboxylic acids due to the presence of three strongly electron-withdrawing nitro groups.

Ortho and para nitrophenols are prepared by treating phenol with nitrous acid ($NaNO_2$ + dil. H_2SO_4) to yield a mixture of ortho and para nitrosophenols followed by their oxidation to give a mixture of ortho and para nitro phenols.

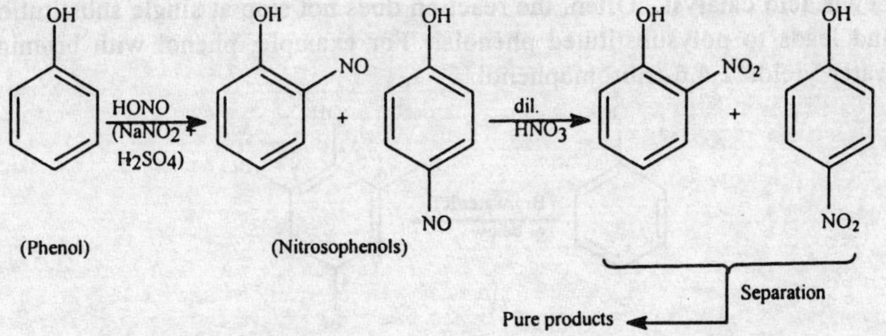

(Phenol) —HONO (NaNO2 + H2SO4)→ (Nitrosophenols) o-nitrosophenol + p-nitrosophenol —dil. HNO3→ o-nitrophenol + p-nitrophenol —Separation→ Pure products

m-Nitrobenzene cannot to preparad by nitration of phenol. It is prepared by an indirect method shown below;

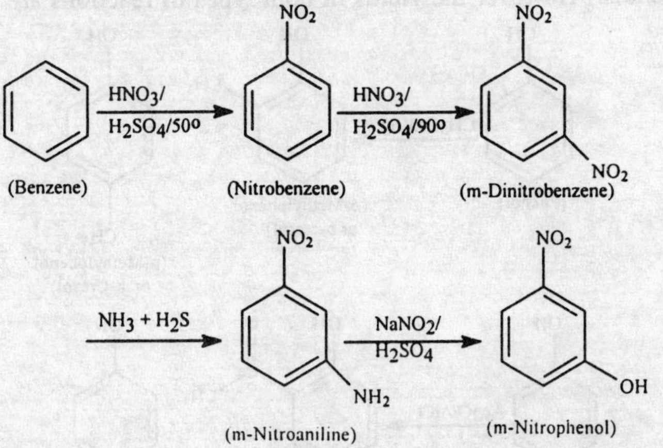

(Benzene) —HNO_3/H_2SO_4/50°→ (Nitrobenzene) —HNO_3/H_2SO_4/90°→ (m-Dinitrobenzene)

—$NH_3 + H_2S$→ (m-Nitroaniline) —$NaNO_2$/H_2SO_4→ (m-Nitrophenol)

(b) Sulphonation: Phenol undergoes sulphonation with conc. H_2SO_4 at RT to yield a mixture of ortho and para sulphonated phenols.

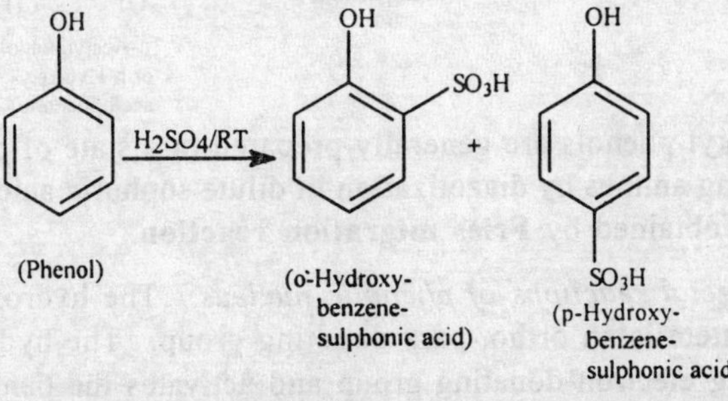

(Phenol) —H_2SO_4/RT→ (o-Hydroxy-benzene-sulphonic acid) + (p-Hydroxy-benzene-sulphonic acid)

The ortho isomer is the kinetically controlled product and predominates at lower temperatures whereas the para isomer is the thermodynamically controlled and predominates at higher temperatures.

(c) Halogenation : Phenols react with chlorine or bromine at RT or below to yield electrophilically substituted products without the aid of any

Lewis acid catalyst. Often, the reaction does not stop at single substitution and leads to polysubstituted phenols. For example, phenol with bromine water yields 2,4,6-tribromophenol.

Br_2/water/RT or below

(2, 4, 6 - Tribromo-phenol)

(d) Friedel-Crafts alkylation and acylation : Phenols undergo Friedel-Crafts alkylation and acylation to yield ring substituted alkyl phenols and acyl phenols. However the yields in both types of reactions are very poor.

$CH_3I/AlCl_3$

(Phenol) (o-Methylphenol or o-cresol) + (p-Methylphenol or p-Cresol)

$Ac_2O/AlCl_3$

(o-Acetylphenol or o-Hydroxy-acetophenone) + (p-Acetylphenol or p-Hydroxy-acetophenone)

The alkyl phenols are generally prepared in a state of purity from the corresponding amines by diazotization in dilute suphuric acid whereas acyl phenols are obtained by **Fries migration reaction.**

(ii) Special reactions of phenolic nucleus: The hydroxyl group of a phenolic system is an ortho, para directing group. The hydroxyl group is a very strong electron-donating group and activates the benzene ring very greatly towards electrophilic aromatic substitution, bypassing even the phenolic ethers/alkoxy benzenes. In fact, the phenoxide ion is the strongest electron-releasing and ring activating group known so far. As a consequence, the benzene ring of a phenol undergoes substitution even with very weak electrophiles — a property special to the phenolic nucleus and *not* exhibited

by any other class of benzene derivatives. The activity of phenols is paralleled only by highly reactive, π-excessive heterocyclic systems like furan, pyrrole and thiophene (to be studied later). The following are some of the special reactions undergone by phenols;

(a) Kolbe reaction: Treatment of the salt of a phenol with CO_2 under pressure leads to the formation of phenol carboxylic acids. The major product in this reaction is the ortho-isomer together with smaller amounts of the para-isomer.

(Sodium phenoxide) —O=C=O→ (Sodium-salicylate) + (Sodium p-hydroxy-benzoate) —$H_3\overset{\oplus}{O}$→ Salicylic acid (major) + (p-Hydroxy-benzoic acid) (minor)

The reaction follows the mechanism given below;

The Kolbe reaction is synthetically useful since it gives in one step, hydroxy benzoic acids which are otherwise obtained by a circuitous process.

(b) Riemer-Tiemann reacion: When phenol is heated with chloroform and potassium *t*-butoxide/conc. *aq.* NaOH, an aldehyde group is introduced into the ring yielding a mixture of ortho and para substituted products. The reaction has synthetic application and is known as Riemer-Tiemann reaction. For example;

OH → $CHCl_3$/base/Δ and then H_3O^+ → OH, CHO + OH, CHO

(Phenol) (o-Hydroxy benzaldehyde) (p-Hydroxy-benzaldehyde)

The mechanism of the reaction is as shown below;

$O^{\ominus}$ ⟷ O, $\ominus$ H; $\ddot{C}Cl_2$ → O, H, $\ominus CCl_2$ → $O^{\ominus}$, $CHCl_2$

$\overset{\ominus}{OH}$ → $O^{\ominus}$, $CH(OH)_2$; $-H_2O$ → $O^{\ominus}$, CHO; $H_3O^{\oplus}$ During quenching → OH, CHO

An initial reaction of chloroform with the strong base leads to the loss of elements of HCl from the chloroform resulting in the generation of dichlorocarbene ($\dot{:}CCl_2$). The latter, although neutral, is a highly reactive and electrophilic species with only a sextet of electrons around carbon. The dichlorocarbene attacks the ortho or para position of the phenoxide ion yielding an electrophilically substituted product, which ultimately gives the mixture of ortho-and parahydroxybenzaldehydes. This mixture can readily be separated into its constituents by steam distillation or by fractionation due to wide variations in their water miscibilities and volatilities.

(c) Gattermann-Koch formylation : Phenols react with HCN in the presence of anh.HCl to yield, after hydrolysis, mixture of ortho and para hydroxybenzaldehydes.

OH → i) HCN/HCl ii) H_2O → OH, CHO + OH, CHO

(o-Hydroxy benzaldehyde or salicylaldehyde) (p-Hydroxy-benzaldehyde)

The reaction proceeds through an electrophilic attack on the phenol nucleus involving the species $H\text{-}C^{+}\text{=}NH$ generated *in situ* in the reaction medium.

OH, H-C=NH, CH=NH, $H_3O^{\oplus}$, CH=O

+ The para - isomer

(d) Coupling with diazonium salts: Phenoxide ions couple with aromatic diazonium salts (which are weak electrophiles) to give brightly colored diazo products. β-Naphthol has been extensively used as the phenolic component in these reactions which find wide application in dye-manufacture. For example;

HONO (Nitrous acid) ($NaNO_2$ + HCl)

(Aniline or aminobenzene) → (Benzenediazonium chloride)

(β-Naphthol or 2-Naphthol) —NaOH→ (Sodium 2-naphthote)

N=N—Ph, $H_3O^{\oplus}$, OH

(e) Formation of phthaleins: When phenols are heated with phthalic anhydride in the presence of an acid catalyst, phthaleins are produced. These phthaleins give characteristic colours in alkaline solution and are used as *indicators*, the familiar example being phenolphthalein.

OH + OH (2 moles of phenol) + (Phthalic anhydride) —Conc. H_2SO_4 or anh. $ZnCl_2$ or anh. $AlCl_3$→ (Phenol phthalein produces pink colour)

(Colourless species- in acid medium) —$\overset{\ominus}{OH}$→ ⟷ (Resonance stabilised pink coloured species- alkaline medium)

(f) Reaction with Formaldehyde -Formation of Bakelite: Phenol reacts with active aldehydes to form condensation products. Notable in this series is the reaction of the phenol with formaldehyde in the presence of an acid or base catalyst. If mild conditions are used and reaction time kept small,a monomeric product is obtained. However,if conditions are sufficiently strong,like higher temperature (>40°C) and reaction time quite large (above 3-4hr), a polymeric resinous product is formed, known as *Bakelite*. The latter, after suitable processing, has been used for many years as a structural,sheet material in various domestic and industrial applications,one of the general commercial product being decolam laminations.

OH —HCHO/RT/1-2hrs. (mild) $\overset{\oplus}{H}$ or $\overset{\ominus}{OH}$→ (CH_2OH) + (CH_2OH) —HCHO/>40° > 4hrs. $\overset{\oplus}{H}$ or $\overset{\ominus}{OH}$ (rigorous)→

The mechanism of initial condensation is as follows:

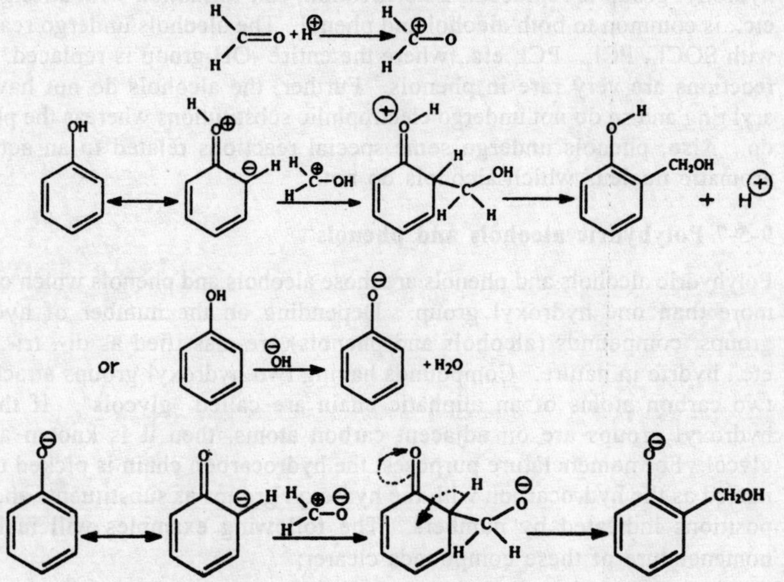

The initial condensation product undergoes further reaction on heating and/or on increasing the reaction time.

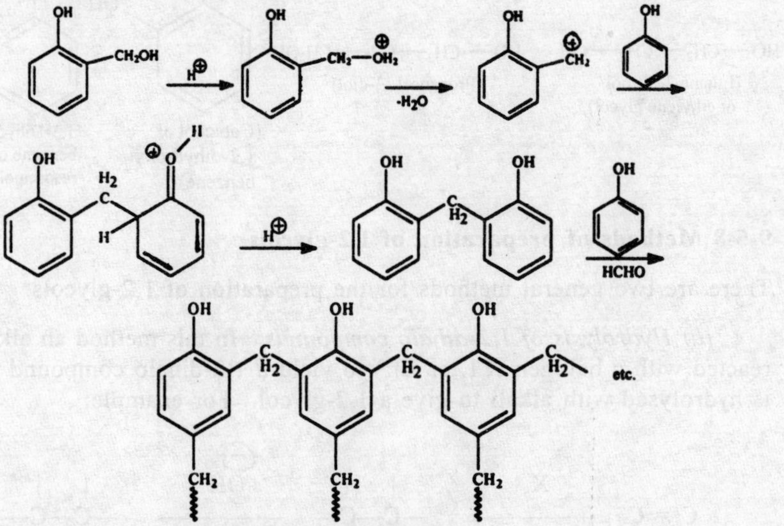

9-5-6 Comparison of alcohols and phenols

From a study of the group properties of alcohols and phenols, it follows that alcohols and phenols are both highly associated substances. Alcohols are much weaker acids ($pK_a \approx 16$-18), compared to phenols ($pK_a \approx 10$). Alcohols do not dissolve in aq. NaOH solutions to form salts whereas phenols do. The alcohols and phenols undergo reactions, wherein the H from the do. The alcohols and phenols undergo reactions, wherein the H from the hydroxyl group is replaced. Esterification, salt formation with strong bases etc., is common to both alcohol and phenol. The alcohols undergo reactions with $SOCl_2$, PCl_3. PCl_5 etc., where the entire -OH group is replaced. Such reactions are very rare in phenols. Further, the alcohols do not have any aryl ring and so do not undergo electrophilic substitutions whereas the phenols do. Also, phenols undergo some special reactions related to an activated aromatic nucleus which alcohols do not.

9-5-7 Polyhydric alcohols and phenols

Polyhydric alcohols and phenols are those alcohols and phenols which contain more than one hydroxyl group. Depending on the number of hydroxyl groups, compounds (alcohols and phenols) are classified as di-, tri-, tetra, etc., hydric in nature. Compounds having two hydroxyl groups attached to two carbon atoms of an aliphatic chain are called "glycols". If the two hydroxyl groups are on adjacent carbon atoms, then it is known as 1,2-glycol. For nomenclature purposes, the hydrocarbon chain is picked up and named as the hydrocarbon with the hydroxyl groups as substituents and their positions indicated by numbers. The following examples will make the nomenclature of these compounds clearer;

HO—CH₂—CH₂—OH (Ethane-1, 2-diol or ethylene glycol)

HO—CH₂—CH₂—CH₂OH (Propane-1, 3-diol)

(Catechol or 1,2-dihydroxybenzene)

(1,3-Dihydroxybenzene or resorcinol)

9-5-8 Methods of preparation of 1,2-glycols

There are two general methods for the preparation of 1,2-glycols.

(a) Hydrolysis of 1,2-dihalo compounds: In this method an alkene is reacted with a halogen (Cl_2 or Br_2) to yield a 1,2-dihalo compound which is hydrolysed with alkali to give a 1,2-glycol. For example;

$$\underset{\text{(Alkene)}}{C=C} \xrightarrow[(Cl_2 \text{ or } Br_2)]{X_2} \underset{\text{(1,2-Dihalo compd.)}}{-\underset{X}{\overset{|}{\underset{|}{C}}}-\underset{X}{\overset{|}{\underset{|}{C}}}-} \xrightarrow{\overset{\ominus}{OH}} \underset{\text{(1,2-Glycol)}}{-\underset{HO}{\overset{|}{\underset{|}{C}}}-\underset{OH}{\overset{|}{\underset{|}{C}}}-}$$

(b) Oxidation of alkene: Controlled oxidation of an alkene with $KMnO_4$ or OsO_4 yields a 1,2 glycol.

$$>C=C< \xrightarrow{KMnO_4} -\underset{HO}{\underset{|}{\overset{|}{C}}}-\underset{OH}{\underset{|}{\overset{|}{C}}}-$$

$$>C=C< \xrightarrow{OsO_4} \text{(Osmium tetroxide complex)} \xrightarrow{H_2O} -\underset{HO}{\underset{|}{\overset{|}{C}}}-\underset{OH}{\underset{|}{\overset{|}{C}}}- \text{ (1,2-Glycol)}$$

(Osmium tetroxide complex) (1,2-Glycol)

Alternatively, the alkene may also be oxidized with peracid to yield a cyclic oxide which on hydrolysis yields the 1,2-glycol.

$$>C=C< + R-\underset{O}{\underset{||}{C}}-O-OH \longrightarrow \text{(Cyclic oxide)} + R-\underset{O}{\underset{||}{C}}-O-H \xrightarrow{H_2O} -\underset{HO}{\underset{|}{\overset{|}{C}}}-\underset{OH}{\underset{|}{\overset{|}{C}}}- \text{ (1,2-Glycol)}$$

(Cyclic oxide) (1,2-Glycol)

9-5-9 Properties of 1,2-glycols

Due to the presence of two hydroxyl groups, glycols are very strongly associated substances. Futher, the glycols form very strong (rather doubly stronger) H-bonds with water and so are readily soluble in water.

As far as chemical properties are concerned, most glycols show the usual reaction of alcohols except double the amount of reagent is required to react with both hydroxyl groups. For example;

$$R-\underset{OH}{\underset{|}{CH}}-\underset{OH}{\underset{|}{CH_2}} \xrightarrow{2\ Na} R-\underset{Na^{\oplus}\ O^{\ominus}}{\underset{|}{CH}}-\underset{O^{\ominus}Na^{\oplus}}{\underset{|}{CH_2}} + H_2\uparrow$$

$$R-\underset{OH}{\underset{|}{CH}}-\underset{OH}{\underset{|}{CH_2}} \xrightarrow[ZnCl_2]{HCl/anh.} R-\underset{Cl}{\underset{|}{CH}}-\underset{Cl}{\underset{|}{CH_2}}$$

$$R-\underset{OH}{\underset{|}{CH}}-\underset{OH}{\underset{|}{CH_2}} \xrightarrow{2\ X\ PCl_5} R-\underset{Cl}{\underset{|}{CH}}-\underset{Cl}{\underset{|}{CH_2}}$$

Noteworthy among the reactions of 1,2-glycols is the acid catalysed *pinacol-pinacolone rearrangement* reaction. Pinacol is the classical, special name for 1,2-glycols. When a pinacol or 1,2-glycol is heated with an acid, a ketone is obtained by rearrangement. The ketone is given the special name of pinacolone and hence the name pinacol to pinacolone rearrangement. The reaction is;

```
      R  R           ⊕              R
      |  |          H /Δ            |
    R—C——C—R      ———————→      R—C——C—R
      |  |                          |  ||
     HO  OH                         R  O
  (a 1,2-glycol                 (a ketone-
   or pinacol)                   pinacolone)
```

The reaction follows the mechanism shown below;

```
   R  R         ⊕         R  R                  R  R
   |  |         H         |  |        -H₂O      |  |
 R—C——C—R    ——————→   R—C——C—R     ——————→   R—C——C—OH
   |  |                   |  |                  ⊕  |
  HO  OH               H₂O⊕  OH                    R

                          R          ⊕          R
                          |  ⊕       -H         |
             ——————→   R—C——C—R     ——————→   R—C——C—OH
                          |  |                  |  ||
                          R  OH                 R  O
```

Formation of a highly stable tertiary carbonium ion and its subsequent rearrangement to the more stable hydroxy carbonium ion is the driving force for this reaction.

9-5-10 Glycerol

It is 1,2,3-trihydroxy propane and industrially very important chemical. Glycerine is its common name. Glycerol is prepared by two methods;

(a) From propene : In this method, propene is subjected to allylic chlorination at higher temperatures to yield allyl chloride, which is hydrolysed to allyl alcohol. The latter on treatment with Cl_2/NaOH followed by hydrolysis results in glycerol. The reactions can be formulated as shown below;

$$H_3C{-}CH{=}CH_2 \xrightarrow{Cl_2/500^\circ} ClCH_2{-}CH{=}CH_2 \xrightarrow[Na_2CO_3]{Aq} HO{-}CH_2{-}CH{=}CH_2$$

(Propene) (Allyl chloride) (Allyl alcohol)

$$\xrightarrow[(+\ HOCl)]{Cl_2/NaOH} \left.\begin{array}{c} HO{-}CH_2{-}\underset{\displaystyle Cl}{\underset{|}{CH}}{-}CH_2OH \\ + \\ HO{-}CH_2{-}\underset{\displaystyle OH}{\underset{|}{CH}}{-}CH_2Cl \end{array}\right\} \xrightarrow{NaOH} \underset{\displaystyle HO}{\underset{|}{H_2C}}{-}\underset{\displaystyle OH}{\underset{|}{CH}}{-}\underset{\displaystyle OH}{\underset{|}{CH_2}}$$

(Glycerol)

(b) From oils and fats by saponification: Oils and fats are the naturally occurring triesters of glycerol with higher fatty acids such as stearic acid [R-$(CH_2)_{14}$-COOH]. When reasonably pure, oils and fats are subjected to hydrolysis with alkali, the sodium salt of fatty acid (which is the soap) and glycerol are obtained. Hydrolysis with aq. NaOH of an oil or fat, in commercial terminology, is known as saponification.

$$\begin{array}{l} CH_2{-}O{-}\underset{\displaystyle O}{\underset{||}{C}}{-}R \\ | \\ CH{-}O{-}\underset{\displaystyle O}{\underset{||}{C}}{-}R \\ | \\ CH_2{-}O{-}\underset{\displaystyle O}{\underset{||}{C}}{-}R \end{array} \xrightarrow[\Delta]{H_2O/OH^-} \begin{array}{l} CH_2{-}OH \\ | \\ CH{-}OH \\ | \\ CH_2{-}OH \end{array} + \overset{+}{Na}\ \overset{-}{O}{-}\underset{\displaystyle O}{\underset{||}{C}}{-}R \times 3$$

(Oil or fat) (where R'-Stearyl group etc.) (Soap)

9-5-11 Nitroglycerine

The word nitroglycerine is misleading in that the compound does not contain any C-NO_2 grouping/s. On the other hand, nitroglycerine is simply glyceroltrinitrate. It is the triester of glycerol with nitric acid and contains three C-O-NO_2 groupings. Being the triester, glycerol trinitrate is prepared by esterification of glycerol with nitric acid.

$$\begin{array}{l} CH_2{-}O{-}H \\ | \\ CH{-}O{-}H \\ | \\ CH_2{-}O{-}H \end{array} + 3 \times HO{-}NO_2 \longrightarrow \begin{array}{l} CH_2{-}O{-}NO_2 \\ | \\ CH{-}O{-}NO_2 \\ | \\ CH_2{-}O{-}NO_2 \end{array}$$

(Glycerol) (Nitric acid) (Glycerol trinitrate or nitroglycerine)

The production process involves treatment of purified glycerol with a mixed acid containing nitric acid and sulphuric acid in water. The temperature, during reaction is carefully controlled and the product when separated from the refuse acid is washed free of acid.

Nitroglycerine was discoverd by Sobrero in 1846 but its preparation developed on a commercially scale by Alfred Nobel, around 1865. Nitroglycerine is a powerful explosive but it is highly sensitive. The sensitive explosive is set off very easily by most forms of friction and impact. During the early years of its developement, many accidents took place, taking a large toll of innocent lives including those of Alfred Nobel's father and brother. However, by an ingenious mind, Alfred Nobel found out that when nitroglycerine (liquid) is absorbed in certain inert substances like kieselguhr, diatomaceous earth, etc., a less sensitive material results without loss of explosive power. This new composition was called *dynamite* for which Alfred Nobel became famous and extremely rich. It was this money which was placed by Nobel in a trust foundation and Nobel awards came from it year after year for outstanding discoveries in Science, Medicine, Economics and Peace.

Nitroglycerine is used as a high explosive for civilian purposes. It is also used in certain initiatory compositions for high explosives. Along with nitrocellulose, nitroglycerin finds application in the manufacture of propellants and rockets. Nitroglycerine is a viscous yellow liquid which freezes at 13.2°C to a sensitive solid explosive. This property of nitroglycerine made its storage in cold countries very dangerous and risky. However, the difficulty can be overcome by adding glycoldinitrate to nitroglycerine thereby reducing the freezing point of the latter.

9-6 INDUSTRIAL SIGNIFICANCE OF HYDROXY COMPOUNDS

9-6-1 Methyl alcohol

Prior to world war II, wood was the principal source of methyl alcohol and so it was originally called as *wood alcohol.* When wood is heated at 400°C, a carbonaceous residue *"charcoal"* is left behind, a liquid condensate and gaseous mixture are obtained. The liquid condensate contains 5-8%MeOH. The latter is also obtained by controlled oxidation of natural gas.

$$\underset{\text{(From naturalgas)}}{CH_4} \xrightarrow{O_2/600°} \underset{\text{(Carbon monoxide)}}{CO} \xrightarrow[ZnO\text{–}Cr_2O_3]{H_2/300°} \underset{\text{(Methanol)}}{H_3C\text{—}OH}$$

Methanol is a clear, colourless liquid with a characteristic smell. It has a boiling point of 65°C and is miscible with water.

The most important use of methanol is as solvent in industry particularly of plastics and textile ones. Methanol is used in manufacturing operations in industry for cleansing purposes. Besides, methanol is used in the preparation of a large number of industrial chemicals like HCHO (required in the manufacture of polymers and resins, notably Bakelite), formic acid, ethyl chloroformate, etc. Methanol is also used as an anti-freeze.

9-6-2 Ethyl alcohol

Among the top ten organic chemicals produced commercially, ethanol is second or third in terms of annual production. Ethanol is prepared by two methods:

In oil-rich countries, where a large reservoir of petroleum is available, ethanol is obtained by hydration of ethylene. The latter is produced by dehydrogenation of ethane obtained from natural gas.

$$\underset{\text{(Ethane from natural gas)}}{H_3C{-}CH_3} \xrightarrow{-H_2/\Delta} \underset{\text{(Ethylene)}}{H_2C = CH_2} \xrightarrow{H_2O/H^{\oplus}} \underset{\text{(Ethanol)}}{H_3C{-}CH_2{-}OH}$$

In tropical countries, ethanol is manufactured by biochemical degradation of molasses. The latter is obtained as mother liquor during cane sugar manufacture from sugar-cane. Molasses contain largely water and a few chemicals together with 18-20% cane sugar. Fermentation of molasses yields alcohol which is isolated by distillation and purified.

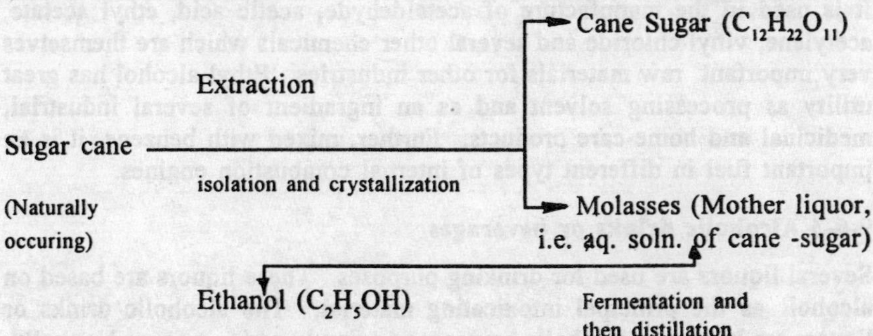

Ethyl alcohol is a clear, colourless liquid with a pleasant, fruity smell. Its boiling point is 78.1°C and is miscible with water as well as most organic solvents. Ethanol, for most purposes (industrial or laboratory) is supplied in any one of the following three forms, namely;

(i) *Industrial alcohol or commercial spirit or denatured spirit:* Sold under various names, this is the commercial grade alcohol which is around 92-94% ethanol, ≈ 5% water and 1-2% denaturant. The latter is usually methanol, pyridine or monoethyl oxalate, etc. The denaturant is added mostly with the intention of making the commercial grade alcohol unfit for human consumption.

(ii) *Rectified spirit :* This is the industrial grade alcohol minus the denaturant which has been subjected to distillation. Thus, rectified spirit is ≈ 95% ethanol and ≈ 5% water. Since ethanol and water form an azeotropic mixture of approximately 95% : 5% composition, further purification of rectified spirit by distillation is not possible.

(iii) *Absolute alcohol:* This is 99 to 99.5% pure ethanol and 0.5 to 1.0% water. It is usually prepared by refluxing rectified spirit with anhydrous lime (CaO) for a few hours followed by distillation over a small amount of sodium metal. Anhydrous lime (CaO) required for this purpose is specially prepared by heating commercial lime in a furnace at 800 - 1000°C and then cooling it to RT in the absence of moisture.

Commercially, absolute alcohol is prepared by mixing it with benzene and then distilling the mixture. Since benzene forms a ternary azeotrope with alcohol and water, boiling at 64.5°C, the first fraction collected at 64.5°C will be benzene + water + alcohol. When all the water (corresponding to 5% in the original lot) is removed, the boiling point of mixture rises to 68° when ethanol + benzene, a binary azetrope mixture, distills over. After all the benzene is removed by distillation, the boiling point of liquid rises to 78.1° when pure ethanol distills over. The absolute alcohol thus obtained is highly hygroscopic and should be kept away from moisture using suitable anh.$CaCl_2$ guard tubes etc. Otherwise, there will be risk of absolute alcohol absorbing moisture from atomosphere and slowly changing to rectified spirit.

Ethyl alcohol is a very important and highly useful organic raw material. It is used in the manufacture of acetaldehyde, acetic acid, ethyl acetate, acetylene, vinyl chloride and several other chemicals which are themselves very important raw materiàls for other industries. Ethyl alcohol has great utility as processing solvent and as an ingradient of several industrial, medicinal and home-care products. Further, mixed with benzene, it is an important fuel in different types of internal combustion engines.

9-6-3 Alcoholic drinks or beverages

Several liquors are used for drinking purposes. These liquors are based on alcohol as the principal intoxicating material. The alcoholic drinks or liquors are known as alcoholic beverages or simply as *beverages*. Normally, alcoholicbeveragescontain, besides alcohol and water, coloring and flavoring materials as other constituents. Beverages are of two types : (a) distilled and (b) undistilled.

(a) Distilled beverages : These are very strong drinks and are prepared by fermentation of molasses, maize, barley, etc. The fermented liquor is then subjected to distillation and the distillate is bottled. The distillate consists of 40-45% alcohol, low molecular weight volatile organic substances,

water and characteristic material from raw material imparting particular taste to the drink.

(b) Undistilled beverages : These are light drinks and are prepared by fermentation of fruit juices or grains. The fermented liquor is used as such without distillation. In the undistilled material, the alcohol percentage rarely exceeds 10.0. The names, sources and alcohol content of some common beverages are given below;

Undistilled				Distilled			
S.No.	Name	Source	% EtOH	S.No.	Name	Source	% EtOH
(1)	Beer	Barley	3-5	(1)	Brandy	Grapes	40-50
(2)	Cider	Apples	3-5	(2)	Whisky	Barley	40-50
(3)	Port & Sherry	Grapes	10-15	(3)	Rum	Molasses	45-48
(4)	Champagne	Grapes	8-10	(4)	Gin	Maize	40-45

Ethyl alcohol is an instant source of energy for the human body. When consumed, ethyl alcohol is very quickly metabolized, releases entire amount of energy in one stroke and in the shortest possible time. The release of energy by alcohol is even faster than the one by glucose. However, alcohol has a great addiction liability whereas glucose has almost none. It is for this reason that students, sports men and hospital patients are given glucose and *not* alcohol. When addiction occurs, larger doses of alcohol are required to get the same amount of energy and the alcohol acts directly on brain as well as on the liver or stomach, setting in a chain of damages in the human body.

9-6-4 Ethylene glycol

This is a powerful solvent and is used in several commercial processes. In the laboratory, glycol is used for ketalisation reaction. Ethylene glycol is used, commercially, in the manufacture of dacron required for synthetic fibremaking. Glycol is used in the commercial preparation of dioxane (a versatile solvent), glycol dinitrate (explosive), glycol stearate (lubricant), etc.

9-6-5 Phenol

The largest volume use of phenols is in the manufacture of plastics and resins, notably, Bakelites. A great amount of phenol is used in the preparation of a variety of intermediates required for the manufacture of drugs, pharmaceuticals, antiseptics, disinfectants, agro-chemicals and pesticides.

9-7 SELF-TEST QUETIONS

Choose the correct answer for the following questions.

9-7-1 2,2,-Dimethyl-1-hydroxy propane is an example of
(A) Secondary alcohol (B) Tertiary alcohol
(C) Primary alcohol (D) Aromatic alcohol

9-7-2 n-Butyl alcohol is prepared by
(A) Reaction of *n*-butyl chloride with aq.NaOH at RT
(B) Reaction of *n*-butane with anh. NaOH at 100°C
(C) Reaction of 1-butene with aq.NaOH at 100°C
(D) Reaction of 1-butyne with steam at 100°C

9-7-3 When styrene (1-phenyl ethylene) is treated with conc.H_2SO_4 followed by water at RT, the major product obtained is
(A) Benzyl alcohol (B) 2-Phenyl ethanol
(C) 1-Hydroxy -1-phenylethane (D) none of the three.

9-7-4 Reaction of formaldehyde with benzyl magnesium chloride results in the formation of
(A) 1-Phenyl-1-hydroxy ethane (B) 2-Phenylethanol
(C) Benzyl alcohol (D) none of the three

9-7-5 Reaction of formaldehyde with benzylmagesium chloride results in the formation of
(A) 1-Phenyl-1-hydroxyethene (B) 2-Phenylethanol
(C) Benzyl alcohol (D) All the three compounds

9-7-6 Reduction of benzaldehyde with $LiAlH_4$ yields
(A) Toluene (B) Benzoic acid
(C) Benzene (D) Benzyl alcohol

9-7-7 Among the following compounds, the one which has the highest boiling point is
(A) Styrene (B) Phenyl acetylene
(C) 2-Phenylethanol (D) Ethyl benzene

9-7-8 Phenol can be prepared in the laboratory by
(A) Hydrolysis of chlorobenzene with NaOH at RT
(B) Diazotization of aniline with $NaNO_2$ and dil.H_2SO_4 at RT
(C) Reaction of toluene with NaOH at 100°C
(D) All the three methods

9-7-9 Complete oxidation of primary alcohols with retention of carbon-chain leads to
(A) Carboxylic acids (B) Carboxaldehydes
(C) Ketones (D) none of the three

9-7-10 Intermolecular H-bonding in a liquid
(A) Increases its boiling point
(B) Decreases its boiling point
(C) Does not cause any change in boiling point
(D) Causes all the above three changes

9-7-11 Reaction of phenol with $CHCl_3$ in the presence of potassium *t*-butoxide is known as
(A) Gattermann.Koch formylation
(B) Kolbe-Schmidt reaction
(C) Villsmeyer-Haack formylation
(D) Riemer-Tiemann reaction.

9-8 SHORT AND ESSAY TYPE QUESTIONS

9-8-1 Describe any two methods for the preparation of primary alcohols giving at least one example for each method.

9-8-2 How can alcohols be converted into alkyl halides? Give two methods for this conversion and mention at least one example for each method.

9-8-3 What is Lucas reagent? How can it be used to distinguish between primary, secondary and tertiary alcohols? (O.U., Mar., 1994)

9-8-4 What is a Grignard reagent? What happens when methyl magnesium iodide is treated with HCHO,CH_3—CHO and CH_3COCH_3 individually? (N.U., Sept., 1994)

9-8-5 Describe a simple method for the laboratory preparation of phenols? How is phenol prepared commercially?

9-8-6 Explain why phenols are more acidic than alcohols? (N.U., Mar., 1995)

9-8-7 What are glycols? Describe any two methods for the preparation of 1,2-glycols.

9-8-8 Write short notes on the following. (O.U., Mar., 1996)
(i) Carbinols (ii) Oxidation (iii) Nitroglycerine (iv) Picric acid

9-8-9 Explain how the following transformations can be carried out?

(i) Benzene to phenol

(ii) Benzene to benzyl alcohol

(iii) *n*-propyl chloride to 2-hydroxy pentane

(iv) Phenol to salicylic acid (N.U., Sept., 1995), (N.U., Mar., 1996)

(v) Toluene to benzyl acetate

(vi) Acetone to *t*-butyl alcohol.

(vii) Methyl alcohol to ethyl alcohol

9-8-10 Using qualitative test-tube experiments how can the following be distinguished?

(a) *p*-Cresol from benzyl alcohol

(b) *n*-Butyl chloride from *n*-butyl alcohol

(c) 1-Butanol from 2-butanol

(d) Phenol from picric acid (N.U., Mar., 1996)

(e) Propylene from *n*-propyl alcohol

9-8-11 Give the structures of products formed in the following trasformations?

(i) $H_3C—CH_2—CH_2—OH \xrightarrow{PCl_5} A \xrightarrow{KCN} B \xrightarrow{H_3O^+} C$
(n-propanol)

(ii) $H_3C—CH_2—OH \xrightarrow[\text{oxidation } (CrO_3)]{\text{Careful}} A \xrightarrow{MeMgI} B \xrightarrow{\text{Oxidation}} C$
(Ethanol) (an aldehyde) (a ketone)

(iii) $\text{Toluene} \xrightarrow{Cl_2/AlCl_3} A + B$

$A \xrightarrow{Cl_2/h\nu} C \xrightarrow{HO^-} D \xrightarrow[KMnO_4]{[O]} E$ (p-disubstitued benzene)

$B \xrightarrow{Cl_2/h\nu} F \xrightarrow{HO^-} G \xrightarrow[{[O]}]{KMnO_4} H$
(alcohol) (o-disubstituted benzene)

(iv) $\text{Naphthalene} \xrightarrow[160°]{H_2SO_4} A \xrightarrow[\text{fusion \& then } H_3O^+]{NaOH} B \xrightarrow{Ph\overset{+}{N_2}\overset{-}{Cl}} C$ (An AZO compound)

(v) $\text{1-propylene} \xrightarrow{Cl_2/RT/CHCl_3} A \xrightarrow{NaOH} B \xrightarrow[KMnO_4]{\text{Oxidation}} C$
(Dihydroxy compound) (An α-keto acid)

(vi) $\text{Tetramethyl ethylene} \xrightarrow{CH_3COOOH} A \xrightarrow{H_2O/H^+} B \xrightarrow{H_3O/\Delta} C$
(An expoxide) (a glycol) (a ketone)

(vii) $\text{Amyl alcohol } or \text{ n-pentanol} \xrightarrow[\text{anh.}ZnCl_2]{HCl/} A \xrightarrow[\text{base}/\Delta]{\text{Pyridine}/} B \xrightarrow[\text{ii)}H_2O]{\text{i) }OsO_4} C \xrightarrow[HNO_3]{\text{Conc.}} C$

9-8-12 Give any two methods of preparation of primary, secondary and tertiary alcohols. (O.U.,Oct.1991)

9-8-13 Give equations for all the steps involved in the following reactions?

(i) Ethyl alcohol,iodine and Sod. hydroxide when warmed (N.U., Sept., 1995), (N.U., Mar., 1996)

(ii) n-Butyl alcohol and conc. H_2SO_4 on heating (A.U., Mar., 1994)

(iii) Phenol and bromine water. (O.U., Oct.1991)

9-8-14 Compare the reactivities of alcohols and phenols. Write about the industrial significance of ethanol and phenol (A.U.,Dec.1989)

9-8-15 How is glycerol prepared? Give its important properties and mention its uses. (A.U., June 1990), (A.U., Nov., 1994), (N.U., Sept., 1995)

9-8-16 How will you differentiate between the following pairs of compounds using test-tube experimetns.

(i) i-propyl alcohol and n-propyl alcohol.

(ii) Ethyl alcohol and phenol.

What is saponification? Mention two uses of glycerol.

(K.U., Apr., 1990), (O.U., Mar., 1995)

9-8-17 How do phenols differ from alcohols in their properties.

(N.U., Apr., 1992), (N.U., Mar., 1995)

SOLUTIONS TO PROBLEMS

9-7-1 **C**	9-7-2 **A**	9-7-3 **C**	9-7-4 **B**	9-7-5 **D**	9-7-6 **C**
9-7-7 **B**	9-7-8 **A**	9-7-9 **A**	9-7-10 **D**	9-7-11 **D**	

9-8-9

(i) (Benzene) $\xrightarrow[H_2SO_4]{HNO_3/}$ (Nitro benzene) $\xrightarrow[Fe/H2O/H+]{Reduction}$ (Aniline or aminobenzene) $\xrightarrow[dil\ H_2SO_4/RT]{NaNO_2}$ (Phenol)

(ii) (Benzene) $\xrightarrow[AlCl_3]{CH_3I/}$ (Toluene) $\xrightarrow{Cl2/h\nu}$ (Benzyl chloride) $\xrightarrow{\overset{\ominus}{O}H}$ (Benzyl alcohol

(iii) $H_3C-CH_2-CH_2Cl \xrightarrow[Et_2O]{Mg/} H_3C-CH_2-CH_2-MgCl$

$\xrightarrow[\&\ then\ H_3O^{\oplus}]{CH_3-\overset{O}{\overset{\|}{C}}-H} H_3C-\underset{OH}{\underset{|}{\overset{H}{\overset{|}{C}}}}-CH_2-CH_2-CH_3$

(iv) (Phenol) $\xrightarrow[(Riemer - Tiemann)]{CHCl3/\overset{\oplus}{K}\ t - Bu\overset{\ominus}{O}}$ (Salicylal-dehyde) $\xrightarrow{Ac_2O}$ (o-Acetoxy-salicylaldehyde)

OAc / COOH on benzene ring $\xrightarrow[\text{[Ag (NH}_3)_2]^+]{\text{KMnO}_4 \text{ or}}$ (Acetylsalicylic acid or aspirin) $\xrightarrow{H_3O^{\oplus}}$ OH / COOH on benzene ring (Salicylic acid or o-Hydroxy benzoic acid)

(v) CH_3 on benzene ring (Toluene) $\xrightarrow{Cl_2/h\nu}$ CH_2Cl on benzene ring (Benzyl chloride) $\xrightarrow{H_2O/\overset{\cdot}{O}H}$ CH_2OH on benzene ring (Benzyl alcohol) $\xrightarrow{CH_3COOH}$ $CH_2—O—CO—CH_3$ on benzene ring (Benzoyl acetate)

(vi) $H_3C-C(=O)-CH_3$ (Acetone) $\xrightarrow{H_3C-MgI}$ $(H_3C)_2C(CH_3)-O-Mg-I$ (Grignard adduct) $\xrightarrow{H_2O}$ $(H_3C)_2C(CH_3)-OH$ (t-Butanol)

(vii) H_3C-OH (Methanol) $\xrightarrow[\text{(Careful oxidation)}]{CrO_3}$ $H-C(=O)-H$ (Formaldehyde) $\xrightarrow{Me-MgI}$ $H_2C(CH_3)-O-MgI$

$\xrightarrow{H_2O}$ H_3C-CH_2-OH (Ethanol)

10

ETHERS

10-1 DEFINITION

Ethers are a group of compounds containing a C—O—C linkage. Ethers can formally be derived from alcohols or phenols by replacing the hydrogen of the hydroxyl group with alkyl or aryl groups. Ethers may be aliphatic, alicyclic, aromatic or aliphatic-aromatic in nature. Thus,ethers have the general formula R—O—R, Ar—O—Ar or R—O—Ar etc.

10-2 NOMENCLATURE

Ethers are of two types, namely symmetrical ethers and unsymmetrical ethers. Symmetrical ethers are those in which the substituents are identical. Ethers in which the substituent groups are different, are known as mixed or unsymmetrical ethers. Most ethers are called by their common names or trivial names. Where common names do not exist, the ethers were named in the older system by mentioning the names of the two alkyl groups and adding the ending "ether".

For example:

$H_3C—O—C_2H_5$	$C_2H_5—O—C_6H_5$	$C_6H_5—O—C_6H_5$
(Methyl ethyl ether)	(Ethyl phenyl ether)	(Diphenyl ether)

The most common of all ethers, diethyl ether, $C_2H_5—O—C_2H_5$, is called ethyl ether or simply just "ether".

According to the IUPAC system, ethers are named as alkoxy alkanes. The larger of the alkyl groups is considered to be the parent alkane.

For example:

$$\overset{1}{H_3C}–O–\overset{2}{CH_2}–\overset{3}{CH_2}–\overset{4}{CH_2}–CH_3$$

(1-Methoxybutane)

$$H_3C–CH_2–O–\overset{2}{CH}(\overset{1}{CH_3})(\overset{3}{CH_3})$$

(2-Ethoxypropane),

$$Ph–O–\overset{1}{CH_2}–\overset{2}{CH}(\overset{3}{CH_3})(CH_3)$$

(2-Methyl-1-phenoxypropane)

10-3 STRUCTUE OF ETHERS

Structurally, ethers are very similar to organic carbon compounds wherein the oxygen atom is isosteric with a tetrahedral carbon atom. The oxygen atom in ethers is sp^3 hybridized and the C—O—C bond angle is 110° which is very close to the tetrahedral angle of 109°28'. Thus, the oxygen atom of ethers is bonded to two carbon atoms and the remaining two of the four sp^3 hybrid orbitals are occupied by two pairs of non-bonding electrons as shown below:

(Ethers) (Ethers) (Tetrahedral carbon) (Water molecule)

It may be mentioned here that in water molecule, the H—O—H bond angle is 105°, much less than the tetrahedral bond angle of 109°28'. This is believed to be due to the repulsion between the two pairs of non-bonding electrons on oxygen of the water molecule. On the other hand, in ethers, the C—O—C bond angle (110°) is very close to that of tetrahedral angle (109°28') because there is believed to be repulsion between the bulky alkyl groups forcing increase of C—O—C bond angle.

10-4 FUNCTIONAL ISOMERISM

Ethers are isomeric with corresponding same-numbers-of-carbons-containing alcohols. Since ethers and alcohols constitute two widely different functional groups,the isomerism between ethers and alcohols is known as functional isomerism. For example:

$H_3C—O—CH_3$ is isomeric with $CH_3—CH_2—OH$
(dimethyl ether) (Ethanol)

$C_2H_5—O—C_2H_5$ is isomeric with $CH_3—CH_2—CH_2—CH_2—OH$
(Diethyl ether) (*n*-Butanol)

$C_6H_5—O—CH_3$ is isomeric with $C_6H_4(OH)CH_3$
(Anisole or Methyl phenl ether) (*o*-Cresol)

10-5 PREPARATION OF ETHERS

Unlike in the case of alcohols, there are not very many methods available in literature for the preparation of ethers. However, the following two methods are very important for making ethers in the laboratory.

10-5-1 Williamson's Method

In this method,an alkyl halide is treated with the sodium or potassium salt of a phenol or an alcohol.

$$R^1\text{—OH} \xrightarrow{Na} R^1\text{—ONa} + \tfrac{1}{2}\,H_2$$

$$R\text{—X} + R^1\text{—ONa} \longrightarrow R\text{—O—}R^1 + \text{Nax}$$

One of the disadvantages in this method is the lower yield of ethers in some reactions. This is because alkoxide, being a strong base, engineers dehydrohalogentaion in the alkyl halide leading to elimination as a side reaction.

$$R^1\text{—O}^- + R^2\text{—}CH_2\text{—}CH_2\text{—X} \longrightarrow R^2\text{—CH=CH}^2 + R^1\text{—O—H} + X^-$$

10-5-2 By dehydration of alcohols

This method involves, heating an equimolar mixture of alcohol and sulphuric acid with excess of alcohol at ≈ 135°C to yield an ethereal compound. For example:

$$C_2H_5\text{—OH} + H\text{—O—}SO_2\text{—OH} + H\text{—O—}C_2H_5 \xrightarrow{135°C} C_2H_5\text{—O—}C_2H_5 + H_2SO_4.H_2O$$

The above method gives satisfactory yields of ethers only when a primary alcohol is used. Secondary and tertiary alcohols usually undergo dehydration. yielding alkenes. For the formation of ethers, the reaction proceeds through the following mechanism:

$$C_2H_5\text{—O—H} + H\text{—O—}\overset{\overset{O}{||}}{\underset{\underset{O}{||}}{S}}\text{—O—H} \xrightarrow{RT} C_2H_5\text{—O—}\overset{\overset{O}{||}}{\underset{\underset{O}{||}}{S}}\text{—O—H}$$

(An alkyl hydrogen sulphate)

$$C_2H_5\text{—O—}\overset{\overset{O}{||}}{\underset{\underset{O}{||}}{S}}\text{—OH} \xrightarrow{C_2H_5\text{–O–H}/135°} C_2H_5\text{—}\overset{\oplus}{\underset{\underset{H}{|}}{O}}\text{—}C_2H_5 + \overset{-}{O}\text{—}\overset{\overset{O}{||}}{\underset{\underset{O}{||}}{S}}\text{—O—H}$$

$$C_2H_5\text{—}\overset{\oplus}{\underset{\underset{H}{|}}{O}}\text{—}C_2H_5 + \overset{-}{O}\text{—}\overset{\overset{O}{||}}{\underset{\underset{O}{||}}{S}}\text{—OH} \longrightarrow C_2H_5\text{—O—}C_2H_5 + HO\text{—}\overset{\overset{O}{||}}{\underset{\underset{O}{||}}{S}}\text{—OH}$$

(Diethyl ether)

10-6 PHYSICAL PROPERTIES OF ETHERS

The ethers are highly non-polar, non-associated substances. Their melting and boiling points are very close to those of hydrocarbons and are as expected on the basis of molecular weights. Table 10.1 shows the boiling points of some ethers and compares them with those of isomeric alcohols, which being strongly associated substances,possess quite high boiling points.

Table 10.1 Comparison of the boiling points of ethers and isomeric alcohols.

S.No.	Name of ether	B.Pt.°C	Name of alcohol	B.Pt.°C
1.	Dimethıyl ether	-24	Ethanol	78
2.	Diethyl ether	36	*n*-Butanol	118
3.	Di-n-propyl ether	90	*n*-Hexanol	158
4.	Di-i-propyl ether	70	-	-
5.	Di-n-butyl ether	142	Octanol	194
6.	Anisole	154	*o*-Cresol	191
7.	1,4-Dioxane	101	1,4-Dihydroxy butene *	30
8.	Tetrahydrofuran	66	*n*-Butanol*	118

(* not isomeric in nature with the ether but very close in molecular weight)

Since ethers are non-associated substances, they cannot form hydrogen bonds with water and so do not show any tendency to dissolve in water. On the other hand, ethers are freely soluble in hydrocarbons, halohydrocarbons, etc. However, sometimes ethers dissolve in hydroxylic, but strongly acidic solvents like H_2SO_4, HI, etc. due to protonation on oxygen leading to oxonium-ion formation. For example:

$$C_2H_5\text{–O–}C_2H_5 + \text{HO–}\overset{\overset{\large O}{||}}{\underset{\underset{\large O}{||}}{S}}\text{–OH} \longrightarrow C_2H_5\text{–}\overset{+}{\underset{\underset{\large H}{|}}{O}}\text{–}C_2H_5 + \overset{-}{O}\text{–}\overset{\overset{\large O}{||}}{\underset{\underset{\large O}{||}}{S}}\text{–OH}$$

10-7 CHEMICAL PROPERTIES

The ethereal oxygen is one of the highly unreactive functional groups. Consequently, ethers are one of the most inert compounds resembling very closely the saturated hydrocarbons in reactivity. For example, ethers are inert to alkali and aqueous acids and to most reducing and oxidizing agents. On heating with strong, anhydrous acids, ethers are cleaved due to oxonium-ion formation folowed by C-O bond breaking.

$$R\text{—O—}R^1 + HI \longrightarrow R\text{—}\overset{+}{\underset{\underset{\large H}{|}}{O}}\text{—}R' + \overset{-}{I} \longrightarrow R\text{—O—H} + R^1\text{—I}$$

$$R—\overset{+}{\underset{|\atop H}{O}}—H \text{ ... }$$

$$R—O—H + HI \longrightarrow R—\overset{+}{\underset{H}{O}}—H + \overset{-}{I} \longrightarrow HOH + R—I$$

Overall, the reaction may be written as:

$$R—O—R^{1} + 2HI \longrightarrow R—I + R^{1}—I + H_2O$$

This reaction forms the basis of the **Zeisel's method** for the estimation of ethers. In this method, an ether is heated with excess HI acid and the cleaved components consisting of alkyl iodides determined by titration against standardized solution of $AgNO_3$.

10-8 USES OF ETHERS

Diethyl ether, di-isopropyl ether, ethylene glycol dimethyl ether, dioxane, tetrahydrofuran, etc., are some of the most important ethers in use today. Ethylene oxide may also be considered to be an ethereal compound by virtue of its possession of a C—O—C linkage. However, it is normally not regarded as an ether due to its high reactivity arising as a consequence of the ring strain in the small three-membered ring.

Diethyl ether or solvent ether or simply ether is an outstanding solvent for chemical extractions of organic compunds from aqueous solutions in the laboratory. Diethyl ether is used extensively as a solvent in Grignard reactions. It is also used as an anaesthetic for performing medical surgery operations.

Ethylene glycol dimethyl ether or glyme is a solvent of choice in reactions where ether cannot be used. One great disadvantage of ether is its low boiling point (36°C) and high volatility. Furthermore, its highly inflammable nature makes it a vulnerable solvent for chemical reactions. In such cases, glyme is the solvent of choice, since glyme is more polar than ether. The former is preferred as a solvent in reactions where polar reactants are involved. Moreover, glyme is used as a solvent in industrial processes such as production of polymers and plastics.

Dioxane,by virtue of its highly polar but inert nature, is an excellent solvent for reactions where polar reagents such as salts or ionic reagents are involved. However,its great disadvantage is its toxicity. Dioxane is prepared commercially by heating a mixture of ethylene glycol and sulphuric acid in a ration of 2:1.

$$\begin{matrix} HO–CH_2–CH_2–OH \\ + \\ HO–CH_2–CH_2–OH \end{matrix} \xrightarrow[\Delta]{H_2SO_4} \begin{matrix} H_2C—CH_2 \\ O \qquad O \\ H_2C—CH_2 \end{matrix} + 2\,H_2O$$

(Dioxane)

The high solvent capacity of dioxane is believed to be due to its high polarity which arises due to the existence of a puckered /cyclohexane-type chair system in its structure.

H_2C—CH_2, O, O, H_2C—CH_2 ≡ [chair structures]

A further advantage of dioxane is its miscibility with water and most of the common organic solvents.

Tetrahydrofuran, is probably the most important ethereal solvent used in organic sysnthesis. THF, as it is commonly known, is a reasonably polar, water miscible and fairly non-toxic ethereal solvent. It has a much higher boiling point (66°C) when compared to ether (36°C) and is liquid even at sufficiently low temperatures,as low as -90°C. These properties make THF the alternative solvent in several Grignard reactions, the only solvent of choice in metallations involving organometallic bases like phenyllithium, methyllithium, n-butyllithium, lithium diisopropyl amide, etc.,

10-9 SELF TEST QUESTIONS

Choose the correct answer for the following questions:

10-9-1 The systematic IUPAC name for anisole is

(A) Methyl phenyl ether (A) Methoxy benzene

(C) Anisole (D) None of the above three.

10-9-2 Anisole is an example of

(A) Aromatic ethers (B) Aliphatic ethers

(C) Alicyclic ethers (D) Aromatic aliphatic ethers

10-9-3 Which one of the following compounds does not show the phenomenon of H-bonding?

(A) Acetic acid (B) Ethyl alcohol

(C) Diethyl ether (D) Ethyl amine

10-9-4 Which of the following compounds give two moles of n-butyl iodide on boiling with HI?

(A) n-Butyl ether (B) n-Butyl alcohol

(C) n-Butane (D) All the above three

10-9-5 Which of the following compounds is isomeric with anisole?

(A) Benzyl alcohol (B) Cyclohexyl methyl ether

(C) Benzaldehyde (D) All the above three.

10-9-6 Williamson's method for syntheses of ethers involves

(A) Reaction of an alcohol with alkyl halide

(B) Reaction of an acid with alkyl halide

(C) Reaction of sodium salt of an alcohol with alkyl halide

(D) Reaction of an alcohol with alkyl halide in the presence of sodium hydroxide.

10-10 SHORT AND ESSAY TYPE QUESTIONS

10-10-1 How is diethyl ether prepared and what is its reaction with anhydrous sulphuric acid?

10-10-2 What is anisole?How is it prepared and what is its reaction with anhydrous HI?

10-10-3 Describe the Williamson's method for sysnthesis of ethers? What is its main drawback? (O.U., Mar., 1995)

10-10-4 Explain why ethers generally have lower melting and boiling points than the isomeric alcohols?

10-10-5 Explain why the boiling point of dimethyl ether (-24°C) is considerably lower than that of ethanol (78°C)? (N.U., Sept., 1994)

10-10-6 Nerolin is a pleasant smelling compound used in several commercial perfumes. Chemically, it is β-napthyl methyl ether. Describe a simple synthesis of it based on naphthalene as starting material.

10-10-7 Describe a synthesis for each one of the following compounds from readily available starting materials?

(a) Ethyl *n*-propyl ether

(b) *p*-Chloroanisole

(c) *o*-methyl phenyl ethyl ether

(d) *p*-Nitrophenyl n-propyl ether

(e) Di-*n*-butyl ether

10-10-8 Using qualitative experiments, how can one distinguish between

(i) Ether and *n*-butanol

(ii) Anisole and *o*-methyl phenol (0-Cresol)

(iii) Toluene and ethyl phenyl ether

(iv) Methyl *n*-propyl ether and *n*-pentanol

(v) *p*-Chlorobenzyl alcohol and *p*-chloroanisole.

10-10-9 Give two functional isomers with the molecular formula C_2H_6O? which one of them has the higher boiling point and why? (O.U.,Apr.1991)

10-10-10 What are "simple ethers" "mixed ethers" and "cyclic ethers"? Explain with a suitable example of each. Write a note on Williamson's synthesis of ethers. (O.U., Apr.1990)

10-10-11 How is diethyl ether prepared in the laboratory? How does it react with chlorine, hydriodic acid and phosphorus pentachloride? (SKU., Sep.1991)

ANSWER TO PROBLEMS AND EXERCISES

10-9-1 (i) B (ii) D (iii) C (iv) A (v) A (vi) C

10-10-6 Nerolin can be prepared starting from naphthalene using the following sequence of reactions;

(Naphthalene) $\xrightarrow[160^oC]{H_2SO_4}$ (SO_3H) $\xrightarrow[\Delta]{NaOH}$ ($\bar{O}\overset{+}{Na}$) (β-Naphthol sodium salt)

$\xrightarrow{Me_2SO_4}$ (OCH_3) (β-Naphthyl methyl ether or Nerolin)

10-10-7 (a) Ethyl *n*-propyl ether: This can be prepared in two ways:

(i) $C_2H_5—OH \xrightarrow{Na} C_2H_5–ONa \xrightarrow{CH_3—CH_2—CH_2Cl} C_2H_5—O—CH_2CH_2CH_3$

(ii) $CH_3CH_2CH_2OH \xrightarrow{Na} CH_3CH_2CH_2ONa \xrightarrow{C_2H_5—I} CH_3CH_2CH_2\text{- O- }C_2H_5$

(b) *p*-Chloroanisole:

OH, Cl $\xrightarrow[RT]{Me_2SO_4/Aq.NaOH}$ OMe, Cl

(c) *o*-Methyl phenyl ethyl ether.

OH, CH_3 (o-Cresol) $\xrightarrow[Aq.NaOH/RT]{Et_2SO_4}$ OEt, CH_3 (o-Methyl phenyl ethyl ether)

(d) *p*-Nitrophenyl *n*-propyl ether

(Benzene) $\longrightarrow$ Cl (Chlorobenzene) $\xrightarrow{HNO_3}$ Cl, NO_2 (p-Nitrochloro benzene) + Cl, NO_2 (o-Nitrochloro benzene)

Separate

O—$CH_2CH_2CH_3$ ... $CH_2CH_2CH_2$—Cl ... $\overset{-}{O}\overset{+}{Na}$ ... NaOH ... Cl

NO_2
(p-Nitrophenyl-n-propyl ether)

(e) Di-*n*-butyl ether: There are two methods of preparing this compound:

(i) $CH_3\text{-}CH_2\text{-}CH_2\text{-}\dot{C}H_2\text{- OH}$ (*n*-butanol) $\xrightarrow{H_2SO_4/RT}$ $CH_3\text{-}CH_2\text{-}CH_2\text{-}CH_2\text{-O-}SO_2\text{-OH}$ (*n*-Butyl hydrogen sulphate)

$\xrightarrow[\text{above } 120°]{n\text{—BuOH} / \Delta}$ $H_3C\text{-}CH_2\text{-}CH_2\text{-}CH_2\text{-O-}CH_2\text{-}CH_2\text{-}CH_3$

(ii) $CH_3\text{-}CH_2\text{-}CH_2\text{-}CH_2\text{-OH}$ (*n*-Butanol) $\xrightarrow{Na}$ $CH_3\text{-}CH_2\text{-}CH_2\text{-}CH_2\text{-}\bar{O}\overset{+}{Na}$

$\xrightarrow{CH_3\text{-}CH_2\text{-}CH_2\text{-}CH_2\text{-Cl}}$ $H_3C\text{-}CH_2\text{-}CH_2\text{-}CH_2\text{-}CH_2\text{-O-}CH_2\text{-}CH_2\text{-}CH_2\text{-}CH_3$
(Di-*n*-butyl ether)

10-10-8 (i) Ether and n-butanol

$CH_3\text{-}CH_2\text{-O-}CH_2\text{-}CH_3$ (Diethyl ether or ether) $CH_3\text{-}CH_2\text{-}CH_2\text{-}CH_2\text{-OH}$ (*n*-Butanol)

n-Butanol will react with sodium metal at RT liberating a gas (hydrogen) from it wheras diethyl ether will not respond to this treatment and will remain unaffected by it.

(ii) Anisole and *o*-Cresol:

O—CH_3 (Anisole) OH, CH_3 (o-Cresol)

o-Cresol will react with Aq.NaOH giving a clear Aq.solution and will be regenerated from the aq.alkali on addition of aq.HCl; whereas anisole will remain unaffected by aq.alkali at RT.

(iii) Toluene and ethyl phenyl ether:

CH_3 (Toluene) O—C_2H_5 (Ethyl phenyl ether)

Toluene is a typical aromatic hydrocarbon wheras ethyl phenyl ether is an alkyl aryl ether. The latter will dissolve in conc. H_2SO_4 at RT due to oxonium ion formation whereas toluene will remain unaffected by it.

(iv) Methyl *n*-propyl ether and *n*-pentanol:

$H_3C\text{-}O\text{-}CH_2\text{-}CH_2\text{-}CH_3$	$CH_3\text{-}CH_2\text{-}CH_2\text{-}CH_2\text{-}CH_2\text{-}OH$
(Methyl *n*-propyl ether)	(*n*-pentanol)

The ether will not react with sodium metal at RT liberating any gas (hydrogen) from it wheas the latter, by virtue of its being a hydroxy compound, will respond to this test.

(v) *p*-Chlorobenzyl alcohol and *p*-chloroanisole:

CH_2OH, Cl — (p-Chlorobenzyl alcohol)

O, CH_3, Cl — (p-Chloroanisole)

The alcohol will react with sodium metal at RT to give the corresponding alkoxide and at the same time liberating a gas (H_2) whereas the ether will remain silent in this test unaffected by it.

11

ALDEHYDES AND KETONES

11-1 INTRODUCTION

Aldehydes and ketones contain the common functionality C = O group, known as carbonyl group. The carbonyl group is one of the most important functional groups in organic chemistry. Aldehydes have the general formula R—CO—H and ketones have the general formula R—CO—R', where R and R' may be different or identical alkyl or aryl groups. Formaldehyde is the simplest and the parent member of the group of carbonyl compounds wherein a carbonyl group is attached to two hydrogen atoms as shown below;

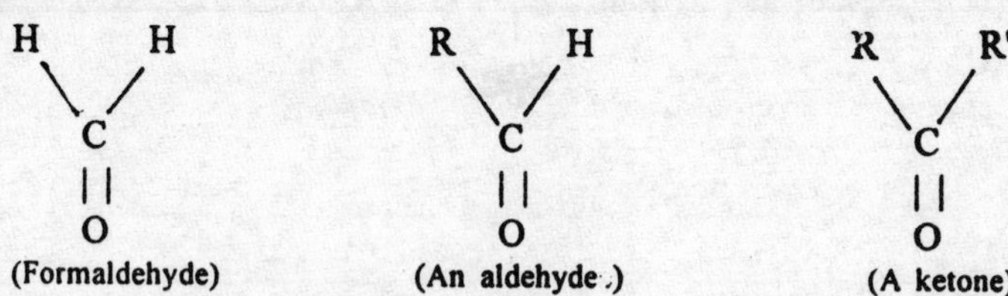

(Formaldehyde) (An aldehyde) (A ketone)

Aldehydes and ketones are collectively referred to as ***carbonyl compounds.*** It is the carbonyl group that determines the chemistry of aldehydes and ketones.

11-2 CLASIFICATION

Aldehydes and ketones may be classified as aliphatic or aromatic depending upon the attachment of alkyl or aryl groups to the carbonyl functionality. In the case of ketones, R may be alkyl and R' may be aryl group. Alternatively, the two alkyl groups of a ketone may be part of an alicyclic ring in which case it is known as cyclic ketone. These classifications are shown below;

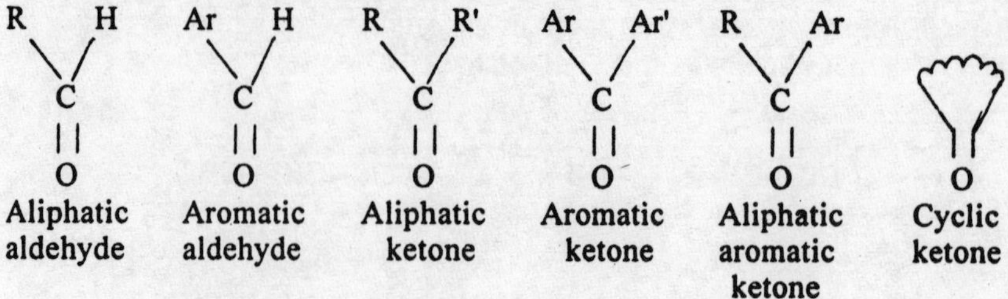

Aliphatic aldehyde · Aromatic aldehyde · Aliphatic ketone · Aromatic ketone · Aliphatic aromatic ketone · Cyclic ketone

11-3 NOMENCLATURE

In the IUPAC system, aldehydes are named as derivatives of alkanes using the suffix *al* instead of *e* in the end. For example, the first three members of the aldehyde family are known as methanal, ethanal and propanal. However, most aldehydes have common names derived from the names of immediate carboxylic acid produced on their oxidation. For example, CH_3—CHO, was called acetaldehyde, CH_3—CH_2—CHO, propionaldehyde, CH_3—CH_2—CH_2—CHO, n-butyraldehyde, etc. In complicated systems, the aldehyde group is treated as a substituent and in the IUPAC system, it is referred to as formyl group.

The class ending for ketones in the IUPAC system is "one" and when the keto group becomes a substitutent it is known as *oxo* group. Aromatic ketones are named as derivatives of *phenones* and aromatic-aliphatic ketones as *acyl* derivatives of benzene. These rules of nomenclature become clear considering the following examples;

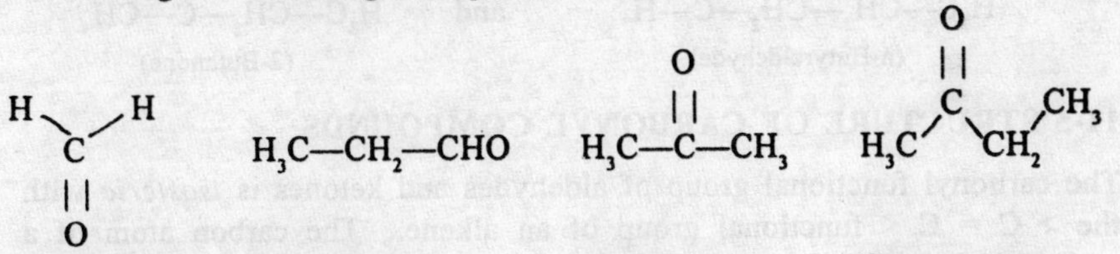

H—C(=O)—H (Common name: Formaldehyde; IUPAC name: Methanal)

H_3C—CH_2—CHO (Common name: Propionaldehyde; IUPAC name: Propanal)

H_3C—C(=O)—CH_3 (Common name: acetone; IUPAC name: Propanone)

H_3C—C(=O)—CH_2—CH_3 (Common name: Ethylmethyl ketone; IUPAC name: 2-Butanone)

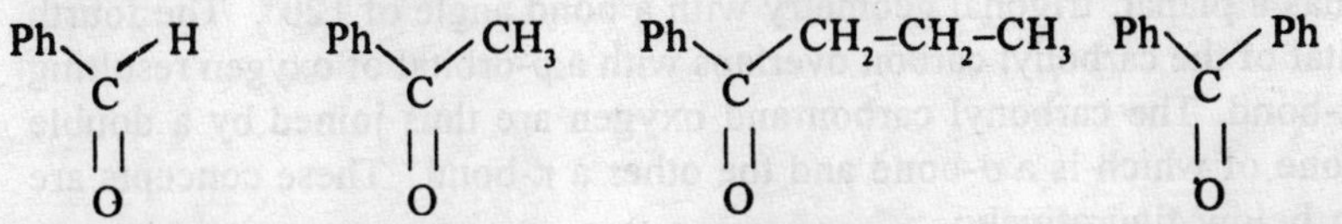

Ph—C(=O)—H (Common name: Benzaldehyde; IUPAC name Formylobenzene

Ph—C(=O)—CH_3 (Common name: acetophenone; IUPAC name 1-phenylethanone)

Ph—C(=O)—CH_2-CH_2-CH_3 (Common name: Butyrophenone; IUPAC name 1-phenylbutanone

Ph—C(=O)—Ph (Common name: Benzophenone; IUPAC name 1-phenylphenone)

11-4 ISOMERISM IN CARBONYL COMPOUNDS

Aldehydes and ketones exhibit the phenomenon of isomerism. The isomerism of carbonyl compounds can be any one of the following three types;

11-4-1 Chain isomerism

This type of isomerism arises as a result of variations in the nature of chains attached to carbonyl carbon atom. For example;

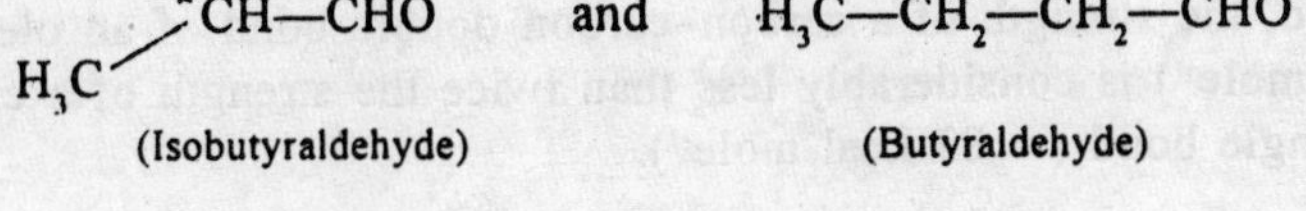

$(H_3C)_2$CH—CHO (Isobutyraldehyde) and H_3C—CH_2—CH_2—CHO (Butyraldehyde)

11-4-2 Position isomerism

This type of isomerism arises as a result of variations in the position or point of attachment of the carbonyl group to a particular carbon chain. For example;

$$H_3C{-}\overset{\overset{\large O}{||}}{C}{-}CH_2{-}CH_2{-}CH_2{-}CH_3 \quad \text{and} \quad H_3C{-}CH_2{-}\overset{\overset{\large O}{||}}{C}{-}CH_2{-}CH_2{-}CH_3$$

(2-Hexanone) (3-Hexanone)

11-4-3 Functional isomerism

This type of isomerism arises due to the presence of different (aldehyde or ketone) functional groups in isomeric compounds. For example;

$$H_3C{-}CH_2{-}CH_2{-}\overset{\overset{\large O}{||}}{C}{-}H \quad \text{and} \quad H_3C{-}CH_2{-}\overset{\overset{\large O}{||}}{C}{-}CH_3$$

(n-Butyraldehyde) (2-Butanone)

11-5 STRUCTURE OF CARBONYL COMPOUNDS

The carbonyl functional group of aldehydes and ketones is *isosteric* with the > C = C < functional group of an alkene. The carbon atom of a carbonyl group is sp^2 hybridized. The carbonyl carbon atom uses the three sp^2 hybridized orbitals for the formation of three σ-bonds, one each with the two alkyl groups and one with the oxygen atom. The carbonyl carbon atom has a planar, trigonal geometry with a bond angle of 120°. The fourth p-orbital of the carbonyl carbon overlaps with a p-orbital of oxygen resulting in a π-bond. The carbonyl carbon and oxygen are thus joined by a double bond one of which is a σ-bond and the other a π-bond. These concepts are shown below figuratively;

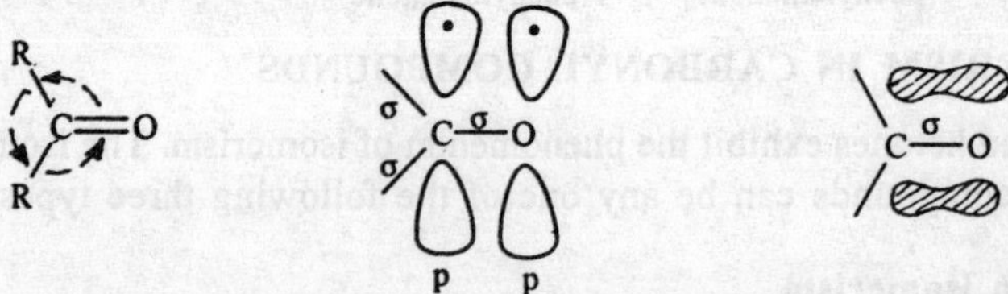

The carbon-oxygen double bond of a carbonyl group is a very strong bond with a bond energy of 176-179 kcal mol^{-1}, which is more than twice the bond energy of carbon-oxygen single bond (86 Kcal $mole^{-1}$). On the other hand, the strength of a carbon–carbon double bond of an olefin (≈ 146 kcal $mole^{-1}$) is considerably less than twice the strength of a carbon–carbon single bond (≈ 83 kcal $mole^{-1}$).

Although, the carbon-oxygen double bond of a carbonyl group is considered to be fairly covalent, the pairs of electrons constituting this bond are never shared equally between carbon and oxygen. This is because the oxygen atom is more electronegative than carbon resulting in the pair of electrons being drawn more towards carbon. This leads to the placement of a partial negative charge on oxygen and partial positive charge on carbon. Thus, the carbonyl group is considered to be a resonance hybrid of the following two contributing structures.

$$>C=O \longleftrightarrow >\overset{+}{C}-\overset{-}{O}$$

It is the fairy large contribution by the charge separated structure that makes the carbon-oxygen double bond of the carbonyl group to be very strong and that explains the high value of > C = O bond energy (176-179 kcal mole^{-1}) when compared to a > C=C< bond energy (≈ 146 kcal mole^{-1}). Evidence for the existence of a charge-seperated structure comes from the high value of dipole moment (2.3–2.7 D units) for carbonyl compounds and the direction of dipole moment being such that the negative pole is towards oxygen, positive pole towards carbon.

It is evident from the above discussion that an important feature of chemical reactivity of carbonyl compounds would involve addition to the carbon-oxygen double bond with the nucleophilic part of the reagent adding on to the positively polarized carbon of the carbonyl group, which is indeed found to be the case.

11-6 GENERAL METHODS OF PREPARATION OF ALDEHYDES AND KETONES

There are several methods described in literature for the preparation of aldehydes and ketones. However, the following are some of the important methods employed for the preparation of carbonyl compounds.

11-6-1 Oxidation of alcohols

Alcohols when subjected to oxidation give aldehydes and ketones. Primary alcohols yield aldehydes whereas secondary alcohols yield ketones. The oxidation may be carried out either by using an oxidizing agent or by using a dehydrogenating agent. For example;

$$R-CH_2-OH \xrightarrow{CrO_3/H^+} R-CHO$$

$$R-\underset{\substack{|\\OH}}{CH}-R^1 \xrightarrow[MnO_2 \text{ or } SeO_2 \text{ etc.}]{KMnO_4/\text{acetone or}} R-\underset{\substack{||\\O}}{C}-R^1$$

$$\text{or} \quad R{-}CH_2{-}OH \xrightarrow{Cu/\Delta} R{-}\underset{\underset{O}{\|}}{C}{-}H$$

$$R{-}\underset{\underset{OH}{|}}{CH}{-}R^1 \xrightarrow[Ph(OAc)_4]{Cu/\Delta \text{ or}} R{-}\underset{\underset{O}{\|}}{C}{-}R$$

In the oxidation of a primary alcohol to an aldehyde using an oxidizing agent like potassium dichromate, care should be taken to remove the aldehyde from the reaction mixture as soon as it is formed. Otherwise, there is risk of the aldehyde getting oxidized to the carboxylic acid. This is usually achieved by adding the alcohol to the hot solution of the dichromate. The aldehyde, that is produced has a much lower boiling point (about 30-40° less) than that of alcohol and is immediately removed from the reaction mixture as vapor. The vapors of aldehyde are condensed and collected. Aldehydes like acetaldehyde, propionaldehyde, butyraldehyde are produced this way.

An alternative method for the *oxidation of alcohols to ketones* is the *Oppenauer oxidation*. In this method, a secondary alcohol is heated with aluminium alkoxide in the presence of acetone.

$$(H_3C)_2C{=}O + (R)(R^1)CH{-}OH \xrightleftharpoons{Al(OR^3)_3} (R)(R^1)C{=}O + (H_3C)_2CH{-}OH$$

The reaction is reversible and appropriate conditions have to be used to shift the reaction into forward direction by using excess of acetone or acetone itself as solvent. Only a trace of aluminium alkoxide catalyst, is required which is usually prepared *in situ* by reacting finely powdered aluminium with the alcohol first and then adding acetone followed by refluxation of the reaction mixture.

$$3\,(R)(R^1)CH{-}OH + Al \longrightarrow Al\left[O{-}CH(R)(R^1)\right]_3 + 1\tfrac{1}{2}\,H_2\uparrow$$

$$(R)(R^1)(H)C{-}O^{\ominus} + (H_3C)_2\overset{\oplus}{C}{-}O^{\ominus} \longrightarrow (R)(R^1)C{=}O + (H_3C)_2(H)C{-}O^{\ominus}$$

$$\underset{R'}{\overset{R}{>}}CH{-}OH + \underset{H_3C\quad H}{\overset{H_3C\quad O^{\ominus}}{>C<}} \longrightarrow \underset{R'}{\overset{R}{>}}CH{-}\overset{\ominus}{O} + \underset{H_3C}{\overset{H_3C}{>}}CH{-}OH$$

Since the reaction is reversible, the back reaction can be used for *the reduction of the ketone to get the corresponding alcohol out of it.* In that case, the reaction is known as *Meerwein-Pondorff-Verley reduction.* The mechanism of the reaction is given above;

11-6-2 Reduction of acid chlorides

In this method, acid chlorides are treated with H_2 gas over partially poisoned palladium catalyst to yield aldehydes. The method is known as *Rosenmaund reduction.*

$$R{-}\overset{\overset{\displaystyle O}{||}}{C}{-}Cl + H_2 \xrightarrow[\text{poisoned)}]{\text{Pd (partially}} R{-}CHO + HCl$$

The reaction proceeds by a simple mechanism involving palladium catalysed addition of hydrogen to the carbonyl group followd by loss of elements of HCl to yield the aldehyde.

$$R{-}\underset{\underset{\displaystyle O}{||}}{C}{-}Cl + H_2 \longrightarrow \left[R{-}\overset{H}{\underset{O{-}H}{C}}{-}Cl \right] \longrightarrow R{-}\underset{\underset{\displaystyle O}{||}}{C}{-}H + HCl$$

Since acid chlorides themselves are obtained from carboxylic acids, the method effectively involves reduction of carboxylic acids to aldehydes.

$$R{-}\underset{\underset{\displaystyle O}{||}}{C}{-}OH \xrightarrow[\text{PCl}_5\text{ or PCl}_3]{\text{SOCl}_2\text{ or}} R{-}\underset{\underset{\displaystyle O}{||}}{C}{-}Cl \xrightarrow[\text{(poisoned)}]{H_2/Pd} R{-}\underset{\underset{\displaystyle O}{||}}{C}{-}H$$

Direct reduction with H_2 over Pd catalyst of carboxylic acids or acid chlorides leads to the formation of alcohols, seldom does the reaction stop at aldehyde stage.

$$R{-}\underset{\underset{O}{||}}{C}{-}OH \xrightarrow{H_2/Cat.} \left[R{-}\underset{\underset{OH}{|}}{CH}{-}OH\right] \xrightarrow{-H_2O} \left[R{-}\underset{\underset{O}{||}}{C}{-}H\right] \xrightarrow{H_2/Pd} R{-}CH_2OH$$

$$\downarrow SOCl_2$$

$$\underset{\text{(Acid chloride)}}{R{-}\underset{\underset{O}{||}}{C}{-}Cl} \xrightarrow{H_2/Pd} \left[R{-}\underset{\underset{OH}{|}}{CH}{-}Cl\right] \xrightarrow{-HCl} \underset{\text{(Aldehyde)}}{\left[R{-}\underset{\underset{O}{||}}{C}{-}H\right]} \xrightarrow{H_2/Pd} \underset{\text{(Alcohol)}}{R{-}CH_2OH}$$

Similar results are obtained when lithium aluminium hydride is used for reduction of carboxylic acids or acid chlorides. However, when $LiAlH_4$ is treated with controlled amounts of alcohols, alkoxy derivatives of $LiAlH_4$ are obtained which are less powerful reducing agents. The latter permit selective reduction of acid chlorides to aldehydes. For example;

$$R{-}\underset{\underset{O}{||}}{C}{-}OH \xrightarrow{LiAlH_4} R{-}CH_2{-}OH$$

$$R{-}\underset{\underset{O}{||}}{C}{-}OH \xrightarrow{SOCl_2} R{-}\underset{\underset{O}{||}}{C}{-}Cl \xrightarrow{LiAlH_4} R{-}CH_2{-}OH$$

$$R{-}\underset{\underset{O}{||}}{C}{-}Cl \xrightarrow{LiAlH(OEt)_3} R{-}\underset{\underset{O}{||}}{C}{-}H$$

The reaction appears to follow the mechanism given below;

$$R{-}\overset{\oplus}{C}{-}Cl \ (\text{with } {-}O^{\ominus}) \xrightarrow{LiAlH(OEt)_3} \left[R{-}\overset{H}{C}{-}Cl \ \ (O{\cdots}\overset{\oplus}{Al}(OEt)_3)\right] \overset{\oplus}{Li} \longrightarrow R{-}\underset{\underset{O}{||}}{C}{-}H + LiCl + Al(OEt)_3$$

11-6-3 From Grignard reagents

Othoesters condense with Grignard reagents to yield acetals or ketals as intermediates which on hydrolysis give aldehydes or ketones respectively. For example;

$$\underset{\text{(Grignard reagent)}}{R{-}Mg{-}X} + \underset{\text{(Ethyl orthoformate)}}{H{-}C(OEt)_3} \longrightarrow \left[H{-}\overset{R}{C}(OEt)_2 + X{-}Mg{-}OEt\right] \xrightarrow{H_2O} \left[\begin{matrix}R & & H\\ & C & \\ HO & & OH\end{matrix}\right] + X{-}Mg{-}OH \longrightarrow \underset{\text{(an aldehyde)}}{R{-}\underset{\underset{O}{||}}{C}{-}H}$$

In an alternative reaction, Grignard reagents can be condensed with nitriles to yield imino intermediates which on hydrolysis lead to ketones.

$$\underset{\text{(Grignard reagent)}}{R{-}MgX} + \underset{\text{(Alkyl or Aryl nitrite)}}{R'{-}C{\equiv}N} \longrightarrow \left[R{-}\underset{\underset{N{-}MgX}{||}}{C}{-}R'\right] \xrightarrow{H_2O} \left[\underset{\underset{NH}{||}}{\overset{R\diagdown\;\diagup R'}{C}}\right] + Mg\begin{matrix}\diagup X\\ \diagdown OH\end{matrix} \longrightarrow \underset{\text{(a ketone)}}{R{-}\overset{\overset{O}{||}}{C}{-}R'} + NH_3$$

The two condensations proceed by the mechanisms shown below:

With Ortho esters:

$$\overset{\delta-\;\;\delta+}{R{-}MgX} + R'{-}C(OEt)_3 \longrightarrow \left[\overset{R\diagdown\;\diagup R'}{C}\underset{EtO\diagup\;\diagdown OEt}{}\right] \xrightarrow{H\ddot{O}H} \left[\overset{R\diagdown\;\diagup R'}{C}\underset{EtO\diagup\;\diagdown \overset{\oplus}{O}H_2}{} + EtO^-\right]$$

$$\longrightarrow \overset{R\diagdown\;\diagup R'}{C}\underset{EtO\diagup\;\diagdown OH}{} + EtOH \xrightarrow{H\ddot{O}H} \left[\overset{R\diagdown\;\diagup R'}{C}\underset{H{-}\overset{\oplus}{O}(H)\diagup\;\diagdown OH}{}\right] \xrightarrow{EtO^{\ominus}} \left[\overset{R\diagdown\;\diagup R'}{C}\underset{HO\diagup\;\diagdown OH}{}\right] + EtOH \longrightarrow \underset{\underset{O}{||}}{\overset{R\diagdown\;\diagup R'}{C}} + H_2O$$

With nitriles:

$$\overset{\delta-\;\;\delta+}{R{-}MgX} + R{-}C{\equiv}\ddot{N} \longrightarrow \left[\underset{\underset{\ddot{N}{-}MgX}{||}}{\overset{R\diagdown\;\diagup R'}{C}}\right] \xrightarrow{H\ddot{O}H} \underset{\underset{\underset{\oplus}{N{-}H}}{||}}{\overset{R\diagdown\;\diagup R'}{C}} + Mg\begin{matrix}\diagup X\\ \diagdown OH\end{matrix}$$

$$\underset{\underset{N{-}H}{||}}{\overset{R\diagdown\;\diagup R'}{C}} \xrightarrow{H\ddot{O}H} \overset{R\diagdown\;\diagup R'}{C}\underset{H\overset{\ominus}{N}\diagup\;\diagdown \overset{\oplus}{O}{-}H(H)}{} \longrightarrow \left[\overset{R\diagdown\;\diagup R'}{C}\underset{H_2N\diagup\;\diagdown O{-}H}{}\right] \longrightarrow \underset{\underset{O}{||}}{\overset{R\diagdown\;\diagup R'}{C}} + NH_3$$

11-6-4 From Ca salts of fatty acids

Distillation of calcium salts of fatty acids leads to the formation of ketones. However, *the reaction is limited to the preparation of aliphatic ketones only.*

$$(RCOO^-)_2\, Ca^{+2} \xrightarrow{\Delta} \underset{\underset{O}{||}}{\overset{R\diagdown\;\diagup R}{C}} + CaCO_3$$

11-6-5 From alkynes

Simple or substituted alkynes add on elements of water in the presence of conc. H_2SO_4 and a trace of $HgSO_4$ as catalyst to yield derivatives of vinyl alcohol which tautomerize giving ketones as products. For example;

$$\underset{\text{(1,2-diphenylacetylene)}}{Ph-C \equiv C-Ph} \xrightarrow[\substack{Hg^{+2} \\ (+H_2O)}]{H^{\oplus}} \left[Ph-\overset{H}{\overset{|}{C}} = \underset{\underset{OH}{|}}{C}-Ph \right] \longrightarrow Ph-CH_2-\underset{\underset{O}{||}}{C}-Ph$$

The mechanism of the reaction is as follows;

$$Ph-C \equiv C-Ph \xrightarrow{Hg^{+2}} Ph-C \equiv C-Ph\ (Hg^{+2}\text{ complex}) \xrightarrow{H^{\oplus}} Ph-\overset{H}{\overset{|}{C}} = \overset{\oplus}{C}-Ph + Hg^{+2}$$

$$\xrightarrow{H-\ddot{O}-H} Ph-\overset{H}{\overset{|}{C}} = C\begin{matrix} Ph \\ \overset{\oplus}{O}-H \\ | \\ H \end{matrix} \longrightarrow Ph-\overset{H}{\overset{|}{C}} = C\begin{matrix} Ph \\ O-H \end{matrix} \longrightarrow PhCH_2-\underset{\underset{O}{||}}{C}-Ph$$

However, the method has limited scope due to non availability of suitably substituted alkynes.

11-7 PREPARATION OF AROMATIC ALDEHYDES AND KETONES

The methods described above are fairly general for the preparation of aliphatic and aromatic carbonyl compounds unless stated otherwise. However, in addition to those given above, there are several methods specific to the preparation of aromatic carbonyl compounds alone. Some of these methods are given below in brief:

11-7-1 Oxidation of aliphatic side chains

Aromatic hydrocarbons containing alkyl side chains can be suitably oxidized to yield aromatic aldehydes or ketones. For example;

$$C_6H_5-CH_3 \xrightarrow[Ac_2O/0\text{-}5^{o}]{CrO_3/} C_6H_5-CH(OAc)_2 \xrightarrow{H_2O} C_6H_5-CHO$$

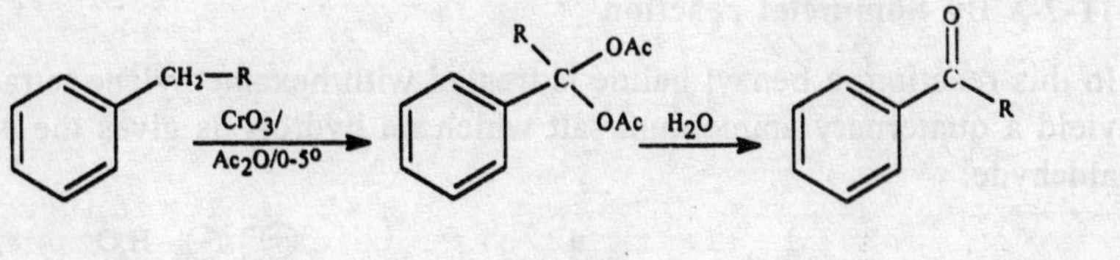

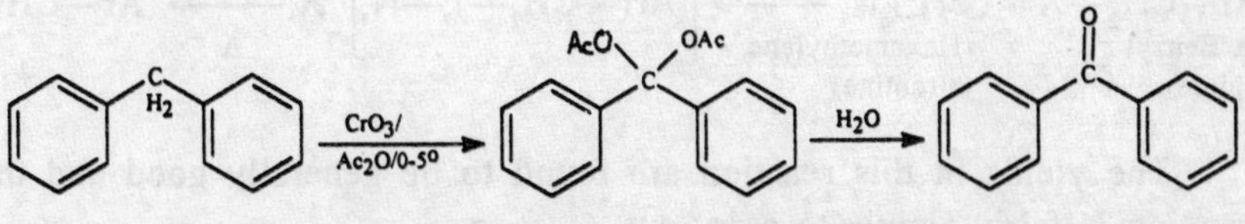

Chromium trioxide (or chromic acid anhydride) oxidizes the C—H bonds of the α-carbon to C—OH bonds which react with acetic anhydride to yield C-OAc bond-containing derivatives. The reaction usually stops at the dihydroxy stage and the diacetate precipitates out of the reaction mixture. The diacetate is separated and hydrolysed to yield aldehyde or ketone as the case may be. If permanganate is used in place of chromium trioxide, the oxidation proceeds further to give benzoic acid.

11-7-2 Hydrolysis of benzal halides

Aromatic hydrocarbons containing suitable alkyl side chains undergo free radical halogenation in the presence of heat and UV radiation to yield dihaloalkyl benzenes. The latter on hydrolysis with aq. acid result in the formation of aromatic carbonyl compounds. For example;

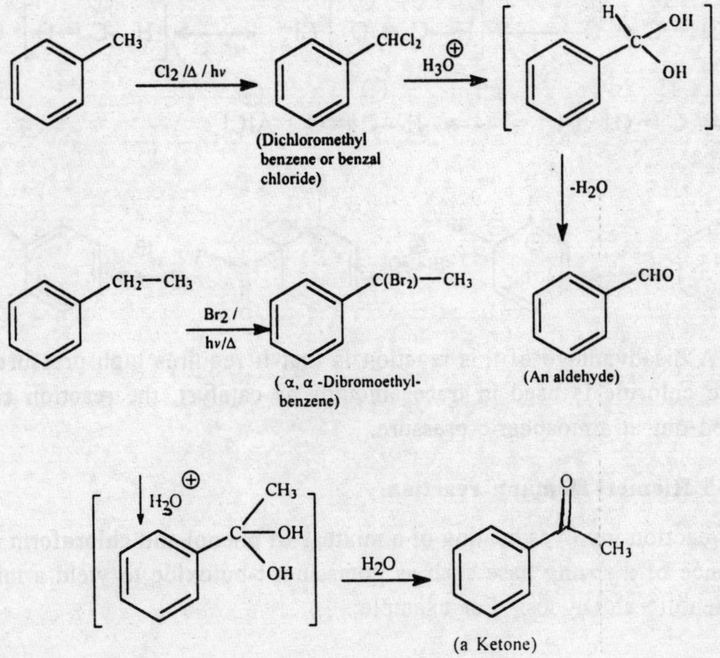

(Dichloromethyl benzene or benzal chloride)

(An aldehyde)

(α, α -Dibromoethyl-benzene)

(a Ketone)

11-7-3 By Sommelet reaction

In this reaction, a benzyl halide is treated with hexamethylene tetramine to yield a quaternary ammonium salt which on hydrolysis gives the aromatic aldehyde.

$$\underset{\text{(a Benzyl halide)}}{Ar{-}CH_2{-}X} + \underset{\text{(Hexamethylene tetramine)}}{(CH_2)_6N_4} \longrightarrow \left[Ar({-}CH_2{-})_7{-}N_4\right]^{\oplus} \overset{\ominus}{X} \xrightarrow[\Delta]{H_2O} Ar{-}CHO$$

The yields in this reaction are found to be generally good and the reaction is fairly simple to carry out.

11-7-4 By Gattermann-Koch formylation reaction

In this reaction, an aromatic hydrocarbon is treated with a mixture of carbon monoxide and HCl in the presence of a trace of anhydrous aluminium chloride to yield an aromatic aldehyde. For example;

$$C_6H_6 \xrightarrow{CO + HCl / AlCl_3} C_6H_5CHO$$

The mechanism of this reaction involves simple electrophilic aromatic substitution of the benzene ring, the effective electrophilic species being H—C⁺ = O as shown below;

$$HCl + \overset{\ominus}{C} \equiv \overset{\oplus}{O} \longrightarrow \left[H{-}C \equiv \overset{\oplus}{O}\right] Cl^- \longleftrightarrow \left[H{-}\overset{\oplus}{C} = O\right] Cl^-$$

$$\left[H{-}\overset{+}{C} = O\right] Cl^- \xrightarrow{AlCl_3} H{-}\overset{\oplus}{C} = O \quad \overset{\ominus}{AlCl_4}$$

$$C_6H_6 \longleftrightarrow [C_6H_6]^{\oplus\ominus} \xrightarrow{[H{-}\overset{\oplus}{C}{=}O]} [C_6H_6{-}CHO]^{\oplus} \xrightarrow{-\overset{\oplus}{H}} C_6H_5{-}CHO$$

A disadvantage of this reaction is that it requires high pressures. If cupric chloride is used in trace amounts as catalyst, the reaction can be carried out at atmospheric pressure.

11-7-5 Riemer-Tiemann reaction

This reaction involves heating of a mixture of phenol and chloroform in the presence of a strong base such as potassium t-butoxide to yield a mixture of phenolic aldehydes. For example;

(Phenol) $\xrightarrow[\text{and then } H_3O^{\oplus}]{CHCl_3/K^{\oplus}\ t\text{-}BuO^{\ominus}/\Delta}$ (Salicylaldehyde) or o-hydroxybenzaldehyde) + (p-Hydroxybenzaldehyde)

The details of this reaction have already been discussed in Chapter 9, on hydroxy compounds.

11-7-6 Friedel-Crafts acylation

In this method, a mixture of an aromatic hydrocarbon and an acid anhydride/ an acid chloride is heated in the presence of anhydrous aluminium chloride to yield acyl/aroyl derivatives of aromatic hydrocarbons.

$$C_6H_6 \xrightarrow[AlCl_3/\Delta]{(RCO)_2 \text{ or } R\text{-CO-Cl} /} C_6H_5\text{-CO-R}$$

$$C_6H_6 \xrightarrow[\Delta/AlCl_3]{(ArCO)_2 \text{ or } Ar\text{-CO-Cl} /} C_6H_5\text{-CO-Ar}$$

Large amounts of Lewis acid catalyst (mole per mole) are required and the details of this reaction have already been discussed in Chapter 7, on arenes.

11-8 PHYSICAL PROPERTIES OF ALDEHYDES AND KETONES

The aldehydes and ketones are not strongly associated substances since they do not contain any—OH or other groups capable of H-bonding. On the other hand, aldehydes and ketones contain the carbonyl group which has a dipolar structure and so aldehydes and ketones are capable of dipole-dipole intermolecular interactions. Consequently, aldehydes and ketones prossess normal boiling and melting points expected from their molecular weights and molecular structures. Thus, aldehydes and ketones possess boiling and melting points lower than the corresponding alcohols and carboxylic acids but higher than the corresponding hydrocarbons and ethers. Except for lower molecular weight compounds, carbonyl compounds do not dissolve in water. They are, however, freely soluble in organic polar and non-polar solvents at room temperature. The physical constants of some of the carbonyl compounds are shown below in Table 11-1.

Table 11-1: Physical constants of some cabonyl compounds

S.No.	Name of the carbonyl compound	Structure of the carbonyl compound	B.P °C	M.P °C
1.	Formaldehyde	HCHO	−21	—
2.	Acetaldehyde	CH_3-CHO	20	—
3.	Propionaldehyde	CH_3-CH_2-CHO	49	—
4.	n-Butyraldehyde	CH_3-CH_2-CH_2-CHO	75	—
5.	Valeraldehyde	CH_3-CH_2-CH_2-CH_2-CHO	104	—
6.	n-Hexanal	CH_3-$(CH_2)_4$-CHO	131	—
7.	n-Heptanal	CH_3-$(CH_2)_5$-CHO	155	—
8.	Benzaldehyde	C_6H_5-CHO	179	—
9.	Salicylaldehyde	o-(OH)-C_6H_4-CHO	197	—
10.	o-Tolualdehyde	o-(H_3C)-C_6H_4-CHO	200	—
11.	p-Tolualdehyde	p-H_3C-C_6H_4-CHO	204	—
12.	o-Methoxybenzaldehyde	o-(CH_3O)-C_6H_4-CHO	236	38
13.	p-Methoxybenzaldehyde	p-(CH_3O)-C_6H_4-CHO	248	20
14.	o-Chlorobenzaldehyde	o-(Cl)-C_6H_4-CHO	213	11
15.	p-Chlorobenzaldehyde	p-(Cl)-C_6H_4-CHO	214	48
16.	Acetone	CH_3-CO-CH_3	56	—
17.	Ethyl methyl ketone	C_2H_5-CO-CH_3	80	—
18.	Ethyl propyl ketone	C_2H_5-CO-CH_2CH_2-CH_3	124	—
19.	Cyclopentanone	$(CH_2)_4$CO	131	—
20.	Cyclohxanone	$(CH_2)_5$ CO	156	—
21.	Acetophenone	C_6H_5-CO-CH_3	202	20
22.	Benzyl methyl ketone	$C_6H_5CH_2$-CO-CH_3	216	27
23.	Propiophenone	C_6H_5-CO-CH_2-CH_3	218	20
24.	Dibenzyl ketone	C_6H_5-CH_2-CO-CH_2-C_6H_5	331	35
25.	Deoxybenzoin	C_6H_5-CO-CH_2-C_6H_5	320	60

11-9 CHEMICAL PROPERTIES OF ALDEHYDES AND KETONES

The carbonyl group is central to the reactivity of aldehydes and ketones. Geometrywise, the carbonyl group is a flat, planar, trigonal unit with the C—C—O, C'—C—O and C—C—C' bond angle of 120° each. The carbonyl group of aldehydes and ketones is a resonance hybrid of uncharged and charge-separated structures with the latter making a contribution of about 40% to the actual structure of the molecule.

R C=O R¹ 120° 120° 120°

R—C(=O)—R¹ (uncharged structure) ⟷ $R-\overset{\oplus}{C}-\overset{\ominus}{O}$ R¹ (Charge-separated structure)

In addition to the carbonyl-resonance, aromatic aldehydes and aromatic ketones will show Kekule resonance of the aromatic rings. Further, there will be resonance interaction between the aromatic ring and the carbonyl group too.

The existence of a charge-separated structure is responsible for the carbonyl group exhibiting three types of chemical reactivity.

11-9-1 Nucleophilic additions

Nucleophilic reagents add on to the carbonyl group at the positively charged carbon atom bringing about addition reactions.

$$>C=O \longleftrightarrow >\overset{\oplus}{C}-\overset{\ominus}{O} \xleftrightarrow{\overset{\ominus\oplus}{Y\ Z}} >C(Y)-O-Z$$

The mechanism of nucleophilic addition throws light on the reactivity of different types of carbonyl compounds and useful conclusions relating the reactivity with structures of aldehydes and ketones can be drawn. When a nucleophile (Y^-) adds on to the carbonyl carbon, the intermediary product is an oxyanion. This rate determining step in the reaction sequence is followed by a fast, nonratedetermining attack of the cationic part (Z^+) leading to the formation of final addition product. The transition state leading to the formation of oxyanion intermediate can be depicted as one having partial negative charge on oxygen and partial bond formation between carbonyl carbon and nucleophile Y^-.

$$\begin{matrix}R\\ \;\;\diagdown\\ \;\;\;C=O\\ \;\;\diagup\\ R^{1}\end{matrix} \xrightarrow{\overset{\ominus}{Y}} \left[\begin{matrix}R & & Y^{\delta-}\\ & C & \\ R^{1} & & O^{\delta-}\end{matrix}\right] \longrightarrow \begin{matrix}R & & Y\\ & C & \\ R^{1} & & O^{\ominus}\end{matrix} \xrightarrow{\overset{\oplus}{Z}} \begin{matrix}R & & Y\\ & C & \\ R^{1} & & OZ\end{matrix}$$

(Reactant) (Trigonal) — (Transition state) (Close to tetrahedrat) — (Product) (Tetrahedral) — (End product) (Tetrahedral)

Any structural factor/s which stabilize the oxyanion will also stabilize the transition state leading to its formation thereby reducing the activation energy and helping its quicker formation. Attachment of electron-withdrawing groups to the carbonyl carbon will stabilize the negative charge on oxygen in the oxyanion by electron-withdrawal and by charge-dispersal. Consequently, electron-withdrawing groups attached to carbonyl groups increase the rate of reaction. Such carbonyl compounds will, therefore, display higher reactivity in nucleophilic additions.

Aldehydes, generally, react faster than ketones towards nucleophilic additions. This is due to a combination of electronic and steric factors. Since alkyl groups (as in a ketone) are more electron-releasing when compared to hydrogen (as in an aldehyde), the alkyl groups destabilize the negative charge, developing on the oxygen in the transition state thereby delaying its formation. Furthermore, carbonyl group in the carbonyl compound is a flat, planar, trigonal structure and the product oxyanion has a tetrahedral geometry around the central carbon atom leading to steric crowding between the two alkyl groups. Hence, large alkyl groups, or aryl groups as it happens in aromatic ketones, will hinder the formation of oxyanion as well as the transition state.

The above discussion serves to rationalize the following observed order of reactivity for carbonyl compounds in nucleophilic additions;

$$H-\underset{\|}{\overset{}{}}\!\!\begin{matrix}O\\ \|\\ C\end{matrix}\!\!-H \;>\; R-\begin{matrix}O\\ \|\\ C\end{matrix}-H \;>\; H-\begin{matrix}O\\ \|\\ C\end{matrix}-R^{1} \;\geqslant\; Ar-\begin{matrix}O\\ \|\\ C\end{matrix}-H \;>\; Ar-\begin{matrix}O\\ \|\\ C\end{matrix}-Ar$$

11-9-2 Base-promoted condensations

The placement of a positive charge on the carbonyl carbon atom makes the hydrogen atom attached to α-carbon of the carbonyl group highly acidic in nature. Consequently the α-hydrogen can be easily removed using a strong base resulting in a carbanion which is a resonance stabilized nucleophilic species. The carbanion can then react with a positively-charged/electron-deficient species generating a series of base-promoted condensations.

```
  α                          ⊕    ⊖OH              ⊖ ⊕
R–CH₂–C–R'   ⟷   R–CH–C–R'  ——→   R–CH–C–R' + H₂O
      ||                |     |                      |
      O                 H    O⊖                     O⊖

⊖ ⊕                        ⊖
R–CH–C–R'   ⟷   R–CH–C–R'   ⟷   R–CH = C–R'
       |                  ||                  |
      O⊖                  O                  O⊖

                         ⊕
⊖                        Y
R–CH–C–R'  ——→   R–CH–C–R'
       ||                |    ||
       O                 Y    O
```

11-9-3 Tautomerism

Carbonyl compounds having an α-hydrogen atom exhibit the phenomenon of *tautomerism*. The latter involves migration of α-hydrogen to the carbonyl oxygen atom with the simultaneous shift of the carbon-oxygen double bond to the carbonyl carbon-α-carbon position.

```
                                      ⊕
R—CH₂—C—R'   ⟷   R—CH—C—R'   ⟷   R—CH = C—R'
       ||                 |      |                    |
       O                  H     O⊖                   O—H
 (Keto form)                                   (Enol form)
```

The carbonyl compound having the keto group is known as *keto form* and the carbonyl compound having enol group (a structure having an alkene group and alcoholic group) is known as *enol form*. It should be noted that tautomerism or *proto-tropic shift* involves an equilibrium between keto and enol forms. At a given temperature, most carbonyl compounds exist in the keto form and there are hardly any amounts of enol form. However, under certain conditions that stabilize hydrogen-bonding and/or resonance, there will be appreciable amounts of enol form for a given carbonyl compound.

(A) Some Nucleophilic Addition Reactions of the Carbonyl Group

(i) Addition of oxygen nucleophiles : Water can add on to the electrophilic carbon of a carbonyl compound resulting in a *gem-dihydroxy compound*. However, the reaction is reversible and with most ketones, the equilibrium is on the side of the carbonyl compound alone.

$$\begin{matrix} R \\ \quad \diagdown \\ C=O \\ \diagup \\ R^1 \end{matrix} \longleftrightarrow \begin{matrix} R \\ \diagdown \\ \overset{\oplus}{C}-\overset{\ominus}{O} \\ \diagup \\ R^1 \end{matrix} \overset{H\ddot{O}H}{\rightleftharpoons} \begin{matrix} R \quad \overset{\oplus}{OH_2} \\ C \\ R^1 \quad \overset{\ominus}{O} \end{matrix} \rightleftharpoons \begin{matrix} R \quad OH \\ C \\ R \quad OH \end{matrix}$$

(Carbonyl compound) (Gem-Dihydroxy compound)

The gem-dihydroxy compounds will be stable, if there are strongly electron-withdrawing groups attached to the alkyl group/s of the carbonyl moiety. For example, trichloroacetaldehyde readily reacts with water to form a gem dihydroxy compound. In fact, trichloroacetaldehyde exists in the form of its hydrate only, known as chloral hydrate and it is very difficult to dehydrate this compound. The hydrate of chloral is nothing but its gem-dihydroxy derivative. Similar is the case with indantrione. The hydrate of indantrione is commonly known as ninhydrin which is used in amino-acid detection.

$$CCl_3-\overset{O}{\overset{\|}{C}}-H \xrightarrow{H_2O} Cl_3C-C(H)(OH)(OH)$$

(Trichloro-acetaldehyde) (Chloral hydrate)

$$\text{Indantrione} \xrightarrow{H_2O} \text{Indantrione hydrate}$$

(Indantrione) (Indantrione hydrate or Ninhydrin)

Carbonyl compounds react with alcohols in the presence of a trace of an acid catalyst to yield, reversibly, addition compounds known as *hemi-acetals*. The latter on heating with excess alcohol yield irreversibly gem-dialkoxy compounds known as *acetals/ketals*.

$$\begin{matrix} R \\ C=O \\ H \end{matrix} \overset{R^1-OH}{\rightleftharpoons} \begin{matrix} R \quad OR^1 \\ C \\ H \quad OH \end{matrix} \xrightarrow[\text{(trace } H^+)]{R^1-OH/} \begin{matrix} R \quad OR^1 \\ C \\ H \quad OR^1 \end{matrix}$$

(Aldehyde) (Hemi-acetal) (An acetal)

$$\underset{\text{(Ketone)}}{R(R^1)C=O} \underset{}{\overset{R^2-OH}{\rightleftharpoons}} \underset{\text{(Hemi-ketal)}}{R(R^1)C(OR^2)(OH)} \xrightarrow[\text{(trace } H^+)]{R^2-OH} \underset{\text{(A ketal)}}{R(R^1)C(OR^2)_2}$$

Once isolated, the acetals/ketals are stable compounds. They are stable to alkalies but are readily hydrolysed by aqueous acids. This property makes acetals/ketals suitable protecting groups for carbonyl compounds during basepromoted reactions. For example, hydrolysis of α-haloketones with alkali to give α-hydroxyketones is usually a low-yield reaction. The α-hydroxy ketones can, however, be prepared by first ketalising the α-haloketone followed by treatment with the alkali and finally carrying out acid hydrolysis.

$$R-\overset{O}{\overset{||}{C}}-CH_2-Cl \xrightarrow[\text{(Low yields)}]{\overset{\ominus}{O}H} R-\overset{O}{\overset{||}{C}}-CH_2OH$$

$$R-\overset{O}{\overset{||}{C}}-CH_2-Cl \xrightarrow{R^1-OH/\overset{\oplus}{H}} R(CH_2Cl)C(R^1O)(OR^1) \xrightarrow[\text{(High yield)}]{\overset{\ominus}{O}H} R(CH_2OH)C(R^1O)(OR^1) \xrightarrow{H_3\overset{\oplus}{O}} R-\overset{O}{\overset{||}{C}}-CH_2OH$$

In the above reaction, ethylene glycol is commonly used as the alcohol for ketalisation. The usual procedure is to reflux a benzene solution of the α-haloketone and ethylene glycol in the presence of trace of p-toluene sulphonic acid as the acid-catalyst with the Dean-Stark removal of water.

$$\underset{\text{(Ketone)}}{R(R^1)C=O} + \underset{\text{(Ethylene glycol)}}{HO-CH_2-CH_2-OH} \xrightarrow[-H_2O]{p\text{-TsOH}} \underset{\text{(Ethylene glycol Ketal derivative)}}{R(R1)C\langle O-CH_2-CH_2-O\rangle}$$

The Mechanism of the reaction is as follows.

$$R(R^1)C=O \xrightarrow{\overset{\oplus}{H}} R(R^1)\overset{\oplus}{C}-OH \xrightarrow{HO\ \ OH} R(R^1)C(OH)-\overset{\oplus}{O}(H)-CH_2CH_2OH \longrightarrow$$

$$R(R')C(OCH_2CH_2\overset{\oplus}{O}H_2\ldots) \longrightarrow R(R')C(O\text{-}CH_2CH_2\text{-}\overset{\oplus}{O}H) + H_2O \xrightarrow{-H^{\oplus}} R(R')C(OCH_2CH_2O)$$

(ii) Addition of nitrogen nucleophiles: Derivatives of ammonia add on to carbonyl compounds giving initially aminohydrins, which undergo loss of elements of water to yield appropriate products. For example, aromatic primary amines react with carbonyl compounds to yield Schiff's bases.

$$\underset{\text{(Aromatic amine)}}{RR'C=O + H_2N\text{—}Ar} \longrightarrow \left[RR'C(OH)(N\text{-}Ar)H\right] \xrightarrow{-H_2O} \underset{\text{(Schiff's base)}}{RR'C=N\text{—}Ar}$$

Similarly, phenyl hydrazine, hydroxylamine and semicarbazide react with aldehydes/ketones to yield respectively a *phenyl hydrazone*, an *oxime* and a *semicarbazone*.

$$RR'C=O + \underset{\text{(Phenylhydrazine)}}{H_2N\text{–}NH\text{–}Ph} \longrightarrow \left[RR'C(OH)(NH\text{–}NH\text{–}Ph)\right] \xrightarrow{-H_2O} \underset{\text{(a phenylhydrazone)}}{RR'C=N\text{–}NH\text{–}Ph}$$

$$RR'C=O + \underset{\text{(Hydroxylamine)}}{H_2N\text{—}OH} \longrightarrow \left[RR'C(OH)(NH\text{—}OH)\right] \xrightarrow{H_2O} \underset{\text{(an oxime)}}{RR'C=N\text{—}OH}$$

$$RR'C=O + \underset{\text{(Semicarbazide)}}{H_2N\text{-}NH\text{-}CONH_2} \longrightarrow \left[RR'C(OH)(NH\text{-}NHCONH_2)\right] \longrightarrow \underset{\text{(a semicarbazone)}}{RR'C=N\text{-}NHCONH_2}$$

In all the reactions mentioned above, the loss of elements of water from the initially formed aminohydrin derivatives is due to formation of stable, conjugated products containing the C=N functionality. The reactions are reversible and the Schiff's bases or their equivalent products may be treated with aqueous acids to yield back carbonyl compounds and amine salts.

The above reactions are very important from the synthetic and analytical points of view. Particularly noteworthy is the reaction of carbonyl compounds with 2,4-dinitrophenylhydrazine (2,4-DNP) to produce an orange-red precipitate of 2,4-dinitrophenyl hydrazone. This reaction is used as a qualitative test for the detection of an aldehyde or ketone. To an alcoholic solution of the 2,4-DNP reagent, a solution of the carbonyl compound in alcohol is added at room temperature. Prompt formation of an orange to orange-red precipitate from the reaction medium in 1 to 10 minutes time is taken as a positive test. Most aldehydes and ketones respond positively to this test while other functional groups like carboxylic acids, esters, amides, ureas etc., *do not.*

$$RR^1C{=}O + H_2N{-}NH{-}C_6H_3(NO_2)_2 \xrightarrow[(-H_2O)]{RT} RR^1C{=}N{-}NH{-}C_6H_3(NO_2)_2$$

(Colourless solution in alcohol) (Orange-yellow solution in alcohol) (Orange-red or reddish precipitate)

(iii) Addition of sulphur nucleophiles : Sulphur in its lower valent states (II and IV), acts as a powerful nucleophile adding on to the carbonyl carbon yielding addition products. Since sulphur belongs to the second period of the VI group in the periodic table, due to its larger size and lower electronegativity, sulphur acts as a weaker base but as a more stronger nucleophile than oxygen. Thus, sodium bisulphite, thiols, sulphinate salts, etc., add on to carbonyl compounds yielding various products. For example;

$$RR^1C{=}O + HO{-}\ddot{S}(=O){-}O^{\ominus}Na^{\oplus} \longrightarrow RR^1C(O^{\ominus})(S^{\oplus}(=O)(H){-}O^{\ominus}Na^{\oplus}) \longrightarrow RR^1C(O^{\ominus}Na^{\oplus})(SO_2{-}OH)$$

(Carbonyl compound) (Sod. bisulphite) (Sod. bisulphite addition compound)

$$RR^1C{=}O \xrightarrow[H^{\oplus}]{R{-}\ddot{S}H} \left[RR^1C(SR^2)(OH)\right] \xrightarrow{R^2{-}SH/H^{\oplus}} RR^1C(SR^2)_2$$

(An α,α-dithioether)

$$\begin{matrix}R \\ R^1\end{matrix}\!\!>C{=}O + NaO{-}\overset{O}{\overset{\|}{S}}{-}Ph \longrightarrow \left[\begin{matrix}R & & \overset{O\ \ Ph}{S^{\ominus}Na^{\oplus}}\\ & C & \\ R & & O^-\end{matrix}\right] \longrightarrow \begin{matrix}R & & SO_2{-}Ph\\ & C & \\ R^1 & & O^{\ominus}Na^{\oplus}\end{matrix}$$

(Sodiumbenzene sulphinate) (Sod. salt of α-hydroxy sulphone)

$$\xrightarrow{H_2O} \begin{matrix}R & & SO_2{-}Ph\\ & C & \\ R^1 & & OH\end{matrix}$$

(α-Hydroxy sulphone)

The reaction of carbonyl compounds with sodium bisulphite is often used as a qualitative test to distinguish between aldehydes and ketones. Since ketones are sterically more crowded (due to the presence of two bulky alkyl groups) around the carbonyl group than aldehydes, the formation of tetrahedral/bulky bisulphite addition compound with ketones as starting materials would be very difficult. On the other hand, aldehydes react readily with sodium bisulphite to yield the bisulphite additon compound which being the sodium salt precipitates out of ether solution. With the *exception* of a few ketones like *acetone*, most ketones do not respond to this test while aldehydes *do*.

(iv) Addition of carbon nucleophiles: The additon of carbon nucleophiles to carbonyl compounds is synthetically a very important reaction

$$\begin{matrix}R \\ R^1\end{matrix}\!\!>C = O + \overset{\delta+}{H}{-}\overset{\delta-}{CN} \longrightarrow \begin{matrix}R & & OH\\ & C & \\ R^1 & & CN\end{matrix}$$

(a Cyanohydrin)

$$\begin{matrix}R \\ R^1\end{matrix}\!\!>C = O + \overset{\delta-}{R^2}{-}\overset{\delta+}{MgX} \longrightarrow \begin{matrix}R & & R^2\\ & C & \\ R^1 & & OMgX\end{matrix} \xrightarrow{-H_2O} \begin{matrix}R & & R^2\\ & C & \\ R^1 & & R^2\end{matrix}$$

(Grignard reagent) (An alcohol)

since it leads to the formation of carbon-carbon bonds and hence, homologation of the carbon chain. For example, HCN and Grignard reagents add to carbonyl compounds giving useful end products. The reactions are shown above.

The former reaction (with HCN) was used extensively by Emil Fischer and coworkers during the early days of developement of carbohydrate chemistry and it led to the establishment of the structure and configuration of several carbohydrates including glucose. The latter reaction (with

R^2—MgX) results in the syntheses of alcohols and depending on the nature of the starting carbonyl compound gives rise to a primary, secondary or tertiary alcohol.

(v) Addition of hydride ions: Hydride ions from the complex metal hydrides such as $LiAlH_4$, $NaBH_4$, etc., add on to the carbonyl carbon leading to the formation of additon compounds, which are simply the reduction products. Thus, aldehydes yield primary alcohols and ketones yield secondary alcohols.

$$\underset{\text{(Aldehyde)}}{\text{RHC=O}} \xrightarrow[(+H^{\ominus})]{LiAlH_4} \text{RCH}_2\text{–}O^{\ominus}Li^{\oplus} \xrightarrow[\text{(on work up)}]{H_2O} \underset{\text{(a primary alcohol)}}{\text{RCH}_2\text{OH}}$$

$$\underset{\text{(a Ketone)}}{\text{RR}^1\text{C=O}} \xrightarrow[(+H^{\ominus})]{NaBH_4} \text{RR}^1\text{CH–}O^{\ominus}Na^{\oplus} \xrightarrow[\text{(on work up)}]{H_2O} \underset{\text{(a secondary alcohol)}}{\text{RR}^1\text{CHOH}}$$

(B) Base - Promoted Condensation Reactions of Carbonyl Compounds

Carbonyl compounds having a hydrogen attached to an α-carbon react with strong bases to yield resonance stabilized carbanions. These carbanions are strongly nucleophilic species and react with electron-deficient carbon atoms (from alkyl halides or other carbonyl carbon atoms) to give condensation products. The following are some of these base promoted condensation reactions.

(i) Aldol Condensation: In the presence of a base, such as aq alkali, two molecules of an aldehyde or ketone condense to give a β-hydroxyaldehyde or a β-hydroxyketone. This reaction is known as *aldol condensation*, since the product has an aldehyde functionality and an alcoholic (hydroxy) functionality.

$$R\text{—}CH_2\text{—}\overset{O}{\overset{||}{C}}\text{—}R^1 + R\text{—}CH_2\text{—}\overset{O}{\overset{||}{C}}\text{—}R^1 \xrightarrow[HO^-]{\text{Base}} R\text{—}CH_2\text{—}\underset{OH}{\underset{|}{\overset{R}{\overset{|}{C}}}}\text{—}\overset{R}{\overset{|}{C}H}\text{—}\overset{O}{\overset{||}{C}}\text{—}R^1$$

The mechanism of the reaction is as follows:

$$\underset{\text{(Carbonyl compound)}}{R{-}CH_2{-}\overset{\overset{\displaystyle O}{||}}{C}{-}R^1} \xrightarrow{\overset{\ominus}{O}H} \underset{\text{(a Carbanion)}}{R{-}\overset{\ominus}{C}H{-}\overset{\overset{\displaystyle O}{||}}{C}{-}R^1} + HOH$$

$$R{-}\overset{\ominus}{C}H{-}\underset{\underset{\displaystyle O}{||}}{C}{-}R^1 + R{-}CH_2{-}\underset{\underset{\displaystyle O^{\ominus}}{|}}{\overset{\oplus}{C}}{-}R^1 \longrightarrow R{-}CH_2{-}\underset{\underset{\displaystyle O^{\ominus}}{|}}{\overset{\overset{\displaystyle R^1}{|}}{C}}{-}\overset{\overset{\displaystyle R}{|}}{C}H{-}\underset{\underset{\displaystyle O}{||}}{C}{-}R^1$$

$$R{-}CH_2{-}\underset{\underset{\displaystyle O^{\ominus}}{|}}{\overset{\overset{\displaystyle R^1}{|}}{C}}{-}\overset{\overset{\displaystyle R}{|}}{C}H{-}\underset{\underset{\displaystyle O}{||}}{C}{-}R^1 + HOH \longrightarrow \underset{\text{(an aldol)}}{R{-}CH_2{-}\underset{\underset{\displaystyle OH}{|}}{\overset{\overset{\displaystyle R^1}{|}}{C}}{-}\overset{\overset{\displaystyle R}{|}}{C}H{-}\underset{\underset{\displaystyle O}{||}}{C}{-}R^1} + \overset{\ominus}{O}H$$

In concentrated alkali or upon heating the dilute alkaline solution, the aldol loses a molecule of water to yield an *enone*.

$$R{-}CH_2{-}\underset{\underset{\displaystyle OH}{|}}{\overset{\overset{\displaystyle R^1}{|}}{C}}{-}\underset{\underset{\displaystyle H}{|}}{\overset{\overset{\displaystyle R}{|}}{C}}{-}\overset{\overset{\displaystyle O}{||}}{C}{-}R^1 \xrightarrow[(-HOH)]{\overset{\ominus}{O}H} \underset{\text{(Enone)}}{R{-}CH_2{-}\overset{\overset{\displaystyle R^1}{|}}{C}=\overset{\overset{\displaystyle R}{|}}{C}{-}\overset{\overset{\displaystyle O}{||}}{C}{-}R^1}$$

The driving force for the formation of enone is that it has a double bond in conjugation with the keto group which confers stability on the molecule. Obviously, the aldol condensation cannot take place with aldehydes/ketones that do *not* have an α-carbon carrying hydrogen atom.

The aldol condensation is an extremely useful synthetic reaction. The aldol product can be converted into a variety of functional groups such as α,β- unsaturated carbonyl compounds, 1,3-diols, saturated aldehydes/ketones, saturated acids, saturated alcohols, hydrocarbons etc., having the same number of carbon atoms. For example;

$$R{-}CH_2{-}CHO + R{-}CH_2{-}CHO \xrightarrow{\overset{\ominus}{O}H} \underset{\text{(an aldol)}}{R{-}CH_2{-}\underset{\underset{\displaystyle OH}{|}}{\overset{\overset{\displaystyle H}{|}}{C}}{-}\overset{\overset{\displaystyle R}{|}}{C}H{-}CHO}$$

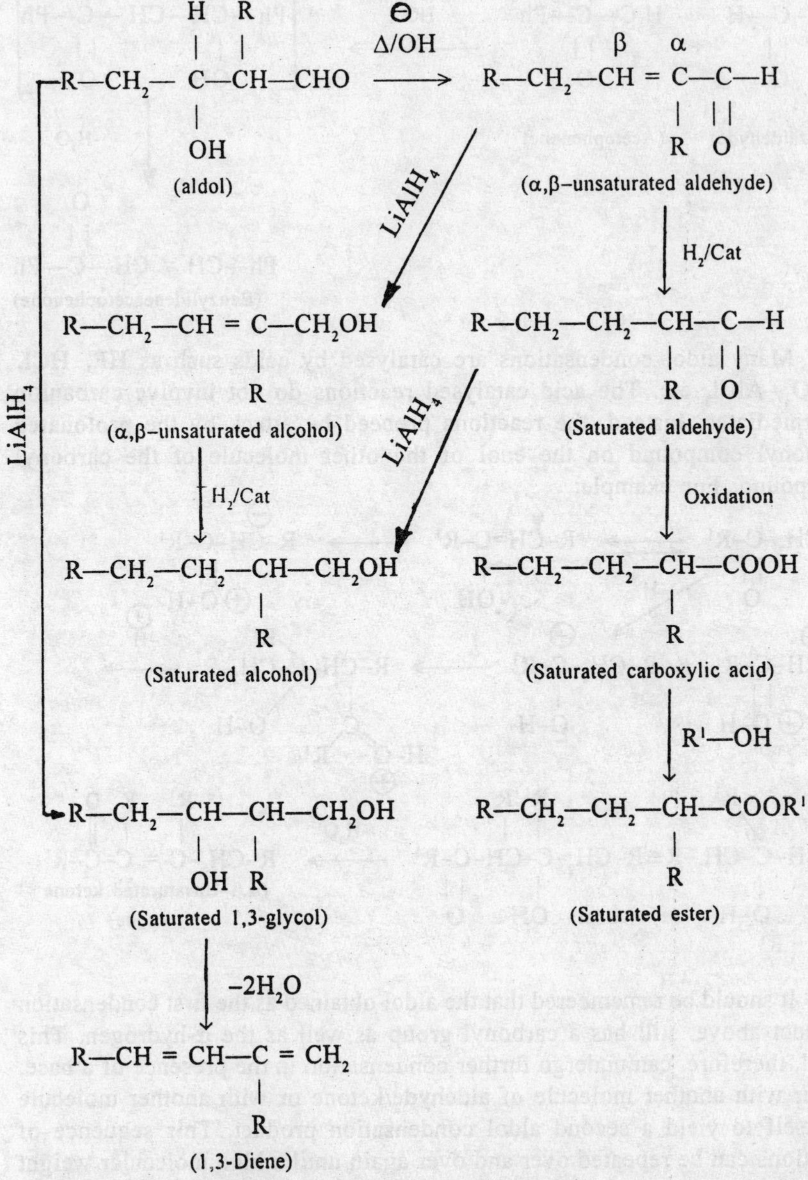

Crossed aldol condensations, involving reaction between a carbonyl compound having an α-hydrogen and a carbonyl compound lacking an α-hydrogen, are also known. For example, benzaldehyde condenses with acetophenone in the presence of alkali to yield, benzylideneacetophenone, also known as *chalcone*. It is a very important starting material for the synthesis of a variety of oxygen heterocyclic compounds, particularly of the flavone family.

$$\underset{\text{(Benzaldehyde)}}{\mathrm{Ph{-}\overset{O}{\overset{||}{C}}{-}H}} + \underset{\text{(Acetophenone)}}{\mathrm{H_3C{-}\overset{O}{\overset{||}{C}}{-}Ph}} \xrightarrow{HO^-} \left[\mathrm{Ph{-}\underset{OH}{\underset{|}{CH}}{-}CH_2{-}\underset{O}{\underset{||}{C}}{-}Ph}\right] \xrightarrow{-H_2O} \underset{\text{(Benzylideneacetophenone)}}{\mathrm{Ph{-}CH{=}CH{-}\overset{O}{\overset{||}{C}}{-}Ph}}$$

Many aldol condensations are catalysed by acids such as HF, HCl, H_2SO_4, $AlCl_3$ etc. The acid catalysed reactions do not involve carbanion intermediates. Instead, the reactions proceed by attack by the protonated carbonyl compound on the enol of the other molecule of the carbonyl compound. For example;

$$\mathrm{R{-}CH_2{-}\underset{O}{\underset{||}{C}}{-}R^1} \rightleftharpoons \mathrm{R{-}CH{=}\underset{\ddot{O}H}{\underset{|}{C}}{-}R^1} \longleftrightarrow \mathrm{R{-}\overset{\ominus}{C}H{-}\underset{\oplus O{-}H}{\underset{||}{C}}{-}R^1}$$

$$\mathrm{R{-}\overset{\ominus}{C}H{-}\underset{\oplus O{-}H}{\underset{||}{C}}{-}R^1} + \mathrm{R{-}CH_2{-}\underset{O{-}H}{\underset{|}{\overset{\oplus}{C}}}{-}R^1} \longrightarrow \mathrm{R{-}\underset{H{-}\overset{\oplus}{O}{=}C{-}R^1}{\underset{|}{CH}}{-}\overset{R^1}{\underset{O{-}H}{C}}{-}CH_2{-}R} \xrightarrow{-\overset{\oplus}{H}}$$

$$\mathrm{R{-}\underset{O{=}C{-}R^1}{\underset{|}{CH}}{-}\overset{R^1}{\underset{O{-}H}{C}}{-}CH_2{-}R} \equiv \underset{\text{(Aldol)}}{\mathrm{R{-}CH_2{-}\underset{OH}{\underset{|}{\overset{R^1}{C}}}{-}\overset{R}{CH}{-}\underset{O}{\underset{||}{C}}{-}R^1}} \xrightarrow{-H_2O} \underset{(\alpha,\beta\text{-Unsaturated ketone or enone})}{\mathrm{R{-}CH_2{-}\overset{R^1}{C}{=}\overset{R}{C}{-}\overset{O}{\overset{||}{C}}{-}R^1}}$$

It should be remembered that the aldol obtained as the first condensation product above, still has a carbonyl group as well as the α-hydrogen. This aldol, therefore, can undergo further condensation in the presence of a base, either with another molecule of aldehyde/ketone or with another molecule of itself to yield a second aldol condensation product. This sequence of reactions can be repeated over and over again until a high molecular weight resin is formed which may not have any use. It is for this reason that prolonged contact of aldehydes/ketones with strong acids/alkalies should be avoided. Even when aldol condensaion is desired, the conditions should be carefully monitored.

(ii) Cannizzaro condensation: Whenever an aldehyde lacking an α-hydrogen atom is treated with a strong base, instead of aldol condensation, another type of condensation called *Cannizzaro condensation* takes place. In a way, it is a disproportionation reaction. For example;

$$\text{Ph—CHO} + \text{Ph—CHO} \xrightarrow{\overset{\ominus}{\text{OH}}} \text{Ph—CH}_2\text{OH} + \text{Ph—COOH}$$

(Benzaldehyde) (Benzaldehyde) (Benzylalcohol) (Benzoicacid)

The mechanism of this reaction is very similar to aldol condensation but somewhat different from it as shown below;

$$\text{Ph—}\overset{\overset{\text{O}}{||}}{\text{C}}\text{—H} \longleftrightarrow \text{Ph—}\underset{\underset{\text{O}^{\ominus}}{|}}{\overset{\oplus}{\text{C}}}\text{—H} \xrightarrow{\overset{\ominus}{\text{OH}}} \text{Ph—}\underset{\underset{\text{O}^{\ominus}}{|}}{\overset{\overset{\text{OH}}{|}}{\text{C}}}\text{—H}$$

$$\text{Ph—}\underset{\underset{\text{O}^{\ominus}}{|}}{\overset{\oplus}{\text{C}}}\text{—H} + \text{Ph—}\underset{\underset{\text{O}^{\ominus}}{|}}{\overset{\overset{\text{H}}{|}}{\text{C}}}\text{—OH} \longrightarrow \left[\text{Ph—}\underset{\underset{\text{O}^{\ominus}}{|}}{\overset{\overset{\text{H}}{|}}{\text{C}}}\text{—H} + \text{Ph—}\overset{\overset{\text{O}}{||}}{\text{C}}\text{—OH}\right]$$

$$\downarrow$$

$$\text{Ph—CH}_2\text{OH} + \text{Ph—}\underset{\underset{\text{O}}{||}}{\text{C}}\text{—OH} \xleftarrow[\text{(During work-up)}]{\overset{\oplus}{\text{H}_3\text{O}}} \left[\text{Ph—}\underset{\underset{\text{OH}}{|}}{\overset{\overset{\text{H}}{|}}{\text{C}}}\text{—H} + \text{Ph—}\underset{\underset{\text{O}}{||}}{\text{C}}\text{—O}^{\ominus}\right]$$

Crossed cannizzaro condensation reactions are also known. However, such reactions are not synthetically very useful since at least four products are fomed. For example;

$$\text{Ar—}\underset{\underset{\text{O}}{||}}{\text{C}}\text{—H} + \text{Ar}^{\text{I}}\text{—}\underset{\underset{\text{O}}{||}}{\text{C}}\text{—H} \longrightarrow \text{Ar—CH}_2\text{OH} + \text{Ar—COOH} + \text{Ar}^{\text{I}}\text{—CH}_2\text{OH} + \text{Ar}^{\text{I}}\text{—COOH}$$

To minimize side-product formations in such reactions, one of the starting materials is usually taken as formaldehyde so that sodium formate

$$\text{Ar—}\underset{\underset{\text{O}}{||}}{\text{C}}\text{—H} + \text{H—}\underset{\underset{\text{O}}{||}}{\text{C}}\text{—H} \xrightarrow{\overset{\ominus}{\text{OH}}} \underset{\text{(Aromatic alkylalcohol)}}{\text{Ar—CH}_2\text{OH}} + \underset{\text{(Formic acid salt)}}{\text{H—}\underset{\underset{\text{O}}{||}}{\text{C}}\text{—O}^{\ominus}}$$

and aromatic alkyl alcohol are obtained as product which can be very easily separated. For example;

The most important pre-requisite of the Cannizzaro reaction is that strongly concentrated solution of alkali (≈50-60%) is required for this condensation to occur.

(iii) Perkin condensation: In this reaction an aromatic aldehyde is treated with acetic anhydride in the presence of an anhydrous base to yield an α,β-unsaturated acid. For example;

$$\underset{\text{(an aromatic aldehyde)}}{Ar{-}\overset{O}{\overset{\|}{C}}{-}H} + \underset{\text{(acetic anhydride)}}{H_3C{-}\overset{O}{\overset{\|}{C}}{-}O{-}COCH_3} \xrightarrow{NaOAc/\Delta} \underset{(\alpha,\beta\text{-unsaturated acid})}{Ar{-}CH = CH{-}COOH}$$

The mechanism of this reaction is believed to be as follows;

$$H_3C{-}\overset{O}{\overset{\|}{C}}{-}OCOCH_3 \xrightarrow{\overset{\oplus\ominus}{Na\,OAc}} \overset{\oplus}{Na}\,\overset{\ominus}{C}H_2{-}\overset{O}{\overset{\|}{C}}{-}OCOCH_3 + AcOH$$

$$Ar{-}\overset{O}{\overset{\|}{C}}{-}H \longleftrightarrow Ar{-}\overset{\oplus}{C}{-}H\ (O^{\ominus}) \xrightarrow{\overset{\ominus}{C}H_2{-}\overset{O}{\overset{\|}{C}}{-}OCOCH_3} Ar{-}\overset{H}{\overset{|}{C}}(O^{\ominus}){-}CH_2{-}\overset{O}{\overset{\|}{C}}{-}OCOCH_3 \xrightarrow{AcOH}$$

$$\underset{(\alpha,\,\beta\text{-hydroxy acid})}{\left[Ar{-}\overset{OH}{\overset{|}{C}H}{-}CH_2{-}\overset{O}{\overset{\|}{C}}{-}OH\right]} + HO{-}\underset{O}{\underset{\|}{C}}{-}H_3C \xleftarrow{H_2O} Ar{-}\underset{OH}{\underset{|}{C}H}{-}CH_2{-}\overset{O}{\overset{\|}{C}}{-}OCOCH_3 + AcOH$$

$$\downarrow -H_2O$$

$$\underset{(\alpha,\,\beta\text{-unsaturated acid})}{Ar{-}CH = CH{-}\underset{O}{\underset{\|}{C}}{-}OH}$$

(iv) Benzoin condensation: Two molecules of an aldehyde (lacking an α-hydrogen) undergo condensation in the presence of a cyanide ion to yield an α-hydroxy ketone derivative commonly known as *benzoin*. For example;

$$\underset{\text{(Benzaldehyde)}}{Ph-\overset{O}{\overset{\|}{C}}-H} + Ph-\overset{O}{\overset{\|}{C}}-H \xrightarrow{\overset{\ominus}{CN}} \underset{\text{(Benzoin)}}{Ph-\overset{HO}{\overset{|}{C}H}-\overset{O}{\overset{\|}{C}}-Ph}$$

The mechanism of this reaction is as follows;

$$Ph-\underset{\underset{O}{\|}}{C}-H \longleftrightarrow Ph-\overset{\oplus}{\underset{\underset{O^{\ominus}}{|}}{C}}-H \xrightarrow{\overset{\ominus}{CN}} Ph-\overset{CN}{\overset{|}{\underset{\underset{O^{\ominus}}{|}}{C}}}-H \longrightarrow Ph-\overset{CN}{\underset{OH}{C^{\ominus}}}$$

$$Ph-\overset{CN}{\underset{OH}{C^{\ominus}}} + Ph-\overset{\oplus}{\underset{\underset{O^{\ominus}}{|}}{C}}-H \longrightarrow Ph-\overset{CN}{\overset{|}{\underset{\underset{HO}{|}}{C}}}-\overset{H}{\overset{|}{\underset{\underset{O^{\ominus}}{|}}{C}}}-Ph$$

$$Ph-\overset{CN}{\overset{|}{\underset{\underset{O}{|}}{C}}}-\overset{H}{\overset{|}{C}}-Ph \; O^{\ominus} \longrightarrow \underset{\text{(Benzoin)}}{Ph-\underset{\underset{O}{\|}}{C}-\overset{H}{\overset{|}{\underset{\underset{OH}{|}}{C}}}-Ph} + \overset{\ominus}{CN}$$

(C) Tautomerism in Carbonyl Compounds

It has been mentioned earlier that carbonyl compounds having an α-hydrogen atom exhibit the phenomena of tautomerism. The two equilibrating forms of tautomers are known as keto and enol forms.

$$\underset{\text{(Keto form)}}{R-CH_2-\underset{\underset{O}{\|}}{C}-R^1} \rightleftarrows \underset{\text{(Enol form)}}{R-CH=\underset{\underset{OH}{|}}{C}-R^1}$$

Under normal conditions of temperature and pressure, the above equilibrium lies entirely to the left side and compound exists predominantly in the keto form. This is due to the fact that the enthalpy of the keto group is quite small and since entropy factor (structural disorderliness) for the two forms is not significantly different, free-energywise the keto form is favored. However, the situation changes drastically, when the enol form is particularly stabilized by hydrogen bonding and/or resonance. In such cases, there will be considerable amounts of enol form in the carbonyl compound. Even in compounds which exist predominantly in the keto form, the compounds may change/tautomerize momentarily into the enol form under dynamical/ reaction conditions react in the enol form and after the reaction is over, revert back to keto form. One such reaction is the α-halogenation of ketones described below.

α-Halogenation of Ketones: Ketones possessing an α-hydrogen atom react with halogens (Cl_2,Br_2 or I_2) or halogenating reagent, PCl_5, $SOCl_2$, Pyridinium bromide perbromide etc., to yield α-halogenated ketones, displacing the hydrogen atom from the α-position. Thus, the reaction can be formulated as follows;

The above reaction is acid catalysed. Since an acid is produced as by-product in the above reaction, it becomes autocatalytic. The mechanism of this reaction is as given below;

$$R{-}CH_2{-}\underset{\|}{\overset{O}{C}}{-}R' \overset{\text{Slow}}{\rightleftharpoons} R{-}CH{=}\underset{\underset{O-H}{|}}{C}{-}R' \longleftrightarrow R{-}\overset{\ominus}{C}H{-}\underset{\underset{\oplus O-H}{\|}}{C}{-}R' \xrightarrow[\text{Fast}]{X-X}$$

$$R{-}\underset{\underset{X}{|}}{CH}{-}\underset{\underset{\oplus O-H}{\|}}{C}{-}R' + \overset{\ominus}{X} \longrightarrow R{-}\underset{\underset{X}{|}}{CH}{-}\underset{\underset{O}{\|}}{C}{-}R' + HX$$

The keto form of the carbonyl compound, slowly, in a rate-determining step, is transformed into the enol form. The enol form is a resonance hybrid of the uncharged and charge-separated (having strongly nucleophilic carbanionic center and weakly electrophilic oxonium ionic center) structures. The enol form,through its strongly nucleophilic carbanionic center, attacks the positively polarized part of halogen molecule, in a fast step to yield the α-haloketone.

Evidence for the above mechanism of the halogenation reaction comes from the following facts;

(i) Ketones lacking an α-hydrogen *do not* undergo halogenation reaction.

(ii) Ketones that cannot enolize under the reaction conditions *do not* under go halogenation.

(iii) The halogenation reaction apparently comes to a halt when monohalogenation has taken place. If the α-haloketone first produced is treated with excess halogen, it does *not* undergo a second halogenation (under acidic conditions) even if an α-hydrogen is still available.

$$\underset{\text{(Ketone)}}{R\text{—}CH_2\text{—}\overset{\overset{O}{||}}{C}\text{—}R'} \xrightarrow{X_2} \underset{\text{(}\alpha\text{-Haloketone)}}{R\text{—}\overset{\overset{X}{|}}{CH}\text{—}\overset{\overset{O}{||}}{C}\text{—}R'} \xrightarrow{X_2} \!\!\!\!\!\!\times\;\; \underset{\text{(}\alpha,\alpha\text{-dihaloketone)}}{R\text{—}\overset{\overset{X}{|}}{C}\text{—}\overset{\overset{O}{||}}{C}\text{—}R'}$$

Explanation for the above observation comes from the fact that α-haloketone first produced does not enolize under acidic conditions and hence further halogenation stops. However, if a base is present in the reaction mixture,such as an alkali, the base abstracts the highly acidic α-proton yielding a highly nucleophilic carbanion which attacks the halogen molecule yielding α,α-dihaloketone.

$$R\text{—}CH_2\text{—}\overset{\overset{O}{||}}{C}\text{—}R' \xrightarrow{X_2/\overset{\oplus}{H}} R\text{—}\overset{\overset{X}{|}}{CH}\text{—}\overset{\overset{O}{||}}{C}\text{—}R' \xrightarrow{X_2/\overset{+}{H}} \!\!\!\!\!\!\times\;\; R\text{—}\underset{\underset{X}{|}}{\overset{\overset{X}{|}}{C}}\text{—}\underset{\underset{O}{||}}{C}\text{—}R'$$

$$R\text{—}\overset{\overset{X}{|}}{CH}\text{—}\overset{\overset{O}{||}}{C}\text{—}R' \xrightarrow{X_2/\overset{\ominus}{O}H} R\text{—}\underset{\underset{X}{|}}{\overset{\overset{X}{|}}{C}}\text{—}\underset{\underset{O}{||}}{C}\text{—}R'$$

$$R\text{—}\underset{\underset{X}{|}}{\overset{\overset{H}{|}}{C}}\text{—}\underset{\underset{O}{||}}{C}\text{—}R' + \overset{\ominus}{O}H \longrightarrow R\text{—}\overset{\overset{X}{|}}{C}=\underset{\underset{O^{\ominus}}{|}}{C}\text{—}R' \longleftrightarrow R\text{—}\underset{\underset{X}{|}}{\overset{\ominus}{C}}\text{—}\overset{\overset{O}{||}}{C}\text{—}R'$$

$$R\text{—}\underset{\underset{X}{|}}{\overset{\ominus}{C}}\text{—}\underset{\underset{O}{||}}{C}\text{—}R' + X\text{—}X \longrightarrow R\text{—}\underset{\underset{X}{|}}{\overset{\overset{X}{|}}{C}}\text{—}\underset{\underset{O}{||}}{C}\text{—}R' + \overset{\ominus}{X}$$

Evidence for the above postulate comes from the fact that when acetone is subjected to halogenation with bromine at room temperature under acidic conditions (trace of HBr) using just one mole of halogen per mole ketone, monobromoacetone is obtained. Further bromination with bromine (one mole) in the presence of a trace of HBr yields α,α-dibromoacetone. Bromination of monobromoacetone with bormine (one mole) in the presence of alkali or proton-sponge as base yields α,α-dibromoacetone.

$$\underset{\text{(Acetone)}}{H_3C{-}\overset{O}{\overset{\|}{C}}{-}CH_3} \xrightarrow[\text{Trace HBr}]{Br_2\text{ (1 mole)/}} \underset{(\alpha\text{–Bromoacetone})}{H_3C{-}\overset{O}{\overset{\|}{C}}{-}CH_2Br} \xrightarrow[\text{Trace HBr}]{Br_2\text{ (1mole)/}} \underset{(\alpha,\alpha^1\text{–Dibromoacetone})}{BrCH_2{-}\overset{O}{\overset{\|}{C}}{-}CH_2Br}$$

$$\underset{(\alpha,\alpha\text{ –Dibromoacetone})}{H_3C{-}\overset{O}{\overset{\|}{C}}{-}CHBr_2} \overset{Br_2\text{ (1 mole)/}}{\underset{\text{Trace HBr}}{\nleftarrow}} \; \alpha\text{-Bromoacetone} \; \xrightarrow[\text{HO}^-\text{ or organic base}]{Br_2\text{ (1mole)/}} \underset{(\alpha,\alpha\text{ –Dibromoacetone})}{H_3C{-}\overset{O}{\overset{\|}{C}}{-}CHBr_2}$$

The above theory of halogenation of ketones finds application in the haloform reaction exhibited by methyl ketones. The haloform reaction, particularly when carried out using iodine as the halogen, is used as a qualitative test to identify/distinguish methyl ketones from other ketones. Thus, the reaction merely involves treating a methyl ketone with excess iodine in the presence of NaOH, when a carboxylic acid and iodoform are obtained as products. The iodoform can be easily identified by its smell and appearance as a yellow precipitate. For example;

$$\underset{\text{(Acetophenone a methylketone)}}{Ph{-}\underset{O}{\underset{\|}{C}}{-}CH_3} + \underset{\text{(Excess)}}{3\,I_2} \xrightarrow{\text{NaOH (aq.)}} \underset{\text{(Sod. Salt of the caboxylic acid)}}{Ph{-}\underset{O}{\underset{\|}{C}}{-}\overset{\ominus}{O}\,\overset{\oplus}{Na}} + \underset{\text{(Iodoform)}}{HCI_3}$$

The mechanism of the haloform/iodoform reaction is as follows;

$$Ph{-}\underset{O}{\underset{\|}{C}}{-}CH_3 \rightleftharpoons Ph{-}\underset{OH}{\underset{|}{C}}=CH_2 \longleftrightarrow Ph{-}\underset{\overset{\oplus}{O}{-}H}{\underset{\|}{C}}{-}\overset{\ominus}{C}H_2 \;\; (I{-}I) \longrightarrow Ph{-}\underset{\overset{\oplus}{O}{-}H}{\underset{\|}{C}}{-}CH_2{-}I + \overset{\ominus}{I}$$

$$\longrightarrow Ph{-}\overset{O}{\overset{\|}{C}}{-}CH_2I \xrightarrow{\overset{\ominus}{OH}} Ph{-}\overset{O}{\overset{\|}{C}}{-}\overset{\ominus}{C}H{-}I \;\; (I{-}I) \longrightarrow Ph{-}\overset{O}{\overset{\|}{C}}{-}CHI_2 + \overset{\ominus}{I}$$

$$\xrightarrow{HO^-} Ph{-}\underset{O}{\underset{\|}{C}}{-}\overset{\ominus}{C}I_2 \;\; (I{-}I) \longrightarrow Ph{-}\underset{O}{\underset{\|}{C}}{-}CI_3 + \overset{\ominus}{I}$$

$$Ph{-}\underset{\underset{O}{\|}}{C}{-}CI_3 + \overset{\ominus}{O}H \longrightarrow \left[Ph{-}\underset{\underset{O}{\|}}{C}{-}\overset{\ominus}{O}H{+}CI_3 \right] \longrightarrow Ph{-}\underset{\underset{O}{\|}}{C}{-}\overset{\ominus}{O} + HCI_3$$

Apart from qualitative analysis, the haloform reaction is at times used in a multi-step synthesis, for the cleavage of methyl ketones to yield corresponding carboxylic acids. For example;

$$R{-}\underset{\underset{O}{\|}}{C}{-}CH_3 + 3\,I_2 + \overset{\ominus}{O}H \longrightarrow R{-}\underset{\underset{O}{\|}}{C}{-}\overset{\ominus}{O} + HCI_3$$

(A methyl Ketone) (Sodi Salt of the corresponding carboxylic acid) (Iodoform)

Compounds like acetaldehyde and ethanol also respond positively to this test. Acetaldehyde may be expected to respond to the iodoform test since it can be considered to be a derivative of methyl ketones. Ethanol responds positively to the test because iodine with HO^- forms hypoiodite which is a mild oxidizing agent. Hypoiodite oxidizes ethanol to acetaldehyde which responds to the iodoform test.

(D) Other Reactions of Aldehydes and ketones

(i) Oxidation: Aldehydes are very readily oxidized to the corresponding carboxylic acids. When aldehydes undergo oxidation, in the process they reduce the oxidizing agent that brings about their oxidation. In other words, aldehydes act as powerful reducing agents.

$$R{-}CHO \xrightarrow{[O]/\text{mild oxidant}} R{-}COOH$$

Aldehydes can be oxidized by very mild oxidizing agents like Tollen's reagent, Fehling's reagent and Benedict's reagent.

(a) Tollen's Reagent as oxidant: Tollen's reagent is an ammoniacal solution of silver nitrate. Silver ion is the effective oxidizing agent and the medium should be alkaline in nature. Since silver will precipitate out as the hydroxide in alkaline medium, thereby preventing its oxidative action, the silver ion is taken in ammoniacal solution where it is present as silver ammonia complex. The Tollen's reagent is readily prepared in the lab. by treating silver nitrate solution with little ammonia when a precipitate of silver hydroxide is first formed which dissolves on adding excess ammonia giving soluble silver amine complex. This solution is treated with a solution

of aldehyde in water or methanol or aqueous methanol. After oxidation of the aldehyde to the carboxylic acid silver ion comes out as beautiful silver mirror or as black precipitate of silver. Thus, the reaction can be formulated as;

$$AgNO_3 \xrightarrow{NH_4OH} Ag(OH)\downarrow + NH_4NO_3 \xrightarrow{2NH_3} [Ag(NH_3)_2]OH \text{ (Soluble)}$$

$$R\text{–}CHO + 2[Ag(NH_3)_2]\,OH \longrightarrow R\text{—}COONH_4 + 2Ag\downarrow + H_2O + 3NH_3$$

(Silver mirror or black ppt.of silver.)

Effectively, the reaction is,

$$R\text{—}CHO + Ag^+ \longrightarrow R\text{—}COO^- + Ag\downarrow$$

(b) Fehling's solution as oxidant: Fehling's solution is blue solution of cupric ions complexed with tartrate ions. It is produced readily in the laboratory by mixing aq. solution of $CuSO_4$ *(Fehling A)* with an aqueous solution of sodium potasium tartrate*(Fehling B)*. When a solution of the aldehyde is mixed with Fehling's solution (Fehling A + Fehling B) and warmed on a steam-bath at 90-95° for 1 minute, a red precipitate of cuprous oxide is formed. Thus, the reaction can be formulated as shown below;

$$R\text{–}CHO + Cu^{+2}\text{—Complex} + HO^- \longrightarrow R\text{—}COO^- + Cu_2O\downarrow + H_2O$$

(Colorless) (Blue in color) (Colorless) (Red.ppt.)

(c) Benedict's solution as oxidant: Benedict's solution is also a cupric ion solution but complexed with citrate ions. Once again, when a solution of the aldehyde is warmed with cupric sulphate sodium citrate complex solution (Benedict's solution), a red precipitate of cuprous oxide is produced in about 30-60 seconds. The test is believed to be sensitive even to the extent of 1 part in million.

Ketones are not oxidized by mild oxidizing agents and they do *not* reduce Tollen's, Fehling's and Benedict's solutions. This difference in behavior of aldehydes and ketones towards mild oxidizing agent is due to the fact that aldehyes contain a hydrogen as one of its substituents. This hydrogen ofaldehydescan be readily extracted out by an oxidizing agent as H or as H^+ leading to oxidation of the aldehyde. On the other hand, a ketone has alkyl/aryl groups as substituents on the carbonyl group and an alkyl/aryl group cannot be extracted out by a mild oxidizing agent. Hence, ketones are very resistant to mild oxidizing agents. Ketones are, however,oxidized by strong oxidizing agents like acidified dichromate and acidified permanganate etc., leading to chain-degraded products.

(ii) Reduction : Aldehydes are reduced to primary alcohols and ketones are reduced to secondary alcohols by catalytic hydrogenation or by $LiAlH_4$. These methods find synthetic use for the preparation of alcohols when the carbonyl compounds are comparatively more readily available such as those from aldol condensation.

$$R—CHO \xrightarrow{H_2/Cat, \text{ or } LiAlH_4} R—CH_2—OH$$

$$R—\underset{\underset{O}{||}}{C}—R' \xrightarrow{H_2/Cat. \text{ or } LiAlH_4} R—\underset{\underset{OH}{|}}{CH}—R'$$

Aldehydes and ketones can be reduced to hydrocarbons by the action of amalgamated zinc and conc. HCl(Clemmensen reduction) or by the action of hydrazine hydrate and potassium t-butoxide (Wolff-Kishner reduction).

$$R—CH_2—CHO \xrightarrow[\text{or } NH_2\text{-}NH_2/K^+ \text{ t-BuO}^-]{Zn(Hg)/HCl} R—CH_2—CH_3$$

$$R—\underset{\underset{O}{||}}{C}—R' \xrightarrow[\text{Wolff-Kishner redn.}]{\text{Clemmensen or}} R—CH_2—R'$$

These methods of reduction are particularly important for getting alkylated derivatives of aromatic hydroxy compounds, since the corresponding carbonyl derivatives are readily obtainable through Friedel-Crafts reaction. For example;

OH, OH (Resorcinol) $\xrightarrow[ZnCl_2]{HOOC\text{-}(CH_2)\text{-}CH_3}$ OH, OH, $C(=O)(CH_2)_4—CH_3$ (Pentano-2, 4-dihydroxy phenone) $\xrightarrow[\text{or w.K. redn.}]{\text{Clemmensen}}$ OH, OH, $(CH_2)_5—CH_3$ (4-n-Hexyl-resorcinol an antiseptic)

(iii) Polymerization: In the presence of an acid, formaldehyde condenses with itself to form a trimer called *trioxane*.

$$3\ H_2C{=}O \rightleftharpoons \text{(CH}_2\text{O)}_3$$

(Three molecules of formaldehyde) (Trioxane)

If the reaction is allowed to go on in the presence of an acid for a longer time, it yields a polymer.

$$n\ (CH_2O) \rightleftharpoons \sim\!\!(\!\sim H_2C{-}O \sim\!)_n\!\sim$$

Formaldehyde Paraformaldehyde

Both the reactions, trioxane formation and *paraformaldehyde* formation, are reversible and catalysed by acids. When the trimer/polymer is treated with acid and the mixture distilled, the reaction is shifted in favor of the monomer.

Acetaldehyde also condenses to a trimer called paracetaldehyde or paraldehyde. This on heating with acid followed by distillation yields the monomer.

$$3\ H_3C{-}CH{=}O \underset{H_3O^{\oplus}/\Delta}{\overset{H^{\oplus}}{\rightleftharpoons}} (CH_3CHO)_3$$

(3 Molecules of acetalbehyde)

Higher aldehydes polymerize only to a small extent and do not yield definite isolable product/s. The trimers or polymers of aldehydes are very sparingly soluble in water and do not show reactions characteristic of aldehydes.

11-10 SELF-TEST QUESTIONS

Choose the correct answer for the following questions.

11-10-1 Acetaldehyde and vinyl alcohol are compounds which exhibit.

(A) Chain isomerism (B) Position isomerism

(C) Functional isomerism (D) All the three types.

11-10-2 Treatment of an acid chloride with H_2 over poisoned palladium catalyst to yield an aldehyde is known as
(A) Clemmensen reduction (B) Rosenmund reduction
(C) Wolff-kishner reduction (D) None of the above three.

11-10-3 Reaction of isopropyl alcohol with aluminium isopropoxide to yield acetone is known as
(A) Oppenauer reaction (B) Williamson's reaction
(C) Wurtz reaction (D) Wurtz-Fittig reaction

11-10-4 Ethyl methyl ketone is obtained by oxidation of
(A) n-Propyl alcohol (B) n-Butyl alcohol
(C) Sec-butyl alcohol (D) Isopropyl alcohol.

11-10-5 Hydrolysis of benzal halides leads to the formation of
(A) Benzoic acids (B) Benzaldehyde
(C) Benzophenones (D) Benzyl alcohols

11-10-6 The boiling point and solubilities of aldehydes and ketones are dependent upon
(A) Resonance of the carbonyl group (B) Hydrogen bonding
(C) Dipole-dipole interactions (D) Van der Waal's interactions.

11-10-7 When phenyl hydrazine reacts with a ketone with the loss of a molecule of water,the product formed is
(A) An oxime (B) A Schiffs base
(C) A semicarbazide (D) A phenyl hydrazone.

11-10-8 Self-condensations of benzaldehyde in the presence of alkali to yield a molecule each of alcohol and acid is known as
(A) Cannizzaro reaction (B) Perkin condensation
(C) Aldol condensation (D) None of the three

11-10-9 Iodoform test is not given by
(A) Ethanol (B) Acetaldehyde (C) Acetone (D) Benzophenone,

11-10-10 In qualitative analysis, an orange red precipitate of a compound with 2,4-D.N.P. is a diagnostic test for
(A) Alcohols (B) Ethers (C) Phenols (D) Aldehydes and ketones.

11-10-11 Which one of the following compounds on reaction with KCN followed by hydrolysis, yields phenylacetic acid
(A) Benzoic acid (B) Benzyl alcohol
(C) Benzyl chloride (D) Ethyl phenyl ether.

11-10-12 Aldehydes and ketones can be distinguished from each other using
(A) Tollen's test (B) 2,4-D.N.P. reagent
(C) Oxime test (D) None of the three.

11-11 SHORT AND ESSAY TYPE QUESTIONS

11-11-1 Explain the isomerism of carbonyl compounds-aldehydes and ketones,taking suitable examples.

11-11-2 Describe any two methods for the preparation of aldehydes giving at least one example of each method.

11-11-3 Explain,with suitable examples any two methods for the preparation of aromatic aldehydes.

11-11-4 Aryl alkyl ketones can be prepared either by oxidation method or by Friedel-Crafts acylation method. Explain this statement with suitable examples.

11-11-5 Describe the nucleophilic addition reactions of carbonyl compounds taking oxygen nucleophiles as examples.

11-11-6 What are the product/s obtained in the addition of 2,4-D.N.P. to acetaldehyde, acetone and ethanol. (O.U., Mar., 1994)

11-11-7 What are the product/s obtained when acetaldehyde and benzaldehyde are treated independently each with saturated $NaHSO_3$ solution.

11-11-8 Explain the reactions - Oppenauer oxidation and Meerwein-Pondorf-Verley reduction with suitable examples. (N.U., Sept., 1995)

11-11-9 Using nucleophilic addition as a specific example, explain the relative reactivity of aldehydes and ketones.

11-11-10 Describe the aldol condensation between benzaldehyde and acetophenone. What is the product formed and what is the mechanism of the reaction?

11-11-11 What is Cannizzaro condensation? How is the reaction carried out and what is the mechanism of the reaction?

11-11-12 Reaction of benzaldehyde with acetic anhydride in the presence of sodium acetate to yield cinnamic acid is known as Perkin condensation. Formulate the reaction and explain the mechanism of it?

11-11-13 What is haloform reaction? What is the use of this reaction in qualitative analysis?

11-11-14 Describe the Tollen's test and Fehling's test with suitable examples.

11-11-15 Using qualitative test-tube experiments, how can one distinguish between the following compounds.

(i) Acetaldehyde and acetone.

(ii) Benzaldehyde and cyclohexane.

(iii) Acetophenone and phenol.

(iv) Benzyl methyl ketone and benzyl alcohol.

(v) Ethyl methyl ketone and ethyl methyl ether.

11-11-16 What happens when

(a) Benzaldehyde is treated with warm Tollen's reagent.

(b) Acetone is treated with 2,4-D.N.P. reagent at RT.

(c) Phenyl acetaldehyde is warmed with Fehling's solution.

(d) Acetaldehyde is heated with SeO_2?

(e) A mixture of benzaldehyde and acetophenone is warmed with alcoholic KOH.

(f) Benzaldehyde is heated with very conc. NaOH?

(g) Formaldehyde and propiophenone are treated with sodium ethoxide.

11-11-17 The empirical molecular formula of a compound A was found to be C_8H_8O. The compound A reacted with 2,4-D.N.P. to form an orange-red precipitate. On treatment with warm aq. Tollen's reagent A gave a beautiful silver mirror. The aq. solution from this reaction mixture on acidification with HCl yielded carboxylic acid B with molecular formula $C_8H_8O_2$. The acid B on decarboxylation with CaO- NaOH yielded toluene. Deduce the structures of A and B and formulate the reactions. Suggest a suitable synthesis of A.

11-11-18 A compound X (C_8H_8O) reacted with 2,4-DNP to yield an orange-red precipitate. This compound X did not give the Tollen's test or Fehling's test. It reacted with I_2/NaOH to yield iodoform and benzoic acid. Derive the structure of X and suggest a suitable synthesis of it.

11-11-19 A compound A ($C_7H_6O_2$) reacted with CH_3I in pyridine to yield a compound B ($C_8H_8O_2$). Both the componds A and B gave orange-red precipitates independently with 2,4-D.N.P. Compound B with Tollen's reagent gave silver mirror aand a compound C ($C_88H_8O_3$). The latter on decarboxylation yielded anisole ($C_7H_8O_3$.) The compound A could be readily prepared from phenol by Riemer-Tiemann reaction and A was found to be non-steam volatile. Deduce the structures of A,B and C and formulate the reactions.

11-11-20 Propiophenone on reaction with CH_3-Mgl yielded a tertiary alcohol. The latter on dehydration gave an alkene which on Aq.$KMnO_4$ oxidation followed by acid-catalysed pinacol rearrangement gave isopropyl phenyl ketone. Formulate the reactions and suggest a synthesis of propiophenone.

11-11-21 What is the reaction of acetophenone with
(i) HCN (ii) Hydroxyl amine (iii) Phenyl hydrazine
(O.U., Oct.1991)

11-11-22 Discuss nulceophilic addition to carbonyl compounds and explain the mechanism of the reaction with examples. Explain why aldehydes are more reactive than ketones. (N.U., Apr., 1991)

11-11-23 Explain why formic acid reduces Tollen's reagent and not acetic acid?
(O.U.,Apr.1991)

11-11-24 Outline the synthesis of acetophenone and benzophenone by Friedel-crafts acylation. Give mechanism for the synthesis of any one of the above. Explain why acetophenone will not undergo further acylation. (O.U.,Nov.1990)

11-11-25 Write the mechanism involved in
a) Aldol condensation (N.U.,Mar.,1996), (O.U.,Mar.,1994), (O.U.,Mar.,1995)
b) Benzoin Condensation (N.U., Sept., 1995), (O.U., Mar., 1996),

SOLUTIONS TO PROBLEMS & EXERCISES

11-10-1 **C**	11-10-2 **B**	11-10-3 **A**	11-10-4 **C**	11-10-5 **B**	11-10-6 **C**
11-10-7 **D**	11-10-8 **A**	11-10-9 **D**	11-10-10 **D**	11-10-11 **C**	11-10-12 **A**

11-11-15

(i) *Acetaldehyde and acetone :* Acetaldehyde will readily react with warm Tollen's reagent to give a silver mirror whereas acetone will not. Similarly acetaldehyde will give a positive Fehling's test and acetone will not.

(ii) *Benzaldehyde and cyclohexanone :* Benzaldehyde will give positive reaction to Tollen's test and Fehling's test whereas cyclohexanone will not.

(iii) *Acetophenone and phenol :* Phenol will dissolve in dil. aq.NaOH at RT and will be regenerated on addition of aq.HCl. Acetophenone will not dissolve in dil.aq.NaOH at RT to give a clear solution. Furthermore, acetophenone will react with 2,4-D.N.P. at RT to give an orange precipitate whereas phenol will not.

(iv) *Benzaldehyde and benzyl alcohol :* Benzaldehyde will react with 2,4-D.N.P. to give an orange precipitate whereas benzyl alcohol will not. Benzaldehyde will give positive reaction to the silver-mirror test (Tollen's reagent). Wheareas benzyl alcohol will not. The benzyl alcohol, on the other hand,will react with sodium metal liberating hydrogen gas, whereas benzaldehyde will not.

(v) *Benzyl methyl ketone and benzyl alcohol :* Benzyl methyl ketone will react with 2,4-D.N.P. to give an orange-red precipitate whereas benzyl alcohol will not. The benzyl alcohol on the other hand,will react with sodium metal liberating a hydrogen gas, wheras benzyl methyl ketone will not.

(iv) *Ethyl methyl ketone and ethyl methyl ether :* Ethyl methyl ketone will give positive reaction to 2,4-D.N.P. reagent test whereas ethyl methyl ether will not.

11-11-16

a) $\text{Ph–CHO} + [\text{Ag}(\text{NH}_3)_2]^+ \longrightarrow \text{Ph—COOH}$

(Benzaldehyde) (Tollen's reagent) (Benzoic acid)

b)

$$\underset{\text{Acetone}}{H_3C-\overset{\overset{O}{\|}}{C}-CH_3} + \underset{\text{(2, 4-D. N. P)}}{H_2N-HN-C_6H_3(NO_2)_2} \longrightarrow \underset{\text{(2, 4-Dinitrophenyl hydrozone of acetone)}}{(H_3C)(CH_3)C{=}N-\overset{H}{N}-C_6H_3(NO_2)_2}$$

c) $\text{Ph—CH}_2\text{—CHO} + \text{Cu}^{+2} \longrightarrow \text{Cu}^{+1}\downarrow + \text{Ph—CH}_2\text{—COOH}$

(Phenyl acetaldehyde) (Fehling's solution) (Red cuprous oxide) (Phenylacetic acid)

d)

$$H_3C-\overset{\overset{O}{\|}}{C}-H \xrightarrow[\text{[O]}]{SeO_2} H_3C-\overset{\overset{O}{\|}}{C}-OH$$

(Acetaldehyde) (Acetic acid)

e)

$$\text{Ph}-\overset{\overset{O}{\|}}{C}-H + H_3C-\overset{\overset{O}{\|}}{C}-\text{Ph} \xrightarrow[\text{NaOH}/\Delta]{\text{Aq. alc.}} \text{Ph—CH} = \text{CH}-\overset{\overset{O}{\|}}{C}-\text{Ph}$$

(Benzaldehyde) (Chalcone product)

f)

$$\text{Ph}-\overset{\overset{O}{\|}}{C}-H \xrightarrow[\text{(Cannizzaro)}]{\text{Conc. NaOH}/\Delta} \text{Ph—CH}_2\text{OH} + \text{Ph—COOH}$$

(Benzylalcohol) (Benzoic acid)

g)

$$H-\overset{\overset{O}{\|}}{C}-H + H_3C-CH_2-\overset{\overset{O}{\|}}{C}-\text{Ph} \xrightarrow[\text{(Aldol)}]{\text{NaOEt}} H-CH = \overset{\overset{H_3C}{|}}{C}-\overset{\overset{O}{\|}}{C}-\text{Ph}$$

(Formaldehyde) (Propiophenone) (α,β-unsaturated ketone)

11-11-17

$Ph—CH_2—CHO$ $\xrightarrow{\text{2, 4 D.N.P}}$ $Ph—CH_2—CH{=}N—NH—C_6H_3(NO_2)_2$ (2,4-dinitrophenyl)
(Orange - Red precipitate)

$A(C_8H_8O)$
(Phenylacetaldehyde)

Tollen's reagent: $Ph—CH_2—CHO \rightarrow Ag\downarrow + Ph—CH_2—COOH$
(Silver mirror) (B, Phenylacetic acid)

Satd. aq. $NaHSO_3$: $Ph—CH_2—CH(ONa)—SO_2—OH$
(Bisulphite addition compound)

(C_8H_8O) $\xrightarrow{-CO_2}$ $Ph—CH_3$
(Toluene C_7H_8)

Synthesis of A

(Toluene) $C_6H_5CH_3$ $\xrightarrow{Cl_2/h\nu}$ $C_6H_5CH_2—Cl$ (Benzyl chloride) $\xrightarrow[\text{R.T.}]{\text{NaOH}}$ $C_6H_5CH_2OH$ (Benzyl alcohol)

(Benzyl alcohol) $\xrightarrow{CrO_3/Ac_2O}$ (Careful oxidation) $Ph—CH_2—CHO$ (Phenyl acetaldehyde)

11-11-18

$C_6H_5—CO—CH_3$ (Compound "X" Acetophenone) $\xrightarrow{I_2/NaOH}$ CHI_3 (Iodoform) + C_6H_5COOH (Benzoic acid)

(Compound "X" Acetophenone) $\xrightarrow{\text{(2,4-D N P)}}$ $Ph(CH_3)C{=}N—HN—C_6H_3(NO_2)_2$
(2,4-Dinitrophenyl hydrazone deriv. of acetophenone)

Tollen's reagent → No silver mirror

11-11-19

(p-Hydroxybenzaldehyde) ("A") ($C_7H_6O_2$) —CH_3I / Pyridine→ (p-Methoxybenzaldehyde) (B) ($C_8H_8O_2$) —$Ag^{\oplus}$ (Tollen's reagent)→ Ag↓ (Silver mirror) + (C) ($C_8H_8O_3$) [COOH, OCH_3]

(C) —$-CO_2$→ (Anisole) (C_7H_8O)

(Phenol) → (p-Hydroxybenzaldehyde)

(Phenol) —$(CH_3)_2SO_4/HO^{\ominus}$→ (Anisole) ($C_7H_8O$)

11-11-20

$$\underset{\text{(Propiophenone)}}{Ph{-}CO{-}CH_2{-}CH_3} \xrightarrow[\text{then } H_2O]{Me{-}MgI} \underset{\text{(Tertiaryalcohol)}}{Ph(HO)C(CH_3)(CH_2{-}CH_3)} \xrightarrow{-H_2O}$$

$$\underset{\text{(An alkene)}}{Ph(H_3C)C{=}C(H)CH_3} \xrightarrow{\text{aq. } KMnO_4} \underset{\text{(A Glycol)}}{H_3C{-}C(Ph)(HO){-}C(H)(OH){-}CH_3} \xrightarrow[\text{(Pinacol)}]{H_3O^{\oplus}} \underset{\text{(Isopropylphenylketone)}}{Ph{-}\underset{\|}{\overset{}{C}}{-}CH(CH_3)_2}$$

12

MONOCARBOXYLIC ACIDS AND THEIR DERIVATIVES

12-1 INTRODUCTION

Organic compounds containing a carbonyl and a hydroxy group on the same carbon atom are called the *carboxylic compounds*. The functional group of carboxylic acids is $-\overset{\overset{\displaystyle O}{||}}{C}-OH$. They are acidic in nature. This class of compounds have the general formula $R-\overset{\overset{\displaystyle O}{||}}{C}-OH$, where R=H, alkyl or aryl group.

$$H-\overset{\overset{\displaystyle O}{||}}{C}-OH \qquad R-\overset{\overset{\displaystyle O}{||}}{C}-OH \qquad Ar-\overset{\overset{\displaystyle O}{||}}{C}-OH$$

Although carboxyl group is virtually a union of one carbonyl and one hydroxyl group, the properties of the functional group as a whole are profoundly modified due to their mutual interaction. Thus *the carboxyl group does not give the reactions shown by aldehydes and ketones; and its hydroxyl group is definitely not alcoholic.* The carboxyl group taken as a whole shows characteristic properties and in its own is regarded as an independent functional group.

These carboxylic acids are classified as monocarboxylic acids, dicarboxylic acids, tricarboxylic acids, etc., according to the number of carboxylic groups present in the molecule. The acidity of carboxylic acids depend upon the number of carboxylic groups present in a molecule. Thus, a monocarboxylic acid is monobasic and a dicarboxylic acid is dibasic in nature and so on. They are saturated as well as unsaturated. The long chain monocarboxylic acids are called fatty acids as many higher members, for example; stearic acid, palmitic acid, oleic acid, etc., occur as glycerides in oils and fats and also resemble fats in some physical properties.

12-2 NOMENCLATURE

There are two methods of naming monocarboxylic acids.

1. *Common system*: Many of straight chain monocarboxylic acids are isolated from natural sources and were assigned common names. These names were usually derived from latin words indicating the source of the acids. The name formic acid comes from *formica* stands for ants in latin,as in the medieval times, alchemists obtained formic acid by distilling red ants. Acetic acid is from the latin word *acetum* = vinegar. In its pure form, it is called glacial acetic acid. The term *glacial* arises from the fact that pure acetic acid is a viscous liquid that solidifies into an icy-looking solid. The name for propionic acid literally means *first fàt*. Propionic acid is the first lowest molecular weight carboxylic acid to exhibit some of the properties of fatty acids, which are obtained from the hydrolysis of fats. Butyric acid is found in rancid butter, Latin:*butyrum*. It may be noted that such names do not follow any rules except that they end in-*ic acid*. In case of substituted acids the position of the substituent is represented by Greek letters - α, β, γ, δ etc.

$$\overset{\delta}{C}-\overset{\gamma}{C}-\overset{\beta}{C}-\overset{\alpha}{C}-COOH$$

2. *IUPAC System:* According to this system, carboxylic acids are named as alkanoic acids. The systematic name of an individual acid is derived by dropping the final 'e' from the name of the corresponding alkane (having same number of 'C' atoms) and adding the suffix-oic acid as shown in Table 12.1.

Table 12.1: Systematic and common names of some carboxylic acid

Formula	Source	Comon name	IUPAC name
HCOOH	L. *Formica* = ant	Formic acid	Methanoic acid
H_3CCOOH	L. *Acetum* = Vinegar	Acetic acid	Ethanoic acid
H_3CCH_2COOH	Gr. *Propon* = first pion = fat	Propionic acid	Propanoic acid
$H_3CCH_2CH_2COOH$	L. Buthyrum = butter	Butyric acid	Butanoic acid
$H_3CCH_2CH_2CH_2COOH$	Root of valerian plant	Valeric acid	Pentanoic acid
$H_3C-(CH_2)_4-COOH$	L. *Caper* = goat fat	Caproic acid	Hexanoic acid
$H_3C-(CH_2)_{10}-COOH$	Laurel oil	Lauric acid	Dodecanoic acid
$H_3C-(CH_2)_{14}-COOH$	Palm oil	Palmitic acid	Hexadecanoic acid
$H_3C-(CH_2)_{16}-COOH$	Gr. *Steur* = tallow	Stearic acid	Octadecanoic acid

The IUPAC system is frequently used for assigning names to higher branched chain and substituted carboxylic acids according to the rules given below:

1. The longest carbon chain containing the carboxylic group is selected and the carboxylic acid is named by replacing the *e* of the corresponding alkane by - *oic* acid

2. The selected carbon chain is numbered to indicate the position of the side - chains attached to it. The carboxylic carbon is always given the number 1.

$$\overset{\delta}{\underset{5}{C}}-\overset{\gamma}{\underset{4}{C}}-\overset{\beta}{\underset{3}{C}}-\overset{\alpha}{\underset{2}{C}}-\underset{1}{\overset{\overset{O}{||}}{C}}-OH$$

The IUPAC names of some substituted acids are given below.

$$H_3C-CH_2-\overset{\overset{Cl}{|}}{CH}-COOH$$

2-Chloro butanoic acid
(α-Chlorobutyric acid)

$$H_3C-\overset{\overset{CH_3}{|}}{CH}-CH_2-CH_2-COOH$$

4-Methyl pentanoic acid
(γ-Methyl valeric acid)

Aromatic carboxylic acids (ArCOOH) are usually named as derivatives of the parent acid, *benzoic acid*. If the aromatic group is not bonded directly to the carboxyl group, then the aromatic group is named as a substituent.

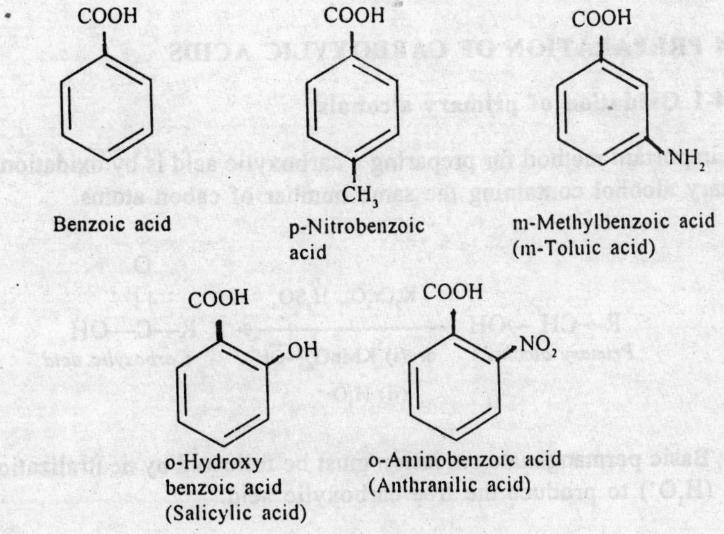

Benzoic acid

p-Nitrobenzoic acid

m-Methylbenzoic acid (m-Toluic acid)

o-Hydroxy benzoic acid (Salicylic acid)

o-Aminobenzoic acid (Anthranilic acid)

$C_6H_5CH_2COOH$
phenylethanoic acid
(phenylacetic acid)

$C_6H_5CH(OH)COOH$
α-Hydroxyphenylacetic acid
(Mandelic acid)

12-3 STRUCTURE

Carboxylic acid functional group is structurally related to both ketones and alcohols and we might therefore expect to see some familiar properties. In fact, carboxylic acids, are indeed similar to both ketones and alcohols in some ways, but there are major differences. Like ketones, the carboxylic carbon is sp^2 hybridized, and carboxylic acid groups are therefore planar, with C–C–O and O–C–O bond angles of approximately 120°. The physical parameters of acetic acid are given below along with its structure.

H_3C—C(=O)—O—H

Bond angle (degrees)		Bond lentth (Å units)	
C—C = O	119	C—C	1.52
C—C—OH	119	C = O	1.25
O = C—OH	122	C—OH	1.31

12-4 PREPARATION OF CARBOXYLIC ACIDS

12-4-1 Oxidation of primary alcohols

One important method for preparing a carboxylic acid is by oxidation of the primary alcohol containing the same number of cabon atoms.

$$\underset{\textit{Primary alcohol}}{R-CH_2-OH} \xrightarrow[\text{or (i) } KMnO_4-\overline{O}H \text{ (ii) } H_3O^+]{K_2Cr_2O_7,\ H_2SO_4} \underset{\textit{Carboxylic acid}}{R-\overset{O}{\overset{||}{C}}-OH}$$

Basic permanganate oxidation must be followed by neutralization with acid (H_3O^+) to produce the free carboxylic acid.

The principal limitation of alcohol oxidation is that the necessary strength of the oxidizing agent *precludes* the presence of another oxidizable functional group in the molecule (unless it is protected by a blocking group). Even with this limitation, the oxidation of primary alcohols is the most common oxidational procedure for obtaining carboxylic acids because alcohols are often available easily.

12-4-2 Oxidation of aldehydes

Aldehydes are very easily oxidized to carboxylic acids. Almost any reagent that oxidizes an alcohol also oxidizes an aldehyde. Permanganate or dichromate salts are the most popular oxidizing agents.

$$\underset{\text{An aldehyde}}{R{-}\overset{\overset{O}{||}}{C}{-}H} \xrightarrow{KMnO_4/OH} \underset{\text{A carboxylic acid}}{R{-}\overset{\overset{O}{||}}{C}{-}OH}$$

In addition to oxidation by permanganate or dichromate, aldehydes are oxidized by very mild oxidizing agents, such as Ag^+ or Cu^{+2}, that do not oxidize other groups. However, aldehydes are not as readily available as primary alchols.

$$R{-}\overset{\overset{O}{||}}{C}{-}H + [Ag(NH_3)_2]^+ \xrightarrow{2HO^-} R{-}\overset{\overset{O}{||}}{C}{-}\bar{O} + Ag\downarrow$$

$$\downarrow H^+$$

$$R{-}\overset{\overset{O}{||}}{C}{-}OH$$

12-4-3 Hydrolysis of nitriles

Aliphatic and aromatic nitriles give the corresponding carboxylic acids on hydrolysis in either acidic or basic solution. The hydrolysis reaction appears to occur in two steps. The nitrile is first converted to an amide, which is then hydrolysed to the corresponding carboxylic acid.

Acidic hydrolysis

$$\underset{\text{(Nitrile)}}{R{-}C\equiv N} \xrightarrow[\Delta]{H_3O^+} R{-}\overset{\overset{O}{||}}{C}{-}NH_2 \xrightarrow{H_3O^+} R{-}\overset{\overset{O}{||}}{C}{-}OH + \overset{+}{N}H_4$$

Basic hydrolysis

$$R{-}C{\equiv}N \xrightarrow[H_2O,\ \Delta]{\bar{O}H} R{-}\overset{\overset{O}{||}}{C}NH_2 \xrightarrow[H_2O]{\bar{O}H} R{-}\overset{\overset{O}{||}}{C}{-}\bar{O} + \dot{N}H_3$$

$$\uparrow\downarrow H_3O^+$$

$$R{-}\overset{\overset{O}{||}}{C}{-}OH$$

In general, basic conditions are chosen to convert a nitrile to an acid because the soluble salt of the carboxylic acid is formed and evidence of complete reaction can be seen visually when the original heterogeneous reaction mixture becomes homogeneous, under acidic aqueous conditions, the reaction mixture remains heterogeneous unless a polar organic solvent is used.

12-4-4 Hydrolysis of acid chlorides, anhydrides, esters and amides

These compounds which are collectively called acid derivatives possess a paired functional group in which a carbonyl is joined to an oxygen (as in carboxylic acid), nitrogen, halogen or any other electronegative atom. All these acid derivatives undergo hydrolysis to produce carboxylic acids with varying degrees of difficulty.

$$R{-}\overset{\overset{O}{||}}{C}{-}Cl + H_2O \longrightarrow R{-}\overset{\overset{O}{||}}{C}{-}OH + HCl$$

$$R{-}\overset{\overset{O}{||}}{C}{-}O{-}\overset{\overset{O}{||}}{C}{-}R + H_2O \longrightarrow 2R{-}\overset{\overset{O}{||}}{C}{-}OH$$

$$R{-}\overset{\overset{O}{||}}{C}{-}OR + H_2O \begin{cases} \xrightarrow{\text{acid}} R{-}\overset{\overset{O}{||}}{C}{-}OH + R{-}OH \\ \xrightarrow{\text{alkali}} R{-}\overset{\overset{O}{||}}{C}{-}OH + R{-}OH \end{cases}$$

$$R-\overset{O}{\overset{||}{C}}-NH_2 + H_2O \begin{cases} \xrightarrow{\text{acid}} R-\overset{O}{\overset{||}{C}}-OH + \overset{+}{N}H_4 \\ \xrightarrow{\text{alkali}} R-\overset{O}{\overset{||}{C}}-\overset{-}{O} + NH_3 \end{cases}$$

Acid chlorides react instantly with water, acid anhydrides react rather slowly, esters on the other hand, require days to react with water at room temperature, although the reaction times can be shortened to a matter of hours by raising the temperature. Amides react at a very slow rate, even at elevated temperatures. Usually, hydrolysis of esters or amides is carried out with the catalytic assistance of either a mineral acid or an alkali.

12-4-5 Carbonation of Grignard reagent

One of the most useful general methods for preparing both aliphatic and aromatic carboxylic acids begins with the corresponding alkyl or arylhalide, which is first converted into the corresponding Grignard reagent and then allowed to react with carbon dioxide. On hydrolysis,a carboxylic acid containing one more carbon atom is obtained.

$$\underset{X=Cl,\ Br,\ I}{R-X + Mg} \xrightarrow[\text{ether}]{\text{anhydrous}} RMgX \xrightarrow{CO_2} R-\overset{O}{\overset{||}{C}}-OMgX \xrightarrow{H_3O^+} R-\overset{O}{\overset{||}{C}}-OH$$

$$Ar-Br + Mg \longrightarrow Ar-Mg-Br \xrightarrow{CO_2} Ar-\overset{O}{\overset{||}{C}}-OMgBr \xrightarrow{H_3O^+} Ar-\overset{O}{\overset{||}{C}}-OH$$

The carbondioxide required for the reaction is usually supplied from dry ice, and this reaction is some times called *carbonation.*

The mechanism of the reaction between the Grignard reagent and carbon dioxide takes the following path.

$$\overset{\delta-}{R}\ \overset{\delta+}{MgX} + O=C=O \longrightarrow R-\overset{O}{\overset{||}{C}}-OMgX$$

$$\downarrow H_3O^+$$

$$R-\overset{O}{\overset{||}{C}}-OH + MgX(OH)$$

This is a typical nucleophilic addition reaction of a carbon-oxygen double bond.

12-4-6 Oxidation of alkylbenzenes

Aromatic rings containing an alkyl group are converted into the corresponding benzoic acid derivative on treatment with strong oxidizing agent under vigorous conditions.

$$\text{Ar—R} \xrightarrow[\Delta]{KMnO_4\,/\,HO^-} \text{Ar—}\overset{\overset{\displaystyle O}{||}}{C}\text{—OH}$$

Recall that methyl groups, as well as more complex side chains such as n-butyl and isopropyl groups, are oxidized so that only the carboxylic group involving the carbon attached directly to the aromatic ring remains. This reaction is limited to aromatic rings that are not highly activated, for example; highly activated phenols or aromatic amines are completely degraded by vigorous oxidation.

12-4-7 Permanganate oxidation of alkenes

Certain alkenes react with basic potassium permanganate under vigorous conditions to produce carboxylic acids.

$$\text{R(H)C=C(H)R'} \xrightarrow[\Delta]{KMnO_4/HO^-} \text{R—}\overset{\overset{\displaystyle O}{||}}{C}\text{—OH} + \text{R'—}\overset{\overset{\displaystyle O}{||}}{C}\text{—OH}$$

12-4-8 Hydrolysis of trihalogen derivatives of alkanes

Trihalogen derivatives of alkanes on the same carbon atom, when hydrolysed with aqueous alkali give carboxylic acids.

$$\text{R—CX}_3 \xrightarrow{\text{aq KOH}} \text{R—C(OH)}_3 \xrightarrow{H_2O} \text{R—}\overset{\overset{\displaystyle O}{||}}{C}\text{—OH}$$

12-5 PHYSICAL PROPERTIES

The first nine monocarboxylic acids are colorless liquids, while the higher ones are colorless wax like solids. The first three members possess pungent odor, butyric acid has rancid butter odor, while the higher ones have goat-like odor. Lower members up to C_4 are completely soluble in water due to the formation of hydrogen bonds between the hydroxy group of acid and oxygen of water and oxygen of acid and hydrogen of water.

```
    O              H              R
    ||             |              |
R—C— OH . . . . . .O——H. . . . .O = C—OH
```

Valeric acid is slightly soluble, while higher members are insoluble in water. All the acids are soluble in organic solvents such as alcohol, benzene, ether, etc.

The carboxylic acids have abnormally high boiling points, even higher than those of alcohols of similar molecular weights. Thus propionic acid (mol.wt., 74) boils at 141°C while n-butanol (mol.wt., 74) boils at 118°C. This indicates greater strengths of the hydrogen bonds in acids than in alcohols. This is so because –O –H of acid is more strongly ionized as $-O^-$ $-H^+$ and is also capable of forming hydrogen bonds to the rather negative oxygen of the carbonyl dipole than just to the oxygen of another hydroxyl as in alcohols. In benzene, the lower acids, occur in dimeric state due to hydrogen bonding.

```
              O.......H——O
             //           \
   H3C—C                   C—CH3
             \            //
              O——H.......O
```

The boiling and melting points of fatty acids show a regular gradation with the increase in molecular weight.

12-5-1 Acidity of carboxylic acid

Carboxylic acids have considerable acidic nature and are the acids of organic chemistry in much the same way that mineral acids are the acids of inorganic chemistry. In an aqueous solution molecules of carboxylic acid interact with water molecules to form carboxylate ions and hydronium ions.

$$\underset{\text{Carboxylic acid}}{R{-}\overset{O}{\overset{\|}{C}}{-}OH} + \underset{\text{Water}}{H_2O} \rightleftharpoons \underset{\text{Carboxylate ion}}{R{-}\overset{O}{\overset{\|}{C}}{-}O^-} + \underset{\text{Hydronium ion}}{H_3O^+}$$

Carboxylic acids are weak acids and the extent of ionization is described by an equilibrium constant K_a, known as acidity constant.

$$K_a = \frac{[RCOO^-]\,[H_3O^+]}{[RCOOH]}$$

The value of K_a which is directly proportional to the concentration of hydrogen ions is thus a measure of acidity of the individual acid under consideration. The larger the value of K_a the greater the concentration of H^+ and stronger the acid. The cause of acidity may be undersstood from the following considerations.

(i) Resonance effect: The reasons for the acidity of carboxylic acids is that they lose a proton readily because the carboxylate ion formed by ionization or reaction with a base is stabilized by resonance. In the case of a carboxylic acid, both the acid and the carboxylate ion formed by dissociation can exist in two canonical forms.

$$\underset{\text{I}}{R\text{—}\overset{\overset{\Large O}{||}}{C}\text{—}OH} \longleftrightarrow \underset{\text{II}}{R\text{—}\overset{\overset{\Large O^{\ominus}}{|}}{C}=\overset{\oplus}{O}\text{—}H} \xrightarrow{-H^{\oplus}} \underset{\text{III}}{R\text{—}\overset{\overset{\Large O}{||}}{C}\text{—}O^-} \longleftrightarrow \underset{\text{IV}}{R\text{—}\overset{\overset{\Large O^{\ominus}}{|}}{C}=O}$$

In canonical form II there is charge separation, therefore,it would contain more energy and hence is less stable than form I. On the other hand, in the canonical forms III and IV, carboxylate ions have equivalent structures which are necessarily of equal stability. Thus both acid and carboxylate ion are resonance stabilized but stabilization is far greater for the carboxylate ion than for the acid. The driving force for ionization of carboxylic acids is, therefore, due to gain in the stability in going from carboxylic acid to the more stable carboxylate ion.

(ii) Stability of carboxylate ion: According to the resonance theory, a carboxylate ion is a hybrid of two canonical forms as described above. These forms being of equal stability, contribute equally and in the hybrid ion, the carbon is joined to each oxygen by equivalent *one-and-a-half* bond. The negative charge is evenly distributed over both oxygen atoms.

$$R\text{—}C\begin{matrix}\nearrow O \\ \searrow O^{\ominus}\end{matrix} \longleftrightarrow R\text{—}C\begin{matrix}\nearrow O^{\ominus} \\ \searrow O\end{matrix} \equiv R\text{—}C\begin{matrix}O \\ \ominus \\ O\end{matrix}$$

Cononical forms of carboxylate ion **Resonance hybrid**

The fact that the anion is indeed a resonance hybrid receives support from the bond length measurements. Formic acid,for example, contains a carbon oxygen double bond and a carbon-oxygen single bond. We would expect these bonds to have different lengths. But on the other hand if sodium formate (existing as formate anion), is a resonance hybrid, it ought to contain two equivalent carbon oxygen bonds or in other words we would expert these to have the same bond length, intermediate between double and single bonds. X-ray and electron diffraction studies showed that these expectations are correct. Formic acid contains one carbon oxygen bond of 1.36A° (single bond) and another 1.23A° (double bond) where as sodium formate contains two equal carbon-oxygen bonds each, 1.27A° long.

Formic acid: H—C(=O, 1.23 A°)—OH (1.36 A°)

Sodium formate: [H—C(O)(O), 1.27 A°, 1.27 A°]$^{\ominus}$ $Na^{\oplus}$

(iii) *Molecular orbital structure:* The stability of carboxylate ion can also be explained on the basis of its molecular orbital structure. Carbon atom of carboxylate is sp^2 hybridized. Two of the sp^2 orbitals of this carbon atom form σ–bonds with the *p*-orbitals of the two oxygen atoms, the third sp^2 orbital of the carbon forms σ–bond with the carbon of the alkyl group of hydrogen atom in the case of formic acid. The unused (fourth) sp^2 and *p*-orbital of both the oxygen atoms form a stable delocalized molecular orbital. Thus the π-electron are delocalized and hence the carboxylate ion is stable.

12-5-2 Effect of substituents on acidity

We know that the carboxylic acids are acidic in nature because of the stabilization of carboxylate ion. Any factor that stabilizes the carboxylate ion should increase the acidity and any factor that makes carboxylate ion less-stable should decrease acidity. Electron withdrawing substituents should disperse the negative charge, stabilize the carboxylate ion and thus increase acidity. Electron releasing substituents should intensify the negative charge thereby destabilize the carboxylate and thus lead to a fall in acidity.

G ← C(O)(O) ⊖

G-withdraws electrons, stabilizes anions and strengthens acid

G → C(O)(O) ⊖

G-releases electrons, destabilizes anions and weakens acid.

Now since alkyl groups are electron releasing, their presence on the —COOH group will decrease the acidity. In general greater the length of the alkyl chain lower shall be the acidity of the acid. Thus formic acid, having no alkyl group is about 10 times stronger than acetic acid, which in turn is stronger than the propionic acid and so on. The electron withdrawing halogens strengthen acids. Chloroacetic acid is about 100 times as strong as acetic acid, dichloroacetic acid is still stronger and tri chloroacetic acid is more than 10,000 times as strong as the unsubstituted acid. The other halogens exert similar effect.

α-Chlorobutyric acid is about as strong as chloroacetic acid. As the chlorine is moved away from the -COOH, however, its effect rapidly dwindles. β-chlorobutyric acid is only six times as strong as butyric acid and γ-chlorobutyric acid is only twice as strong. It is typical of inductive effects that they decrease rapidly with distance, and are seldom important when acting through more than four atoms.

$$Cl \longleftarrow \overset{\gamma}{CH_2} \longleftarrow \overset{\beta}{CH_2} \longleftarrow \overset{\alpha}{CH_2} \longleftarrow C(O)(O)^{\ominus}$$

Inductive effect decreases with distance. Benzoic acid is somewhat stronger than simple aliphatic acids. Here the carboxylate group is attached to a more electronegative carbon (sp^2 hybridized) than in aliphatic acids (sp^3 hybridized).

The aromatic acids are similarly affected by substituents. The presence of an electron withdrawing substituent such as NO_2, CN, Cl, SO_3H, etc., disperses the negative charge of the carboxylate anion and thus stabilizes it and hence increases the acidity of the aromatic acid. On the other hand, presence of an electron releasing substituent such as $-CH_3$, –OH, $-OCH_3$, $-NH_2$, etc., intensify the negative charge of the carboxyte anion and thus destabilizes it and hence decreases the acidity of the aromatic acid.

O C O ⊖ (benzene ring with –G) O C O ⊖ (benzene ring with –G)

12-6 CHEMICAL PROPERTIES

12.6.1 Reactions involving removal of proton from - COOH group

(A) Reaction with metals: Metals like Na, K, Ca and Zn react with carboxylic acids to form salts with the evolution of hydrogen.

$$\underset{\text{Carboxylic acid}}{R{-}\overset{\overset{\displaystyle O}{\|}}{C}{-}OH} + Na \longrightarrow \underset{\text{Sodium salt of carboxylic acid}}{R{-}\overset{\overset{\displaystyle O}{\|}}{C}{-}O^{-}} \; Na^{+} + 1/2\, H_2$$

$$H_3CCOOH + Na \longrightarrow CH_3CO\bar{O}\;\; Na^{+} + 1/2\, H_2$$

(B) Reaction with bases: Carboxylic acids react with inorganic bases to form the corresponding salts

$$\underset{\text{Carboxylic acid}}{RCOOH} + NaOH \longrightarrow \underset{\text{Sodium salt of carboxylic acid}}{RCOONa} + H_2O$$

$$H_3CCOOH + NH_4OH \longrightarrow \underset{\text{Ammonium acetate}}{CH_3COO^{-}NH_4^{+}} + H_2O$$

12-6-2 Reaction involving $-\overset{\overset{\displaystyle O}{\|}}{C}$ part of the –COOH group (Replacement of –OH group of COOH)

These are the reactions of carboxylic acids in which the carbon of the carboxyl group of –COOH is attacked by a nucleophile: Nu to form acid derivatives such as esters, acid halides and acid amides.

$$\underset{\text{Carboxylic acid}}{R{-}\overset{\overset{\displaystyle O}{\|}}{C}{-}OH} + :Nu \rightleftharpoons R{-}\underset{\underset{\displaystyle Nu}{|}}{\overset{\overset{\displaystyle O^{-}}{|}}{C}}{-}OH \rightleftharpoons \underset{\text{Acid derivative}}{R{-}\overset{\overset{\displaystyle O}{\|}}{C}{-}Nu}$$

where Nu = OR, Cl, NH_2, etc.

(A)Esterification : carboxylic acids react with alcohols in the presence of mineral acid (HCl or H_2SO_4) catalyst to form esters. This reaction is known as esterification and is a reversible process.

Mechanism : The mechanism of esterification involves the following steps:

(i) Protonation of the carboxylic group by H^+ of the acid catalyst.

(ii) This makes the carbonyl carbon more strongly electropositive which is attacked by the nuclophile, the alcohol molecule.

(iii) The transfer of proton to one of the –OH groups takes place.

(iv) The –OH group leaves as water molecule giving protonated ester.

(v) The resonance stabilized protonated ester losses a proton to form ester.

$$R{-}\overset{\overset{O}{||}}{C}{-}OH + H^+ \rightleftarrows R{-}\overset{\overset{+OH}{||}}{C}{-}OH \underset{R'OH}{\rightleftarrows} R{-}\underset{\underset{\underset{R}{|}}{{}^+O{-}H}}{\overset{\overset{OH}{|}}{C}}{-}OH \rightleftarrows R{-}\underset{\underset{OR'\;H}{|\quad |}}{\overset{\overset{OH}{|}}{C}}{-}\overset{+}{O}{-}H$$

$$\underset{\text{Ester}}{R{-}\overset{\overset{O}{||}}{C}{-}OR'} \overset{-H^+}{\rightleftarrows} R{-}\overset{\overset{OH}{||}}{C}{-}OR' \rightleftarrows R{-}\overset{\overset{OH}{|}}{C_+}{-}OR' \quad \uparrow\downarrow -H_2O$$

(B) Formation of acid halide : Carboxylic acids can be converted into corresponding acid halides by reaction with PCl_3, PCl_5, PBr_3 or $SOCl_2$

$$3R{-}\overset{\overset{O}{||}}{C}{-}OH + PCl_3 \longrightarrow 3R{-}\overset{\overset{O}{||}}{C}{-}Cl + H_3PO_3$$

$$R{-}\overset{\overset{O}{||}}{C}{-}OH + PCl_5 \longrightarrow R{-}\overset{\overset{O}{||}}{C}{-}Cl + POCl_3 + HCl$$

$$R{-}\overset{\overset{O}{||}}{C}{-}OH + \underset{\text{Thionyl chloride}}{SOCl_2} \longrightarrow R{-}\overset{\overset{O}{||}}{C}{-}Cl + SO_2 + HCl$$

Of all the reagents, thionyl chloride is the best and the mildest reagent for the preparation of acid halides because the products (SO_2 and HCl) being gaseous can be easily removed, leaving pure acid halide as the residue.

(C) Formation of acid anhydride: Carboxylic acids undergo intermolecular anhydride formation in the presence of a strong dehydrating agent like phosphorous pentoxide.

$$\underset{\text{Carboxylic acid molecules}}{R{-}\overset{\overset{O}{||}}{C}{-}OH + HO{-}\overset{\overset{O}{||}}{C}{-}R} \xrightarrow[-H_2O]{P_2O_5} \underset{\text{Acid anhydride}}{R{-}\overset{\overset{O}{||}}{C}{-}O{-}\overset{\overset{O}{||}}{C}{-}R}$$

Carboxylic acids also react with acid halides to form acid anhydrides.

$$R-\overset{O}{\overset{||}{C}}-ONa + Cl-\overset{O}{\overset{||}{C}}-R \xrightarrow{P_2O_5} R-\overset{O}{\overset{||}{C}}-O-\overset{O}{\overset{||}{C}}-R + NaCl$$

12-6-3 Reduction

Carboxylic acids are conveniently reduced to primary alcohols by reaction with lithium aluminium hydride in ether.

$$4R-\overset{O}{\overset{||}{C}}-OH + LiAlH_4 \longrightarrow 4R-CH_2OH$$

Acylchlorides when reduced with H_2 in the presence of Pd supported over $BaSO_4$ as catalyst, form aldehydes. This reaction is called as *Rosenmund reduction.*

For example;

$$H_3C-\overset{O}{\overset{||}{C}}-Cl \xrightarrow[Pd-BaSO_4]{H_2} H_3C-\overset{O}{\overset{||}{C}}-H + HCl$$

The function of $BaSO_4$ is to act as catalytic poison and reduce the activity of the catalyst to the extent that the reaction stops at the aldehyde stage.

Esters can be reduced to alcohols by a variety of reagents such as $LiAlH_4$, Na and ethanol or by catalytic reduction.

$$R-\overset{O}{\overset{||}{C}}-O-R \xrightarrow{LiAlH_4} \underset{1° \text{ Alcohol}}{R-CH_2OH} + ROH$$

$$\underset{\text{Methyl propionate}}{H_3C-CH_2-\overset{O}{\overset{||}{C}}-OCH_3} \xrightarrow{LiAlH_4} \underset{\text{n-Propyl alcohol}}{H_3C-CH_2CH_2OH} + CH_3OH$$

The reduction of esters to alcohols with sodium and alcohol is commonly known as *Bouealt-Blanc Reduction.* This reduction is particularly important because while converting an ester into alcohol, *it does not reduce ethylenic double bond if present in the ester molecule.*

12-6-4 Degradation of carboxylic acids.

(A) Hunsdiecker reaction: Dry silver salt of a carboxylic acid upon refluxing with bromine in carbon tetrachloride, affords the corresponding alkyl bromide wtih the liberation of carbon dioxide

$$RCO\overset{\ominus}{O}\,Ag^{+} + Br_2 \xrightarrow{CCl_4} RBr + CO_2 + AgBr$$

Mechanism: This reaction proceeds by a free radical mechanism.

$$\text{(i)}\quad RCO\overset{\ominus}{O}\,Ag+ \xrightarrow{Br_2} RCOOBr + AgBr$$

$$\text{(ii)}\quad RCOOBr \longrightarrow RCOO^{\bullet} + Br^{\bullet}$$

$$\text{(iii)}\quad RCOO^{\bullet} \longrightarrow R^{\bullet} + CO_2$$

$$\text{(iv)}\quad R^{\bullet} + RCOOBr \longrightarrow RBr + RCOO^{\bullet}$$

The yields of alkyl bromides can be considerably improved by adding bromine to fatty acid in the presence of mercuric oxide in carbon tetrachloride solution at reflux temperature.

$$2\,RCOOH + 2Br_2 + HgO \longrightarrow 2RBr + 2CO_2 + HgBr_2 + H_2O$$

(B) Schmidt reaction: When a solution of an aromatic carboxylic acid in conc. H_2SO_4 is treated with chloroform solution of hydrazoic acid (HN_3), an amine is produced

$$R{-}\overset{\overset{\displaystyle O}{||}}{C}{-}OH + N_3H \xrightarrow{\text{Conc. } H_2SO_4} R{-}NH_2 + N_2 + SO_2$$

(C) Arndt - Eistert synthesis : Acyl chlorides readily react with diazomethane to give diazoketones.

$$R{-}\overset{\overset{\displaystyle O}{||}}{C}{-}Cl + 2H_2\overset{\ominus}{C}{-}\overset{\oplus}{N}\equiv N \longrightarrow R{-}\overset{\overset{\displaystyle O}{||}}{C}{-}\overset{\ominus}{C}H{-}\overset{\oplus}{N}\equiv N + CH_3Cl + N_2$$

The diazoketone thus obtained on warming with water in presence of Ag_2O as catalyst, produces the next higher homologue of the parent acid of acyl chloride. This method of converting a carboxylic acid to its next higher homologue via the acyl chloride is referred to as *Arndt-Estert synthesis.*

$$\underset{\text{diazoketone}}{R{-}\overset{\overset{\displaystyle O}{||}}{C}{-}CHN_2} \xrightarrow[H_2O]{Ag_2O} R{-}CH_2{-}COOH$$

(D) Halogenation : Hell-Volhard-Zelinsky reaction: In the presence of a small amount of phosphorous, aliphatic carboxylic acids react smoothly with chlorine or bromine to yield a compound in which α-hydrogen has been replaced by halogen. Because of its specificity only *alpha* halogenation and the readiness with which it takes place, is of considerable importance in synthesis.

$$\underset{\text{Acetic acid}}{H_3CCOOH} \xrightarrow{Cl_2,\ P} ClCH_2COOH \xrightarrow{Cl_2,\ P} Cl_2CHCOOH \xrightarrow{Cl_2,\ P} \underset{\text{Trichloroacetic acid}}{Cl_3CCOOH}$$

12-6-5 Urea

Urea is the diamide of carbonic acid and is physiologically important. It is the final decomposition product of proteins and occurs in the urine. Its historical importance lies in its being the first organic compound synthesized in the laboratory in 1828 by Wohler.

(i) Wohler synthesis of urea : By evaporating a solution, containing potassium cyanate and ammonium sulphate, ammonium cyanate is formed which undergoes molecular rearrangement to give urea.

$$2\ KCNO + (NH_4)_2SO_4 \xrightarrow{\Delta} 2\ NH_4CNO \xrightarrow{\Delta} \underset{\text{Molecular rearrangement}}{H_2N{-}\overset{\overset{\displaystyle O}{||}}{C}{-}NH_2}$$

(ii) Laboratory method : Urea may be prepared in the laboratory by the action of ammonia on a carbonyl chloride

$$\underset{\text{Ammonia}}{\begin{matrix} H_2N{-}H \\ + \\ H_2N{-}H \end{matrix}} \quad \underset{\text{Carbonylchloride}}{\begin{matrix} Cl \\ \ \ \ \ \ \ \ \ C=O \\ Cl \end{matrix}} \longrightarrow \underset{\text{Urea}}{H_2N{-}\overset{\overset{\displaystyle O}{||}}{C}{-}NH_2}$$

(iii) Industrial manufacture :

(a) by partial hydrolysis of cyanamide in feeble acidic solution urea is obtained. Cyanamide is obtained from calcium cyanamide manufactured by passing nitrogen through heated calcium carbide.

$$CaC_2 \xrightarrow{N_2} CaNC \equiv N \xrightarrow{H_2SO_4} H_2NC \equiv N \xrightarrow{H_2O} \underset{\text{Urea}}{H_2NCOHN_2}$$

(b) Liquid carbon dioxide and liquid ammonia are reacted in an autoclave to form ammonium carbamate, the latter is now heated at 1300 - 1500°C under pressure of about 35 atmosphere to give urea.

Physical properties : Urea is a white crystalline solid, mp. 132°C, readily soluble in water and alcohol, but insoluble in ether.

Chemical properties : Urea responds to all characteristic reactions of an amide. However, in certain respects it behaves like amine also. Some of its important reactions are given below:

(i) Basic nature : Although an aqueous solution of urea is neutral to litmus, it behaves as a very weak mono acid base. It forms crystalline salts with strong acids.

$$H_2N—\overset{\overset{O}{||}}{C}—NH_2 + HNO_3 \longrightarrow \underset{\text{Urea nitrate}}{H_2N—\overset{\overset{O}{||}}{C}—NH_2 \cdot HNO_3}$$

$$2\underset{\text{Urea}}{H_2N—\overset{\overset{O}{||}}{C}—NH_2} + \underset{\text{Oxalic acid}}{H_2C_2O_4} \longrightarrow \underset{\text{Urea oxalate}}{(NH_2CONH_2)_2H_2C_2O_4}$$

(ii) Hydrolysis : When urea is boiled with dilute mineral acids or alkalies, urea undergoes hydrolysis forming carbon dioxide and ammonia.

$$H_2N—\overset{\overset{O}{||}}{C}—NH_2 + H_2O \xrightarrow[\text{alkali}]{\text{acid or}} CO_2 + 2\ NH_3$$

Enzyme urease, also brings about same change. This reaction is used for quantitative estimation of urea by measuring the amount of ammonia liberated.

Smell of ammonia, often observed in public urinals is due to hydrolysis of urea present in urine by micrococcus urease, a micro organism present in atmosphere.

(iii) Action of hydrazine : Urea reacts with hydrazine to produce semi carbazide.

$$H_2N{-}CO{-}NH_2 + H{-}NHNH_2 \longrightarrow H_2NCONHNH_2 + NH_3$$

(iv) Action of heat (Biuret reaction) : When heated, urea melts at 132°C, on further heating it loses ammonia at 150°C and forms a solid white residue known as *Biuret.*

$$H_2NCONH_2 + H_2NCONH_2 \longrightarrow H_2NCONHCONH_2 + NH_3$$

When an aqueous solution of biuret is treated with sodium hydroxide and a drop of very dilute copper sulphate solution, violet color is produced. This reaction is used for detecting the presence of peptide group $-CO-NH_2-$ in a given compound.

(v) Action of alkaline hypobromite : With alkaline solution of sodium hypobromite, urea is decomposed into gaseous nitrogen.

$$H_2NCONH_2 + 3NaOBr + 2NaOH \longrightarrow N_2 + Na_2CO_3$$

(vi) Acylation : Urea reacts with acid chlorides or acid anhydrides to form acyl derivatives known as *ureides.*

$$\underset{\text{Urea}}{H_2NCONH_2} + \underset{\text{Acetylchloride}}{ClCOCH_3} \longrightarrow \underset{\text{Acetyl urea}}{H_2NCONHCOCH_3} + HCl$$

(vii) Formation of cyclic ureides : Urea reacts with oxalic, malonic esters and ethylacetoacetate to form cyclic ureides.

$$O{=}C\begin{matrix}NH_2\\NH_2\end{matrix} + \begin{matrix}EtO{-}\overset{O}{\overset{\|}{C}}\\ |\\ EtO{-}\underset{O}{\underset{\|}{C}}\end{matrix} \longrightarrow \underset{\text{Parabanic acid}}{O{=}C\begin{matrix}\overset{H}{N}{-}C{=}O\\ |\\ \underset{H}{N}{-}C{=}O\end{matrix}}$$

$$O{=}C\begin{matrix}NH_2\\NH_2\end{matrix} + \begin{matrix}EtO{-}\overset{O}{\overset{\|}{C}}\\ \\ EtO{-}\underset{O}{\underset{\|}{C}}\end{matrix}\!\!>CH_2 \longrightarrow \underset{\substack{\text{Malonyl urea}\\ \text{(Barbutric acid)}}}{O{=}C\begin{matrix}\overset{H}{N}{-}\overset{O}{\overset{\|}{C}}\\ \\ \underset{H}{N}{-}\underset{O}{\underset{\|}{C}}\end{matrix}\!\!>CH_2}$$

(viii) Reaction with formaldehyde (Formation of urea-formaldehyde resins) : Urea and formaldehyde react to give urea formaldehyde resins, which are widely used in the manufacture of textiles, adhesives, plywood, fibers, etc.

Formaldehyde reacts with urea in the presence of a base or acid to form methylol urea and dimethylol urea.

$$H_2NCONH_2 + CH_2O \xrightarrow{\text{room temp.}} H_2NCONHCH_2OH$$

Monomethylol urea

$$H_2NCONH_2 + 2CH_2O \longrightarrow HOH_2C\ NH\ CONHCH_2OH$$

Dimethylol urea

Dimethylol urea may condense with urea molecules to form linear polymers or it may condense with formaldehyde molecules to form cross linked polymers or dimethylol urea may condense with each other to form linear polymers. The complex mixture of linear and cross-linked condensation products is known as *urea-formaldehyde resin.*

Uses of urea : It is used

(i) as a fertilizer

(ii) as a stabilizer for nitrocellulose explosives

(iii) in the manufacture of urea formaldehyde resins

(iv) in the manufacture of drugs such as urea stibamine (for treatment of kala-azar fever) and vernal (as valuable hypnotic).

(v) in the manufacture of hydrazine.

12-7 SELF TEST QUESTIONS

12-7-1 Which one of the following carboxylic acids is optically active.

(A) α-Hydroxy acetic acid (B) Mallic acid
(C) Mandelic acid (D) Salicylic acid

12-7-2 Increase in chain length of monocarboxylic acids can be done by

(A) Hunsdiecker reaction (B) Schmidt reaction
(C) Kiliani synthesis (D) Arndt-Eistert synthesis

12-7-3 The most acidic among the following is

(A) α-Chlorobutyric acid (B) β-Chlorobutyric acid
(C) γ-Chlorobutyric acid (D) Butyric acid

12-7-4 Carbonation of benzyl magnesium bromide followed by hydrolysis yields

(A) Benzoic acid (B) Benzylbromide
(C) Toluene (D) Phenylacetic acid

12-7-5 From which of the following urea is manufactured

(A) Calcium cyanamide (B) $CO + NH_3$

(C) Both (D) $HCO_2H + NH_3$

12-7-6 Methanoic acid and Ethanoic acid can be distinguished by

(A) aqueous $NaHCO_3$ (B) $FeCl_3$ test

(C) Tollens test (D) None

12-7-7 Which one of the following has highest K_a value

(A) *p*-Hydroxybenzoic acid (B) *p*-Nitrobenzoic acid

(C) *p*-Methoxybenzoic acid (D) *p*-Aminobenzoic acid

12-7-8 Benzoic acid is obtained by the alkaline hydrolysis of

(A) Benzylchloride (B) Benzalchloride

(C) Benzotrichloride (D) Phenylacetylchloride

12-7-9 The change takes place in Hell-Volhard-Zelinsky reaction is

(A) $CH_3COOH \longrightarrow CH_3CH_2COOH$

(B) $CH_3–CH_2–COOH \longrightarrow ClCH_2CH_2COOH$

(C) $CH_3–CH(Cl)–COOH \longrightarrow CH_3–CH_2COOH$

(D) $CH_3–CH_2–COOH \longrightarrow CH_3–CH(Cl)–COOH$

12-7-10 The product formed by reactiong salicylic acid with acetic anhydride in the presence of H^+ is

(A) Methyl benzoate (B) Acetyl salicyclic acid

(C) Phenylacetate (D) None

12-8 ESSAY AND SHORT TYPE QUESTIONS

12-8-1 How will you distinguish between

(a) Benzoic acid and salicylic acid

(b) Formic acid and acetic acid

12-8-2 Show all the steps, conditions and reagents required to carry out the following transformations.

(a) Benzoic acid $\longrightarrow$ Phenylacetic acid

(b) Propionic acid $\longrightarrow$ acetic acid

(c) Propanoic acid $\longrightarrow$ Propionic anhydride

(d) Ethylbenzene $\longrightarrow$ Benzoic acid

(e) Benzoyl chloride $\longrightarrow$ Benzaldehyde.

12-8-3 Give explanation for the following

(a) *p*-Nitrobenzoic acid is more acidic than *p*-Toluic acid

(b) Chloroacetic acid is less acidic than fluoroacetic acid

(c) Formic acid exhibits some of the aldehydic properties

12-8-4 Give any two methods of preparation of phenylacetic acid.

12-8-5 Write snort notes on
(a) Hunsdiecker reaction
(b) Hell-Volhard Zelinsky reaction

12-8-6 Give the reagents and equations for the conversion of acetic acid to acetamide. (O.U., Apr. 1992)

12-8-7 Give equations for the conversion of acetic acid into ethylacetate. (O.U., Oct. 1987)

12-8-8 Give the structure of the product of the following reaction.

$$CO_2 + CH_3MgCl \xrightarrow{H_3O^+}$$

(O.U., Apr. 1987)

12-8-9 Write the reaction for the following
(a) Methanol esterification with acetic acid
(b) Conversion of acetic acid into chloroacetic acid
(c) Schmidt reaction. (O.U., Oct. 1988), (N.U., Mar. 1995)

12-8-10 Write the equations for the reactions of acetic acid with
(a) Thionyl chloride (N.U., Mar. 1995), (N.U., Mar. 1996)
(b) Cl_2 in the presence of red P
(c) Hydrazoic acid (O.U., Mar. 1989)

12-8-11 How is acetic acid converted into
(a) Methane (b) Ethanal (c) Acetyl chloride (O.U., Mar. 1990)

12-8-12 Explain why chloroacetic is more acidic than acetic acid. (O.U., Mar. 1990)

12-8-13 Explain the following
(a) Propionic acid is weaker than acetic acid
(b) Formic acid reduces Tollens reagent while acetic acid does not.
(c) Chloroacetic acid is stronger than acetic acid. (K.U., Apr. 1990)

12-8-14 Write an account on the following
(a) Schmidt reaction
(b) Arndt-Eistert synthesis (O.U., Oct. 1990)

12-8-15 Write an account of the mechanism of
(a) Ester hydrolysis
(b) How is Urea prepared. (A.U., Jan. 1990), (N.U., Mar. 1995)

12-8-16 Outline the synthesis of α-chloro propionic acid from propionic acid by Hell-Volhard Zelinsky reaction explain why α-chloropropionic acid is more acidic than propionic acid. (O.U., Apr., 1991), (N.U., Mar. 1996)

ANSWERS

12-7-1 **C**	14-7-2 **D**	12-7-3 **A**	12-7-4 **D**	12-7-5 **C**	12-7-6 **C**
12-7-7 **B**	12-7-8 **C**	12-7-9 **D**	12-7-10 **B**		

13
CARBANIONS

13-1 INTRODUCTION

Carbanions are one of the most important intermediates in synthetic organic chemistry. The significance of carbanions is largely due to their ability to form carbon-carbon bonds with a variety of electrophilic reagents such as alkyl halides, aldehydes, ketones, esters etc.

13-2 STRUCTURE

In simple terms, a carbanion may be described as a carbon atom having three alkyl groups attached to it together with an unshared electron pair in an orbital. The central carbon atom in carbanions is believed to be sp^3 hybridized with the unshared electron pair occupying one apex of the tetrahedron. Carbanions would thus have pyramidal structures, similar to those of amines, oscillating between two invert - tetrahedrons.

In certain cases, the carbanion carbon atoms, instead of being bonded to three alkyl groups, may also be bonded to three hydrogens or to an sp carbon atom through a triple bond or to an sp^2 carbon atom or two sp^2 carbons and a hydrogen atom etc. Some examples of the carbanions in various structural environments are shown below;

H—C≡C:

Methyl anion | Acetylide anion | Phenyl anion | Cyclopentadienyl anion

13-3 FORMATION AND STABILITY

Carbanions may, theoretically, be formed by the loss of a hydrogen in the form of proton, attached to a tetravalent carbon atom. In principle, therefore, all compounds containing C-H bonds may be considered to be potential **Carbon-acids**. However, the extent of dissociation in solution of most carbon acids is quite small, being very much less than that of weak organic acids such as acetic acid (pK_a=4.8) or phenol (pK_a=9.9). The pK_a values of some hydrocarbon compounds are given below in Table 13-1.

Table 13-1 : pK_a values of some carbon - acids

S.No.	Compound	pK_a	S.No.	Compound	pK_a
1.	Methane	48	7.	Benzene	43
2.	Ethane	49	8.	Toluene	41
3.	Propane	51	9.	Diphenyl methane	34
4.	2-Methyl-propane	52	10.	Triphenyl methane	31.5
5.	Ethylene	44	11.	Acetylene	25
6.	Propene	40	12.	Cyclopentadiene	15

The extremely high pK_a value of hydrocarbons indicates the extremely weak acidity of its C-H bonds. By definition, a carbanion possesses an unshared pair of electrons and is therefore a base. When a hydrocarbon loses a proton, it is converted into its conjugate base i.e., **the carbanion.** Therefore, weaker the hydrocarbon acid, greater is the basic strength, and hence lower is the stability of the conjugate base, i.e. the carbanion. In general, it can be said, on the basis of relative acidity of hydrocarbons, that the stability of simple carbanions follows the order given below :

Methyl > Primary > Secondary > Tertiary

This order of stability of carbanions may be explained as follows;

According to the laws of physics and electrostatics, charge dispersal in a system leads to stability and charge concentration leads to instability. Alkyl groups are unusual when compared to hydrogens in that they are electron releasing. The electron donating alkyl group increases the already existing electron density on the central carbanion-carbon thereby decreasing its stability. The greater the number of alkyl groups, greater is the electron donation followed by a decrease in stability and hence the above order of stability of carbanions.

13-3-1 Stabilization of Carbanion

The carbanions mentioned above are relatively less stable ones. However, their stability can be increased by suitable structural modifications. For example, one method is the attachment of strongly electronegative elements to the carbanion carbon resulting in greater dispersal of negative charge from the central carbanion-carbon, by inductive and/or field effects. Thus, chloroform ($CHCl_3$) has a pK_a of 24 compared to that of 48 for methane. A second method is the attachment of strongly electron-withdrawing groups to the carbanion carbon which would help in dispersing the negative charge from the central carbon by a resonance interaction.

Table 13-2 : pK Values of some compounds containing electron withdrawing groups

S.No	Compound	pK_a	S.No.	Compound	pK_a
1.	$CHCl_3$	24	7.	CH_3NO_2	11
2.	CH_3COCH_3	20	8.	$CH_2(NO_2)_2$	4
3.	CH_3CHO	19	9.	$CH_3SO_2CH_3$	29
4.	CH_3COOCH_3	24	10.	$CH_3COCH_2COCH_3$	9
5.	CH_3CN	25	11.	$CH_3COCH_2COOC_2H_5$	11
6.	$CH_2(CN)_2$	12	12.	$CH_2(COOC_2H_5)_2$	13

Groups such as carbonyl, nitro, cyano etc., fall into this category. The pK_a values of some compounds containing electron-withdrawing groups are given in Table 13-2.

13-3-2 Enolate anions - Keto - enol Tautomerism :

A carbonyl group increases the acidity of C-H bonds on a carbon directly attached to it by many powers of ten over an inactivated carbon-hydrogen bond. Removal of an α - hydrogen from a carbonyl compound, in the form of proton, leaves behind a carbanion which is a resonance hybrid of two structures; one in which the negative charge is on the carbon and the other in which the negative charge is on the oxygen, the latter being called an **enolate anion.**

$$-\overset{\overset{H}{|}}{\underset{|}{C}}-\underset{\underset{O}{\|}}{C}- + \ddot{B} \longrightarrow \left[-\overset{\ominus}{\underset{|}{C}}-\underset{\underset{O}{\|}}{C}- \longleftrightarrow -\underset{|}{C}=\underset{\underset{O^{\ominus}}{|}}{C}- \right] + \overset{\oplus}{BH}$$

(Carbanion) (Enolate anion)

The anion can be reprotonated either at the carbon, to give back the original keto compound, or at the oxygen to give the enol. The two structures, keto form and enol form, are isomeric with each other and at the same time intercovertible. This phenomenon is known as **tautomerism or prototropic isomerism.**

$$-\overset{\ominus}{C}-\underset{\|}{\overset{}{C}}- + \overset{\oplus}{BH} \longrightarrow -\overset{H}{\overset{|}{C}}-\underset{\underset{O}{\|}}{C}- + :B$$

$$\updownarrow$$

$$-C=\underset{\underset{O^{\ominus}}{|}}{C}- + \overset{\oplus}{BH} \longrightarrow -C=\underset{\underset{OH}{|}}{C}- + :B$$

Acidic substances promote interconversion between keto and enol forms. Since enols are stronger acids (having pK_a in the range 9 to 14) than the corresponding ketone forms, strong bases convert carbonyl compounds into the corresponding enolate anions.

In some cases, it has been found possible by careful exclusion of acids and bases and by using low temperature conditions to isolate keto and enol forms but under ordinary conditions of room temperature, there are sufficient acids/bases present to cause rapid interconversion. There is usually an equilibrium between the two forms of a compound with distinct keto and enol contents. In the case of simple aliphatic and aromatic aldehydes and ketones, the compounds exist almost entirely in the keto form, whereas considerable amounts of enol tautomer are present in ß-diketones and ß-ketoesters. In these latter cases, the enol contains a conjugated p-electron system and an intramolecularly hydrogen bonded chelate system.

$$CH_3-\underset{\underset{O}{\|}}{C}-CH_2-\underset{\underset{O}{\|}}{C}-OEt \rightleftharpoons CH_3-\underset{\underset{OH}{|}}{C}=CH-\underset{\underset{O}{\|}}{C}-OEt$$

[Chelate structure: H,C(C=O···H–O–C(H₃C)) with OEt]

Compounds usually show a tendency to exist in the ketonic form since the latter is believed to have lower enthalpy content, entropy differences being not very significant. Compounds start showing considerable enolic contents when the above factor is largely outweighed by energy gains due to hydrogen-bonding and/or resonance. Acetone exists almost entirely in the ketonic form whereas phenol exists entirely in the enol form, since in the case of phenol, the alternative keto structure would sacrifice the stability associated with the sixπ-electron aromatic system. However, in polyhydroxy phenols (like catechol for example), it has been possible in a number of cases to obtain keto forms and the keto form predominates in heterocyclic systems such as pyridones and quinolones.

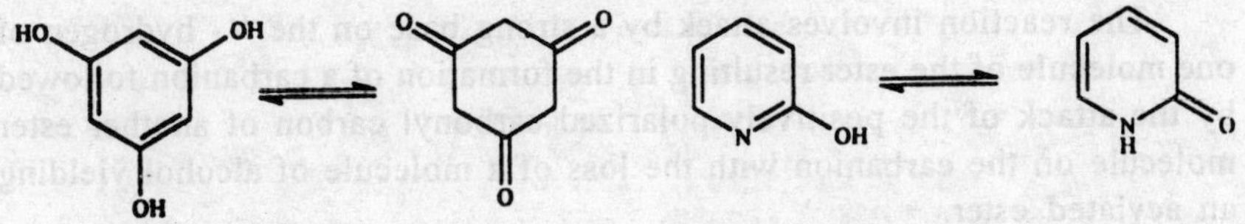

The enol contents of some representative compounds are shown in the Table 13-3.

Table 13-3 : The enol contents of some representative compounds

S.No.	Name	Structure	Enol content (%)
1.	Acetone	CH_3-CO-CH_3	0.0001
2.	Acetaldehyde	CH_3-CO-H	0.00001
3.	Acetophenone	Ph-CO-CH_3	0.04
4.	Ethyl acetoacetate	CH_3-CO-CH_2-$COOC_2H_5$	8.2
5.	Diethyl Malonate	$CH_2(COOC_2H_5)_2$	1.0
6.	2-Pyridone	HN-CH=CH-CH=CH-CO	0.11
7.	Acetylacetone	CH_3-CO-CH_2CO-CH_3	76
8.	Dibenzoylmethane	Ph-CO-CH_2-CO-Ph	100
9.	Phenol	Ph-OH	100

It may be mentioned here that several reactions of carbonyl compounds having one or more α hydrogens occur through the enol form. Reaction of ketones with chlorine, bromine and iodine result in the substitution of α-hydrogens by the respective halogen. Rates of these reactions are characteristically first order with respect to ketone and independent of halogen concentration and irrespective of which halogen is used.

13-4 CLAISEN CONDENSATION

This is one of the most important and synthetically useful reactions of esters. In this reaction, two molecules of an ester condense in the presence of a strong base such as sodium amide, sodium hydride or sodium ethoxide or sodium metal to form an α-acylated ester.

$$R{-}CH_2{-}\underset{\underset{O}{||}}{C}{-}OR' + R{-}CH_2{-}\underset{\underset{O}{||}}{C}{-}OR' \xrightarrow{\text{Base}} R{-}\underset{\underset{\underset{\underset{CH_2-R}{|}}{C=O}}{|}}{CH}{-}\overset{\overset{O}{||}}{C}{-}OR' + R'{-}OH$$

The reaction involves attack by a strong base on the α- hydrogen of one molecule of the ester resulting in the formation of a carbanion followed by the attack of the positively polarized carbonyl carbon of another ester molecule on the carbanion with the loss of a molecule of alcohol yielding an acylated ester.

$$R{-}CH_2{-}\underset{\underset{O}{||}}{C}{-}OR' + :B \longrightarrow R{-}\overset{\ominus}{C}H{-}\underset{\underset{O}{||}}{C}{-}OR' + \overset{\oplus}{B}H$$

$$R{-}\overset{\ominus}{C}H{-}\underset{\underset{O}{||}}{C}{-}OR' + R{-}CH_2{-}\overset{\oplus}{\underset{\underset{O^{\ominus}}{|}}{C}}{-}OR' \longrightarrow R{-}\underset{\underset{R'O\ \ \underset{CH_2-R}{C}\ \ O^{\ominus}}{|}}{CH}{-}\overset{\overset{O}{||}}{C}{-}OR'$$

$$\longrightarrow R{-}\underset{\underset{\underset{\underset{CH_2-R}{|}}{C=O}}{|}}{CH}{-}\overset{\overset{O}{||}}{C}{-}OR' + R'{-}\overset{\ominus}{O}$$

$$R'{-}\overset{\ominus}{O} + \overset{\oplus}{B}H \longrightarrow R'{-}OH + :B$$

When the above reaction is carried out using ethyl acetate as the ester, sodium ethoxide usually being the base, the product obtained is the ethyl acetoacetate.

13-4-1 Crossed Claisen condensation

The Claisen condensation is a reaction which involves self condensation between two molecules of the same ester in the presence of a base. However, when the reaction is carried out between the molecules of two different esters, it is known as **crossed Claisen condensation**. It is advantageous in reactions of the latter type to have one of the esters without an α-hydrogen to avoid complication by multiplicity of the products. The following examples will illustrate this point clearly:

$$\underset{\text{(Methyl benzoate)}}{Ph-\overset{O}{\overset{||}{C}}-OCH_3} + \underset{\text{(Methyl acetate)}}{H_3C-\overset{O}{\overset{||}{C}}-OCH_3} \xrightarrow{\overset{\ominus}{O}CH_3} \underset{\text{(Methyl benzoyloacetate)}}{Ph-\overset{O}{\overset{||}{C}}-CH_2-\overset{O}{\overset{||}{C}}-OCH_3} + CH_3-OH$$

$$\underset{\text{(Ethyl formate)}}{H-\overset{O}{\overset{||}{C}}-OC_2H_5} + \underset{\text{(Ethyl acetate)}}{H_3C-\overset{O}{\overset{||}{C}}-OC_2H_5} \xrightarrow{\overset{\ominus}{O}C_2H_5} \underset{\text{(Ethyl formyloacetate)}}{H-\overset{O}{\overset{||}{C}}-CH_2-\overset{O}{\overset{||}{C}}-OC_2H_5} + C_2H_5OH$$

$$\underset{\text{(Ethyl carbonate)}}{H_5C_2-O-\overset{O}{\overset{||}{C}}-OC_2H_5} + \underset{\text{(Ethyl acetate)}}{H_3C-\overset{O}{\overset{||}{C}}-OC_2H_5} \xrightarrow{\overset{\ominus}{O}C_2H_5} \underset{\text{(Ethyl malonate)}}{H_5C_2O-\overset{O}{\overset{||}{C}}-CH_2-\overset{O}{\overset{||}{C}}-OC_2H_5}$$

$$\underset{\text{(Ethyl oxalate)}}{\begin{matrix}COOC_2H_5\\|\\COOC_2H_5\end{matrix}} + \underset{\text{(Ethyl phenylacetate)}}{Ph-CH_2-\overset{O}{\overset{||}{C}}-OC_2H_5} \xrightarrow{\overset{\ominus}{O}C_2H_5} \underset{\text{(Ethyl oxalophenylacetate)}}{H_5C_2OOC-\overset{O}{\overset{||}{C}}-\underset{Ph}{\underset{|}{CH}}-COOC_2H_5}$$

13-4-2 Dieckmann condensation

The self-condensation of a diester in the presence of a strong base, involving reaction between the carbanion alpha to an ester grouping and the carbonyl carbon of second ester grouping, is known as **Dieckmann condensation**.Thus Dieckmann condensation is an **intramolecular Claisen condensation.** For example,

$$CH_2(CH_2-COOC_2H_5)(CH_2-CH_2-COOC_2H_5) \xrightarrow{\overset{\ominus}{O}C_2H_5} \left[H_2C(CH_2-COOC_2H_5)(CH_2-\overset{\ominus}{C}H-COOC_2H_5) \right] \xrightarrow{-C_2H_5OH} \text{2-oxocyclopentane-}COOC_2H_5$$

13-5 ACYLATED ESTERS

The general structural formula for an acylated ester may be written as follows :

$$R-\underset{\underset{O}{\|}}{C}-C(R^1R^2)-\underset{\underset{O}{\|}}{C}-OR^3$$

In the above formula, R^3 is usually methyl or ethyl group and R^1 and R^2 are hydrogens. R may be alkyl or aryl group. For naming purpose, these compounds may be regarded as derivatives of ester, the alkyl/aryl group also being treated as a substituent on the α - carbon of the ester functionality. Table 13-4 gives a few examples of nomenclature system for acylated esters:

Table 13-4 : Example of nomenclature system for acylated esters

S.No	Structure	Name of the compound
1.	CH_3-CO-CH_2-$COOC_2H_5$	Ethyl α-acetoacetate
2.	CH_3-CO-$C(CH_3)_2$-$COOC_2H_5$	Ethyl α,α-dimethylacetoacetate
3.	Ph-CO-CH_2-$COOCH_3$	Methyl α-benzoyloacetate
4.	H-CO-$C(C_2H_5)_2$-$COOCH_3$	Methyl α,α-diethyl formyloacetate
5.	Ph-CO-CH(Pr^i)-$COOC_2H_5$	Ethyl α-isopropylbenzoyloacetate.

Very often the acyl group in these compounds is the acetyl group. Hence, the acylated esters are referred to very frequently as **acetoacetic esters**. These acetoacetic esters exhibit the phenomenon of keto-enol tautomerism in that they will undergo reactions characteristic of both the ketone and enol functionalities. For example, a sample of an acetoacetic ester will form an oxime, a bisulfite addition compound, a cyanohydrin etc., properties typical of ketones. That, these reactions are not due to the ester C=O group, is proved by the fact that simple esters do not exhibit these ketonic properties. On the other hand, a sample of the same acetoacetic ester will add halogens, will give deep coloration with neutral ferric chloride, will dissolve in alkali etc., a behavior attributed to enols.

The separate keto and enol forms of an acetoacetic ester were first isolated by Ludwig Knorr in the year 1911. An ethereal solution of ethyl acetoacetate was cooled to -78°C and then evaporated under reduced pressure to yield a crystalline solid which was proved to be the keto form of the ß-ketoester. In ether solvent, essentially only the keto form existed. In another experiment, the sodium salt of ethyl acetoacetate in light petrol was cooled to -78°C and then treated with just enough hydrogen chloride to yield the liquid enol. The tautomers thus obtained were found to be stable for long periods when kept in quartz vessels at -78°C but at room temperature they rapidly change to equilibrate with one another.

13-6-1 Acetoacetic ester synthesis

The acetoacetic esters serve as significant intermediates in synthetic organic chemistry. They are used in the syntheses of several substituted ketones and substituted acetic acids which are not, otherwise, easily preparable. The use of acetoacetic esters in these reactions depends on two important chemical properties, namely

(i) Alkylation and

(ii) Hydrolysis.

These properties are discussed below in detail. The reactions are perfectly general although most often they are carried out with ethyl acetoacetate.

(i) Alkylation : When treated with a strong base such as sodium hydride, sodium ethoxide or sodium metal, the acetoacetic ester forms the corresponding sodium salt. The latter is a resonance hybrid of two anionic forms and reacts readily with primary / secondary alkyl halides to form alkyl derivatives of acetoacetic ester in which the alkyl group is attached to carbon.

$$H_3C\text{-}\underset{\underset{O}{\|}}{C}\text{-}CH_2\text{-}\underset{\underset{O}{\|}}{C}\text{-}OC_2H_5 \xrightarrow{\overset{\ominus}{O}C_2H_5} H_3C\text{-}\underset{\underset{O^{\ominus}}{|}}{C} = CH\text{-}\underset{\underset{O}{\|}}{C}\text{-}OC_2H_5 + C_2H_5OH$$

$$H_3C\text{-}\underset{\underset{O^{\ominus}}{|}}{C} = CH\text{-}\underset{\underset{O}{\|}}{C}\text{-}OC_2H_5 \rightleftharpoons H_3C\text{-}\underset{\underset{O}{\|}}{C}\text{-}\overset{\ominus}{C}H\text{-}\underset{\underset{O}{\|}}{C}\text{-}OC_2H_5$$

$$H_3C\text{-}\underset{\underset{O}{\|}}{C}\text{-}\overset{\ominus}{C}H\text{-}\underset{\underset{O}{\|}}{C}\text{-}OC_2H_5 + R'\text{–}X \longrightarrow H_3C\text{-}\underset{\underset{O}{\|}}{C}\text{-}CH(R')\text{-}\underset{\underset{O}{\|}}{C}\text{-}OC_2H_5 + X^{\ominus}$$

The reaction of the alkyl halides with the anion occurs at the carbon (ie. in the carbanionic form) rather than at the oxygen (ie. in the enolate anionic form). This is due to the fact that the positively charged/polarized carbon of the alkyl halides constitutes a soft acidic center, the carbon of the anion-the soft basic center, whereas the oxygen of the anion the hard basic center. Since **hard-hard or soft-soft interactions** would be favored in a given reaction system, the alkyl halide preferentially attacks the carbanion rather than the enolate anion.

After the introduction of one alkyl group, the entire process may be repeated to yield the dialkyl derivative of the acetoacetic ester;

$$H_3C\text{-}\underset{\|}{\overset{}{C}}\text{-}CH(R^1)\text{-}\underset{\|}{C}\text{-}OC_2H_5 \xrightarrow{NaOC_2H_5} [H_3C\text{-}\underset{\|}{C}\text{-}\overset{\ominus}{C}R^1 - COOC_2H_5]\ \overset{+}{Na}$$
(with =O on the marked carbons)

$$[H_3C\text{-}\underset{\|}{C}\text{-}\overset{\ominus}{C}(R^1)\text{-}COOC_2H_5]\ \overset{+}{Na} \xrightarrow{R^2\text{-}X} H_3C\text{-}\underset{\|}{C}\text{-}C(R^1R^2)\text{-}COOC_2H_5 + NaX$$
(with =O on the marked carbons)

(ii) **Hydorlysis** : The substituted acetoacetic ester can be suitably hydrolysed to obtain the desired products. This is usually done in two ways:

(a) ***Ketonic hydrolysis*** : It is so-called because a ketone is the main product in this reaction and is carried out by boiling the substituted acetoacetic ester with dilute aqueous alcoholic KOH or with dilute acid.

$$H_3C\text{-}CO\text{-}CR^1R^2\text{-}COOC_2H_5 + H_2O \xrightarrow{\bar{O}H/\overset{+}{H}} H_3C\text{-}CO\text{-}CHR^1R^2 + CO_2 + C_2H_5OH$$

The first step in the above reaction is a simple hydrolysis of the ester group yielding the unstable acetoacetic acid which loses CO_2 under the reaction conditions resulting in the ketone. Thus, the hydrolysis sequence can be formulated as follows;

$$H_3C\text{-}CO\text{-}CR^1R^2\text{-}\underset{\underset{O}{\|}}{C}\text{-}OC_2H_5 + H_2O \longrightarrow [H_3C\text{-}CO\text{-}CR^1R^2\text{-}\underset{\underset{O}{\|}}{C}\text{-}OH] + C_2H_5\text{-}OH$$

$$H_3C\text{-}\underset{}{\overset{\overset{O}{\|}}{C}}\text{-}C(R^1R^2)\text{-}\overset{\overset{O}{|}}{C}\text{-}O\text{-}H \longrightarrow H_3C\text{-}\overset{\overset{OH}{|}}{C}{=}C(R^1R^2) + CO_2 \longrightarrow H_3C\text{-}\overset{\overset{O}{\|}}{C}\text{-}CHR^1R^2$$

A large number of substituted ketones can be prepared in the above reaction by varying the structures of R^1 and R^2.

(b) ***Acid Hydrolysis*** **:** It is so-called because an acid is the main product in this reaction and is carried out by boiling the substituted acetoacetic ester with conc. alcoholic KOH.

$$H_3C\text{-}\underset{\parallel}{\underset{O}{CO}}\text{-}CR^1R^2\text{-}\underset{\parallel}{\underset{O}{C}}\text{-}OC_2H_5 + 2\bar{O}H \longrightarrow H_3C\text{-}\underset{\parallel}{\underset{O}{C}}\text{-}O^{\ominus} + CHR^1R^2\text{-}\underset{\parallel}{\underset{O}{C}}\text{-}O^{\ominus} + C_2H_5OH$$

The first step in the above reaction is the reductive cleavage of the $\text{-}CO\text{-}CR^1R^2\text{-}$ bond followed by hydrolysis of the resulting ester to yield the substituted acid salt.

$$H_3C\text{-}\underset{\parallel}{\underset{O}{C}}\text{-}CR^1R^2\text{-}\underset{\parallel}{\underset{O}{C}}\text{-}OC_2H_5 \xrightarrow{\bar{O}H}$$

$$[H_3C\text{-}\overset{OH}{\overset{|}{\underset{|}{\underset{O^{\ominus}}{C}}}}\text{-}CR^1R^2\text{-}\underset{\parallel}{\underset{O}{C}}\text{-}OC_2H_5] \longrightarrow [H_3C\text{-}\underset{\parallel}{\underset{O}{C}}\text{-}OH + \overset{\ominus}{C}R^1R^2\text{-}\underset{\parallel}{\underset{O}{C}}\text{-}OC_2H_5]$$

$$\longrightarrow H_3C\text{-}\underset{\parallel}{\underset{O}{C}}\text{-}O^{\ominus} + HCR^1R^2\text{-}\underset{\parallel}{\underset{O}{C}}\text{-}OC_2H_5$$

$$HCR^1R^2\text{-}\underset{\parallel}{\underset{O}{C}}\text{-}OC_2H_5 + \overset{\ominus}{O}H \longrightarrow [CHR^1R^2\text{-}\underset{\parallel}{\underset{O}{C}}\text{-}OH + \overset{\ominus}{O}C_2H_5]$$

$$[CHR^1R^2\text{-}\underset{\parallel}{\underset{O}{C}}\text{-}OH + \overset{\ominus}{O}C_2H_5] \longrightarrow CHR^1R^2\text{-}\underset{\parallel}{\underset{O}{C}}\text{-}O^{\ominus} + C_2H_5\text{-}OH$$

The free acid is generated from the acid salt by treatment with dilute mineral acid.

Examples of the Ketone synthesis :

(i) ***1-Benzyl -1-Methyl -2-Propanone*** **:** In planning a synthesis of this compound, the structure of the ketone, to be prepared, is first written down. Since ethyl acetoacetate is the β -Ketoester most commonly employed in these type of syntheses, the portion corresponding to $CH_3CO\text{-}CH$ in the ketone is first identified; then, the appropriate alkyl groups are added to the

α^1 - carbon followed by attachment of the carbethoxy group. The resulting dialkyl substituted β-ketoester may be considered to be the target of synthesis whose hydrolysis will yield the desired ketone. Thus,

$$\underbrace{\overset{\alpha\quad\alpha}{CH_3\text{-}\underset{\underset{O}{||}}{C}\text{-}CH\text{-}}}(CH_3)(\text{-}CH_2Ph) \Longleftarrow H_3C\text{-}\underset{\underset{O}{||}}{C}\text{-}\overset{\overset{CH_3}{|}}{\underset{\underset{CH_2Ph}{|}}{C}}\text{-}COOC_2H_5$$

1-Benzyl-1-methyl-2-propanone (a-disubstituted ketone)

(Ethyl-a-benzyl-b-methyl- (a-acetoacetate)

Then, the synthetic sequence may be formulated as shown below:

$$H_3C\text{-}\underset{\underset{O}{||}}{C}\text{-}CH_2\text{-}COOC_2H_5 \xrightarrow[\text{then } H_3O^+]{CH_3I/\bar{O}C_2H_5} H_3C\text{-}\underset{\underset{O}{||}}{C}\text{-}CH(CH_3)\text{-}COOC_2H_5$$

$$H_3C\text{-}\underset{\underset{O}{||}}{C}\text{-}CH(CH_3)\text{-}COOC_2H_5 \xrightarrow[\text{then } H_3O^+]{PhCH_2Cl/\bar{O}C_2H_5} H_3C\text{-}\underset{\underset{O}{||}}{C}\text{-}C(CH_3)(CH_2Ph)\text{-}COOC_2H_5$$

$$H_3C\text{-}\underset{\underset{O}{||}}{C}\text{-}C(CH_3)(CH_2Ph)\text{-}COOC_2H_5 \xrightarrow{\bar{O}H\ /\ \Delta} [H_3C\text{-}\underset{\underset{O}{||}}{C}\text{-}C(CH_3)(CH_2Ph)\text{-}COOH]$$

$$[H_3C\text{-}\underset{\underset{O}{||}}{C}\text{-}C(CH_3)(CH_2Ph)\text{-}COOH] \longrightarrow H_3C\text{-}\underset{\underset{O}{||}}{C}\text{-}CH(CH_3)(CH_2Ph) + CO_2$$

(ii) ***3-Ethyl-hexan-2-one*** : Analogous to the above case,

$$\overset{\alpha}{CH_3\text{-}\underset{\underset{O}{||}}{C}\text{-}CH\text{-}}\ \overset{\alpha'}{(\text{-}C_2H_5)(\text{-}CH_2CH_2CH_3)} \Longleftarrow H_3C\text{-}\underset{\underset{O}{||}}{C}\text{-}\overset{\overset{C_2H_5}{|}}{\underset{\underset{CH_2CH_2CH_3}{|}}{C}}\text{-}COOC_2H_5$$

3-Ethylhexane-2-one (α,αDisubstituted ketone)

Ethyl α-ethyl-α-n-propyl-β-acetoacetate

This ketone may be synthesized as follows:

$$\underset{\underset{O}{||}}{H_3C\text{-}C}\text{-}CH_2\text{-}COOC_2H_5 \xrightarrow[\text{then } H_3O^+]{C_2H_5I/\bar{O}C_2H_5} \underset{\underset{O}{||}}{H_3C\text{-}C}\text{-}CH(C_2H_5)\text{-}COOC_2H_5$$

$$\underset{\underset{O}{||}}{H_3C\text{-}C}\text{-}CH(C_2H_5)\text{-}COOC_2H_5 \xrightarrow[\text{then } H_3O^+]{CH_3CH_2CH_2\text{-}Cl/\bar{O}C_2H_5} \underset{\underset{O}{||}}{H_3C\text{-}C}\text{-}\underset{\underset{(CH_2CH_2CH_3)}{|}}{C}(\text{-}C_2H_5)COOC_2H_5$$

$$\underset{\underset{O}{||}}{H_3C\text{-}C}\text{-}C(\text{-}C_2H_5)(\text{-}CH_2CH_2CH_3)\text{-}COOC_2H_5 \xrightarrow{H^+/\Delta} [\underset{\underset{O}{||}}{H_3C\text{-}C}\text{-}\underset{\underset{(CH_2CH_2CH_3)}{|}}{C}(\text{-}C_2H_5)\ COOH]$$

$$[\underset{\underset{O}{||}}{H_3C\text{-}C}\text{-}C(\text{-}C_2H_5)(\text{-}CH_2CH_2CH_3)\text{-}COOH] \longrightarrow \underset{\underset{O}{||}}{H_3C\text{-}C}\text{-}\underset{\underset{(CH_2CH_2CH_3)}{|}}{CH}(\text{-}C_2H_5) + CO_2$$

Examples of the acid synthesis:

(i) ***2-Methylbutanoic acid*** : In planning a synthesis of mono carboxylic acids, the approach is very similar to that of ketone synthesis. First, the structure of the monocarboxylic acid, to be prepared, is written down. The portion of this acid corresponding to -C-$COOC_2H_5$ of the ß-ketoester is identified. Then, the alkyl groups are attached to the α-carbon of ethyl acetoacetate whose cleavage with hot concentrated alkali will yield the substituted acetic acid as desired.

$$H_3C\text{-}CH_2\text{-}\boxed{\overset{\overset{CH_3}{|}}{CH}\text{-}COOH} \Longleftarrow \underset{\underset{O}{||}}{H_3C\text{-}C}\text{-}\underset{\underset{CH_2CH_3}{|}}{\overset{\overset{CH_3}{|}}{C}}\text{-}COOC_2H_5$$

2-Methylbutanoic acid — Ethyl α-Methyl-α-ethylacetoacetate

Thus, the synthesis may be formulated as shown below;

$$\underset{\underset{O}{||}}{H_3C\text{-}C}\text{-}CH_2\text{-}COOC_2H_5 \xrightarrow[\text{then } H_3O^+]{CH_3I/\bar{O}C_2H_5} \underset{\underset{O}{||}}{H_3C\text{-}C}\text{-}CH(\text{-}CH_3)\text{-}COOC_2H_5$$

$$H_3C\text{-}\underset{\underset{O}{||}}{C}\text{-}CH(\text{-}CH_3)\text{-}COOC_2H_5 \xrightarrow[\text{then } H_3O^+]{C_2H_5I/\bar{O}C_2H_5} H_3C\text{-}\underset{\underset{O}{||}}{C}\text{-}C(\text{-}CH_3)(\text{-}C_2H_5)\text{-}COOC_2H_5$$

$$H_3C\text{-}\overset{\overset{O}{||}}{C}\text{-}C(\text{-}CH_3)(\text{-}C_2H_5)\text{-}COOC_2H_5 \xrightarrow{\bar{O}H} H_3C\text{-}\underset{\underset{O}{||}}{C}\text{-}OH + \bar{C}(\text{-}CH_3)(\text{-}C_2H_5)\text{-}COOC_2H_5$$

$$H_3C\text{-}\underset{\underset{O}{||}}{C}\text{-}OH + \overset{\ominus}{C}(\text{-}CH_3)(\text{-}C_2H_5)\text{-}COOC_2H_5 \longrightarrow H_3C\text{-}\underset{\underset{O}{||}}{C}\text{-}\bar{O} + HC(\text{-}CH_3)\underset{\underset{C_2H_5}{|}}{}\text{-}COOC_2H_5$$

$$HC(\text{-}CH_3)(\text{-}C_2H_5)\text{-}COOC_2H_5 \xrightarrow{\bar{O}H} [HC(\text{-}CH_3)(\text{-}C_2H_5)\text{-}COOH + C_2H_5\bar{O}]$$

$$HC(\text{-}CH_3)(\text{-}C_2H_5)\text{-}COOH + \bar{O}C_2H_5 \longrightarrow HC(\text{-}CH_3)(\text{-}C_2H_5)\text{-}COO^- + C_2H_5OH$$

$$H_5C_2\text{-}\overset{\overset{CH_3}{|}}{CH}\text{-}COO^- + H_3C\text{-}\underset{\underset{O}{||}}{C}\text{-}O^- \xrightarrow[\text{(in the end)}]{H_3O^{\oplus}} H_3C\text{-}CH_2\text{-}\overset{\overset{CH_3}{|}}{CH}\text{-}COOH + H_3C\text{-}\underset{\underset{O}{||}}{C}\text{-}OH$$

(ii) <u>2-Ethylpentanoic acid</u> : Analogous to the above case,

$$CH_3\text{-}CH_2\text{-}CH_2\text{-}\boxed{\overset{\overset{CH_2CH_3}{|}}{CH}\text{-}COOH} \Longleftarrow H_3C\text{-}\underset{\underset{O}{||}}{C}\text{-}\overset{\overset{CH_2CH_3}{|}}{\underset{\underset{CH_2CH_2CH_3}{|}}{C}}\text{-}COOC_2H_5$$

This acid may be synthesized as follows;

$$H_3C\text{-}\underset{\underset{O}{||}}{C}\text{-}CH_2\text{-}COOC_2H_5 \xrightarrow{C_2H_5I/\bar{O}C_2H_5} H_3C\text{-}\underset{\underset{O}{||}}{C}\text{-}CH(\text{-}C_2H_5)\text{-}COOC_2H_5$$

$$H_3C\text{-}\underset{\underset{O}{||}}{C}\text{-}CH(\text{-}C_2H_5)\text{-}COOC_2H_5 \xrightarrow{CH_3CH_2CH_2Cl/\bar{O}C_2H_5}$$

$$[H_3C\text{-}\underset{\underset{O}{||}}{C}\text{-}C(\text{-}C_2H_5)(\text{-}CH_2CH_2CH_3)\text{-}COOC_2H_5]$$

$$H_3C\text{-}\underset{\underset{O}{||}}{C}\text{-}C(\text{-}C_2H_5)(\text{-}CH_2CH_2CH_3)\text{-}COOC_2H_5 \xrightarrow{\overset{\ominus}{O}H} [H_3C\text{-}\underset{\underset{O}{||}}{C}\text{-}OH + \bar{C}(\text{-}C_2H_5)(\text{-}C_3H_7)COOC_2H_5]$$

$$H_3C\text{-}\underset{\underset{O}{||}}{C}\text{-}OH + \bar{C}(\text{-}C_2H_5)(\text{-}C_3H_7)\text{-}COOC_2H_5 \longrightarrow H_3C\text{-}\underset{\underset{O}{||}}{C}\text{-}\bar{O} + HC(\text{-}C_2H_5)(\text{-}C_3H_7)\text{-}COOC_2H_5$$

$$HC(\text{-}C_2H_5)(\text{-}C_3H_7)\text{-}COOC_2H_5 \longrightarrow [HC(\text{-}C_2H_5)(\text{-}C_3H_7)\text{-}COOH + \bar{O}C_2H_5]$$

$$[HC(\text{-}C_2H_5)(\text{-}C_3H_7)\text{-}COOH + \bar{O}C_2H_5] \longrightarrow HC(\text{-}C_2H_5)(\text{-}C_3H_7)\text{-}CO\bar{O} + C_2H_5OH$$

$$HC(\text{-}C_2H_5)(\text{-}C_3H_7)\text{-}CO\bar{O} + H_3C\text{-}\underset{\underset{O}{||}}{C}\text{-}\bar{O} \xrightarrow[\text{(in the end)}]{H_3O^+} H\text{-}\underset{\underset{CH_2CH_2CH_3}{|}}{\overset{\overset{C_2H_5}{|}}{C}}\text{-}COOH + H_3C\text{-}\underset{\underset{O}{||}}{C}\text{-}OH$$

In the above reactions of **acid hydrolysis**, the substituted acetic acids formed are usually water-insoluble and so can be readily isolated on acid work-up. However the yields of substituted acetic acids in these reactions are sometimes very low due to **ketonic hydrolysis** occurring simultaneously. Thus, in the preparation of acids, ketone is formed as a bye-product. For this reason, malonic ester synthesis. (discussed in the next section) is used to synthesize acids in which the yields are always higher.

13-6-2 Malonic ester synthesis

Malonic acid is a dicarboxylic acid with the structural formula $CH_2(COOH)_2$. Its di-ethyl ester is commonly referred to as malonic ester. It is prepared from chloroacetic acid by treatment with excess KCN followed by boiling with ethanolic HCl.

$$\underset{\text{(Chloroacetic acid)}}{Cl\text{-}CH_2\text{-}COOH} \xrightarrow{KCN} \underset{\text{(Potassium cyanoacetate)}}{NC\text{-}CH_2\text{-}CO\bar{O}\ \overset{+}{K}} \xrightarrow[HCl]{EtOH/} \underset{\text{(Ethyl malonate)}}{CH_2\begin{matrix} \diagup COOC_2H_5 \\ \diagdown COOC_2H_5 \end{matrix}}$$

Malonic ester is a pleasant smelling, clear, colorless liquid with a boiling point of 199°C. It is a significant intermediate for the syntheses of substituted acetic acids. The synthetic use of malonic ester depends on its two important chemical properties, namely

(i) Alkylation and

(ii) Hydrolysis.

(i) ALKYLATION : Diethyl malonate is a reasonably acidic compound with a pK_a of around 13.0 (see Table 13-2). On treatment with a strong base, such as sodium hydride, sodium ethoxide or sodium metal, malonic ester is transformed completely to the corresponding sodium salt. The latter is a nucleophile, combines readily with primary/secondary alkyl halides to form alkyl derivatives of malonic ester in which the alkyl group is attached to carbon.

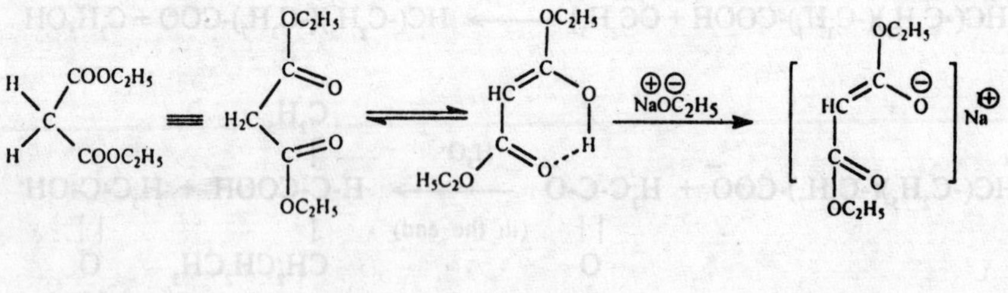

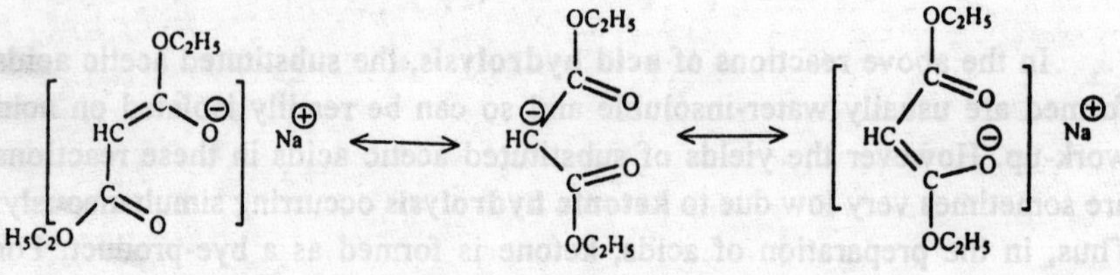

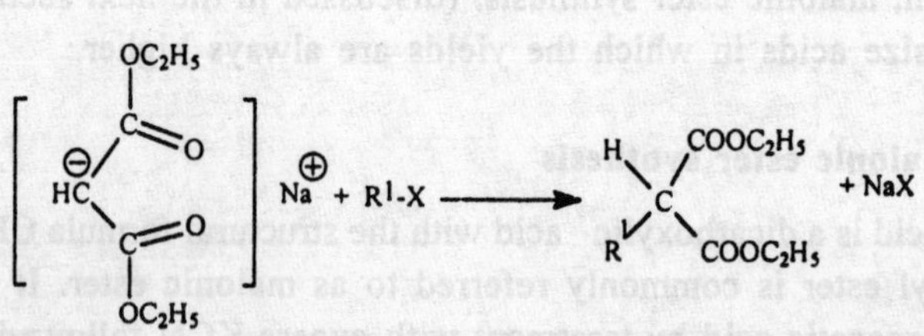

$$H_2C(COOC_2H_5)_2 \rightleftharpoons HC(OC_2H_5)C{=}C(OH)OC_2H_5 \xrightarrow{\overset{\oplus\ominus}{NaOC_2H_5}} [\overset{\ominus}{C}H(COOC_2H_5)_2]\,\overset{\oplus}{Na}$$

$$[\overset{\ominus}{C}H(COOC_2H_5)_2]\,\overset{\oplus}{Na} + R^1\text{-}X \longrightarrow R\text{-}CH(COOC_2H_5)_2 + NaX$$

The sodium derivatives of malonic esters are ambident nucleophiles and reaction with alkyl halides can occur either at carbon or at oxygen. However, for reasons given in the section on acetoacetic ester synthesis above, reaction occurs preferentially at carbon. The whole process of alkylation can be repeated to yield di-alkyl derivatives of malonic ester.

$$\begin{array}{c} H \\ R^1 \end{array} > C < \begin{array}{c} COOC_2H_5 \\ COOC_2H_5 \end{array} \xrightarrow{\overset{\oplus\ominus}{NaOC_2H_5}} \left[R'C(COOC_2H_5)_2 \right]^{\ominus} \overset{\oplus}{Na} \xrightarrow{R^2-X} \begin{array}{c} R^2 \\ R' \end{array} > C < \begin{array}{c} COOC_2H_5 \\ COOC_2H_5 \end{array}$$

(ii) HYDROLYSIS : The dialkyl derivatives of malonic esters undergo hydrolysis on heating with dilute aqueous acids, yielding the di-alkyl derivatives of malonic acid. The latter, on heating in sulfuric acid solution or on heating above the melting point of the compound, lose CO_2 to form derivatives of acetic acid.

$$\begin{array}{c} R^2 \\ R' \end{array} > C < \begin{array}{c} COOC_2H_5 \\ COOC_2H_5 \end{array} \xrightarrow{H_3\overset{\oplus}{O}/\Delta} \begin{array}{c} R^2 \\ R' \end{array} > C < \begin{array}{c} COOH \\ COOH \end{array} \xrightarrow{-CO_2} \begin{array}{c} R^2 \\ R' \end{array} > CH-COOH$$

Examples of Malonic ester synthesis :

(a) ***2-Methylpentanoic acid*** : In attempting a synthesis of substituted acetic acid, the structure of the acid to be prepared is first written down. Then, the portion of the acid corresponding to [-CH-COOH] of the malonic ester is identified. Substitution of the appropriate alkyl groups in place of hydrogens on diethyl malonate will indicate the desired compound to be synthesized; thus

$$H_3C-CH_2-CH_2-\underset{\text{(2-Methylpentanoic acid)}}{\overset{CH_3}{\overset{|}{[CH-COOH]}}} \Longleftarrow \begin{array}{c} H_3C-CH_2-CH_2 \\ H_3C \end{array} > \left[C < \begin{array}{c} COOC_2H_5 \\ COOC_2H_5 \end{array} \right]$$

The reaction sequence can be written down as follows:

$$\begin{array}{c} H \\ H \end{array} > C < \begin{array}{c} COOC_2H_5 \\ COOC_2H_5 \end{array} \xrightarrow[\text{then } H_3O^{\oplus}]{CH_3I/\overset{\ominus}{O}C_2H_5} \begin{array}{c} H \\ H_3C \end{array} > C < \begin{array}{c} COOC_2H_5 \\ COOC_2H_5 \end{array} \xrightarrow[\overset{\ominus}{O}C_2H_5 \text{ and then } H_3O^+]{H_3CCH_2CH_2Cl} \begin{array}{c} H_3CCH_2CH_2 \\ H_3C \end{array} > C < \begin{array}{c} COOC_2H_5 \\ COOC_2H_5 \end{array}$$

$$\begin{array}{c} H_3CCH_2CH_2 \\ H_3C \end{array} > C < \begin{array}{c} COOC_2H_5 \\ COOC_2H_5 \end{array} \xrightarrow{H_3\overset{\oplus}{O}/\Delta} \begin{array}{c} H_3CCH_2CH_2 \\ H_3C \end{array} > C < \begin{array}{c} COOH \\ COOH \end{array} \xrightarrow[(-CO_2)]{\Delta} \begin{array}{c} H_3CCH_2CH_2 \\ H_3C \end{array} > C < \begin{array}{c} H \\ COOH \end{array}$$

$$\begin{array}{c} H_3CCH_2CH_2 \\ H_3C \end{array} > C < \begin{array}{c} H \\ COOH \end{array} \equiv H_3C-CH_2-CH_2-\overset{CH_3}{\overset{|}{C}H}-COOH$$

(2-Methylpentanoic acid)

(b) 2-Ethylhexanoic acid : Just as in the above case,

$$\underset{\text{(2-Ethylexanoic acid)}}{H_3C-CH_2CH_2CH_2-\overset{C_2H_5}{\overset{|}{CH}}-COOH} \Longleftarrow (H_5C_2)(H_3CH_2CH_2CH_2C)C(COOC_2H_5)_2$$

this acid can be prepared as follows:

$$H_2C(COOC_2H_5)_2 \xrightarrow[\text{then } H_3O^+]{C_2H_5I/\bar{O}C_2H_5} (H_5C_2)(H)C(COOC_2H_5)_2 \xrightarrow[\bar{O}C_2H_5 \text{ and then } H_3O^{\oplus}]{CH_3CH_2CH_2CH_2Cl} (H_5C_2)(H_3CCH_2CH_2CH_2)C(COOC_2H_5)_2$$

$$(H_5C_2)(H_3CCH_2CH_2CH_2)C(COOC_2H_5)_2 \xrightarrow{H_3O^{\oplus}/\Delta} (H_5C_2)(H_3CCH_2CH_2CH_2)C(COOH)_2 \xrightarrow[(-CO_2)]{\Delta} (H_5C_2)(H_3CCH_2CH_2CH_2)C(H)(COOH)$$

$$(H_5C_2)(CH_3CH_2CH_2CH_2)C(H)(COOH) \equiv CH_3-CH_2-CH_2-CH_2-\overset{C_2H_5}{\overset{|}{CH}}-COOH$$

(2-Ethylhexanoic acid)

In several cases, ethyl cyanoacetate can be used instead of malonic ester. The ethyl cyanoacetate can be decarboxylated to yield substituted acetic acid.

$$H_2C(COOC_2H_5)(CN) \xrightarrow[\text{then } H_3O^{\oplus}]{R^1-X/\bar{O}C_2H_5} (H)(R^1)C(COOC_2H_5)(CN) \xrightarrow[\text{then } H_3O^{\oplus}]{R^2-X/\bar{O}C_2H_5} (R^2)(R^1)C(COOC_2H_5)(CN)$$

$$(R^2)(R^1)C(COOC_2H_5)(CN) \xrightarrow{H_3O^{\oplus}/\Delta} (R^2)(R^1)C(COOH)_2 \xrightarrow{\Delta} (R^2)(R^1)C(H)(COOH)$$

The ethyl cyanoacetate required in the above reaction can be readily prepared by esterification of chloroacetic acid with ethanol followed by treatment with KCN.

13-7 PROBLEMS AND EXERCISES

Choose the correct answer for each of the following questions.

13-7-1 A carbanion is

A) a free radical
B) a Zwitterion
C) an Electrophile
D) a Nucleophile

13-7-2 The carbanion stability follows the order

A) Tertiary > Secondary > Primary
B) Secondary > Primary > Tertiary
C) Primary > Secondary > Tertiary
D) Primary > Tertiary > Secondary

13-7-3 Carbanions are isoelectronic with

A) Carbonium ions
B) Amines
C) Free-radicals
D) Carbenes

13-7-4 Carbanions are formed by the loss of an acidic hydrogen containing.

A) -CH- grouping
B) -NH-grouping
C) -O-H grouping
D) -SH grouping

13-7-5 Which of the following compounds will form carbanion most readily on treatment with a base?

A) $CH_3COCH_2COCH_3$
B) CH_3COCH_3
C) CH_3COOCH_3
D) CH_3COOH

13-7-6 Tautomerism can be readily observed in

A) Esters
B) ß-Ketoesters
C) α - Ke[illegible]ters
D) Ketones

13-7-7 ß-Diketones are well-known examples for

A) Resonance
B) Hyperconjugation
C) Tautomerism
D) Polarizability effects in organic molecules.

13-7-8 Reaction between two molecules of an ester in the presence of a strong base leading to the formation of a ß-Keto ester in known as

A) Cannizzaro
B) Perkin condensation
C) Michael condensation
D) Claisen condensation.

13-7-9 Acetoacetic ester synthesis is used in the preparation of
A) Substituted acetones
B) Substituted alcohols
C) Substituted aldehydes
D) Substituted phenols.

13-7-10 Malonic ester can be prepared by the hydrolysis of the following acid followed by esterification:
A) Chloroacetic acid
B) Cyanoacetic acid
C) Bromoacetic acid
D) Methoxyacetic acid.

13-7-11 Malonic ester synthesis is used in the preparation of
A) Substituted ketones
B) Substituted aldehydes
C) Substituted alcohols
D) Substituted acetic acids

13-7-12 Diethyl malonate undergoes alkylation in the α- position because it has
A) Acidic hydrogens
B) Non-ionizable hydrogens
C) Hydrogens that act as good leaving groups
D) None of the above three properties.

13-7-13 Dieckmamn condensation is
A) Analogous to Michael condensation
B) Similar to Perkin condensation
C) An internal claisen condensation
D) None of these reactions

13-7-14 Acetoacetic ester synthesis is an important example of
A) Carbanion formation
B) Alkylation of carbanions.
C) Reaction of carbonium ions with carbanions
D) None of the above three

13-7-15 Diethyl malonate on hydrolysis and decarboxylation yields
A) Acetone
B) Acetaldehyde
C) Acetic acid
D) Ethylacetate

13-8 SHORT AND ESSAY TYPE QUESTION

13-8-1 Based on Claisen condensation, outline a synthesis for each of the following:
(a) Acetylacetone
(b) Dibenzoylmethane
(c) Ethyl benzoyloacetate
(d) Ethyl α-benzoylophenylacetate
(e) 2-Benzoylcyclohexanone

13-8-2 Using ethyl acetoacetate, describe a synthesis for the following:
(a) 4-Methyl-2-Pentanone
(b) 4-Phenylbutane-2-one
(c) Methylisopropyl ketone
(d) 3,4 -Dimethyl-2-pentanone
(e) 3-Ethyl-2-pentanone

13-8-3 How can the following carboxylic acids be prepared starting from ethyl acetoacetate?
(a) 3-Methyl butanoic acid
(b) Caproic acid
(c) ß-Phenylpropionic acid
(d) 2-Ethyl-3-phenylpropanoic acid
(e) α-Cyclohexyl-α-methylacetic acid

13-8-4 Using malonic ester synthesis, formulate a reaction sequence to prepare the following;
(a) 2-Methylbutanoic acid
(b) α-Methyl-ß-phenyl propionic acid
(c) α-Ethylvaleric acid
(d) α,α-Dibenzylacetic acid
(e) p-Nitrophenylacetic acid.

13-8-5 Explain how cyclopentanone can be synthesized starting from caproic acid?

13-8-6 Describe a method for the preparation of cyclohexanone using 1,5-dibromopentane as starting material.

13-8-7 Using qualitative test-tube experiments how can the following be distinguished?
(a) Ethyl acetoacetate from ethyl α,α-dimethylacetoacetate.
(b) Ethyl α-acetoacetate from methyl ß-acetopropionate.
(c) Ethyl malonate from methyl glutarate.
(d) Methyl butyrate from acetylacetone
(e) Hexane 2,5-dienone from ethyl butyrate.

13-8-8 How can ethyl acetoacetate be prepared from ethyl alcohol as the only source of organic compound?

13-8-9 How acetoacetic ester is prepared? Discuss the mechanism involved in its preparation. Mention two of its applications. (S.K.U.,Apr 1990)

13-8-10 What is meant by reactive methylene group? How dimethyl malonate is prepared? Discuss any two of its applications. (S.K.U.,Apr 1990)

13-8-11 Ethyl acetoacetate gives a deep purple color with neutral ferric chloride. Write the structure responsible for this. Which other structure is possible for ethyl acetoacetate? Which class of isomers do they belong to?
(K.U.,Apr 1990)

13-8-12 (a) Explain with the help of a suitable example the term active methylene group.
(b) Provide evidence for the presence of keto-enol tautomerism in acetoacetic ester.
(c) How is butanoic acid prepared from (i) diethylmalonate (ii) acetoacetic ester?
(O.U.,Apr 1990)

13-8-13 Discuss the keto-enol tautomerism of acetoacetic ester. (S.K.U.,Apr 1990)

13-8-14 Outline the synthesis of α-methylsuccinic acid from diethylmalonate.
(K.U.,Apr 1990)

13-9 SOLUTIONS TO PROBLEMS

13-7-1 D **13-7-2 C** **13-7-3 B** **13-7-4 A** **13-7-5 A** **13-7-6 B**
13-7-7 C **13-7-8 A** **13-7-9 A** **13-7-10 B** **13-7-11 D** **13-7-12 A**
13-7-13 C **13-7-14 B**

13-8-7

(a)

$$CH_3{-}C({=}O){-}CH_2{-}COOC_2H_5 \quad \text{and} \quad H_3C{-}C({=}O){-}C(Me)_2{-}COOC_2H_5$$

(Ethyl acetoacetate) (Ethyl α,α-dimethylacetoacetate)

The first compound is capable of enolization (since it has two α-hydrogens), whereas the second one is not (since it has no α-hydrogens). Therefore, the first compound will show characteristic properties of an enol such as coloration with neutral $FeCl_3$ and dissolution in aq.alkali. The second compound will give negative response to both these tests.

(b)

$$CH_3{-}C({=}O){-}CH_2{-}COOC_2H_5 \quad \text{and} \quad H_3C{-}C({=}O){-}\overset{\beta}{C}H_2{-}\overset{\alpha}{C}H_2{-}C({=}O){-}O{-}CH_3$$

(Ethyl acetoacetate) (Methyl β-acetopropionate)

The first compound is a β-ketoester whereas the second compound is the γ-ketoester. The former is capable of enolisation but the latter is not. Therefore, the first compound will show characteristic properties of an enol such as coloration with neutral $FeCl_3$ and dissolution in aqueous alkali. The second compound will not exhibit these properties.

(c)

$$H_2C(COOC_2H_5)_2 \quad \text{and} \quad H_2C(CH_2{-}COOCH_3)_2$$

(Ethyl malonate) (Methyl glutarate)

Although the two compounds are isomeric with each other, the first one is a β-dicarbonyl compound whereas the second one is a δ-dicarbonyl compound. The former can enolise exhibiting typical chemical properties like dissolution in aq.alkali and coloration with neutral $FeCl_3$. The latter cannot enolize and so will not dissolve in dil.alkali, nor give coloration with neutral $FeCl_3$.

(d)

$$H_3C{-}CH_2{-}CH_2{-}C({=}O){-}OCH_3 \quad \text{and} \quad H_3C{-}C({=}O){-}CH_2{-}C({=}O){-}CH_3$$

(Methyl butyrate) (Acetyl acetone)

The former is a simple carboxylic acid ester whereas the latter is a β-diketone. The latter will dissolve in dil. alkali and produce coloration with neutral $FeCl_3$ whereas the former will not give any of these tests.

(e)

$$H_3C{-}C({=}O){-}CH_2{-}CH_2{-}C({=}O){-}CH_3 \quad \text{and} \quad H_3C{-}CH_2{-}CH_2{-}C({=}O){-}OC_2H_5$$

(Hexane-2,5-dione) (Ethyl butyrate)

The first compound is a simple diketone whereas the second one is a simple carboxylic acid ester. The former will give positive tests with carbonyl reagents like 2,4 dinitrophenylhydrazine, hydroxylamine, schiff's base etc., while the latter will not show any of these tests.

13-8-8

$$\underset{\text{(Ethyl alcohol)}}{H_3C{-}CH_2{-}OH} \xrightarrow{\text{oxidation}} \underset{\text{(Acetic Acid)}}{H_3C{-}COOH}$$

$$H_3C{-}\underset{\underset{O}{\|}}{C}{-}OH + HO{-}CH_2{-}CH_3 \xrightarrow{H^+/-H_2O} \underset{\text{(Ethyl acetate)}}{H_3C{-}\underset{\underset{O}{\|}}{C}{-}O{-}C_2H_5}$$

$$\underset{\text{(Ethyl acetate)}}{H_3C{-}\underset{\underset{O}{\|}}{C}{-}O{-}C_2H_5} \xrightarrow{C_2H_5O^-} \underset{\text{(Ethyl acetoaccetate)}}{H_3C{-}\underset{\underset{O}{\|}}{C}{-}CH_2{-}\underset{\underset{O}{\|}}{C}{-}OC_2H_5}$$

14

NITROGEN COMPOUNDS

14-1 ALKYL NITRITES

Four classes of compounds, namely alkyl nitrates, alkyl nitrites, nitroalkanes and nitrosoalkanes, can be considered to be derived from nitric and nitrous acids. Of these, alkyl nitrates and alkyl nitrites are regarded as esters of nitric acid and nitrous acid with the corresponding alcohols respectively.

$$\text{H–O–}\overset{\oplus}{\text{N}}(\text{=O})\text{–O}^{\ominus} \qquad \text{H–O–}\ddot{\text{N}}\text{=O}$$

(Nitric acid) (Nitrous acid)

$$\text{R–O–}\overset{\oplus}{\text{N}}(\text{=O})\text{–O}^{\ominus} \qquad \text{R–O–}\ddot{\text{N}}\text{=O}$$

(Alkyl nitrate) (Alkyl nitrite)

$$\text{R–}\overset{\oplus}{\text{N}}(\text{=O})\text{–O}^{\ominus} \qquad \text{R–}\ddot{\text{N}}\text{=O}$$

(Nitroalkane) (Nitrosoalkane)

The alkyl nitrites are isomeric with nitroalkanes. However, the former contain a –C–O–N– bonding whereas the latter contain a –C–N– bond. The alkyl nitrites are usually prepared by adding concentrated hydrochloric or sulfuric acid to a solution of sodium nitrite in the alcohol.

$$\text{R–O–H} + \text{HCl} + \text{NaNO}_2 \longrightarrow \text{R–O–N=O} + \text{H}_2\text{O} + \text{NaCl}$$

The reaction probably involves in–situ generation of nitrous acid which attacks the alcohol through the nitrosonium ion as follows;

$$\text{HCl} + \text{NaNO}_2 \longrightarrow \text{NaCl} + [\text{H–O–N=O}]$$

$$[\text{H–O–N=O}] + \text{H}^+ \longrightarrow \text{H}_2\text{O} + \overset{+}{\text{N}}\text{=O}$$

$$R\text{–}O\text{–}H + \overset{+}{N}{=}O \longrightarrow R\text{–}\overset{+}{O}\text{–}H \longrightarrow R\text{–}O\text{–}N{=}O + H^+$$
$$\quad\quad\quad\quad\quad\quad\quad\quad\quad\quad\quad\quad | $$
$$\quad\quad\quad\quad\quad\quad\quad\quad\quad\quad\quad\quad N{=}O$$

The alkyl nitrites, on hydrolysis with dilute acids, yield back the starting alcohol. On reduction with complex metal hydrides, alkyl nitrites yield the corresponding alcohol and ammonia whereas catalytic reduction affords alcohol and hydroxylamine.

$$R\text{–}O\text{–}N{=}O \xrightarrow{LiAlH_4} R\text{–}O\text{–}H + NH_3$$
$$R\text{–}O\text{–}N{=}O \xrightarrow{H_2\ /\ Catalyst} R\text{–}O\text{–}H + NH_2\text{–}OH$$

The only important alkyl nitrites are the ethyl nitrite and iso amyl nitrite. The latter is prepared by treating isoamyl alcohol (or isopentanol) with $NaNO_2$ in the presence of hydrochloric acid. Ethyl nitrite, b.p 17°c and isoamyl nitrite, b.p. 99°C, are pleasant smelling liquids. Both these compounds are used as sources of nitrous acid in anhydrous media and are commonly employed in syntheses of α–dicarbonyl compounds. For example;

$$\underset{(I)}{R^1\text{–}CH_2\text{–}\underset{\underset{O}{||}}{C}\text{–}R^2} \xrightarrow[R\text{–}O\text{–}N=O/H^+]{(\overset{+}{N}O)} \underset{(II)}{R^1\text{–}\underset{\underset{O=N}{|}}{CH}\text{–}\underset{\underset{O}{||}}{C}\text{–}R^2} \longleftrightarrow \underset{(III)}{R^1\text{–}\underset{\underset{HO\text{–}N}{||}}{C}\text{—}\underset{\underset{O}{||}}{C}\text{–}R^2}$$

$$(III) \xrightarrow{\overset{+}{H_3O}} R^1\text{–}\underset{\underset{O}{||}}{C}\text{—}\underset{\underset{O}{||}}{C}\text{–}R$$

In the above reaction, the carbonyl compound (I) reacts, through its enol form, with the nitrosonium ion (From the alkyl nitrite and HCl) to yield a C–nitroso derivative (II), which tautomerizes to the more stable α–keto oxime, (III) in which the C=N bond is in conjugation with the > C= O group. The α-keto oxime, III, is then hydrolysed during reaction work–up to give α-dicarbonyl compound, IV.

14–2 NITROALKANES

Nitroalkanes are the nitro derivatives of alkanes corresponding to a general formula $R\text{–}NO_2$. Usually, these compounds are named by adding the prefix "nitro" to the name of the alkane, for example, nitromethane, nitroethane, 1–nitrobutane etc., wherein the nitro group is treated as a substituent.

The structure of a nitro compound is considered to be the resonance hybrid of the following;

$$R{-}N(=O)(\rightarrow O) \longleftrightarrow R{-}N(\rightarrow O)(=O) \longleftrightarrow R{-}\overset{\oplus}{N}(=O)(-O^{-}) \longleftrightarrow R{-}\overset{\oplus}{N}(-O^{\ominus})(=O)$$

According to the molecular orbital theory, the nitro group is completely conjugated, the delocalization of electrons having been spread over the surface of O–N–O grouping. In this respect, it is isoelectronic with the carboxylate anion.

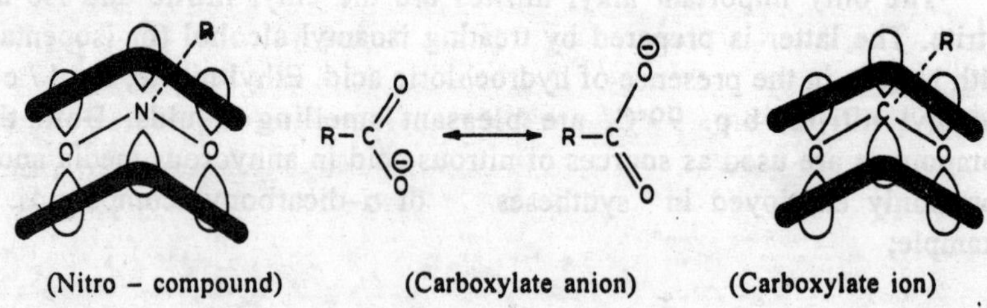

(Nitro – compound) (Carboxylate anion) (Carboxylate ion)

This view is supported by the fact that both the N– O bond distances in a nitro–compound are equal, being 1.21A° close to a normal N= O bond distance, 1.18A°, as different from a N– O bond distance, 1.36A°.

14–2–1 Methods of preparation

Commercially, nitroalkanes are prepared by the vapor–phase nitration of alkanes.

$$R{-}H + HO{-}NO_2 \xrightarrow{400^\circ / \Delta} R{-}NO_2 + HOH \text{ (Tert. > Sec. > Pri.)}$$

Normally, several nitration products are formed in this type of reaction which are usually separated by fractionation. In the laboratory, nitroalkanes may be prepared by reacting alkyl halides with silver nitrite.

$$R{-}X + AgNO_2 \longrightarrow R{-}NO_2 + AgX \downarrow$$

In the above reaction, it has been observed that an important by–product is the alkyl nitrite. This is due to the fact that the nitrite ion is an ambident nucleophile and can react with alkyl halide either at oxygen or at nitrogen. If the reaction with alkyl halide occurs at oxygen of the metal nitrite, an alkyl nitrite is the product whereas if the alkyl halide reacts at nitrogen of the metal nitrite, a nitroalkane results. Use of sodium nitrite or potassium nitrite in place of silver nitrite results in a higher proportion of the alkyl nitrite in the above reaction. Furthermore, it is observed that with a given metal nitrite, the yield of nitro compound decreases from primary to secondary to tertiary alkyl halides. With a given metal nitrite and with a given alkyl halide, the yield of nitroalkane can be remarkably improved by using a polar but non–hydroxylic solvent like dimethyl formamide, dimethyl sulfoxide etc., or by using phase–transfer catalytic conditions. In any case, the mixture of nitroalkane and alkyl nitrite thus obtained can be readily separated by physical methods (such as distillation) due to wide variations in the physical properties (such as boiling points) of the two types of compounds.

The simplest nitroalkane, that is nitromethane, is prepared by boiling an aqueous solution of sodium nitrite with chloroacetic acid.

$$ClCH_2\text{–}COOH + NaNO_2 \longrightarrow [O_2N\text{–}CH_2\text{–}COOH] + NaCl$$

$$[O_2N\text{–}CH_2\text{–}COOH] \longrightarrow CH_3\text{–}NO_2 + CO_2 \uparrow$$

The intermediary α-nitroacetic acid readily loses a molecule of CO_2 due to the attachment of a strongly electron withdrawing group (nitro group) to the carbon carrying a carboxylic acid substituent.

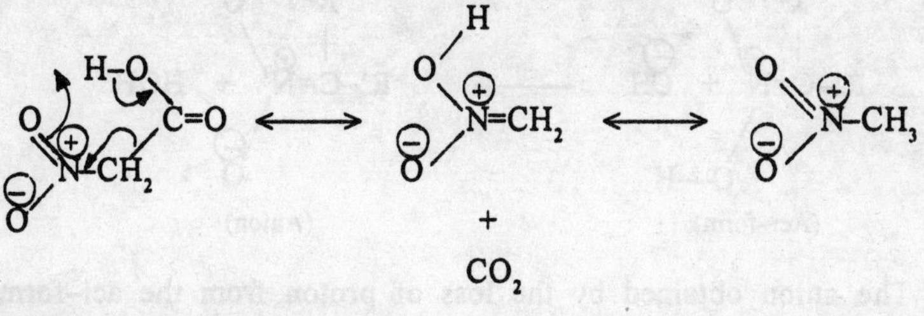

H–O, C=O, O, N⊕, CH₂, O⊖ ⟷ H–O, N⊕=CH₂, O⊖ + CO_2 ⟷ O, N⊕–CH₃, O⊖

14–2–2 Physical Properties

The simple nitroalkanes are, usually, colorless liquids with a pleasant smell. With the exception of nitromethane (which is about 10% soluble), most nitroalkanes are very largely insoluble in water. Due to the presence of charges in the nitro group, the nitroalkanes are highly polar compounds. Consequently, they possess large dipole moments (>3.0D), high boiling points and a great affinity for dissolution in polar solvents. In chromatography, when adsorbed on a suitable support, the nitroalkanes require highly polar solvents for ellution. The nitroalkanes serve as good solvents for various types of organic compounds. The physical constants of some representative nitroalkanes and alkyl nitrites are shown in table-14-1 for comparison purposes.

Table 14–1 : Physical constants of nitroalkanes and alkylnitrites.

S.NO	Nitroalkane	B.P.°C	Alkyl nitrite	B.P °C
1.	Nitromethane	101	Methyl nitrite	–12
2.	Nitroethane	114	Ethyl nitrite	+17
3.	1–Nitropropane	132	n–Propyl nitrite	+57
4.	1–Nitrobutane	152	n–Butyl nitrite	+79
5.	1–Nitropentane	172	Amyl nitrite	104

14–2–3 Chemical properties

(a) Tautomerism : Nitroalkanes exhibit the phenomenon of tautomerism similar to ketones.

$$R^1R^2CH-\overset{\oplus}{N}(=O)-\overset{\ominus}{O} \rightleftharpoons R^1R^2C=\overset{\oplus}{N}(-\overset{\ominus}{O})-O-H$$

(Nitro form) (Aci–form)

The aci–form is the equivalent of an enol and can form salts with bases due to the loss of hydrogen as proton.

$$R^1R^2C=\overset{\oplus}{N}(-\overset{\ominus}{O})-O-H + \overset{\ominus}{O}H \longrightarrow R^1R^2C=\overset{\oplus}{N}(-\overset{\ominus}{O})-\overset{\ominus}{O} + HOH$$

(Aci–form) (Anion)

The anion obtained by the loss of proton from the aci–form of a nitroalkane is a resonance hybrid of the following structures:

$$R^1R^2C=\overset{\oplus}{N}(-\overset{\ominus}{O})-\overset{\ominus}{O} \longleftrightarrow R^1R^2\overset{\ominus}{C}-\overset{\oplus}{N}(-\overset{\ominus}{O})=O \longleftrightarrow$$

$$R^1R^2C=\overset{\oplus}{N}(-O)-\overset{\ominus}{O} \longleftrightarrow R^1R^2\overset{\ominus}{C}-\overset{\oplus}{N}(=O)-\overset{\ominus}{O}$$

Obviously, only the primary and secondary nitroalkanes, which contain at least one alpha hydrogen, can tautomerize and thus dissolve in alkaline solutions. The tertiary nitroalkanes, $R^1R^2R^3C–NO_2$, which do **not** contain an alpha hydrogen, cannot tautomerize and so will remain unaffected by alkali.

(b) Hydrolysis: Primary nitro–compounds, when treated with boiling HCl or 85% H_2SO_4, undergo hydrolysis yielding a carboxylic acid and hydroxylamine.

$$R–CH_2–NO_2 + H_2O \xrightarrow{H^+ / \Delta} R–COOH + H_2N–OH$$

The mechanism probably involves addition of water to the aci–form of the nitro compound, then acid catalysed cleavage of N–O bond followed by hydride transfer from carbon to nitrogen. The reaction is used for the commercial production of hydroxylamine. Secondary nitro compounds under similar conditions yield ketones and nitrous oxide.

$$R^1–\underset{}{\overset{R^2}{\overset{|}{C}}}H–NO_2 + H_2O \xrightarrow{H^+ / \Delta} R^1–\underset{\underset{O}{||}}{C}–R^2 + NO_2$$

The tertiary nitroalkanes, generally, remain unchanged during these conditions.

If the sodium salt of the aci–form of a primary or secondary nitro compound is treated with conc. H_2SO_4, an aldehyde or a ketone and nitrous acid are produced. This is known as the Nef reaction.

$$R^1–CH{=}NO_2Na + H_2SO_4 \longrightarrow R^1–CHO + \tfrac{1}{2} N_2O + NaHSO_4 + H_2O$$

$$R^1R^2C{=}NO_2Na + H_2SO_4 \longrightarrow R^1R^2CO + \tfrac{1}{2} N_2O + NaHSO_4 + H_2O$$

(c) Halogenation : Like aldehydes and ketones,primary and secondary nitroalkanes are halogenated readily in alkaline solution.

$$R^1CH_2–NO_2 + \bar{O}X \longrightarrow R^1CX_2–NO_2 + H_2O$$

$$R^1R^2CH–NO_2 + \bar{O}X \longrightarrow R^1R^2CX–NO_2 + H_2O$$

Only α–halogenation takes place. Primary nitro compounds yield the dihalo derivatives, the reaction seldom stopping at the monohalo stage. Secondary nitro compounds must of necessity give monohalo products. The reaction occurs through the aci–form of the nitro compounds and hence tertiary nitroalkanes remain unreactive to these conditions.

Chloropicrin is the trichloro derivative of nitromethane. It is obtained by the action of Cl_2 gas and NaOH on nitromethane. It can be prepared by the action of nitric acid on chloroform.

$$H_3C{-}NO_2 + 3Cl_2 + 3NaOH \longrightarrow CCl_3{-}NO_2 + 3NaCl + 3H_2O$$

$$\text{or} \quad CHCl_3 + HNO_3 \longrightarrow CCl_3{-}NO_2 + H_2O$$

Chloropicrin is used in the manufacture of tear–bombs and chemical–weapons.

(d) Reduction : The nitro compounds undergo reduction with hydrogen over a catalyst to yield amines. With complexed metal hydrides, like $LiAlH_4$ or $NaBH_4$, alkyl hydroxylamines are usually obtained as products.

$$R{-}NO_2 \xrightarrow{H_2/Cat} R{-}NH_2$$

$$R{-}NO_2 \xrightarrow{LiAlH_4} R{-}NH{-}OH$$

(e) Reaction with nitrous acid : Primary and secondary nitroalkanes react with nitrous acid to yield C–nitroso derivatives which are blue in color.

$$R^1{-}CH_2{-}NO_2 + HONO \longrightarrow \underset{\substack{|\\ N=O \\ \text{(Blue)}}}{R^1{-}CH{-}NO_2} + H_2O$$

$$R^1R^2CH{-}NO_2 + HONO \longrightarrow \underset{\substack{|\\ N=O \\ \text{(Blue)}}}{R^1R^2C{-}NO_2} + H_2O$$

The reaction involves attack at carbon by nitrosonium ion on the aci–form of the nitro compound and hence tertiary nitro–alkanes do not take part in this reaction.

The nitroso derivative obtained from a primary nitro compound, called a nitrolic acid, contains an alpha hydrogen which is reasonably acidic in nature. Therefore, it can be easily extracted out by an alkali resulting in an anion, which being a resonance hybrid of several conjugated, delocalised structures,is intensely colored usually red in appearance;

$$\underset{\substack{|\\ N=O \\ \text{(Blue)}}}{R^1{-}CH{-}NO_2} \xrightarrow{\overset{\ominus}{O}H} \underset{\substack{|\\ N=O \\ \text{(Red)}}}{R^1{-}\overset{\ominus}{C}{-}NO_2}$$

$$R^1-\overset{\ominus}{C}(N{=}O)-\overset{\oplus}{N}(=O)-\overset{\ominus}{O} \longleftrightarrow R^1-C(-N{-}\overset{\ominus}{O})=\overset{\oplus}{N}(=O)-\overset{\ominus}{O} \longleftrightarrow R^1-C(N{=}O)=\overset{\oplus}{N}(-\overset{\ominus}{O})-\overset{\ominus}{O} \longleftrightarrow R^1-\overset{\ominus}{C}(N{=}O)-\overset{\oplus}{N}(-\overset{\ominus}{O})=O$$

The nitroso derivative obtained from a secondary nitroalkane lacks an alpha hydrogen and so is unaffected by alkali. Sometimes, this reaction is used as a means of distinguishing qualitatively between the three types of nitroalkanes by visual examination of colors and is referred to as the **Red, Blue and White reaction.**

(f) Michael condensation : The primary nitroalkanes contain alpha hydrogens which are acidic in nature. The pK_a value of these compounds usually falls in the range 10 to 12. On treatment with a strong base, these nitro compounds yield carbanions that are strongly nucleophilic in character. The latter combine with aldehydes and ketones to give a,ß–unsaturated nitro compounds.

$$R^1-CH_2-NO_2 \xrightarrow{\text{Strong base}} R^1-\bar{C}H-NO_2$$

$$R^1-\bar{C}H-NO_2 + (R^2)(R^3)C{=}O \xrightarrow{-H_2O} R^2-C(R^3)(OH)-C(R^1)(H)-NO_2 \longrightarrow R^2R^3C{=}C(R^1)NO_2$$

The α,β–unsaturated nitro compounds, obtained above, can be further condensed with a primary nitroalkane in the presence of a base. In the latter case, the reaction is known as Michael Condensation and the product is known as Michael adduct.

$$R^2R^3C{=}C(R^1)NO_2 + R-\bar{C}H-NO_2 \longrightarrow R^2R^3C(-CH(R)-NO_2)-CH(R^1)-NO_2$$

(Michael adduct)

Traditionally, Michael condensation has been carried out by condensing carbanions with a,ß–unsaturated carbonyl compounds.

$$R^2R^3C{=}CH-\underset{\|}{\overset{}{C}}(O)-R^4 + R^1-CH-NO_2 \longrightarrow R^2R^3C(-R^1CH-NO_2)-CH_2-C(=O)-R^4$$

(Michael adduct)

The reaction is fairly general and any activated olefin may be condensed with a carbanion to give Michael adducts.

14–2–4 Applications of nitroalkanes

Simple aliphatic nitro compounds find extensive use in organic synthesis as sources of carbanions; especially in reactions of the type Michael addition. Nitroalkanes are used in plastics industry and in the manufacture of insecticides, propellants, explosives, initiators etc. Chloropicrin, $Cl_3C–NO_2$, is a powerful lachrymator and is used in the manufactured of tear–gas, chemical weapons etc. Some hydroxy nitroalkanes find use as raw materials for the manufactures of drugs and pharmaceuticals.

14–3 NITROBENZENES

The nitrobenzenes belong to the broad category of aromatic nitro compounds. The nitrobenzenes are one of the most important intermediates in the syntheses of a variety of substituted benzenes.

14–3–1 Preparation

The methods used for the preparation of nitrobenzenes may be divided into three types and these are as follows :

(i) Direct introduction of nitro groups : In this method an aromatic compound is subjected to nitration wherein a mixture of o–and p–isomers are obtained which can generally be separated to yield the pure isomers. Some of the reagents commonly employed for the direct introduction of nitro groups into the benzene rings are given below;

(A) A mixture of conc.HNO_3 and conc.H_2SO_4 (commonly known as nitrating mixture).

(B) Fuming nitric acid in organic solvents like acetic acid, nitromethane etc.

(C) Acyl nitrates like acetyl nitrate and benzoyl nitrate.

(D) Nitronium salts like $\overset{+}{N}O_2\overset{-}{B}F_4$, $\overset{+}{N}O_2Cl\overset{-}{O}_4$, etc., in organic solvents.

(E) Sodium nitrite and dil. H_2SO_4 followed by oxidation.

The nature of reagent employed in a given reaction will depend upon the substituent already present in the benzene ring and the product ratio (i.e., o–/p–) desired. The parent compound, nitrobenzene,is normally prepared, both in the laboratory and on a commercial scale, by treatment of benzene with the nitrating mixture.

(ii) Replacement of the diazonium group : The reaction of an aromatic primary amine with nitrous acid results in a diazonium salt, which on treatment with cuprous nitrite yields the nitro compound. For example;

$$Ar–NH_2 \xrightarrow{[HONO]} Ar–\overset{+}{N}_2 \xrightarrow{Cu(NO_2)} Ar–NO_2$$

(iii) Oxidation of amino groups : Aromatic primary amines may be oxidized with peroxytrifluoroacetic acid to yield nitro derivatives. For example;

$$\text{Ar–NH}_2 \xrightarrow{\text{CF}_3\text{–CO–O–OH}} \text{Ar–NO}_2$$

14–3–2 Properties

Most of the nitrobenzenes are yellow crystalline solids although some, like the parent nitrobenzene, are high boiling yellow liquids. They are stable compounds at room temperature, inert to oxidizing agents and denser than water. Almost all of them are highly insoluble in water and so they can be advantageously separated from reaction mixture at the boiling point of water by steam distillation. The most important property of nitrobenzenes is their reduction, by a variety of reagents, to aromatic primary amines which are valuable intermediates in organic syntheses. The different reduction products obtained from nitrobenzene and the different reagents that reduce nitrobenzene to aniline are shown in the chart 14–1 .

CHART–14–1

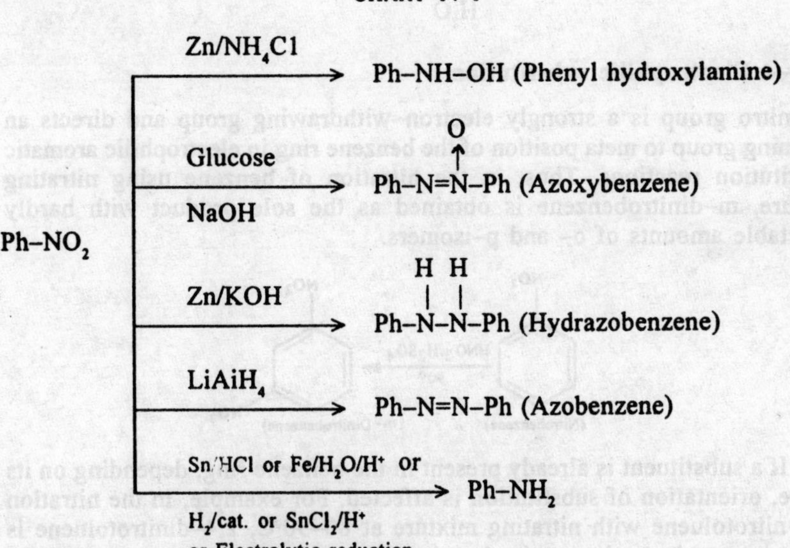

The following mechanistic scheme has been proposed for the complete reduction of nitrobenzene, or in general of any aromatic nitro compound, to the corresponding amine using metal–acid reagents;

$$\text{Ar–}\overset{\oplus}{\text{N}}(=\text{O})\text{–O}^{\ominus} \xrightarrow{e^-} \left[\text{Ar–}\overset{\oplus}{\text{N}}(\text{–O}^{\ominus})\text{–}\dot{\text{O}}^{\ominus} \longleftrightarrow \text{Ar–}\overset{+}{\text{N}}(=\text{O})\text{–O}^{\ominus}\right]^{\bullet}$$

$$Ar\text{-}\overset{+\bullet}{N}(=O)O^{\ominus} \xrightarrow{H^{\oplus}} Ar\text{-}\overset{\cdot +}{N}(=O)OH \xrightarrow{e^-} Ar\text{-}N(O^{\ominus})OH$$

$$\downarrow H^+$$

$$Ar\text{-}\overset{+\bullet}{N}\text{-}\overset{-}{O} \xleftarrow{e-} Ar\text{-}N{=}O \xleftarrow{-H_2O} Ar\text{-}N(OH)_2$$

$$\downarrow H^{\oplus}$$

$$Ar\text{-}\overset{+\bullet}{N}\text{-}OH \xrightarrow{e^-} Ar\text{-}\overset{\ominus}{N}{=}O \xrightarrow{H^{\oplus}} Ar\text{-}NH\text{-}OH$$

$$\downarrow H^+$$

$$Ar\text{-}NH_2 \xleftarrow{e^-,\ H^{\oplus}} Ar\text{-}\overset{+\bullet}{NH} + H_2O \xleftarrow{e^-} Ar\text{-}NH\text{-}\overset{\oplus}{O}H_2$$

14–3–4 Electrophilic substitution

The nitro group is a strongly electron–withdrawing group and directs an incoming group to meta position of the benzene ring in electrophilic aromatic substitution reactions. Thus, in the nitration of benzene using nitrating mixture, m–dinitrobenzene is obtained as the sole product with hardly detectable amounts of o– and p–isomers.

$$\text{(Nitrobenzene)} \xrightarrow[90^{o}]{HNO_3/H_2SO_4} (m\text{-Dinitrobenzene})$$

If a substituent is already present in the benzene ring, depending on its nature, orientation of substitution is affected. For example, in the nitration of o–nitrotoluene with nitrating mixture at 80–90°C, 2,4–dinitrotoluene is the major product whereas in the nitration of m-Nitrobenzoic. acid, 3,5–dinitrobenzoic acid is the only identifiable product obtained.

$$\text{o-nitrotoluene} \xrightarrow[80\text{-}90^{o}]{HNO_3/H_2SO_4/} \text{2,4-dinitrotoluene (Major product)}$$

COOH, NO_2 (m-Nitrobenzoic acid) $\xrightarrow[100^0]{HNO_3+H_2SO_4/}$ COOH, O_2N, NO_2 (3,5-Dinitrobenzoic acid)

The strongly directive influence of nitro group is seen from the fact that in the nitration of o–nitrobenzoic acid, 2,4–dinitrobenzoic acid is the sole product isolated and none of the other isomers.

COOH, NO_2 $\xrightarrow[100^0]{HNO_3/H_2SO_4}$ COOH, NO_2, NO_2

14–3–5 Nucleophilic substitution :

By virtue of its strongly electron withdrawing nature, the nitro group activates the benzene ring towards nucleophilic aromatic substitution. Nitrobenzene, when fused with KOH, yields a mixture of o– and p–nitrophenols.

NO_2 $\xrightarrow{KOH/\Delta}$ NO_2, OH + NO_2, OH

The presence of two nitro groups, in m–position to one another, further activates the ring to nucleophilic substitution. Treatment of m–dinitrobenzene with alkaline potassium ferricyanide yields 2,4–dinitrophenol.

NO_2, NO_2 $\xrightarrow{KOH+K_3Fe(CN)_6}$ NO_2, NO_2, OH

The reaction probably involves nucleophilic attack by hydroxyl ion on the benzene ring. The purpose of ferricyanide is to facilitate loss of hydrogen as hydride ion which is normally a poor leaving group.

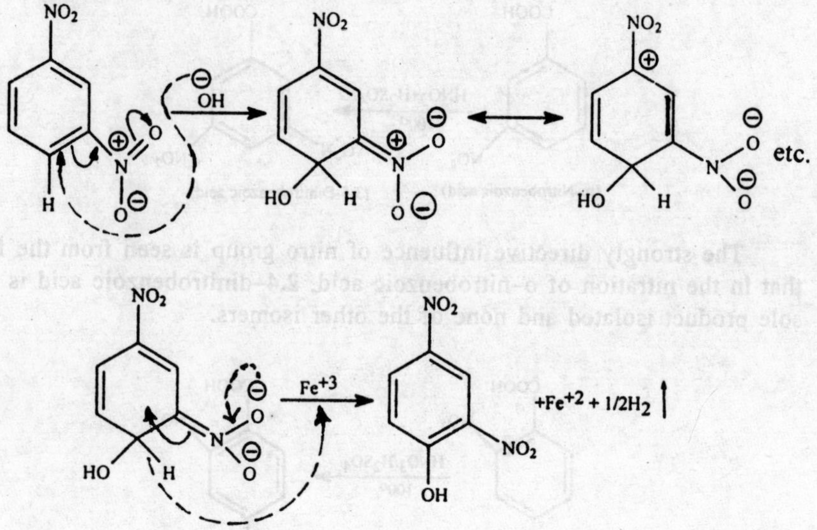

The nitro groups also facilitate displacement of good leaving groups by nucleophiles. For example;

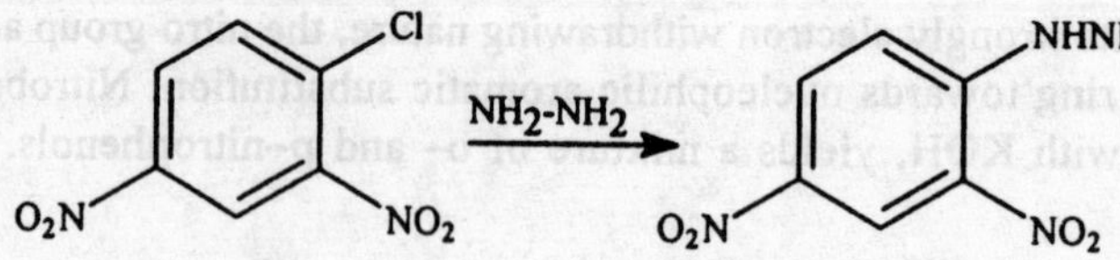

NH_2-NH_2

14–3–6 Uses of Nitrobenzenes :

Apart from their use as important intermediates in organic syntheses, the largest volume is used as explosives. 2,4,6 Trinitrotoluene, commonly known as TNT, is a standard tertiary explosive used for military purposes. Nitrobenzene is used as a solvent and at higher temperatures as a good dehydrogenating agent.

14–4 AMINES :

Amines are the derivatives of ammonia in which one, two or three of the hydrogen atoms have been replaced by alkyl or aryl groups.

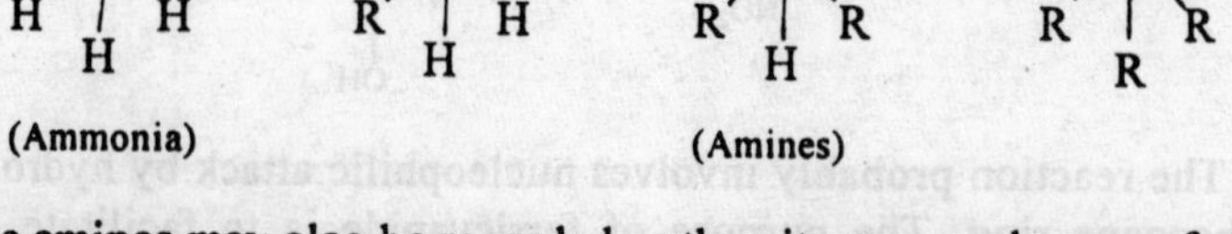

(Ammonia) (Amines)

The amines may also be regarded as the nitrogen analogues of alcohols and ethers. Thus, for example;

$$R-\ddot{O}-H \quad R-\ddot{O}-H \quad R-\ddot{N}(H)-H \quad R-\ddot{N}(H)-H$$

(Alcohol) (Ether) (Amines)

14–4–1 Classification

The amines are classified as primary, secondary or tertiary depending on the number of hydrogen atoms of ammonia molecule that have been successively replaced by one,two or three alkyl groups. Thus, RNH_2 is a primary amine, R_2NH is a secondary amine and R_3N is a tertiary amine. When one or more of substituents is an aryl group, the amine is an aromatic amine. for example;

Ar–NH_2 (Primary) Ar–NH–R or Ar–NH–Ar (Secondary) Ar–NR_2 or Ar–N(R)–Ar or Ar–N(Ar)–Ar (Tertiary)

It should be noted that the method of classification of amines as primary, secondary or tertiary is opposite to that for alcohols or halogen compounds. For example, $(CH_3)_3C–OH$ is a tertiary alcohol, $(CH_3)_3$ $C–NH_2$ is a primary alkylamine.

In addition to the amines, the tetraalkyl derivatives of ammonium salts are also known and these are commonly referred to as quaternary ammonium salts, $[R_4N^+]X^-$

14–4–2 Nomenclature

The amines are named as derivatives of ammonia by calling them with the ending amine after mentioning the name of the alkyl groups attached to the nitrogen.

Thus;

$CH_3–NH_2$ is called methylamine.

$(CH_3)_2$ NH is called dimethylamine.

$(C_2H_5)_3N$ is called triethylamine.

Ph–$N(CH_3)(C_2H_5)$ is called phenyl methyl ethylamine.

However, in more complex compounds, the amino group is regarded as a substituent. Thus,the compound(I) is called as 3–amino–4–methylcycloheptene,

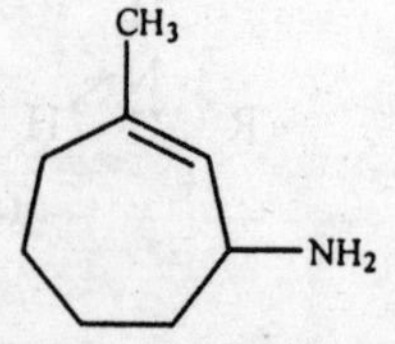

(3–Amino–5–methylcycloheptene)

The quaternary ammonium salts are named by adding the ending ammonium after mentioning the four alkyl groups attached to the nitrogen atom and then mentioning the ammonia part. For example;

$(CH_3)_4N^+Cl^-$ is named as tetramethylammonium chloride.

$PhCH_2{-}\overset{+}{N}(C_2H_5)_3Br^-$ is named as benzyltriethylammonium bromide.

$(C_2H_5)_3\overset{+}{N}(CH_3)I^-$ is named as methyl triethylammonium iodide.

14–4–3 Importance

The amines are probably the most important class of organic compounds. A number of biologically significant classes of compounds, such as nucleic acids, proteins, polypeptides, aminoacids and alkaloids are all derivatives of amines. Several amines, of natural occurrence and also of synthetic origin, are physiologically active, are commercially important and find great applications in our day to day life. Some examples of these are given below:

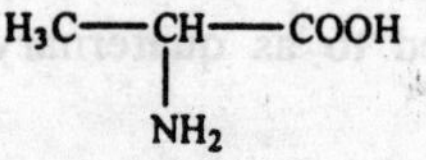

(alanine – an essential amino acid)

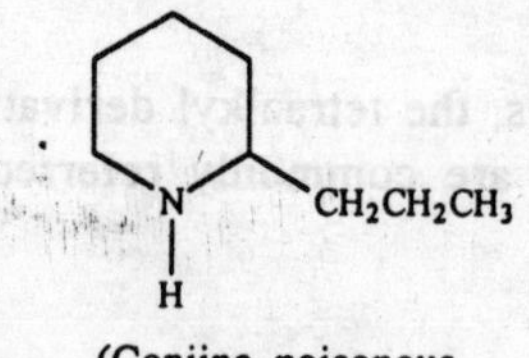

(Coniine–poisonous hemlock alkaloid)

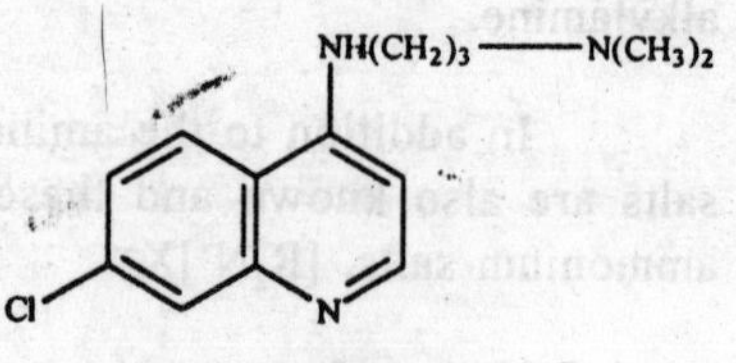

(Chloroquine–an antimalarial drug)

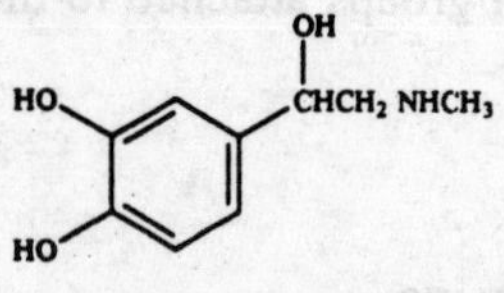

(Adrenaline–an adrenal hormone)

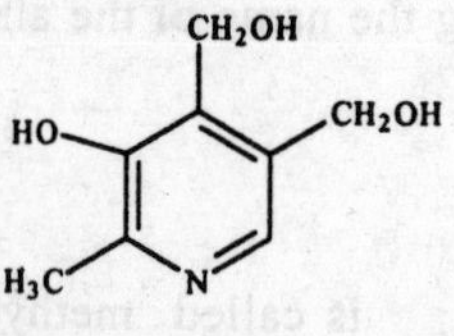

(Pyridoxine–vitamin– B_6)

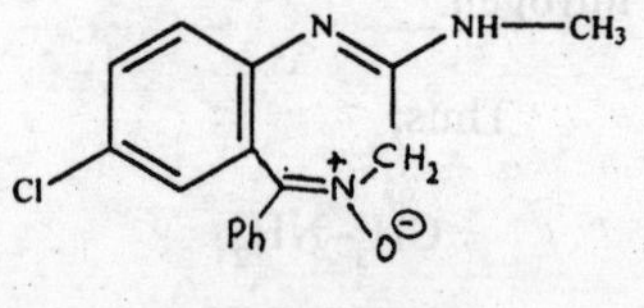

(Chlordiazepoxide – a widely used tranquilizer)

14–4–4 Stereochemistry of Amines

The nitrogen atom has an atomic number of 7 and a ground state electronic configuration of $1s^2\ 2s^2\ 2p_x^1\ 2p_y^1\ 2p_z^1$. In the formation of ammonia molecule, the nitrogen atom's valency orbitals undergo sp^3 hybridization which are used for overlapping with the 1s orbitals of three hydrogen atoms. The result is the formation of three hybrid covalent bonds and a fourth hybrid orbital containing a pair of non–bonding electrons. The shape of the ammonia molecule may, therefore, be expected to be a regular tetrahedron with the nitrogen atom at its center. In fact, the H–N–H bond angle in ammonia has been found to be 107° which is very close to the tetrahedral angle of $109^{\circ} 28^{1}$ in carbon compounds. The slight deviation from the normal tetrahedral angle is believed to be due to repulsion between non–bonding electron–pair in the fourth hybrid sp^3 orbital. Thus, the ammonia molecule is a slightly distorted tetrahedron or rather pyramidal in shape.

Since amines are derivatives of ammonia obtained by successive replacements of hydrogens of alkyl groups, the amines also possess (distorted) tetrahedral or pyramidal geometry. In the case of tertiary amines, when three different groups are attached to the nitrogen atom, the compounds possess non–superimposable mirror images and therefore, they may be expected to show optical activity.

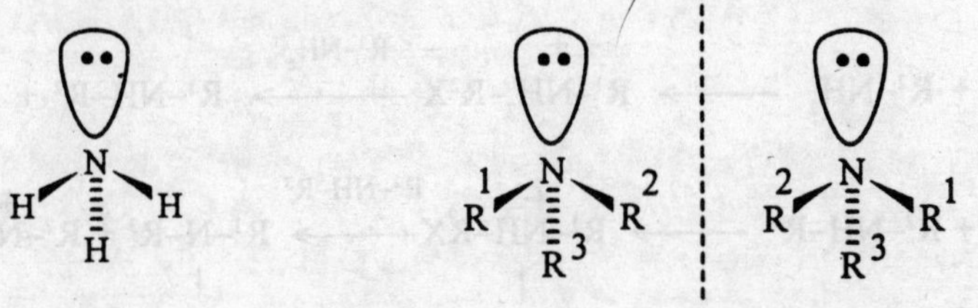

(Ammonia) (Tertiary amines–non superimposable mirror images)

However, all attempts to observe optical activity or to achieve resolution in tertiary amines have been unsuccessful. This is due to the fact that the two mirror–image forms of the tertiary amine are inter–convertible and their inter–conversion is caused by configurational inversion. The electron–pair on nitrogen rapidly oscillate from one side to another and in so doing it inverts the configuration of the entire molecule thereby inter converting the mirror–image structures.

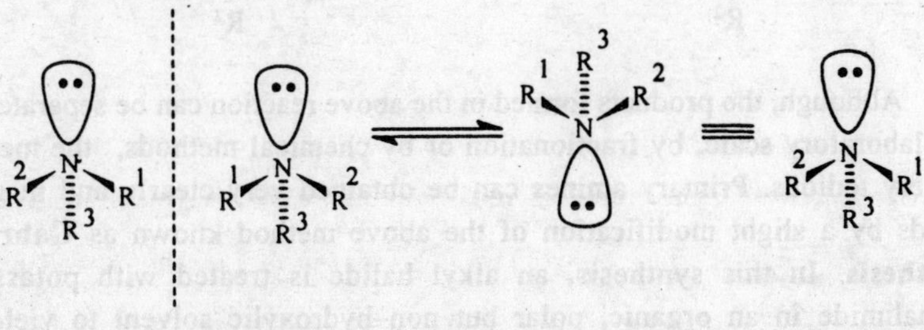

It is estimated from spectroscopic studies that the ammonia molecule undergoes inversion at the rate of 2×10^{11} times per second.

14–4–5 Preparation

The preparation of amines may be divided into two types; namely

(i) Preparation of aliphatic amines and

(ii) Preparation of aromatic amines.

Each of these types is discussed below:

(i) PREPARATION OF ALIPHATIC AMINES : There are a number of methods described in literature for the preparation of aliphatic amines. However, the following four are significant;

(a) From alkyl halides : In this method, an alkyl halide is treated with ammonia or an amine to yield a primary or secondary or tertiary amine.

$$R{-}X + NH_3 \longrightarrow R{-}\overset{+}{N}H_3\overset{-}{X} \xrightarrow{NH_3} R{-}NH_2 + \overset{+}{N}H_4\overset{-}{X}$$

$$R^1{-}X + R^2{-}NH_2 \longrightarrow R^1{-}\overset{+}{N}H_2{-}R^2\overset{-}{X} \xrightarrow{R^2{-}NH_2} R^1{-}NH{-}R^2 + R^2{-}\overset{+}{N}H_3\overset{-}{X}$$

$$R^1{-}X + R^2{-}NH{-}R^3 \longrightarrow R^1{-}\overset{+}{N}H(R^2){-}R^3X \xrightarrow{R^2{-}NH{-}R^3} R^1{-}N(R^2){-}R^3 + R^2{-}\overset{+}{N}H_2{-}R^3X^-$$

The disadvantage of this reaction is that usually mixtures of products are formed as successive alkylation goes on. At times, even quaternary salts are produced by reaction of the first formed tertiary amine with remaining alkyl halide.

$$R^1{-}N(R^2){-}R^3 + R^4{-}X \longrightarrow R^1{-}N^+(R^2)(R^4){-}R^3X^-$$

Although, the products formed in the above reaction can be separated,on the laboratory scale, by fractionation or by chemical methods, the method is very tedious. Primary amines can be obtained very clearly and in good yields by a slight modification of the above method known as **Gabriel's synthesis**. In this synthesis, an alkyl halide is treated with potassium phthalimide in an organic, polar but non–hydroxylic solvent to yield an alkyl phthalimide which on hydrolysis with acid or alkali results in the primary aliphatic amine.

(Potassium phthalimide) (An alkylphthalimide)

$\xrightarrow{R-X}$ N—R $\xrightarrow[\text{or } \overset{\ominus}{O}H]{H_3O^{\oplus}}$

$\longrightarrow$ (Phthalic acid) + $H_2N—R$ (Aliphatic primary amine)

(b) By reduction : Nitrites, nitroalkanes, amides or alkyl azides on reduction, with suitable reducing agents, yield primary amines.

$R–CN \xrightarrow{\text{Reducing agent}} R–CH_2–NH_2$ Reducing agent =Na/EtOH, or $LiAlH_4$, or H_2/catalyst

$R–NO_2 \xrightarrow{\text{Reducing agent}} R–NH_2$ Reducing agent =Fe/H_2O, or Sn/HCl, or Zn–Hg/HCl, or H_2/catalyst

$R–\underset{\underset{O}{||}}{C}–NH_2 \xrightarrow{\text{Reducing agent}} R–NH_2$ Reducing agent = $LiAlH_4$

$R–N_3 \xrightarrow{\text{Reducing agent}} R–NH_2$ Reducing agent = $LiAlH_4$, H_2/catalyst etc.

The alkyl azides required in the above reaction for reduction purpose may be obtained by treating appropriate alkyl halides with sodium or potassium azide.

(c) By rearrangement : There are two rearrangement reactions known as **Hofmann rearrangement** and **Schmidt rearrangement** which yield primary amines. In the former reaction, an acid amide is treated with bromine/NaOH and in the latter an acid–azide is pyrolyzed under hydrolytic conditions. The two rearrangement reactions may be formulated as shown below;

Hofmann rearrangement :

$$\underset{\text{(An acid amide)}}{R{-}CO{-}NH_2} + Br_2 + 4NaOH \longrightarrow R{-}NH_2 + 2NaBr + Na_2CO_3 + 2H_2O$$

Schmidt rearrangement :

$$\underset{\text{(An acid azide)}}{R{-}CO{-}N_3} \xrightarrow{\Delta \text{ and then } H_2O} R{-}NH_2 + CO_2 + N_2$$

Both the reactions proceed through the intermediacy of an alkyl isocyanate which is formed by rearrangement of the first generated nitrene. The alkyl isocyanate hydrolyses under these reaction conditions resulting in alkylamine and CO_2. The mechanisms for the two rearrangements are as shown below;

$$Br_2 + 2\,NaOH \longrightarrow NaBr + NaOBr + H_2O$$

$$R{-}CO{-}NH_2 + \bar{O}Br \longrightarrow [R{-}CO{-}\bar{N}H + H{-}O{-}Br] \longrightarrow R{-}CO{-}NHBr + OH$$

$$R{-}CO{-}NH{-}Br + \bar{O}H \longrightarrow R{-}CO{-}\bar{N}{-}Br + H_2O$$

$$R{-}CO{-}\bar{N}{-}Br \xrightarrow{-Br^-} \underset{\text{(A Nitrene)}}{R{-}CO{-}\ddot{N}} \xrightarrow{\text{Rearrangement}} R{-}N{=}C{=}O$$

$$R{-}N{=}C{=}O + H_2O \longrightarrow [R{-}NH{-}COOH] \longrightarrow R{-}NH_2 + CO_2$$

$$R{-}CO{-}N_3 = R{-}CO{-}N{=}\overset{+}{N}{=}\overset{-}{N} \xrightarrow{\Delta} \underset{\text{(A nitrene)}}{[R{-}CO{-}\ddot{N}]} + N_2$$

$$[R{-}CO{-}\ddot{N}] \xrightarrow{\text{Rearrangement}} R{-}N{=}C{=}O \xrightarrow{H_2O} R{-}NH_2 + CO_2$$

It may be mentioned here that although aliphatic amines can be prepared by the two, above–mentioned, rearrangement reactions, however, they are more suitable for the preparation of aromatic amines since in the latter case the starting materials (acid amide and acid azide respectively) are readily available from the aromatic carboxylic acids. For example;

$$Ar{-}COOH \xrightarrow{SOCl_2} Ar{-}CO{-}Cl \begin{cases} \xrightarrow{NH_3} Ar{-}CO{-}NH_2 \\ \xrightarrow[NaN_3]{} Ar{-}CO{-}N_3 \end{cases}$$

(d) From carbonyl compounds : Carbonyl compounds, like aldehydes and ketones, react with ammonia or amines under reducing conditions to yield higher substituted amines. The reaction is known as reductive aminolysis and proceeds through the imino intermediates. For example;

$$\underset{R^2}{\overset{R^1}{\diagdown}}C{=}O + NH_3 \longrightarrow \left[\underset{R^2}{\overset{R^1}{\diagdown}}C{=}NH\right] \xrightarrow[\text{[+2H]}]{\text{Reduction}} \underset{R^2}{\overset{R^1}{\diagdown}}CH{-}NH_2$$

Imine

Usually, the reaction is carried out by heating the carbonyl compound with ammonium formate which decomposes to give ammonia and formic acid. Ammonia combines with the carbonyl compound to yield the imine which gets reduced by the hydrogen liberated due to the decomposition of formic acid to hydrogen and carbon dioxide.

(ii) Preparation of aromatic amines

Aromatic primary amines are most often prepared by the reduction of corresponding aromatic nitro compounds with metal/acid, H_2/Catalyst reducing agents.

$$Ar - NO_2 \xrightarrow{\text{Reduction}} Ar - NH_2$$

Aryl–alkyl secondary amines and aryl–dialkyl tertiary amines may be prepared by simple alkylation or successive alkylation of aromatic primary amines.

$$Ar{-}NH_2 \xrightarrow[\text{base}]{R^1{-}X/} Ar{-}NH{-}R^1 \xrightarrow[\text{base}]{R^2{-}X/} Ar{-}\underset{\underset{R^2}{|}}{N}{-}R^1$$

Diarylamines can be prepared by heating aromatic primary amines with substituted phenols in the presence of anhydrous zinc chloride.

$$Ar^1{-}OH + Ar^2{-}NH_2 \xrightarrow{ZnCl_2} Ar^1{-}NH{-}Ar^2 + H_2O$$

Preparation of Quaternary Ammonium Salts

These are usually prepared by heating tertiary amines with alkyl halides in a non-polar solvent like benzene, toluene, xylene etc.

$$R_3N + R'X \xrightarrow{\Delta} [R_3\overset{\oplus}{N}\text{–}R']\overset{\ominus}{X}$$

Both the reactants, the tertiary amine and alkyl halide, are usually soluble in the non–polar solvent where as the quaternary ammonium salt precipitates out. Upon treatment with alc. KOH or with silver oxide, quaternary ammonium salts are converted to the corresponding hydroxides which are strong bases comparable in strength to the alkali–metal hydroxides. For example;

$$R_4\overset{\oplus}{N}\overset{\ominus}{X} + KOH \longrightarrow R_4\overset{\oplus}{N}\overset{\ominus}{OH} + \overset{\oplus}{K}\overset{\ominus}{X}$$

14–4–6 Physical properties of Amines

The amines are reasonably polar compounds and exhibit the phenomenon of hydrogen bonding. The hydrogen bonding of amines is of the intermolecular type and arises as a consequence of –NH.....NH– interaction between different molecules of the same compound.

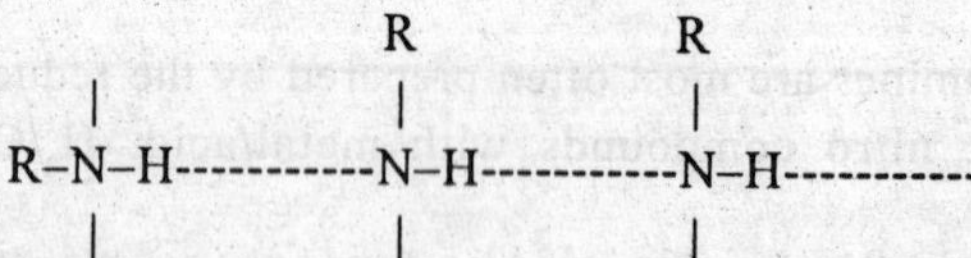

As a consequence, the amines possess higher melting and boiling points than the corresponding non–polar compounds of comparable molecular weights. However, the degree of intermolecular hydrogen bonding exhibited by amino groups is lower than that shown by hydroxyl groups and hence, amines show lower melting and boiling points than the corresponding alcohols or carboxylic acids. The phenomenon of hydrogen bonding is restricted to primary and secondary amines only. Tertiary amines, which lack hydrogen/hydrogens on the amino nitrogen, are incapable of showing hydrogen–bonding effects and, therefore, have physical constants close to hydrocarbons of comparable molecular weights.

All the three types of amines are capable of forming hydrogen bonds with water and, as a result, possess appreciable solubilities in water. The lower amines are completely soluble in water with borderline solubility at about five carbons. Amines with six or more carbons are either sparingly soluble or insoluble in water. This is due to the insolubilizing effect of the hydrocarbon portion outweighing the solubilizing effect of the amino portion of the molecules. The aromatic amines are still less soluble in water for the same reason. Table 14–2 shows the physical constants of some simple amino compounds.

Table 14–2 : Physical constants of some simple amino compounds

S.NO.	Name	Structure	B.P. °C	M.P. °C	Solubility in water(g lit^{-1})
1.	Methylamine	CH_3-NH_2	-7.5	-91.5	Completely soluble
2.	Ethylamine	$CH_3CH_2-NH_2$	17	-84	"
3.	n–Propylamine	$CH_3CH_2CH_2NH_2$	49	-83	"
4.	n–Butylamine	$CH_3CH_2CH_2CH_2NH_2$	78	-51	"
5.	Isobutylamine	$(CH_3)_2CHCH_2NH_2$	68	-86	"
6.	sec–Butylamine	$CH_3CH_2CH(CH_3)NH_2$	63	-104	"
7.	tert–Butylamine	$(CH_3)_3CNH_2$	45	-68	"
8.	Dimethylamine	$(CH_3)_2NH$	7.4	95	"
9	Trimethylamine	$(CH_3)_3N$	3.5	-117	"
10.	Diethylamine	$(C_2H_5)_2NH$	56	-48	"
11.	Triethylamine	$(C_2H_5)_3N$	90	-115	"
12.	Cyclohexylamine	$C_6H_{11}-NH_2$	134	-18	Sparingly soluble
13.	Benzylamine	$C_6H_5-CH_2-NH_2$	185		"
14.	N,N–Dimethylaniline	$C_6H_5-N(CH_3)_2$	154	2	"
15.	Aniline	$C_6H_5-NH_2$	184	-6	"
16.	p–Toluidine	$p-H_3C-C_6H_4-NH_2$	200	49	"

Simple aliphatic amines have the smell resembling that of ammonia. However, higher aliphatic and aromatic amines possess a characteristic, unpleasant odor sometimes referred to as fishy smell. Aromatic amines are generally very toxic in nature and prolonged exposure to vapors should be avoided. Most aromatic amines are colorless when pure or when freshly distilled. They, however, develop color (reddish to reddish black) on standing due to the formation of quinonoidal compounds by aerial oxidation.

14–4–7 Chemical properties of amines

The chemical properties may be divided into the following types;

(i) Basicity : Amines, being derivatives of ammonia, are basic in nature forming salts with acids in aqueous solution. However, the amines are relatively weak bases when compared with hydroxide ions, alkoxide ions or carbanions but moderately stronger when compared with ammonia. According to the Bronsted & Lowry concept of acids and bases, the basic strength of a compound depends on its ability to generate hydroxyl ions in solution or on its ability to extract a proton from an acid. Thus, in the following equilibrium reaction,

$$R-\ddot{N}H_2 + H_2O \rightleftharpoons R-\overset{\oplus}{N}H_3 + \overset{\ominus}{O}H$$

The equilibrium constant is given by the equation,

$$K = \frac{[R\text{-}\overset{\oplus}{N}H_3]\ [\overset{\ominus}{O}H]}{[R\text{-}NH_2]\ [H_2O]}$$

Since water is in far excess in the reaction so that its concentration does not change as a result of reaction, the equation becomes,

$$K_b = \frac{[R\text{-}\overset{\oplus}{N}H_3]\ [\overset{\ominus}{O}H]}{[R\text{-}NH_2]}$$

Where K_b is known as the basicity constant and its value is a measure of the basic strength of the amine. The amine ($R\text{-}NH_2$) acting as a base can acquire a proton changing itself to conjugated acid ($R\text{-}\overset{+}{N}H_3$) and the water ($H_2O$) acting as an acid, looses a proton changing itself to give a base $(OH)^-$ Thus, the reaction can be written down as a conjugate acid–base equilibrium system;

$$\text{Base}_1 + \text{Acid}_2 \rightleftharpoons \text{Acid}_1 + \text{Base}_2$$

It follows from simple considerations that

$$K_a \times K_b = K_w = 10^{-14}$$

Where K_a is the acidity constant for the acid $R\text{-}\overset{+}{N}H_3$ and K_w is the ion–product constant of water. The basic strength of a base can be expressed in terms of its K_b value or in terms of the acidic strength, K_a value, of its conjugate acid. In recent years, there has been a tendency to express the basic strength of a base in terms of its K_a value or more appropriately in terms of its pK_a value for three reasons; namely

(1) A large number of bases are insoluble in water to enable measurement of K_b values

(2) It is easier to prepare acid salts ($R\text{-}NH_3X$) from bases and then measure the K_a or pK_a values for the acid salts and

(3) Uniformity of scale for the acidity of acids and bases (that is, conjugate acids of bases) so that comparisons can be readily made. Table 14–3 shows the basicity constants of some simple amino compounds including ammonia.

TABLE 14–3 : pK_a VALUES OF SOME BASES

S.NO.	Name	Structure	K_b at 25°C	pK_a at 25°C
1.	Ammonia	NH_3	1.8×10^{-5}	9.25
2.	Methylamine	$CH_3{-}NH_2$	4.4×10^{-4}	10.65
3.	Dimethylamine	$(CH_3)_2NH$	5.2×10^{-4}	10.72
4.	Trimethylamine	$(CH_3)_3N$	5.0×10^{-4}	10.70
5.	Ethylamine	$CH_3CH_2{-}NH_2$	5.6×10^{-4}	10.75
6.	Diethylamine	$(C_2H_5)_2NH$	9.6×10^{-4}	10.98
7.	Triethylamine	$(C_2H_5)_3N$	5.7×10^{-4}	10.76
8.	Aniline	$C_6H_5{-}NH_2$	3.8×10^{-10}	4.58
9.	p–Toluidine	$p{-}H_3C{-}C_6H_4{-}NH_2$	1.2×10^{-9}	5.08
10.	p–Nitroaniline	$p{-}O_2N{-}C_6H_4{-}NH_2$	0.1×10^{-12}	1.00
11.	Guanidine	$H_2NC(=NH)NH_2$	5.0×10^{-11}	3.70

It is obvious from the Table 14–3 that primary amines are stronger bases than ammonia and that secondary amines are stronger than primary amines.

$$R_2NH \quad > \quad R-NH_2 \quad > \quad NH_3$$

This is understandable in view of the fact that alkyl groups are electron–donating when compared with hydrogen and so increase availability of electron–density on the basic nitrogen thus facilitating ready attack by an approaching proton. Furthermore, the ammonium ion thus formed is stabilized by electron–release from the alkyl groups.

$$R_2NH + \overset{+}{H} \longrightarrow R_2\overset{\oplus}{NH_2} \equiv R \rightarrow \overset{\overset{H}{|}}{\underset{\underset{R}{\uparrow}}{N}}-H$$

The case of tertiary amines is a little more complicated since according to the reasons given above, a tertiary amine should be the strongest base in the series, secondary > Primary > Ammonia. This is probably due to the steric bulk of the alkyl groups of the tertiary amine hindering approach of the proton and due to non–bonded repulsive interactions between the alkyl groups of the sterically crowded trialkylammonium ion.

It is also evident from the Table 14–3 that aniline is a far weaker base when compared to ammonia and other aliphatic amines. This is due to the fact that the non–bonding, lone pair of electrons on the nitrogen are drawn into a resonance with the benzene ring thereby making it less available to a proton. Further, the anilinium ion formed after protonation lacks this stabilization due to resonance. The different resonance structures involving interaction between the amino nitrogen and the benzene ring are shown below;

$$\ddot{N}H_2\text{-}C_6H_5 \longleftrightarrow \overset{\oplus}{N}H_2{=}C_6H_5^{\ominus}\,(o) \longleftrightarrow \overset{\oplus}{N}H_2{=}C_6H_5^{\ominus}\,(p) \longleftrightarrow \overset{\oplus}{N}H_2{=}C_6H_5^{\ominus}\,(o') \longleftrightarrow \ddot{N}H_2\text{-}C_6H_5$$

Based on the above reasoning it is logical to conclude that electron-donating groups in the nuclear substituted anilines will increase the basicity (as in the case of p–toluidine) and electron withdrawing groups will decrease the basicity (as in the case of p–nitroaniline). Guanidine is probably one of the strongest organic bases known since the guanidinium ion, produced by protonation of guanidine, is enormously stabilized by resonance. Further, the complete equivalence of different resonance structures increases the degree of resonance in the guanidinium ion as shown below;

$$\underset{\text{(Guanidine)}}{H_2N\text{–}\underset{\underset{NH}{\|}}{C}\text{–}NH_2} \xrightarrow{H^{\oplus}} H_2N\text{–}\overset{\oplus}{\underset{\underset{NH_2}{|}}{C}}\text{–}NH_2 \longleftrightarrow \overset{\oplus}{H_2N}{=}\underset{\underset{NH_2}{|}}{C}\text{–}NH_2$$

$$\longleftrightarrow H_2N\text{–}\underset{\underset{\overset{\oplus}{N}H_2}{\|}}{C}\text{–}NH_2 \longleftrightarrow H_2N\text{–}\underset{\underset{NH_2}{|}}{C}{=}\overset{\oplus}{N}H_2$$

(ii) Alkylation : Primary, secondary and tertiary amines react with alkyl halides to give secondary amines, tertiary amines and quaternary ammonium halides respectively. The reactions presumably involve nucleophilic attack (of the S_N2 type) by the amino group (through the non-bonding, lone–pair of electrons on nitrogen) on the positively polarized carbon–atom of the alkyl halide.

For example;

$$R^1\text{–}\ddot{N}H_2 + R \rightarrow X \longrightarrow R^1\text{–}\overset{\oplus}{N}H_2\text{–}R\ \overset{\ominus}{X} \xrightarrow{(-HX)} R^1\text{–}NH\text{–}R$$

$$R^1\text{–}\ddot{N}H\text{–}R^2 + R \rightarrow X \longrightarrow R^1\text{–}\underset{\underset{R}{|}}{\overset{\oplus}{N}H}\text{–}R^2\ \overset{\ominus}{X} \xrightarrow{(-HX)} R^1\text{–}\underset{\underset{R}{|}}{N}\text{–}R^2$$

$$R^1\text{–}\ddot{N}(R^3)\text{–}R^2 + R\text{–}X \longrightarrow R^1\text{–}\overset{\oplus}{N}(R)(R^3)\text{–}R^2\ X^{\ominus}$$

This reaction i.e., nucleophilic attack on the alkyl halide through the non–bonding electron pair, does not occur in the case of oxygenated compounds such as alcohols and ethers because of the higher nucleophilicity of nitrogen (in amines) than the oxygen atoms.

The quaternary ammonium halides react with bases to yield quaternary ammonium hydroxides which, on strong heating, undergo elimination to yield an alkene and a tertiary amine. This sequence of reactions forms the basis of the well–known Hofmann's procedure of exhaustive methylation technique for the structure determination of complex natural products such as alkaloids. Thus, the amine is treated with excess methyl iodide yielding a quaternary ammonium iodide which with moist silver oxide gives the corresponding ammonium hydroxide. The latter on strong heating is converted to the tertiary amine and alkene whose structures are subsequently established. Once the structures of the fragments are known, the structure of the parent compound is also deduced. The above sequence of reactions may be repeated until nitrogen is completely eliminated from the body of the parent molecule.

Pyrrolidine (N–H) $\xrightarrow{2CH_3I}$ N,N-dimethylpyrrolidinium $I^{\ominus}$ $\xrightarrow{Ag_2O/H_2O}$ [H, $\overset{\ominus}{O}H$, H_3C–$\overset{\oplus}{N}$–CH_3]

$\longrightarrow$ $CH_2{=}CH\text{–}CH_2\text{–}CH_2\text{–}N(CH_3)_2$ $\xrightarrow{CH_3I}$ $H_3C\text{–}\overset{\oplus}{N}(CH_3)\text{–}CH_3$ $I^{\ominus}$ $\xrightarrow{Ag_2O/H_2O}$ [H, HO, $\overset{\oplus}{N}$–CH_3, H_3C, CH_3]

$\longrightarrow$ $CH_2{=}CH\text{–}CH{=}CH_2$ + $N(CH_3)_3$

(iii) Acylation : Primary and secondary amines react with acyl halides andwith acid anhydrides to form N–Substituted amides. Tertiary amines do not react with these acylating agents.

$$R\text{–}NH_2 + R^1\text{–}CO\text{–}X \longrightarrow R\text{–}NH\text{–}CO\text{–}R^1 + HX$$

(X = Cl or Br)

$$R\text{–}NH_2 + R^1\text{–}CO\text{–}O\text{–}CO\text{–}R^1 \longrightarrow R\text{–}NH\text{–}CO\text{–}R^1 + R^1\text{–}CO\text{–}OH$$

When acyl halides are used for acylation, a base or an acid scavenger is usually employed in this reaction to eat up the HX liberated so that the latter does not tie up the starting amine. The well-known Schotten–Baumann reaction involves derivatization of amines to amides by treatment with aromatic acid halides in the presence of aqueous NaOH or pyridine. This reaction is often used in the undergraduate laboratory for the identification of amines by preparation of their rational derivatives.

$$R\text{–}NH_2 + Ar\text{–}CO\text{–}Cl \xrightarrow{\overset{\ominus}{O}H \text{ or Pyridine}} R\text{–}NH\text{–}CO\text{–}Ar + Cl^-$$

Aryl sulfonyl chlorides, also, react with primary and secondary amines to yield the corresponding sulfonamides. Tertiary amines do not take part in this reaction.

$$R\text{–}NH_2 + Ar\text{–}SO_2\text{–}Cl \longrightarrow R\text{–}NH\text{–}SO_2\text{–}Ar$$

$$R^1\text{–}NH\text{–}R^2 + Ar\text{–}SO_2\text{–}Cl \longrightarrow R^1\text{–}\underset{\displaystyle R^2}{\underset{|}{N}}\text{–}SO_2\text{–}Ar$$

$$R^1\text{–}\underset{\displaystyle R^3}{\underset{|}{N}}\text{–}R^2 + Ar\text{–}SO_2\text{–}Cl \longrightarrow \text{No reaction}$$

The above reaction forms the basis of the Hinsberg's method for the chemical separation of primary amine, secondary amine and tertiary amine in a mixture of the three. This is possible because the amide ($R\text{–}NH\text{–}SO_2\text{–}Ar$) obtained from a primary amine is soluble in aqueous NaOH due to the formation of the corresponding anion whereas the amide ($R^1\text{–}\underset{\displaystyle R^2}{\underset{|}{N}}\text{–}SO_2\text{–}Ar$) obtained from a secondary amine (lacking an acidic hydrogen on the amide nitrogen) is insoluble in aqueous NaOH. Thus,

$$R\text{–}NH\text{–}SO_2\text{–}Ar \xrightarrow{NaOH} R\text{–}\overset{\ominus}{N}\text{–}SO_2\text{–}Ar\ \overset{\oplus}{Na}$$

(Water soluble)

$$R^1\text{–}\underset{\displaystyle R^2}{\underset{|}{N}}\text{–}SO_2\text{–}Ar \xrightarrow{NaOH} \text{No Salt formation}$$

The usual procedure in Hinsberg's method is to treat a mixture of the three types of amines with arylsulfonyl chloride in pyridine. After completion of the reaction, the mixture is poured in water and extracted with ether. The ether extract is washed with aqueous NaOH. The alkali washing is neutralized with aqueous acid to yield the sulfonamide of primary amine which can be hydrolysed out separately to yield back the starting primary amine. The ether extract left out after alkali washing may then be subjected to aqueous acidic extraction to yield the tertiary amine salt in acidic extract. The left over ether solution may be evaporated to yield the sulfonamide of secondary amine. These reactions are shown in the chart given below;

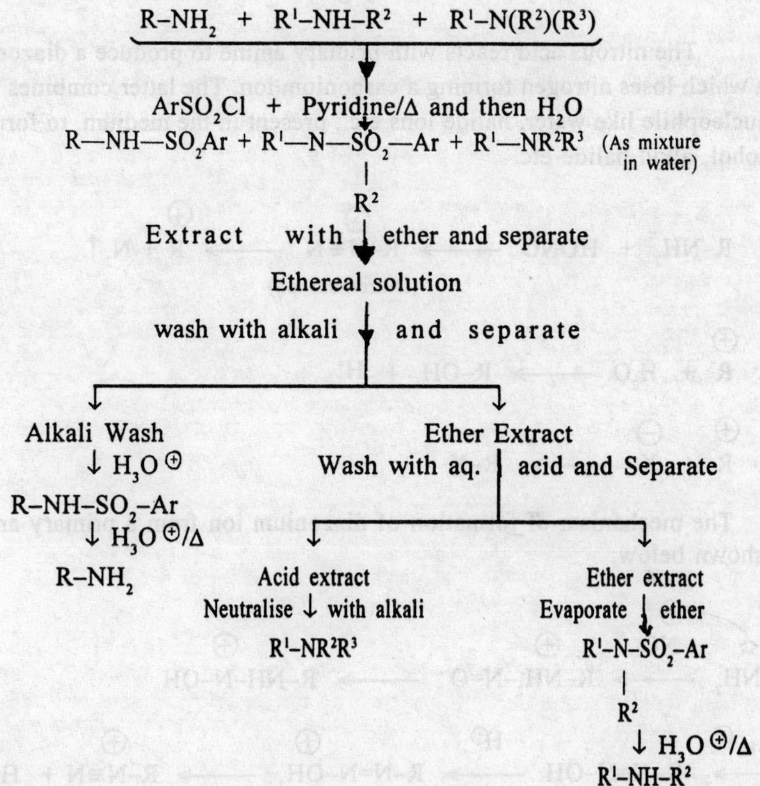

(iv) Reaction with nitrous acid : Primary, secondary and tertiary amines react with nitrous acid in cold conditions (0–5°C) to yield different types of products. The nitrous acid required in this reaction is usually obtained in situ by treating dilute sulfuric acid with sodium nitrite.

$$R\text{–}NH_2 + HONO \longrightarrow R\text{–}OH + H_2O + N_2 \uparrow$$

$$R^1\text{–}NH\text{–}R^2 + HONO \longrightarrow \underset{\substack{|\\ R^2}}{R^1\text{–}N}\text{–}N{=}O + H_2O$$

(N-Nitrosamine)

$$R^1\text{–}\underset{\underset{R_2}{|}}{N}\text{–}R^3 + HONO \longrightarrow \underset{\text{(A Salt)}}{R^1\text{–}\underset{\underset{R_2}{|}}{N}\text{–}R^3 . HNO_2}$$

The reaction of amines with nitrous acid is sometimes used as a test to differentiate between the three types of amines. Thus, a primary amine liberates a gas (N_2), a secondary amine gives a water–insoluble, yellow to green, oil and a tertiary amine merely gives a water–soluble salt.

The nitrous acid reacts with primary amine to produce a diazonium ion which loses nitrogen forming a carbonium ion. The latter combines with a nucleophile like water, halide ions etc., present in the medium, to form an alcohol, alkyl halide etc.

$$R\text{–}NH_2 + HONO \longrightarrow \underset{\text{(Diazonium ion)}}{R\text{–}\overset{\oplus}{N}\equiv N} \longrightarrow \overset{\oplus}{R} + N_2\uparrow$$

$$\overset{\oplus}{R} + H_2O \longrightarrow R\text{–}OH + H^+$$

$$\overset{\oplus}{R} + \overset{\ominus}{X} \longrightarrow R\text{–}X$$

The mechanism of formation of diazonium ion from a primary amine is shown below;

$$R\text{–}\ddot{N}H_2 \xrightarrow{\overset{\oplus}{N}O} R\text{–}\overset{\oplus}{N}H_2\text{–}N{=}O \longrightarrow R\text{–}NH\text{–}\overset{\oplus}{N}\text{–}OH$$

$$\xrightarrow{-H^{\oplus}} R\text{–}N{=}N\text{–}OH \xrightarrow{H^{\oplus}} R\text{–}N{=}N\text{–}\overset{\oplus}{O}H_2 \longrightarrow R\text{–}\overset{\oplus}{N}\equiv N + H_2O$$

(v) Condensation with carbonyl compounds : Primary amines condense with carbonyl compounds such as aldehydes or ketones to yield imino derivatives known as Schiff's bases. For example;

$$R\text{–}NH_2 + R^1\text{–}CO\text{–}R^2 \longrightarrow \underset{\text{(Imines or Schiff bases)}}{\overset{R^1 \quad R^2}{\underset{\underset{N\text{–}R}{||}}{C}}}$$

The reaction involves nucleophilic attack by the amino nitrogen on the carbonyl carbon followed by loss of elements of water.

(vi) Condensation with carbon disulfide : Primary amines react with carbon disulfide under refluxing conditions to yield dithiocarbamic acids. The latter on treatment with silver nitrate or mercuric chloride result in the formation of isothiocyanates.

$$R\text{-}\ddot{N}H_2 + S{=}C{=}S \longrightarrow R\text{-}\overset{\oplus}{N}H_2\text{-}C(=S)\text{-}S^{\ominus}$$

$$\longrightarrow R\text{-}NH\text{-}C(=S)\text{-}SH \xrightarrow[\overset{\oplus}{Hg}]{\overset{\oplus}{Ag}\text{ or}} R\text{-}N{=}C{=}S$$

The above sequence of reaction is known as Hofmann's mustard oil reaction as the product (ie. isothiocyanate) has a characteristic mustard oil smell and the reaction is used for the detection of primary amines.

(vii) Carbylamine test : Warming an alcoholic solution of primary amines (aliphatic or aromatic) with a mixture of aqueous KOH and chloroform leads to the formation of carbylamines (also called isocyanides). The latter compounds possess a characteristic, unpleasant smell akin to that of garlic oil and so the reaction is at times used as a test for the identification of primary amines.

$$R\text{-}NH_2 + CHCl_3 + 3KOH \longrightarrow \underset{\text{(Carbylamine or isocyanide)}}{R\text{-}NC} + 3KCl + H_2O$$

The reaction appears to follow the mechanism given below;

$$CHCl_3 + KOH \longrightarrow :CCl_2 + KCl + H_2O$$

$$R\text{-}\ddot{N}H_2 + :CCl_2 \longrightarrow R\text{-}\overset{\oplus}{N}H_2\text{-}\overset{\ominus}{C}Cl_2 \longrightarrow R\text{-}NH\text{-}CHCl_2$$

$$R\text{-}NH\text{-}CHCl_2 \xrightarrow{\overset{\ominus}{O}H} R\text{-}\ddot{N}{=}CH\text{-}Cl \longrightarrow R\text{-}\overset{\oplus}{N}{\equiv}C\text{-}H$$

$$R{-}\overset{\oplus}{N}{\equiv}C{-}H \xrightarrow{\overset{\ominus}{OH}} R{-}\overset{\oplus}{N}{\equiv}\overset{\ominus}{C} \equiv R{-}NC$$

(vii) Ring substitution reactions : Nitration, sulfonation, halogenation, Friedel–Crafts alkylation or other ring substitution reactions are usually not carried out on aromatic primary amines. This is due to the formation of multisubstituted products since the amino group is a strongly electron-releasing group and activates the aromatic ring towards electrophilic aromatic substitution. This problem is commonly overcome by first acylating the amino group, then carrying out the desired ring substitution reaction followed by deacylation using hydrolysis. A typical reaction sequence may be as follows;

$$C_6H_5NH_2 \xrightarrow{Ac_2O} C_6H_5NHAc \xrightarrow{Br_2} p\text{-}Br{-}C_6H_4NHAc \xrightarrow{H_3O^{\oplus}} p\text{-}Br{-}C_6H_4NH_2$$

$$C_6H_5NH_2 \xrightarrow{Br_2} 2,4,6\text{-}Br_3C_6H_2NH_2$$

(*viii*) **Oxidation :** The oxidation of primary and secondary amines is a very complicated process since a number of products are formed depending upon the nature of amine, oxidizing agent and reaction conditions. The oxidation of tertiary amines with hydrogen peroxide or per acids gives amine oxides.

$$R^1{-}NR^2R^3 \xrightarrow[\text{Peracid}]{H_2O_2 \text{ or}} R^1{-}\underset{\underset{O}{\downarrow}}{\overset{R^2}{\overset{|}{N}}}{-}R^3 \longleftrightarrow R^1{-}\underset{\underset{O^{\ominus}}{|}}{\overset{R^2}{\overset{|}{N^{\oplus}}}}{-}R^3$$

Mechanism :

$$H{-}O{-}O{-}H + \overset{\oplus}{H} \rightleftharpoons H{-}O{-}\overset{\oplus}{O}H_2$$

$$R^1{-}\ddot{N}R^2R^3 + H{-}O{-}\overset{\oplus}{O}H_2 \longrightarrow R^1{-}\underset{\underset{OH}{|}}{\overset{\oplus}{N}}R^2R^3 + OH_2$$

$$R^1\text{-}\overset{\oplus}{N}R^2R^3(\text{-}OH) \longrightarrow R^1\text{-}\overset{\oplus}{N}R^2R^3(\rightarrow O) + \overset{\oplus}{H}$$

14–5 AROMATIC DIAZONIUM SALTS

It was mentioned earlier, that primary amines react with nitrous acid in cold conditions to yield diazonium salts. The reaction is known as diazotization and with an aromatic amine, the following may be shown:–

$$Ar\text{-}NH_2 \xrightarrow[(NaNO_2 + H_3O^+)]{HONO\ (0\text{-}5^\circ)} Ar\text{-}\overset{\oplus}{N}\equiv N\ \overset{\ominus}{X}$$

The diazonium salts obtained from aliphatic primary amines are relatively less stable when compared to those derived from aromatic primary amines. This is due to the stabilizing effect of resonance between the aromatic ring and the diazo group. For example;

$$C_6H_5\text{-}\overset{\oplus}{N}\equiv N \longleftrightarrow \text{(ortho } \overset{\oplus}{C}\text{H)}=\overset{\oplus}{N}=\overset{\ominus}{N} \longleftrightarrow H\text{-}\overset{\oplus}{C}\text{(para)}=\overset{\oplus}{N}=\overset{\ominus}{N} \longleftrightarrow \text{(ortho } \overset{\oplus}{C}\text{H)}=\overset{\oplus}{N}=\overset{\ominus}{N}$$

Consequently, the aromatic diazonium salts can be readily prepared and utilized for reactions whereas the aliphatic diazonium salts decompose spontaneously losing nitrogen and yielding an alcohol/alkyl halide.

14–5–1 Properties of aromatic diazonium salts

The arene diazonium salts such as chloride, bromide, sulfate etc., are water soluble whereas the fluoroborate, naphthalene sulfonate, trifluoroacetate etc. are water insoluble. The water soluble salts ionize and are stable in cold water but decompose and react with water on raising the temperature. The diazonium salts decompose readily in the dry state and some, like nitrate, explode spontaneously when isolated. On the large scale, these salts should be handled very carefully. Usually, these salts are generated in situ under cold conditions and used as such immediately from reactions. The diazonium salts form complexes with metal ions, particularly zinc, and the metal complexes are reasonably stable. Thus, complexation offers a method to stabilize diazonium ions in solution.

$$2[Ph\text{-}\overset{\oplus}{N}\equiv N]\overset{\ominus}{Cl} + \overset{+2}{Zn} + 2\overset{\ominus}{Cl} \longrightarrow \left[\begin{matrix} Cl & & N\equiv\overset{\oplus}{N}\text{-}Ph \\ & Zn & \\ Ph\text{-}\overset{\oplus}{N}\equiv N & & Cl \end{matrix}\right] + 2\,Cl^{\ominus}$$

The arenediazonium salts are significant synthetic intermediates and undergo two important types of reactions. These are;

(A) Those in which the diazo group is replaced by a monovalent nucleophile called replacement reactions and

(B) Those in which the diazo group is retained and it acts as an electrophile attacking electron–rich aromatic systems, called coupling reactions. These reactions are discussed below in detail.

(A) Replacement reactions

(1) Replacement by hydroxyl group : When a solution of the diazonium salt in dil. H_2SO_4 is warmed or boiled, the diazo group is replaced by hydroxyl group and the product is a phenol. for example;

$$Ar{-}\overset{\oplus}{N_2} \xrightarrow[\Delta]{Dil.H_2SO_4/} \overset{\oplus}{[Ar]} + N_2\uparrow \xrightarrow{HOH} [Ar{-}\overset{\oplus}{OH_2}] \longrightarrow Ar{-}OH$$

This reaction is used for the laboratory preparation of phenols.

(2) Replacement by Cl, Br, I or CN groups : When an aqueous solution of a diazonium salt is treated with potassium halide or cyanide in the presence of metallic copper powder, replacement of diazo group occurs to yield a halo or cyano derivative. The reaction is known as **Gattermann reaction.**

$$\underset{(X=\text{halo or CN})}{Ar{-}\overset{\oplus}{N_2} + KX} \xrightarrow{Cu/\Delta} \underset{(X=\text{halo or CN})}{Ar{-}X} + N_2\uparrow + \overset{\oplus}{K}$$

It has been observed that better yields of products are obtained if a cuprous halide or cuprous cyanide is used instead of potassium halide or cyanide + copper combination in the above reaction. In that case, the reaction is known as **Sandmeyer reaction.**

$$\underset{(X=\text{halo or CN})}{Ar{-}\overset{\oplus}{N_2} + CuX} \longrightarrow Ar{-}X + N_2\uparrow + \overset{\oplus}{Cu}$$

(3) Replacement by F : When a solution of diazonium salt in water is treated with sodium fluoroborate,the fluoroborate salt of arenediazonium ion separates out. The latter on filtering, washing, drying and subsequent decomposition yields aryl fluorides, nitrogen and boron trifluoride.

$$Ar{-}\overset{\oplus}{N_2} + NaBH_4 \longrightarrow Ar{-}\overset{\oplus}{N_2}\,\overset{\ominus}{BF_4}\downarrow + \overset{\oplus}{Na}$$

$$Ar{-}\overset{\oplus}{N_2}\,\overset{\ominus}{BF_4} \xrightarrow{\Delta} Ar{-}F + N_2\uparrow + BF_3\uparrow$$

The reaction provides a convenient method for the preparation of aryl fluorides and is known as Shiemann reaction.

(4) Replacement of H : Treatment of a solution of arenediazonium salt with hypophosphorus acid or of an arenediazonium fluoroborate with sodium borohydride yields the parent arene.

$$Ar{-}\overset{\oplus}{N_2} \xrightarrow{H_3PO_2} Ar{-}H + N_2\uparrow$$

$$Ar{-}\overset{\oplus}{N_2}\,\overset{\ominus}{BF_4} \xrightarrow{NaBH_4} Ar{-}H + N_2\uparrow$$

The reaction provides a method for the deamination or reduction of substituted aromatic primary amines yielding arenes otherwise not readily accessible. For example;

NH_2 (p-toluidine) $\xrightarrow{Ac_2O}$ NHAc, CH_3 $\xrightarrow{Cl_2/AcOH}$ NHAc, Cl, CH_3 $\xrightarrow{(i)\ H_3\overset{\oplus}{O}\ (ii)\ NaNO_2\ (iii)\ H_3PO_2}$ Cl, CH_3

(m-Chlorotoluene)

(5) Replacement by NO_2 : The diazo group undergoes replacement with a nitro group when the fluoroborate salt is treated with $NaNO_2$ in the presence of Cu metal powder.

$$Ar{-}\overset{\oplus}{N_2}\,\overset{\ominus}{BF_4} + Na\overset{\oplus\ominus}{NO_2} \xrightarrow{Cu} Ar{-}NO_2 + N_2\uparrow + NaBF_4$$

The reaction seems to be useful for the syntheses of aromatic nitro compounds containing electron–withdrawing groups in the o– and p–positions. For example;

(o-Nitroaniline) $\xrightarrow[\text{(ii) NaBF}_4]{\text{(i) NaNO}_2}$ (o-Nitrophenyl-diazonium fluoroborate) $\xrightarrow[\text{Cu}]{\text{NaNO}_2}$ (o-Dinitro-benzene)

(6) Replacement of aryl : Reaction of an aryldiazonium salt with an activated aromatic ring system, such as anisole, in the presence of a base, like NaOH, NaOAc etc., yields biaryl derivatives.

$$Ar\text{–}\overset{\oplus}{N_2} + Ar^1\text{–}H \xrightarrow[\text{NaOAC}]{\text{NaOH or}} [\overset{\oplus}{Ar}] + N_2\uparrow \longrightarrow Ar\text{–}Ar^1 + H_2O$$

(B) Coupling reactions

These are some of the most important and characteristic reactions of arenediazonium salts. It is these reactions that make the salts crucial intermediates in the syntheses of dyes, drugs and pharmaceuticals. In fact, it is these reactions that distinguish categorically between arene diazonium salts and their aliphatic analogues. The coupling reactions are based on the fundamental property of these salts to act as weak electrophiles and to attack highly activated aromatic nuclei such as the naphthols bringing about electrophilic aromatic substitution. For example;

$$R\text{–}C_6H_4\text{–}NH_2 \xrightarrow{HNO_2} R\text{–}C_6H_4\text{–}\overset{\oplus}{N}\equiv N \xrightarrow{\beta\text{-Naphthol}} R\text{–}C_6H_4\text{–}N{=}N\text{–}C_{10}H_6\text{–}OH$$

(Coupled product)

The azo group (–N = N–) is a good chromophore and continuous conjugation between two aromatic systems through an azo group imparts beautiful color to the compound. Usually the coupled products are brightly colored substances, orange to red colored, and are referred to as azo dyes. Several of these coupled products find applications as dyes in textile industry.

Another important characteristic of diazonium salts is their reduction with sodiumsulfite to arylhydrazines.

$$Ar{-}\overset{\oplus}{N}{\equiv}N \xrightarrow{Na_2SO_3} \underset{\text{(Aryl hydrazine)}}{Ar{-}NH{-}NH_2}$$

The reaction seems to proceed through a radical anionic pathway.

14–6 CYANIDES AND ISOCYANIDES

Hydrogen cyanide or hydrocyanic acid is a very weak acid and is known to exist in two tautomeric forms;

$$\underset{(I)}{H{-}C{\equiv}N} \rightleftharpoons \underset{(II)}{\overset{\ominus}{C}{\equiv}\overset{\oplus}{N}{-}H}$$

In the form(I), hydrogen is bound to carbon whereas in form (II) hydrogen is bound to nitrogen. Alkyl or aryl derivatives of form (I) are known as cyanides whereas alkyl or aryl derivatives of form(II) are known as isocyanides. The cyanides are named by giving the name of the alkyl/aryl group and adding the ending cyanide. Similar is the nomenclature for isocyanides. Cyanides are also called nitriles and for nomenclature purposes, compounds may be named by adding the ending nitrile to the name of the acid which is formed upon its hydrolysis. Analogously, isocyanides are called isonitriles. For example;

$CH_3{-}CN$	Methyl cyanide or acetonitrile/methanonitrile
$CH_3{-}CH_2{-}CN$	Ethyl cyanide or propionitrile.
$Ph{-}CH_2{-}CN$	Benzyl cyanide or phenyl acetonitrile / phenylmetha nonitrile.
$(H_3C)_2{-}CH{-}NC$	Isopropyl isocyanide or isopropyl isonitrile.
$H_3C{-}CH_2{-}CH_2{-}CH_2{-}NC$	n–Butyl isocyanide or n–Butylisonitrile.
$C_6H_{11}{-}\overset{\oplus}{N}{\equiv}\overset{\ominus}{C}$ (cyclohexyl ring)	Cyclohexyl isocyanide or Cyclohexyl isonitrile

14–6–1 Preparation of cyanides and isocyanides

(i) From alkyl halides : Treatment of alkyl halides with KCN leads to the formation of alkyl cyanides as major products. However, use of AgCN in this reaction yields predominantly the isocyanides.

$$R{-}X + KCN \longrightarrow R{-}CN + KX$$

$$R{-}X + AgCN \longrightarrow R{-}NC + AgX$$

The cyanide ion is an ambient nucleophile and, depending upon the reaction conditions (solvent, temperature, counter cation etc), yields alkyl cyanides/alkyl isocyanides. In any case, a minor amount of the other isomer is invariably produced.

(ii) Cyanides from amides/from aldoximes : Dehydration of an acid amide or an aldoxime leads to the formation of cyanides.

$$R\text{–}CO\text{–}NH_2 \xrightarrow{P_2O_5 \text{ or } Ac_2O} R\text{–}C{\equiv}N$$

$$\underset{\substack{||\\ N\text{–}OH}}{R\text{–}C\text{–}H} \xrightarrow{Ac_2O/\Delta} R\text{–}C{\equiv}N$$

(iii) Isocyanides from primary amines : Reaction of primary amines with $CHCl_3/KOH$, under anhydrous conditions, yields isocyanides.

$$R\text{–}NH_2 + CHCl_3 + 3KOH \longrightarrow R\text{–}NC + 3\ KCl + 3\ H_2O$$

14–6–2 Properties of cyanides and isocyanides

The cyanides are colorless, neutral, liquids with not much unpleasant smell. They are reasonably polar compounds, due to the polarity of C≡N bond, but do not possess any H–bonding effects. Consequently, their boiling points are higher than the corresponding hydrocarbons of comparable molecular weights but lower than those of carboxylic acids. Except for lower molecular weight compounds, most nitriles are insoluble in water. The physical constants of some of the nitriles are shown in the table 14–4:

Table 14–4 : Physical constants of nitriles

S.NO	Compound	Name	Dipole Moment (D)	B.P. °C
1.	CH_3–CN	Acetonitrile	3.90	82
2.	CH_3–CH_2–CN	Propionitrile	4.02	97
3.	$CH_3CH_2CH_2$–CN	n–Butyronitrile	4.00	117
4.	C_6H_5–CN	Benzonitrile	4.39	191
5.	C_6H_5–CH_2–CN	Benzyl cyanide	—	234

In contrast to cyanides, the isocyanides are toxic compounds with highly unpleasant odors. They are less polar than the cyanides and are, therefore, lower boiling. Due to the presence of an electron pair on carbon, the isocyanides are less stable and undergo ready oxidation to yield oxidized products.

$$R-\overset{+}{N}\equiv\overset{-}{C} \xrightarrow{[O]} R-N=C=O$$

The cyanides undergo hydrolysis in acidic as well as in basic conditions to give carboxylic acids and ammonia whereas isocyanides yield amines and formic acid. Thus,

$$R-C\equiv N + H_2O \xrightarrow{\overset{+}{H} \text{ or } HO^-} R-COOH + NH_3$$

$$R-\overset{\oplus}{N}\equiv\overset{\ominus}{C} + H_2O \xrightarrow{\overset{+}{H} \text{ or } HO^-} R-NH_2 + H-CO-OH$$

The mechanisms of the two reactions are believed to be as follows;

Under acidic conditions :

$$\underset{\text{(Nitrile)}}{R-C\equiv N} \longleftrightarrow R-\overset{\oplus}{C}=\overset{\ominus}{N} \xrightarrow{H^\oplus} R-\overset{\oplus}{C}=NH \xrightarrow{H_2O} R-C(=NH)-\overset{\oplus}{O}H_2$$

$$\longrightarrow R-\overset{\oplus}{C}(-OH)-NH_2 \longrightarrow \underset{\text{(An amide)}}{R-C(=O)-NH_2} \xrightarrow{H^\oplus} R-\overset{\oplus}{C}(-OH)-NH_2$$

$$\longleftrightarrow R-C(=\overset{\oplus}{O}H)-NH_2 \longleftrightarrow R-C(-OH)=\overset{\oplus}{N}H_2 \xrightarrow{H_2O} R-C(-\overset{\oplus}{O}H_2)(-OH)-NH_2$$

$$\longrightarrow R-C(-OH)(-OH)-\overset{\oplus}{N}H_3 \longrightarrow R-C(=O)-OH + NH_3$$

$$\underset{\text{(Isonitrile)}}{R-\overset{\oplus}{N}\equiv\overset{\ominus}{C}} \xrightarrow[H^\oplus]{} R-\overset{\oplus}{N}\equiv C-H \longleftrightarrow R-N=\overset{\oplus}{C}-H \xrightarrow{H_2O}$$

$$R{-}N{=}\underset{\oplus OH_2}{\underset{|}{C}}{-}H \longrightarrow R{-}\overset{H}{\overset{|}{N}}{-}\overset{\oplus}{\underset{\underset{OH}{|}}{C}}{-}H \longleftrightarrow R{-}\overset{H}{\overset{|}{N}}{=}\underset{\underset{OH}{|}}{C}{-}H\ (\oplus \text{ on } N) \longleftrightarrow$$

$$R{-}\overset{H}{\overset{|}{N}}{-}\underset{\underset{\oplus O{-}H}{||}}{C}{-}H \xrightarrow{H_2O} R{-}\overset{H}{\overset{|}{N}}{-}\overset{H}{\overset{|}{C}}{-}\overset{\oplus}{O}H_2\ (\text{C–OH below}) \longrightarrow R{-}\overset{H}{\overset{|}{\underset{\underset{H}{|}}{\overset{\oplus}{N}}}}{-}\overset{H}{\overset{|}{\underset{\underset{\ominus O{-}H}{|}}{C}}}{-}OH$$

$$\longrightarrow \left[R{-}NH_2 + \overset{H}{\overset{|}{\underset{\underset{O{-}H}{|}}{\overset{\oplus}{C}}}}{-}OH\right] \longrightarrow R{-}\overset{\oplus}{N}H_3 + H{-}\underset{\underset{O}{||}}{C}{-}OH$$

Under alkaline conditions :

$$\underset{\text{(Nitrile)}}{R{-}C{\equiv}N} \longleftrightarrow R{-}\overset{\oplus}{C}{=}\overset{\ominus}{N} \xrightarrow{\overset{\ominus}{O}H} R{-}\overset{O{-}H}{\overset{|}{C}}{=}\overset{\ominus}{N} \xrightarrow{H{-}O{-}H}$$

$$\left[R{-}\underset{\underset{O{-}H}{|}}{C}{=}N{-}H + \overset{\ominus}{O}H\right] \longrightarrow \left[R{-}\underset{\underset{O}{||}}{C}{-}NH_2 + \overset{\ominus}{O}H\right] \longrightarrow$$

$$R{-}\overset{OH}{\overset{|}{\underset{\underset{O^{\ominus}}{|}}{C}}}{-}NH_2 \xrightarrow{H_2O} R{-}\overset{O{-}H}{\overset{|}{\underset{\underset{O{-}H}{|}}{C}}}{-}NH_2 + \overset{\ominus}{O}H \longrightarrow R{-}\overset{OH}{\overset{|}{\underset{\underset{O^{\ominus}}{|}}{C}}}{-}\overset{\oplus}{N}H_3$$

$$\longrightarrow R{-}\underset{\underset{O}{||}}{C}{-}OH + NH_3$$

$$\underset{\text{(Isonitrile)}}{R{-}\overset{\oplus}{N}{\equiv}\overset{\ominus}{C}} \xrightarrow[HOH]{} [R{-}\overset{\oplus}{N}{\equiv}C{-}H + \overset{\ominus}{O}H] \longleftrightarrow [R{-}N{=}\overset{\oplus}{C}{-}H + \overset{\ominus}{O}H]$$

$$\longrightarrow R\text{–}N\text{=}C(H)(OH) \longleftrightarrow R\text{–}\overset{\ominus}{N}\text{–}\overset{\oplus}{C}(H)(OH) \longleftrightarrow R\text{–}\overset{\ominus}{N}\text{–}C(H)(\overset{\oplus}{O}\text{–}H)$$

$$\xrightarrow{HOH} R\text{–}\overset{\ominus}{N}\text{–}C(H)(\overset{\oplus}{O}H_2)\text{–}OH \longrightarrow R\text{–}N(H)\text{–}C(H)(O\text{–}H)\text{–}O\text{–}H \longrightarrow$$

$$R\text{–}\overset{\oplus}{N}H_2\text{–}C(H)(\overset{\ominus}{O})\text{–}O\text{–}H \longrightarrow R\text{–}NH_2 + H\text{–}\underset{\underset{O}{\|}}{C}\text{–}OH$$

Both the cyanides and the isocyanides undergo reduction with suitable reagents to yield amines. For example:

$$R\text{–}C\equiv N \xrightarrow[\text{or } H_2/\text{Raney Ni}]{\text{Na/Et.OH or LiAlH}_4} R\text{–}CH_2\text{–}NH_2$$

$$R\text{–}N\equiv C \xrightarrow[H_2/\text{Raney Ni}]{\text{Na(Hg) or}} R\text{–}NH\text{–}CH_3$$

14–6–3 Uses of nitriles and isonitriles

The cyanides serve as important intermediates in multi step syntheses of compounds in the laboratory. The isocyanides, due to their characteristic unpleasant odor help in carrying out diagnostic tests on primary amines. Hydrogen cyanide is very toxic, but commercially useful compound. Acrylonitrile is used in the manufacture of polyacrylonitrile and other synthetic polymers. Acetonitrile is used commercially as a solvent and in spectroscopy.

14–7 SELF TEST QUESTIONS

Indicate the correct answer for each of the following questions from amongst the choices provided;

14-7-1 The relationship between 1–nitropentane and amylnitrite is that of:
(A) Tautomers (B) Isomers (C) Stereoisomers (D) Mesomers.

14-7-2 The commonly used reagent for generation of nitrous acid in aqueous medium is:
(A) $NaNO_2$+HCl (B) $NaNO_3$+H_2SO_4 (C) KNO_3+HCl (D) $NaNO_2$+HNO_3

14-7-3 1–Nitrobutane dissolves in aqueous alkali because of
(A) Basicity of α–Hydrogens (B) Acidity of α–Hydrogens.
(C) Acidity of β–hydrogens (D) Hydrolysis of nitro group by the alkali

14-7-4 Aliphatic nitro compounds can be prepared by
(A) Vapor phase nitration of alkanes
(B) Sulfonation followed by nitration of alkanes
(C) Treating alkanes with nitrating mixture at room temperature.
(D) Treating alkanes with sodium nitrite.

14-7-5 Nitrobenzene can be prepared from benzene by treatment with Conc.HNO_3 in the presence of.
(A) Conc.H_3PO_4 (B) Conc.H_2SO_4
(C) Dil.H_2SO_4 (D) Friedel–Crafts catalyst like $AlCl_3$.

14-7-6 Reduction of nitrobenzene with hydrogen over Pd–C yields:
(A) Phenyl hydrazine (B) Phenyl hydroxylamine
(C) Azobenzene (D) Aniline.

14-7-7 Which one of the following phenols is most acidic in nature?
(A) p–Methylphenol (B) p–Chlorophenol
(C) p–Aminophenol (D) p–Nitrophenol.

14-7-8 In the preparation of an azo dye, arenediazonium ion is usually coupled with
(A) Phenol (B) Aniline
(C) Sodium salt of β–naphthol (D) α–Naphthol

14-7-9 Aniline is best prepared in the laboratory by
(A) Amination of benzene (B) Nitration of benzene
(C) Reduction of nitrobenzene (D) Nitrosation of benzene

14-7-10 Aliphatic primary amines may be prepared by
(A) Amination of alkyl halides (B) Amination of alcohols
(C) Amination of ethers (D) Amination of acids.

14-7-11 The most basic compounds among the following isomers is:
(A) p–Toluidine (B) Benzylamine
(C) N–Methylaniline (D) o–Toluidine

14-7-12 Aromatic primary amines give positive
(A) Lucas Test (B) Iodoform test
(C) Diazotizationtest (D) Unsaturation test.

14-7-13 Acetanilide is formed from aniline by the action of
(A) Acetic acid (B) Acetic anhydride
(C) Sodium acetate (D) None of these reagents.

14-7-14 Gabriel's phthalimide method is convenient for the synthesis of
(A) Aliphatic primary alkyl halides (B) Aliphatic primary alcohols
(C) Aliphatic primary amines (D) Aliphatic carboxylic acids.

14-7-15 Treatment of benzenediazonium chloride with cuprous chloride to form chlorobenzene is an example of
(A) Diazotization reaction (B) Michael reaction
(C) Coupling reaction (D) Sandmeyer reaction.

14-7-16 Reaction of p–nitrochlorobenzene with hot aqueous NaOH to yield p–nitrophenol is a case of
(A) Electrophilic aromatic substitution (B) Nucleophilic aromatic substitution
(C) Free–radical substitution (D) None of the above reactions.

14-7-17 Organic cyanides on complete hydrolysis yield

(A) Alcohols (B) Esters

(C) Carboxylic acids (D) Amines

14–8 ESSAY AND SHORT TYPE QUESTION

14–8–1 Write short notes on the following:

(a) How are aliphatic nitro compounds prepared?

(b) Describe the preparation of nitrobenzene. What are the different reagents employed for nitration?

(c) What are the different reagents, different conditions and different products obtained on reduction of nitrobenzenes? (A.U. Mar., 1993)

(d) How is the amination of alkyl halides carried out? What are the different products obtained in these reactions in successive stages?

(e) What is Gabriel's method for preparation of amines? How is it advantageous over the other method of amination of alkyl halides?

(f) How is diazotization of aniline carried out? What is its reaction with β–naphthol in alkaline media?

(g) Describe in brief the different replacement reactions undergone by arenediazonium salts

(h) How can a simple mixture of a primary, secondary and tertiary amine be separated by a chemical method? What is the method known as? How does it compare with a simple physical method such as fractional distillation of the mixture of three compounds? (N.U. Mar., 1995), (N.U. Sep., 1995), (O.U. Oct., 1993)

14–8–2 Give explanations for the following observations:

(i) Aniline is a much weaker base than cyclohexylamine.

(ii) Dimethylamine is a stronger base than methylamine.

(iii) Arenediazonium ions are stabler than their aliphatic analogues.

(iv) Alkyl nitrites get hydrolysed by aqueous NaOH whereas nitroalkanes form salts with it.

(v) Diazotization of aniline is usually carried out in strongly acidic media.

(vi) Coupling of arenediazonium salts is usually carried out with β–naphthol and not with α-naphthol.

(vii) The principal product of reaction of aniline with bromine under alkaline conditions is tribromoaniline and not monobromo or dibromo derivative even if less than three moles of bromine are employed in the reaction.

14–8–3 Explain how the following pairs of compounds can be distinguished from one another by qualitative test tube experiments?

(a) Amyl nitrite and 1–nitropentane

(b) Nitrobenzene and aniline

(c) Aniline and triethylamine.

(d) n–Butylamine and aniline

(e) p–Toluidine and benzylamine

(f) Acetanilide and ammonium acetate

(g) Anilinium chloride and p–chloroaniline

(h) 3–Aminopentane and 3–cyanopentane

(i) Phenyl isocyanide and benzylamine

(j) o–Toluidine and N–Methylaniline.

14–8–4 Starting from benzene and using common chemical reagents, how can the following compounds be synthesized in the laboratory?

(i) N–Methylaniline (ii) p–chloroaniline
(iii) p–Nitrophenol (iv) o–Nitroaniline
(v) p–Bromoacetanilide (vi) Benzyl cyanide
(vii) Benzanilide (viii) Phenyl isocyanide
(ix) o–Chlorobenzamide (x) m–Nitrochlorobenzene.

14–8–5 Paracetamol is an analgesic, antipyretic drug. It is manufactured by the following sequence of reactions;

$$\text{(p-Nitrochlorobenzene)} \xrightarrow[\text{then } H_3O^{\oplus}]{NaOH/\Delta} C_6H_5NO_3 \xrightarrow[\Delta]{Fe/H_2O/H^{\oplus}} C_6H_7NO \xrightarrow{Ac_2O} C_8H_9NO_2 \text{ (Paracetamol)}$$

Suggest a suitable structure for paracetamol.

14–8–6 Discuss the Hinsberg's method used for separation of amines. (S.K.U.,Apr 1990)

14–8–7 How nitrobenzene is prepared? Write an account of reduction products of nitrobenzene. (S.K.U. Apr., 1990), (A.U. Mar., 1993)

14–8–8 Discuss the action of acids on amines. (S.K.U.,Apr 1990)

14–8–9 (a) Arrange the following bases in the order of their basic strength and justify your answer.
Aniline, Ammonia, p–Methylaniline, p–Nitroaniline.

(b) Name the reagents required in the following conversions;

(i) $PhNO_2 \longrightarrow Ph-NH-NH-Ph$
(ii) $PhNH_2 \longrightarrow Ph-NC$
(iii) $Ph-H \longrightarrow Ph-NO_2$
(iv) $PhNC \longrightarrow Ph-NH-CH_3$ (K.U.,Apr 1990)

14–8–10 Which one of the following is more basic and why?

(a) Aniline and m–nitroaniline
(b) Methylamine and ammonia (O.U. Apr., 1990), (O.U. Mar., 1996)

14–8–11 (a) How are the following compounds prepared? Give equations with necessary conditions?

(i) Methylamine from acetic acid
(ii) Aniline from benzene
(iii) Nitrobenzene from aniline

(b) What is the action of nitrous acid on the following compounds

(i) Ethylamine (ii) Aniline (iii) Nitromethane (O.U.,Nov 1986)

14–8–12 Arrange the following compounds in order of their decreasing basic strength :

$C_6H_5-NH_2$, $p-O_2N-C_6H_4-NH_2$, $p-H_3CO-C_6H_4-NH_2$

Give justification for the order you have choosen. (O.U.,Nov 1986)

14–8–13 How do cyanides and isocyanides differ towards
(a) hydrolysis and (b) reduction? (O.U. Apr., 1984), (A.U. Sept., 1993)

14-8-14 Describe reagents and conditions required for the preparation of isobutylamine by the following methods :
(i) Hofmann degradation of amides
(i) Reductive amination
(iii) Lithium aluminium hydride reduction (O.U., Oct. 1988)

14-8-15 (a) What is diazotization? How is benezenediazonium chloride prepared in solution? Discuss its structure.
(b) How would you convert benzenediazonium chloride into the following :
(i) Fluorobenzene (ii) Benzeneazo-2-naphthol
(O.U., Oct., 1988), (N.U., Mar., 1996), (A.U., Sept., 1993), (A.U., Mar., 1993)

14-8-16 How does aniline prepared? What is diazotization ? Give three uses of aryl diazonium salts. (A.U. March, 1995), (N.U. Sept., 1994)

14-8-17 Mention two difference of alkyl nitrites and nitroalkanes.
(N.U. Sept. 94), (N.U. March 96)

SOLUTION TO PROBLEMS

14-7-1 **B**	14-7-2 **A**	14-7-3 **B**	14-7-4 **A**	14-7-5 **B**	14-7-6 **D**
14-7-7 **D**	14-7-8 **C**	14-7-9 **C**	14-7-10 **A**	14-7-11 **B**	14-7-12 **C**
14-7-13 **B**	14-7-14 **C**	14-7-15 **D**	14-7-16 **B**	14-7-17 **C**	

14-8-1 Answers to these subjective questions are found in the main text of the chapter.

14-8-2 (i) Aniline is an aromatic primary amine. The non-bonding lone pair of electrons on nitrogen is involved in resonance with the benzene ring and is, therefore, not readily available for donation to an approaching proton. Further more if protonation on amino group occurs resulting in the formation of anilinium ion, the stablization due to the resonance interaction between the amino group and benzene ring is lost. This is therefore resisted. But cyclohexylamine is an alicyclic but primary amine, a simple derivative of ammonia. It is therefore, a stronger base than aniline.

(ii) Dimethylamine is a secondary aliphatic amine and methylamine is a primary aliphatic amine in the same series. Since alkyl groups/methyl groups are electron-releasing, two methyl groups on an amino nitrogen (like dimethylamine) make the base stronger than one methyl group on an amino nitrogen (like methylamine).

(iii) This is due to resonance between the diazo group and the aryl ring.

(iv) Structurally, alkyl nitrites are esters of alcohols with nitrous acid whereas nitroalkanes are nitro derivatives of alkanes. Treatment with alkali leads to hydrolysis of nitrites whereas nitro alkanes are unaffected by it. Since nitroalkanes which carry an α-hydrogen make that α-hydrogen strongly acidic, alkali merely forms salt with these compounds extracting out the acidic hydrogen from them.

(v) This is to ensure that self-coupling of diazotized aniline with unreacted, freely basic aniline does not occur.

$$Ph-NH_2 \xrightarrow{HONO} Ph-\overset{\oplus}{N_2} \xrightarrow{Ph-NH_2} H_2N-C_6H_4-N=N-C_6H_5$$

$$Ph-\overset{\oplus}{N_2} \xrightarrow[\times]{Ph-\overset{\oplus}{N}H_3} H_2N-C_6H_4-N=N-C_6H_5$$

(vi) In the naphthalene skeleton, the α-position is more activated than the β-position towards electrophilic aromatic substitution. Since coupling of arenediazonium ion involved electrophilic aromatic substitution of an activated aromatic nucleus, reaction with β-naphthol would be faster and more complete than with α-naphthol.

(vii) The benzene ring of aniline molecule is a highly activated nucleus towards electrophilic aromatic substitution due to electron–release from the amino group into the benzene ring leading to resonance stabilization. Consequently,

When aniline, which has three equally and highly activated, ortho, para and ortho' positions, is treated with bromine, tri–substitution takes place. The reaction does not stop at mono– or di–substitution stage.

14–8–3 (a) Treatment with dil.aq.NaOH at room temperature will yield a clear solution for nitro compound whereas the nitrite isomer will remain unaffected.

(b) Nitrobenzene will be unaffected by dil.aq.HCl(5%) as an oil whereas aniline will dissolve in it giving a clear solution.

(c) Aniline will give positive diazotization test whereas triethylamine will not.

(d) n–Butylamine when diazotized will not couple with ß–naphthol to give a bright red azo dye whereas aniline will.

(e) p–Toluidine (p–Methylaniline) is an aromatic primary amine whereas benzylamine is an aralkyl primary amine. The former will give positive diazotization test whereas the latter will not.

(f) Acetanilide is an amide/acetyl derivative of an aromatic primary amine i.e aniline, whereas ammonium acetate is a salt of ammonia with acetic acid. Treatment of the former (ie.acetanilide) with dil.aq.NaOH(5–10%) in the cold, no change will occur whereas the latter (ie.ammonium acetate) will immediately liberate ammonia which can be detected.

(g) Anilinium chloride is a salt of aniline with HCl. It will readily dissolve in water at room temperature to yield a clear solution, whereas p–chloroaniline will form an oily immiscible layer with water.

(h) 3–Aminopentane, being a base, will dissolve in dilute aq.HCl(5%) yielding a clear solution whereas 3–cyanopentane, being a neutral compound will remain unaffected in dilute aq.HCl(5%).

(i) Phenyl isocyanide can be recognized by its pungent, garlic–like smell which will disappear on treatment with dilute aq.HNO_3 (due to oxidation) whereas benzylamine will dissolve in it forming a clear solution.

(j) o–Toluidine is an aromatic primary amine whereas N–methyl aniline is an aromatic–aliphatic secondary amine. The former will give positive diazotization test whereas the latter will not.

14–8–4 (i) N–Methylaniline :

$$C_6H_6 \xrightarrow[H_2SO_4/60^\circ]{HNO_3} C_6H_5NO_2 \xrightarrow[\overset{\oplus}{H}/\Delta]{Fe/H_2O/} C_6H_5NH_2 \xrightarrow[\text{or } CH_3I]{(CH_3)_2SO_4} C_6H_5NH{-}CH_3$$

(ii) p–Chloroaniline :

Benzene $\xrightarrow{Cl_2/AlCl_3}$ chlorobenzene $\xrightarrow[\text{fractionate}]{HNO_3/H_2SO_4\&}$ p-chloronitrobenzene $\xrightarrow[\overset{\oplus}{H}/\Delta]{Fe/H_2O/}$ p-chloroaniline

(iii) p–Nitrophenol :

Benzene ⟶ ⟶ p-chloronitrobenzene $\xrightarrow[\text{and then } H_3\overset{\oplus}{O}]{NaOH/\Delta}$ p-nitrophenol

(iv) o–Nitroaniline :

Benzene ⟶ ⟶ aniline (NH_2) $\xrightarrow{AC_2O}$ acetanilide (NHAc) $\xrightarrow[\text{(b) Separation}]{\text{(a) } HNO_3/H_2SO_4}$ (c) $H_3\overset{\oplus}{O}$ o-nitroaniline (NH_2, NO_2)

(v) p–Bromoacetanilide :

Benzene ⟶ ⟶ ⟶ acetanilide (NHAc) $\xrightarrow[0\text{-}5^oC]{Br_2/AcOH}$ p-bromoacetanilide (NHAc, Br)

(vi) Benzyl cyanide :

Benzene $\xrightarrow[AlCl_3]{CH_3I/}$ toluene (CH_3) $\xrightarrow[h\nu]{Br_2/\Delta/}$ benzyl bromide (CH_2Br) $\xrightarrow{KCN}$ benzyl cyanide (CH_2CN)

(vii) Benzanilide :

Benzene ⟶ ⟶ aniline (NH_2) $\xrightarrow[\overset{\ominus}{O}H]{PhCOCl/}$ benzanilide (NHCOPh)

(viii) Phenyl isocyanide :

Benzene ⟶ ⟶ aniline (NH_2) $\xrightarrow{CHCl_3/KOH/\Delta}$ phenyl isocyanide ($\overset{\oplus}{N}\equiv\overset{\ominus}{C}$)

(ix) o–Chlorobenzamide :

Benzene $\xrightarrow[AlCl_3]{CH_3Cl/}$ Toluene (CH_3) $\xrightarrow[\text{and then fractionate}]{Cl_2/AlCl_3}$ o-Chlorotoluene (CH_3, Cl) $\xrightarrow{KMnO_4}$ o-Chlorobenzoic acid (COOH, Cl) $\xrightarrow[b)\ NH_3]{a)\ SOCl_2}$ o-Chlorobenzamide ($CO—NH_2$, Cl)

(x) m–Nitrochlorobenzene :

Benzene $\longrightarrow$ $\longrightarrow$ m-Dinitrobenzene (NO_2, NO_2) $\xrightarrow{(NH_4)_2S}$ m-Nitroaniline (NH_2, NO_2) $\xrightarrow[(b)\ CuCl]{(a)\ NaNO_2/HCl}$ m-Nitrochlorobenzene (Cl, NO_2)

14–8–5 The structure of paracetamol is as given below:–

Benzene ring with OH and $NHCOCH_3$ in para positions

(p–Acetamidophenol or p–Hydroxyacetanilide or 1–Acetamido – 4 – hydroxy benzene).

Part III

Physical Chemistry

5

THE SOLID STATE

5-1 INTRODUCTION

As discussed in gaseous state chapter the inter molecular forces enhance as we pass from gaseous state to condensed state. Liquids and solids, the constituents of the condensed state, possess a definite volume, definite density but differ in their rigidity or assuming a definite shape. While liquids take the shape of the vessel in which they are poured, the solids do not change their shape. This rigidity of shape in solids cannot be attributed to any stronger intermolecular forces or compactness of the atoms or molecules in them, as the liquid and solid states have comparable densities. Generally liquids have slightly lower density than solids, with few exceptions like ice and water where the solid ice has lower density than liquid water at a given temperature. Thus the rigidness of solids is ascribed to decrease in randomness of the components rather than any increase in the compact packing of the atoms or molecules.

From Figure 5-1 we see that in both solid and liquid states of a substance the densities are almost same, but the arrangement of particles in the liquid is less orderly when compared to a precise orderly arrangement of the solid state, known as *long range order*. In a small region liquids may show some regularity, termed *short-range order*, but miss the long range order essential for maintaining a definite shape. When we powder a rigid solid, the long range order is broken and the powder assumes the shape of the container, a property of a liquid.

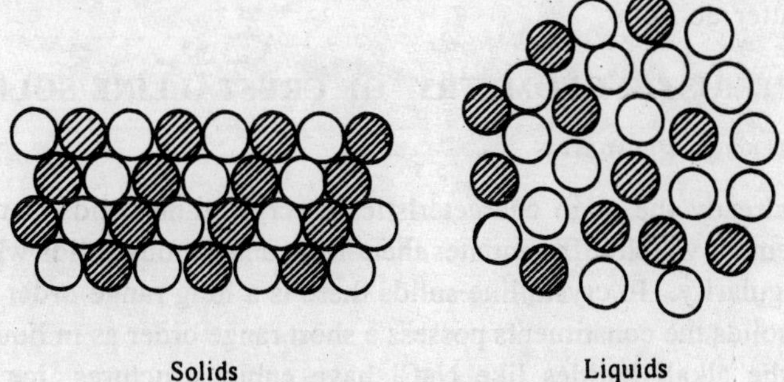

Fig. 5-1 Schematic view of structure in solids and liquids

5-2 CLASSFICATION OF SOLIDS

Solid substances are classified into two classes, depending on the randomness of particle arrangement, as crystalline and amorphous solids.

5-2-1 Crystalline solids

A crystalline solid is one in which the constituents have a long range order and are arranged to give a geometrical form characteristic of the substance. The common salt, mica sheets, ice flakes, sugar particles, etc., are all examples of the crystalline behaviour of solids. On breaking a rock salt lump, it always breaks into small sharp-edged pieces and a mica sheet is comfortably pealed into thin sheets as decided by the inherent crystalline characters. Crystalline forms are obtained when the substances are cooled or created by precipitation reaction and the process of preparing the crystalline forms is called *crystallization. The branch of science which deals with the geometry, properties and structures of crystals and crystalline substances is called* ***crystallography.***

5-2-2 Amorphus solids

An amorphous solid is one in which the constituents are arranged randomly and as such do not have definite geometrical arrangement. Glass, pitch, rubber, plastic, gelatinous precipitate of aluminium hydroxide, etc., are some typical examples of amorphous solids. They have all the attributes of solids like definite shape, hardness, rigidity but only miss the geometric shapes or crystalline behaviour. In true sense, they are not solids and occupy an intermediate position between liquids and solids and are also called super cooled liquids of high viscosity. On long standing or heating or under certian conditions these substances achive certain degree of crystallinity. Thus old glass pieces or heat treated glass show crystalline nature and turns brittle.

A distinguishing property of crystalline and amorphous solids is that the former gives definite diffraction patterns for both crystals and powders whereas the latter do not.

5-3 CHARACTERISTIC GEOMETRY OF CRYSTALLINE SOLIDS

5-3-1 Characteristic geometry

As discussed already the main characteristics of crystalline solids is their regular arrangement which differentiates them from amorphous solids which have partial regularity. In crystalline solids there is a long range order and in amorphous solids the constituents possess a short range order as in liquids. For example the alkali halides like NaCl have cubic structures, ice has hexagonal, while $CaSO_4$ is rhombohedral (trigonal) in geometry.

5-3-2 Crystallization and Fusion

Translational velocities or kinetic energy of the molecules decrease with the decrease in temperature and on approaching freezing point, the attractive forces overcome the translational energy and form crystalline solids. When crystallization starts, heat is evolved, and this evolution arrests any temperature drop. The solid and liquid states are at equilibrium at this temperature known as melting point or fusion temperature of the solid or freezing point of the liquid. Any attempt to lower the temperature only results in the freezing of some more liquid. *The amount of heat evolved for the crystallization of one mole of substance is called* ***heat of crystallization.***

The reverse of crystalllization is *fusion or melting* of the solid. On heating a crystalline solid the vibrational forces increase, breaking completely the attractive forces and converting into liquid. As the liquid and solid are at equilibrium at fusion temperature any extra heat supplied is used for further fusion. *The amount of heat required for the fusion of one mole of a solid substance is called* ***heat of fusion.***

5-3-3 Melting points

Owing to a definite arrangement, the crystal constituents are bound together by same type of cohesive forces throughout the crystal. Hence supply of heat is shared uniformly by all the bonds, followed by an increase of internuclear distance. At a particular temperature, called melting point, the translation forces are sufficient to overcome the internuclear attractions and the solid converts to liquid. Thus *all the crystalline solids have sharp melting points.* On the contrary an amorphous solid has only short range order resulting in different internuclear forces.. Hence heat supplied will be able to snap some weak bonds, while still some other bonds are intact. Consequently *amorphous substances have no sharp melting points and extend over a temperature range.* Supply of heat initially softens followed by gradual conversion into a fluid.

5-3-4 Sublimation

In case of some solid substances like camphor and solid carbondioxide, *the vapour pressure of the solid crystalline substance reaches the* atmospheric pressure before reaching the melting point. In such cases the solid directly passes into vapour phase without touching the liquid form. This phenomenon where the solid and vapour of a substance are at equilibrium is known as *sublimation* and the equilibrium temperature as sublimation temperature. The category of substances can be melted by increasing the external pressure sufficient to arise the vapour pressure above the solid-liquid equilibrium vapour pressure.

5-3-5 Isotropy and anisotropy

All solids that have *same properties, independent of direction are called* ***Isotropic.*** In cubic crystals and amorphous solids, properties like thermal conductivity, elasticity, mechanical compressibility, refractive index etc, do not depend on the direction of measurement or applications. For example light travels with the same velocity through glass, irrespecitve of direction of observation.

The crystalline solids other than cubic crystals are said to be *anisotropic* as *the properties depend on the direction of observations.* When light is passed through an anisotropic crystal, it is split into components which travel with different velocities and follow different paths. This phenomenon is called *double refraction.*

The dependence of properties on direction can be understood by considering a two dimensional, crystal lattice. On focussing light on a lattice with two kinds of atoms, represented in Figure 5-2, experiences at least two different environments.

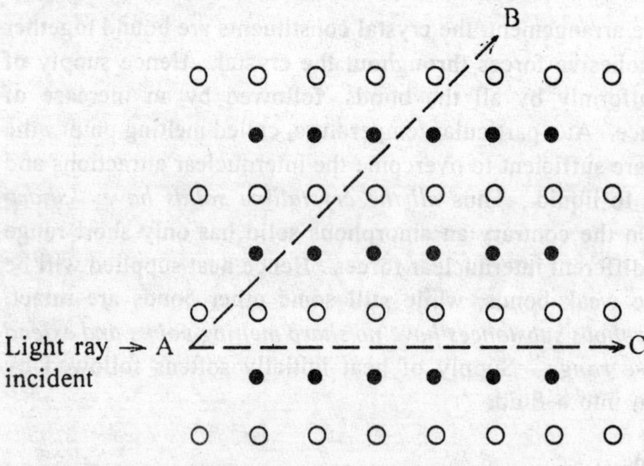

Fig. 5-2 Light passing through an anisotropic lattice

The component AB passes through planes having alternately placed atoms while the component AC passes between two planes each, having one kind of atoms. A lattice made out of one kind of atoms called a cubic lattice. In the cubic lattice components AB and AC experience similar environment, suggesting the absence of directional properties in cubic crystals, Figure 5-3.

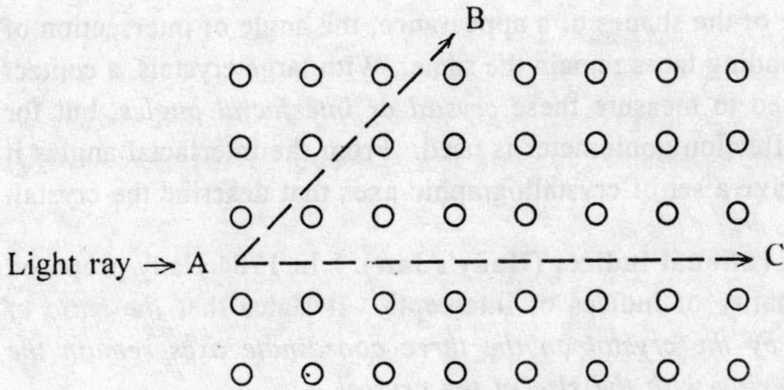

Fig 5-3 Light passing through a cubic crystal lattice.

In case of an amorphous solid, the atoms are placed at random and any ray passing in any direction experiences allways an average environment, explaining their *isotropicity.*

Other characteristics of crystals are covered as three *fundamental laws of crystallography.*

5-4 LAWS OF CRYSTALLOGRAPHY

The size and shapes of crystals depend on the experimental conditions employed for crystallization. For example quick cooling of a hot, saturated solutions of brine gives small crystals but a slow cooling gives crystals of few centimeter length. Traces of impurities, often greatly modify the relative rates of growth of various faces and then lead to different crystalline forms. Irrespective of differences in external geometric shapes in a three dimensional space, the crystal structures are based on three fundamental laws (A) the law of constancy of interfacial angles (B) the law of rational indices and (C) the law of symmetry.

(A) The law of constancy of interfacial angles. *Although the crystal appearance of a substance is variable the angles between the various faces are characteristic and distinctive features of a substance.* Thus, a crystal may appear, in a two dimensional space, as a perfect hexagonal (structure a) or an irregular hexagonal (structures b and c), Figure 5-4.

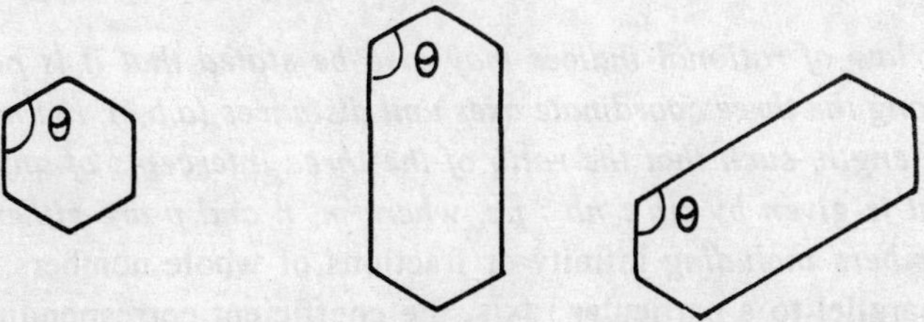

Fig. 5-4 Three crystal structures of the same substance.

Irrespective of the shapes of a appearance, the angle of intersection of any two corresponding faces remain the same. With large crystals, a contact goniometer is used to measure these *crystal or interfacial angles*, but for tiny crystals a reflection goniometer is used. From the interfacial angles it is possible to derive a set of crystallographic axes that describe the crystal.

(B) Law of rational indices (Haüy's law). In 1784 Haüy proposed the law of rationality of indices or intercepts. It states that *the ratio of intercepts made by the crystal on the three coordinate axes remain the same without altering with the size of the crystal.*

Consider a crystal plane intersecting the x, y, z axes at A, B, C respectively such that OA=a, OB=b and OC=c, Figure 5-5. The ratio of the intercepts is $a : b : c$. Considering one and half times larger crystal the same plane intersects the x, y, z axes at K, L, M respectively, such that

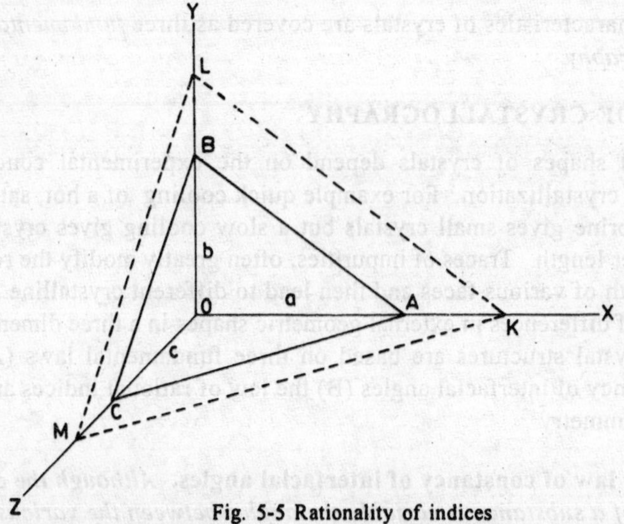

Fig. 5-5 Rationality of indices

OK=3/2a, OL=3/2b and OM=3/2c therefore OK : OL : OM = $a : b : c$. Hence the ratio of the intercepts, OK : OL : OM for larger crystal is same as the intercept of a smaller crystal on the three coordinate axes.

This law of rational indices may also be stated that it is possible to choose along the three coordinate axes unit distances (a,b,c), not necessarily the same length, such that the ratio of the three intercepts of any plane in the crystal is given by ma : nb : pc, where m, n and p are either integral whole numbers including infinity or fractions of whole numbers. When a plane is parallel to a particular axis, the coefficient corresponding to that axis takes a value of infinity.

The Coefficients *m, n* and *p* are called Weiss Indices. In 1839 W.H. Miller, modified the Weiss indices and introduced more convenient *Miller indices* to identify different planes of a crystal. In figure 5-6 let ABC be

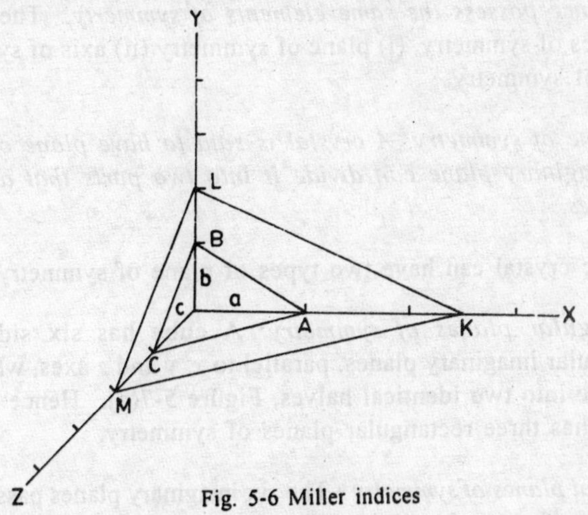

Fig. 5-6 Miller indices

a *unit plane* with distance *a, b* and *c* along *x, y* and *z* axes respectively. In order to describe any other lattice plane KLM in terms of the unit plane, first we measure the distances OK, OL and OM as multiples of *a, b* and *c.* In present example OK = $^5/_2a$, OL = $2b$ and OM = $2c$. The Weiss indices of the plane KLM are 5/2, 2, 2. ***The Miller indices or hkl values of a plane are obtained by taking reciprocals of the Weiss indices and multiplying with some smallest number that will express all the reciprocals as integers.*** Hence Miller indices for KLM plane are obtained as follows.

(a) by taking reciprocals of Weiss indices : 2/5, 1/2, 1/2

(b) Multiplying with 10, the least common multiple (LCM) of the denominators to get integers : 4, 5, 5. The KLM plane is therefore designated as (4, 5, 5) plane. Similarly for the unit plane ABC, the Miller indices are (1 1 1). In case of the plane ABC cutting *x* and *y* axes but parallel to *z* axis is identified by Miller indices as (1 1 0) plane. If the plane ABC just intersects x axis, but parallel to other axes it is designated as (1 0 0) plane.

In addition to designating the planes, Miller indices could also be utilized to calculate the distance, d_{hkl}, between the parallel planes of a cubic crystal using the general formula

$$d_{hkl} = \frac{a}{\sqrt{h^2 + k^2 + l^2}} \qquad (5.1)$$

where a is the length of the side of a cube and h, k, l are the Miller indices of the plane.

(C) The law of symmetry : This law states that *all crystals of the same substance possess the same elements of symmerty.* There are three possible types of symmetry, (i) plane of symmetry (ii) axis of symmetry and (iii) center of symmetry.

(i) Plane of symmetry: A crystal is said to have plane of symmetry when an imaginary plane can divide it into two parts that are identical mirror images.

A cubic crystal can have two types of plane of symmetry.

(a) *Rectangular planes of symmetry:* A cube has six sides. Three rectangular imaginary planes, parallel to x, y and z axes, which can cut the cube into two identical halves, Figure 5-7(a). Hence we say that a cube has three rectangular planes of symmetry.

(b) *Diagonal planes of symmetry* : The six imaginary planes passing through opposite edges and splitting the cube into two identical halves are called diagonal planes of symmetry, Figure 5-7 (b).

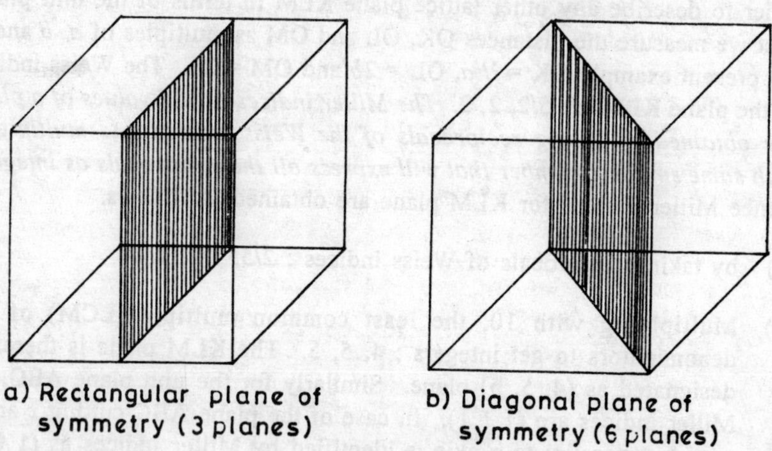

Fig 5-7 Planes of symmetry

(ii) Axis of symmetry: A crystal is said to have an axis of symmetry, when an imaginary line passing through the crystal about which it may be rotated presents exactly the same appearance more than once in the course of one complete (360°) revolution.

(a) *Four fold axis of symmetry:* When a cube is rotated about an imaginary line passing through middle of opposite sides, the faces present themselves four times, Figure 5-8(a). Hence the axis is said to have *four fold symmetry.*

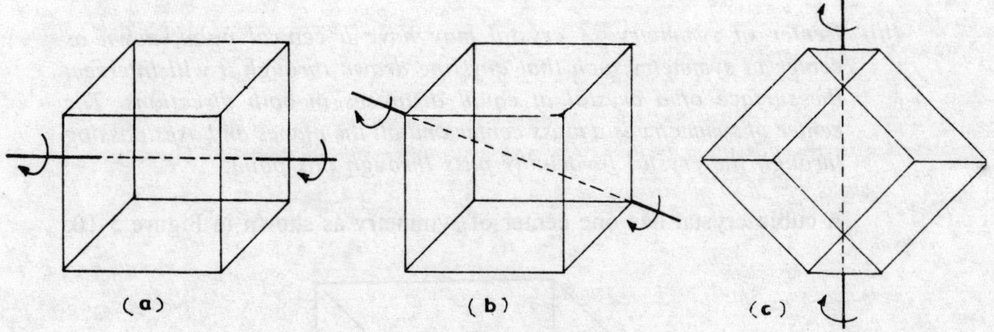

Fig. 5-8 axis of symmetry

(b) *Three fold axis of symmetry:* If a crystal is rotated about an axis that passes through diagonally oppossite corners of a cube the sides repeat thrice in the course of rotation by 360°. Such axis is termed as axis of *three fold (triad) symmetry (Fig.5-8b).*

(c) *Two fold axis of symmetry:* In case, a crystal is rotated about an axis passing through centers of the two diagonally opposite edges of a cube, the cube presents the same appearance twice in the course of rotation of 360°. This axis is said to be one of *two fold (diad) symmetry.*

A cubical crystal has only the above three types of axes of symmetry. A regular cube has three axes of four-fold symmetry, passing through three pairs of opposite sides, four axes of three-fold symmetry passing through opposite corners and six axes of two-fold symmetry emerging from opposite edges making a total of thirteen (3 + 4 + 6) axes of symmetry.

(d) *Multi-fold axis of symmetry:* Depending on the geometry, a crystal is allowed to have even more than four-fold axis of symmetry.

Thus a hexagon can have six-fold axis of symmetry, Figure 5-9(a), and a pentagon can have five-fold axis of symmetry, Figure 5-9(b), etc.,

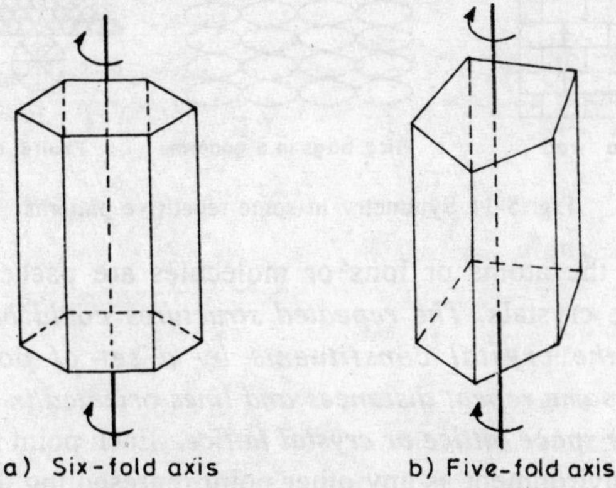

Fig. 5-9 Multi-fold axis of symmetry

(iii) *Center of symmetry: A crystal may have a central point, known as center of symmetry, such that any line drawn through it will intersect the surface of a crystal at equal distances in both directions. The center of symmetry is a mass center and all the planes and axes passing through the crystal invariably pass through this point.*

A cubic crystal has one center of symmetry as shown in Figure 5-10.

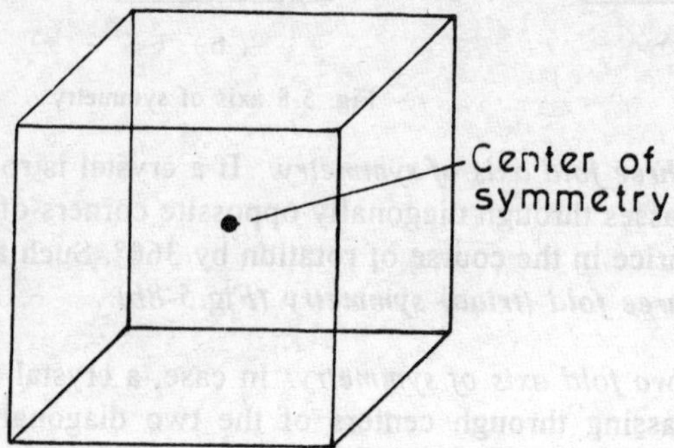

Fig. 5-10 Centre of sysmmetry

All the three types plane, axis and center of symmetry, are collectively called elements of symmetry. A perfect cube has a total of twenty three (9 planes + 13 axes + 1 center) elements of symmetry.

5-5 CRYSTAL SYSTEMS

Any repetitive pattern has a symmetry element associated with it, whether it is stacking of bricks in a wall or stacking of rice bags in a godown or stacking of fruits on a push cart (Figure 5-11).

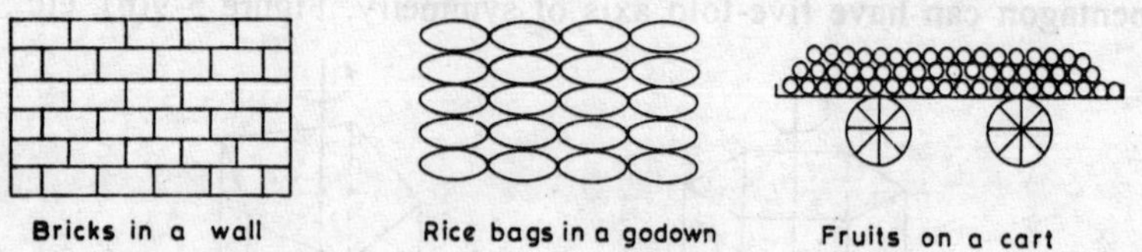

Fig. 5-11 Symmetry in some repetitive patterns

Similarly the atoms or ions or molecules are packed in an orderly fashion to give crystals. *The repeated structures could be designated by representing the crystal constituents by a set of points that have proportionally same repeat distances and lines oriented in the same angles called **lattice or space lattice or crystal lattice.*** Each point in the lattice has exactly same environment as any other point representing the same atom or ion.

5-5-1 Unit Cell

In a crystal the number of particles is enormous and describing the positions of all these particles or their lattice points is impossible. Fortunately *it is sufficient if we describe the basic repeating unit of the lattice to completely understand the space lattice. This basic repeating unit is called* **unit cell.**

Figure 5-12 illustrates a two dimensional latice. In this a square lattice forms the basic unit and entire space lattice could be generated by repeated movement of this unit lattice in a two dimensional space.

O O O O O O O O

O O O O O O O O

O O O O O O O O

O O O O O O O O

Fig. 5-12 A square lattice

Extension of above cited two dimensional lattice model, to a three dimensional space we get a simple cubic lattice as the unit cell, represented by dark lines in Figure 5-13.

Irrespective of shapes of the crystals, the unit cells of all three dimensional lattices are similar in that they all have eight corners (Figure

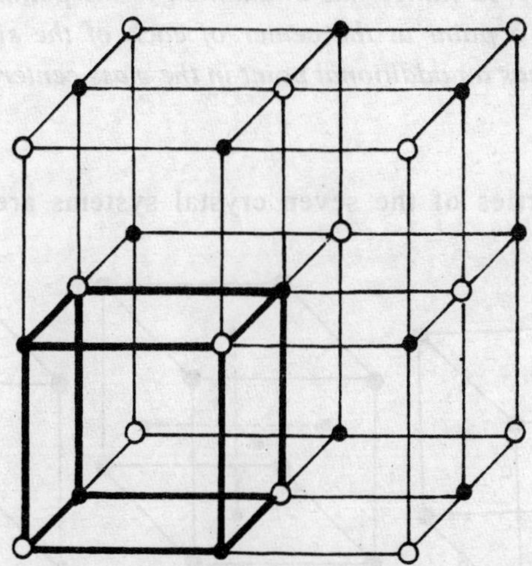

Fig. 5-13 A simple cubic lattice

5-14.). Unit cells differ by the lengths of their edges (a, b and c) and angles between them (α, β and γ), Figure 5-14.

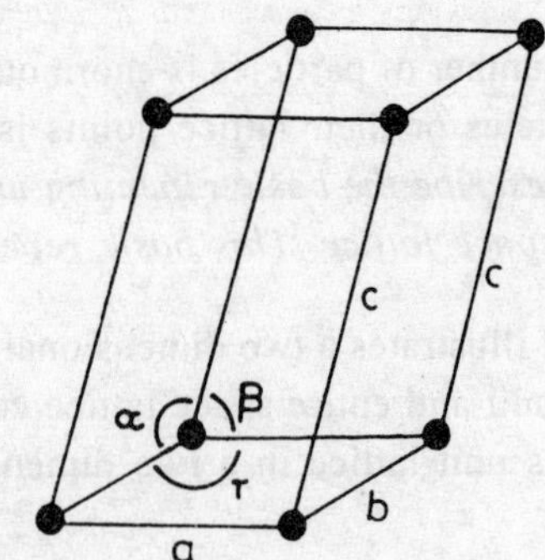

Fig. 5-14 A typical unit cell

5-5-2 Bravais lattice

In 1848 August Bravais showed that only 14 different kinds of lattices are possible, known as *Bravais lattices*. These were further divided into seven basic *crystal systems*. For example a cubic crystal system has one or more of three kinds of lattices known as (i) The simple cube (Fig. 5-15a), (ii) The face centered cube (Fig. 5-15b) and (iii) The body centered cube (Fig. 5-15c). The other non-cubic Bravais lattices are also represented in Figure 5-15.

A simple cube lattice has points only at the eight corners of a unit cell. A face centered lattice has in addition to the points at eight corners of the unit cell, a point at the center of each of the six faces. A body centered lattice has an additional point in the mass center of a simple cubic lattice.

The properties of the seven crystal systems are summarized in Table 5-1.

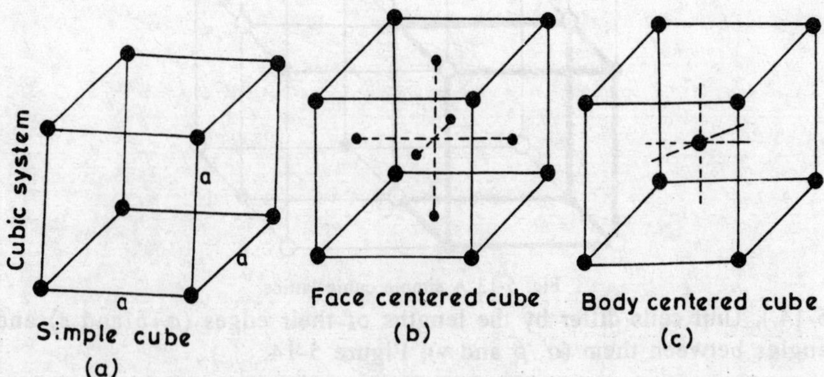

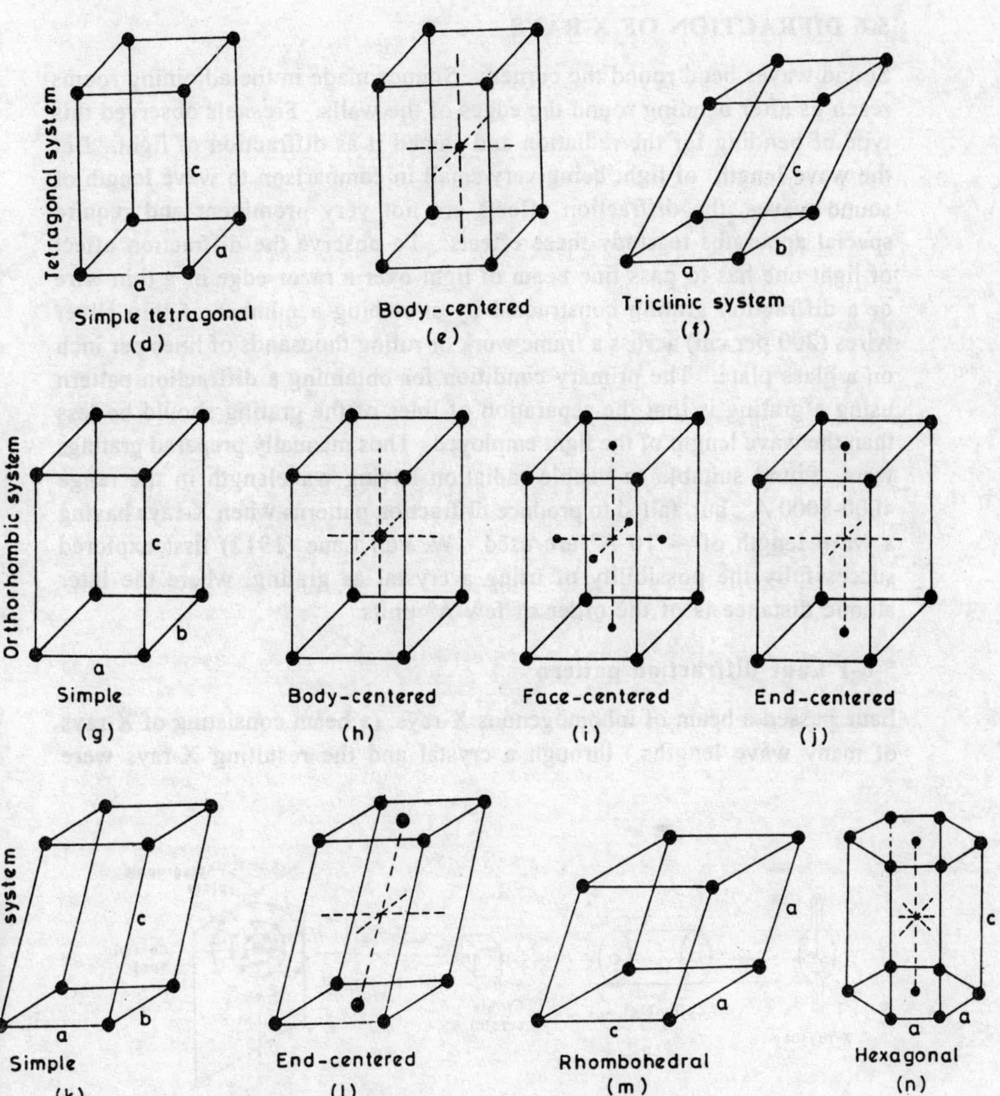

Fig. 5-15 The Bravais lattices

Table 5-1 Properties of the crystal systems

System	Edge Lengths	Angles	Examples
Cubic	a = b = c	$\alpha = \beta = \gamma = 90°$	NaC, CaF2, ZnSS, Diamond
Tetragonal	a = b ≠ c	$\alpha = \beta = \gamma = 90°$	SnO_2, TiO_2, $NiSO_4$
Orthorhombic	a ≠ b ≠ c	$\alpha = \beta = \gamma = 90°$	KNO_3, K_2SO_4, S (Rhombic)
Monoclinic	a ≠ b ≠ c	$\alpha = \beta = 90° \neq \gamma$	$NaHCO_3$, $CaSO_4$, $2H_2O$, S (Monoclinic)
Rhombohedral	a = b = c	$\alpha = \beta = \gamma \neq 90°$	$CaSO_4$, $CaCO_3$, $NaNO_3$
Hexagonal	a = b ≠ c	$\alpha = \beta = 90°, \gamma = 120°$	SiO_2, HgS, Graphite, Ice
Triclinic	a ≠ b ≠ c	$\alpha \neq \beta \neq \gamma \neq 90°$	$K_2Cr_2O_7$, $CuSO_4$. 5 H_2O

5-6 DIFRACTION OF X-RAYS

Sound waves bend round the corners. Sounds made in the adjoining rooms reach us after bending round the edges of the walls. Fresnels observed this type of bending for the radiation and named it as diffraction of light. But the wave length of light being very small in comparison to wave length of sound waves, the diffraction effects are not very prominent and require special apparatus to study these effects. To observe the diffraction effect of light one has to pass fine beam of light over a razor edge or a thin wire or a diffraction grating constructed by stretching a number of thin silver wires (200 per cm) across a frame work or ruling thousands of lines per inch on a glass plate. The primary condition for obtaining a diffraction pattern using a grating is that the separation of lines of the grating should be less than the wave length of the light employed. Thus manually prepared gratings were found suitable to visible radiation having wavelength in the range 4000-8000 A°, but, failed to produce diffraction patterns when X-rays having a wave length of $\approx$ 10 A° are used. W. von Laue (1912) first explored successfully the possibility of using a crystal as grating, where the inter atomic distance is of the order of few A° units.

5-6-1 Laue diffraction pattern

Laue passed a beam of inhomogenous X-rays, (a beam consisting of X-rays of many wave lengths,) through a crystal and the resulting X-rays were

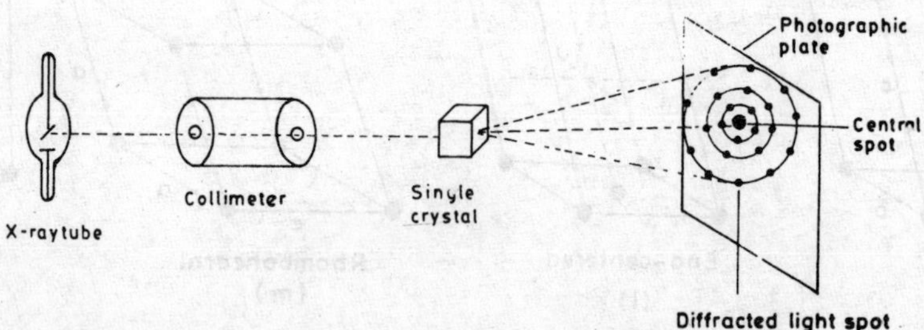

fig. 5-16 The Lave method

allowed to fall on a photographic plate. A pattern indicated by Figure 5-16 was observed, which is called Laue photograph or Laue pattern. Each type of crystal exhibits a characteristic pattern, but interpretation needs advanced mathematical and geometrical knowledge.

The fact that the light coming out of the single crystal forms Laue pattern, consisting of spots lying on equally space circles indicates that the atoms in a crystal are spaced in a definite, repetitive manner.

5-6-2 Bragg's diffraction pattern

A clue from Laue pattern was taken by W.L. Bragg and W.H. Bragg, who obtained a reflection diffraction pattern, which is rather simple in the interpretation. They used homogenous X-rays (X-rays of single wavelength), and crystal as the diffraction grating. Consider that a crystal consists of different planes *AA, BB* etc., parallel to each other, Fig.5-17.

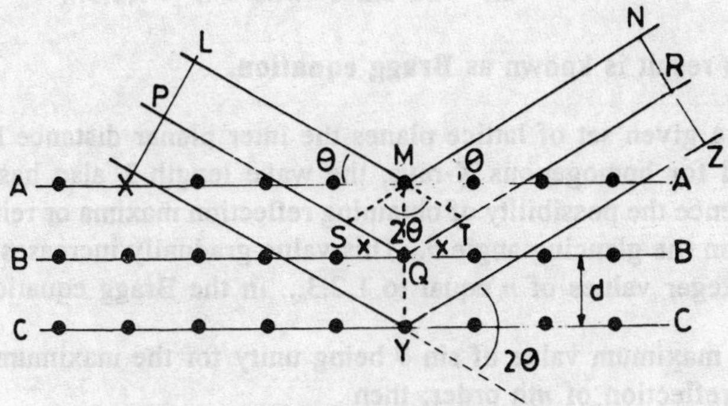

Fig.5-17 Derivation of Bragg's Law

The collimated X-ray beam strikes the crystal surface at a glancing angle θ. Part of the beam *LM* is reflected by the atoms of the first plane, *AA*,to yield reflected ray *MN*. Similarly another part *PQ* that could pass through the interatomic spacing of first plane is reflected by the second plane *BB* to yield *QR*. If the rays *LMN* and *PQR* have to reinforce one another, they must have the same phase. This condition is met when the difference in lengths of the path traced by these two rays is equal to an integral number of wave lengths. The difference in the path length can be obtained by the drawing perpendicular *MS* and *MT* to *PQ* and *QR* respectively.

$$\text{If } SQ = QT = x$$

$$\text{Difference in path length} = SQ + QT = 2x$$

From the above stated condition for reinformcement, the path difference;

$$2x = n\lambda \tag{5.2}$$

Where n is an integer.

As $\widehat{LMS} = \widehat{AMQ} = 90°$, we get the glancing angle

$\widehat{LMA} = \widehat{SMQ} = \theta$

Similarly $\widehat{TMQ}$ is also equal to θ

In the right angle triangle $\widehat{SMQ}$, the length of the side

$$SQ = MQ \sin \theta$$

If the inter planar distance MQ is taken as "*d*" then:

$$x = d \sin \theta \tag{5.3}$$

Substituting the value of *x* in equation 5.2 we have,

$$n\lambda = 2d \sin \theta \quad where\ n = 1,2,3... \tag{5.4}$$

This result is known as **Bragg equation.**

For a given set of lattice planes the inter planar distance has a fixed value and for homogenous X-rays, the wave length λ also has a definite value. Hence the possibility of obtaining reflection maxima or reinforcement depends on the glancing angle θ. This value gradually increases as defined by the integer values of *n* equal to 1,2,3... in the Bragg equation.

The maximum value of sin θ being unity for the maximum reflection angle or reflection of *n*th order; then

the inter planar distance $d = n \ \lambda/2$ (5.5)

5.6-3 Bragg X-ray diffractometer

In the Bragg X-ray diffractometer (or spectrometer), a collimated beam of X-rays are allowed to fall upon a crystal mounted on a rotating table and the angle of rotation may be measured on the scale instrument, Figure 5-18.

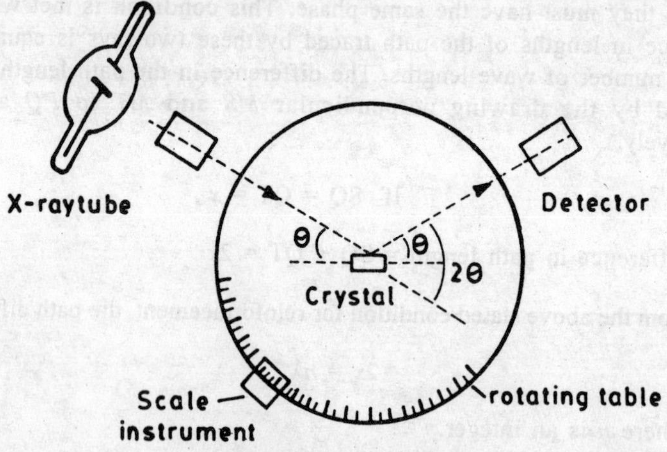

Fig. 5-18 Bragg X-ray diffractometer

After reflection on the crystal surface the X-ray beam is allowed to pass into a detector, an ionizing chamber or a Gieger counter, the response of the detector being directly proportional to the intesity of the reflected X-ray beam. Reinforcement of the beam is indicated by the strongest reflection correspond to the different integral values. By measuring the values of θ, the inter planar spacing d can be computed from Bragg equation which thus defines the plane of a crystal. This procedure may be repeated to identify all important planes of a crystal. This helps in the determination of crystal structure of the compound.

A typical diffraction pattern of tungsten is shown in Figure 5-19. The number on the peaks are Miller indices of the planes which produce the peak.

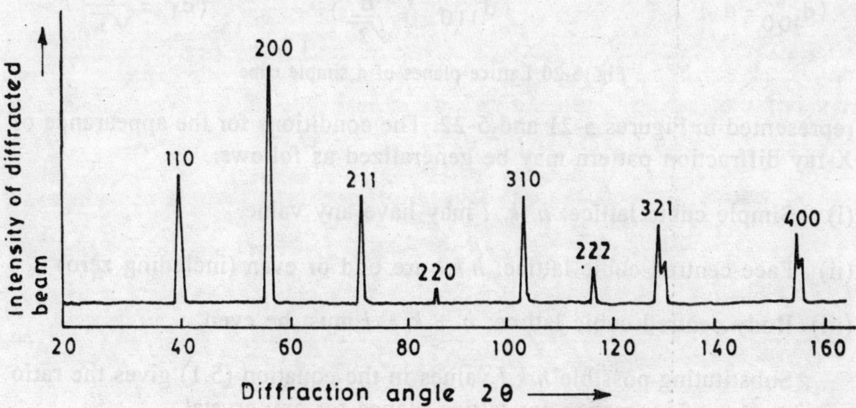

Fig.5-19 Diffraction pattern of tungsten

The fact that some planes are densely placed than others, produces a variation in intensity of the diffracted beam. The planes of high atomic density scatter X-rays better and produce the more intense peak. example 200 peak in Figure 5-19.

5-6-4 Interpretation of X-ray diffraction results

From the knowledge of X-ray diffraction studies we get a very useful quantity d, the distance between the lattice planes. As already discussed under Miller indices, a comparison of experimentally obtained d valuess and the values reached by theoretical consideration of various lattice planes help in arriving at the crystal structure of a substance.

For example :

A simple cube has three lattice planes described by the Miller indices (100),(110) and (111), Figure 5-20. Similarly the lattice planes of a face centred cube are described mainly by Miller indices (200) (220) and (111) and a body centred cubic lattice planes by (110), (200) etc. These are

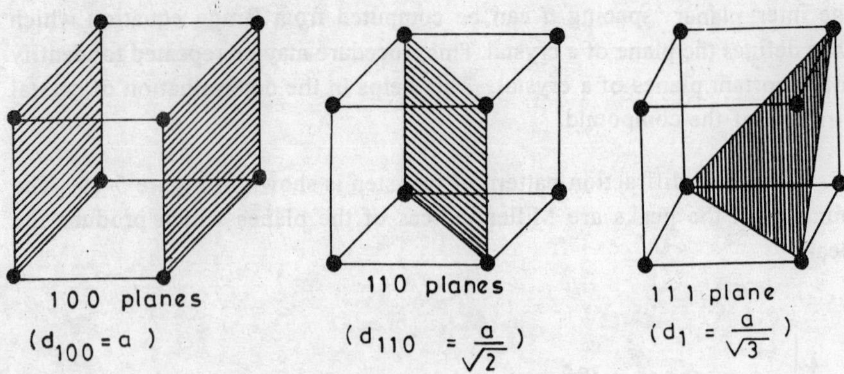

Fig. 5-20 Lattice planes of a simple cube

represented in Figures 5-21 and 5-22. The conditions for the appearance of X-ray diffraction pattern may be generalized as follows:

(i) Simple cubic lattice: h, k, l may have any value.

(ii) Face-centred cubic lattice: h,k,l are odd or even (including zero)

(iii) Body centred cubic lattice: $h + k + l$ must be even.

Substituting possible $h\ k\ l$ values in the equation (5.1) gives the ratio of distances of separation for lattice planes for any crystal.

For simple cube it is ;

$$d_{100} : d_{110} : d_{111} = \frac{a}{\sqrt{1}} : \frac{a}{\sqrt{2}} : \frac{a}{\sqrt{3}}$$

$$= 1 : 0.707 : 0.577$$

The ratio for face centred cubic lattices is ;

$$d_{200} : d_{220} : d_{111} = \frac{a}{2} : \frac{a}{2\sqrt{2}} : \frac{a}{\sqrt{3}}$$

$$= 0.5 : 0.354 : 0.577$$

For body centred cube;

$$d_{200} : d_{110} : d_{222} = \frac{a}{2} : \frac{a}{\sqrt{2}} : \frac{a}{2\sqrt{3}}$$

$$= 0.5 : 0.707 : 0.289$$

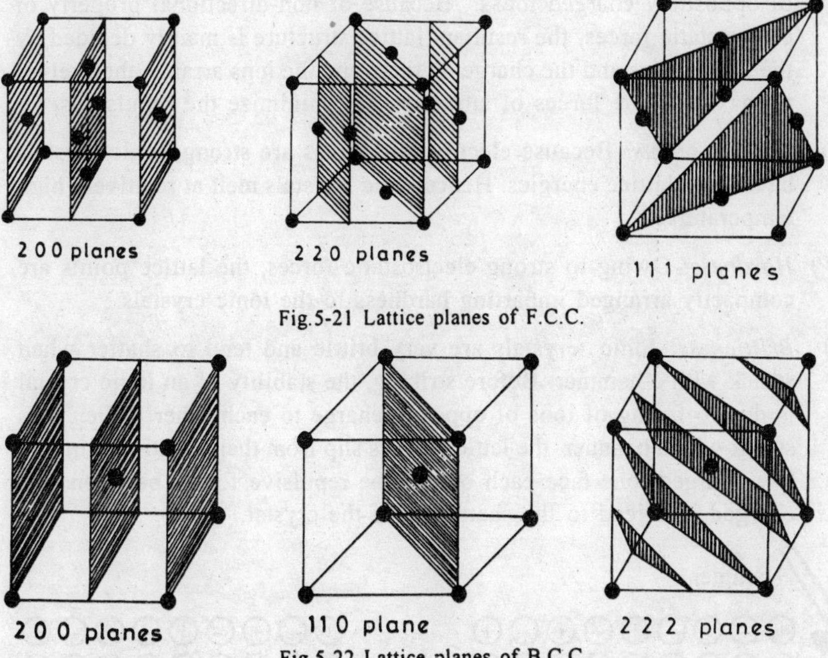

Fig.5-21 Lattice planes of F.C.C.

Fig.5-22 Lattice planes of B.C.C.

Thus a comparision of experimental *d* values, and *d* values obtained from geometrical considerations lead to assignment of crystal structure.

5-7 CRYSTAL TYPES AND PROPERTIES

The geometry and physical properities of crystalline solids are determined by the types of particles present at the lattice points and the nature of attractive forces between them. Crystals,based on the nature of particles that form the space lattice, are classified into four categories, (A) Ionic crystals, (B) Covalent crystals, (C) Molecular crystals and (D) Metallic crystals.

5-7-1 Ionic Crystals

In an ionic crystal, the lattice points are occupied by positive and negative ions and the binding force between them is electrostatic in nature.

Ion compounds like NaCl, $CaCl_2$, ZnS, TiO_2, etc., fall in this category. The ion crystals dissolve in polar solvents, almost 100% ionizing to give respective cations and anions.

(i) *Chemical units at lattice points :* As the definition states positive and negative ions are the chemical units at lattice points.

(ii) *Nature of forces operative between lattice points:* The electrostatic attractions are non-directional and each ion maintains an environment

of oppositely charged ionss. Because of non-directional property of electrostatic forces, the resultant lattice structure is mainly decided by the size of ions and the charge on them and the ions arrange themselves to maximize the forces of attraction and minimize the repulsions.

(iii) *Melting points:* Because electrostatic forces are strong, ionic crystals have large lattice energies. Hence ionic crystals melt at relatively high temperatures.

(IV) *Hardness:* Owing to strong electrostatic forces, the lattice points are compactly arranged imparting hardness to the ionic crystals.

(v) *Brittleness:* Ionic crystals are very brittle and tend to shatter when struck with a hammer. Before striking, the stability of an ionic crystal is due to facing of ions of opposite charge to each other. When it is struck with a hammer, the lattice planes slip from their position bringing like charged ions face each other. The repulsive forces between like charged ions lead to the shattering of the crystal.

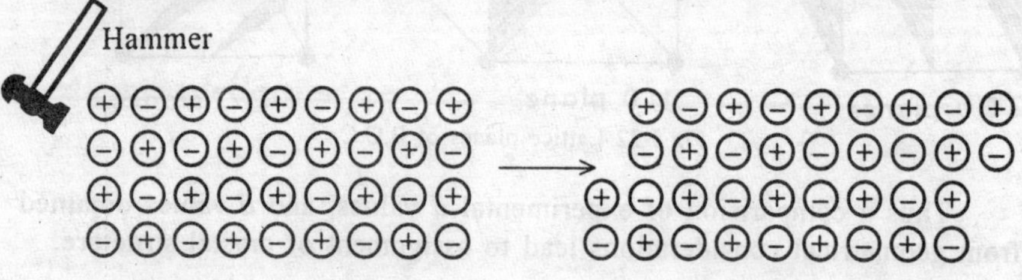

Fig.5.23 Ionic crystals

(vi) *Electrical Conductivity:* In crystalline state the ionic crystals are poor conductors of electricity because the valence electrons are immobile and held rigidly. In the molten state or when dissolved in a solvent, the ions are free to move about and ionic crystals become good conductors of electricity.

5-7-2 Covalent crystals

In a covalent crystal the lattice points are occupied by atoms that are bound to the neighbouring atoms by covalent bonds.

All the group IV elements except lead, boran, phosphorous, arsenic, selenium and tellurium exhibit covalent crystal structures. Some of the elements have allotropes that have metallic and covalent crystal structures. For example tin exhibits metallic properties in white tin form and covalent properties in grey tin form.

(i) *Chemical units at the lattice points:* Covalently bound atoms form the chemicals units at the lattice points.

(ii) *Nature of forces operative between lattice points:* Covalent bonds, which are highly directional, are operative between lattice points. Diamond is an example of this category where elemental carbon atoms are covalently bonded to four nearest, neighbouring carbon atoms. Carborundum(Silicon carbide) and quartz (silicon dioxide) are other typical examples of covalent crystals. The directional properties of these crystals result in the development of specific faces of a crystal.

(iii) *Melting points:* Because of strong covalent bonds covalent crystals have relatively high melting points.

(iv) *Hardness:* Because of interlocking network of the covalent bonds covalent crystals are very hard and used as abrasive materials and cutting tools.

(v) *Electrical conductivity:* The electrons in the crystals are tied up in covalent bonds and hence except graphite they are bad conductors of electricity.

(vi) *Solubility:* Covalent crystals are not easily soluble because it involves breaking of high energy covalent bonds.

5-7-3 Molecular crystals

In molecular crystals,the lattice points are occupied either by molecules or atoms and are bonded by weak forces of attraction like random forces or dipole-dipole attractions or hydrogen bonds.

(i) *Chemicals units at lattice points:* Atoms and molecules occupy the lattice points.

(ii) *Nature of forces operative between lattice points:* The atoms or molecules are bonded by forces much weaker than covalent bonds. Non-polar substances like O_2, naphthalene, etc., are bonded by random forces. Polar molecules like SO_2 in their solid crystalline state are together due to dipole-dipole interactions. Polar molecules like H_2O (in ice), NH_3, HF etc., which have a posssibility of formation of hydrogen bond are held together in the solid state by hydrogen bonds.

(iii) *Melting points:* The weak intermolecular forces present in the molecular crystals are responsible for low melting points of these solids.

(iv) *Hardness:* Molecular crystals are soft due to less compact structures resulted by weak molecular forces.

(v) *Electrical conductivity:* Molecular crystals are also poor conductors of electricity as the electrons bound to molecules are not able to move freely. In the molten state polar molecules conduct electricity but the non-polar solids remain poor conductors.

5-7-4 Metallic crystals

In metallic crystals the positive metal ions are situated at lattice points, with valence electrons that belong to the whole crystal instead of any single element, around them. The negatively charged electrons hold the positively charged ions together. To maintain the electrical neutrality the number of positive and negative charges are exactly balanced. They assume one of the three following arrangements.

1. Hexagonal Close Packed - HCP
2. Cubic Close Packed (Face Centred Cubic) - CCP or FCC
3 Body Centred Cubic - BCC

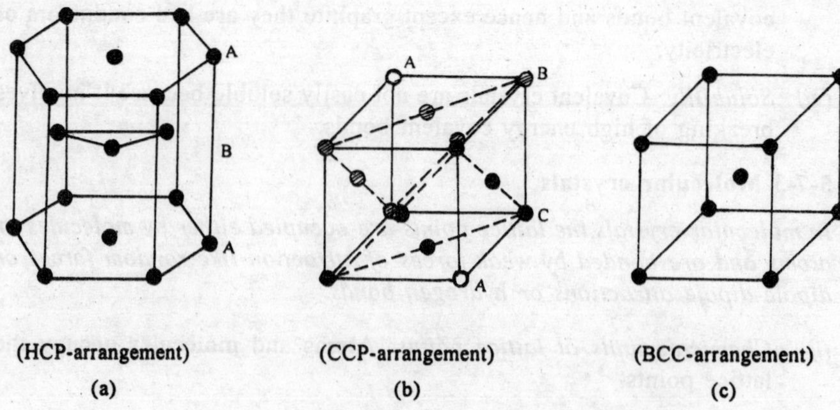

Fig. 5-24 Packing arrangements in metallic crystals

In hexagonal close packed arrangement each atom has 12 closest neighbours (coordination number = 12) and arranged in two different layers of atoms, each type appearing alternatively. This repeating pattern is called AB AB AB ...... pattern. The atoms occupy 74% of the available space.

In CCP or FCC arrangement the crystal is generated by three different layers A, B, C repeated and the pattern is represented by (ABC.ABC......). This crystal too has a coordination number of 12. This class of crystals occupy 74% of the available space.

The BCC crystals are relatively less compact and has 8 neighbouring atoms. They fill 68% of the available space. The characters are summarized in Table 5-2.

Table 5-2 Comparison of hcp, ccp and bcc structures

Property	hcp	ccp	bcc
Coordination No.	12	12	8
Volume occupied	74%	74%	68%
Type of packing	(AB AB AB ...)	(ABC.ABC......)	---
Melleability & ductitity	Less malleable hard & ductile	Malleable & ductile	Malleable & ductile
Metals	Be, Mg, Ca, Cr V, Mo & Zn	Cu, Ag, Au, Pt.	Alkali metals & Fe

Bonding in metals is discussed later.

(i) *Chemical units at lattice points:* The positive metal ions occupy the lattice points and give metallic crystals.

(ii) *Nature of forces opertive between the lattice points:* The positive ions in the crystal lattice are held together by electrostatic attractions with a sea of electrons moving around them. The electrons can move freely in between the lattice planes and they act as a glue in holding any two positive lion lattice points.

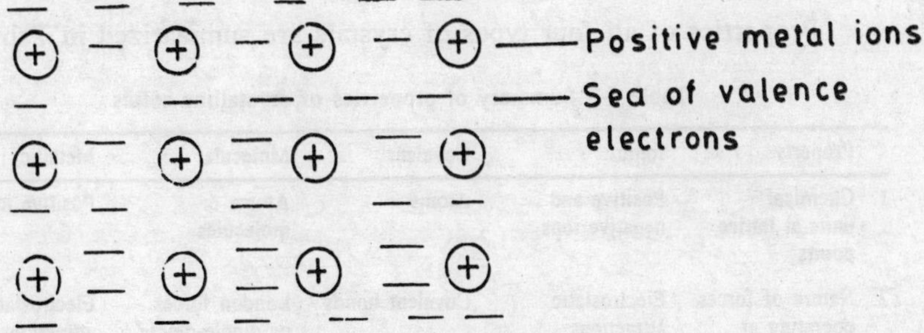

Fig.5-25 Metallic crystals

The formation of sea of mobile electrons can be understood by considering the fact that the valence electrons possess almost identical lattice energy. When two metallic atoms come into contact, due to identical energy states all the valence electrons move around them without any labelling of electrons to a particular atom.The strength of the metallic bonds (electrostatic attractions) depend on the number of free electrons available per atom, the atomic radii and nature of valence orbital (*s, p, d* or *f*).

(iii) *Melting points:*As the electrostatic attractions depend on nature of valence orbital, atomic radii and mobile electrons available per atom; the bond strength vary from very weak to very strong.Hence metallic crystals show low and high melting points. For example sodium metallic

crystal utilizing 2s valence electrons melt at a low temperature while copper or iron utilizing 3d valence electrons have high melting point.

(iv) *Hardness :* Due to dependence of electrostatic attractions on valence electrons the hardness of the metallic crystal also varies from soft to very hard. Sodium metal is soft and can be cut with a knife while iron requires special cutting tools.

(v) *Electrical conductance:* Because of free mobile electrons all the metallic crystals are good conductors of electricity.

(vi) *Metallic lustre:* The free electrons on the external surface of a metallic crystal are excited when light falls on them.The excitation energy is reemmitted in the form of light giving an appearance of reflection of light which we call the metallic lustre.

(vii) *Malleability and ductility:* The metallic bond is non-directional and the lattice planes, containing the positive metal ions are held together by a mobile electron film.These two properties allow the flattening of metal by hammering into thin sheets, known as malleability, and drawing into thin wires,known as ductility.

(viii) *Tensile strength:* Even thin wires of metal bear heavy loads and resist stretching and break after crossing a particular load. This behaviours is termed as tensile strength.The high tensile strength is due to strong electrostatic attractions operating between metallic planes.

Properties of all four types of crystals are summarized in Table 5-3.

Table 5-3 Summary of properties of crystalline solids

Property	Ionic	Covalent	Molecular	Metallic
1. Chemical units at lattice points	Positive and negative ions	Atoms	Atoms or molecules	Positive ions
2. Nature of forces operating at lattice points	Electrostatic attractions betwee the ions	Covalent bonds	London forces or dipole-dipole attractions or hydrogen bonds	Electrostatic attractions between positive ions and valence electron sea
3. Melting point	High melting	High melting	Low melting	Low to high melting
4. Mechanical Properties	Hard and brittle	Very hard	Soft	Soft to hard, Malleable, ductile and high lustre
5. Electrical conductance	Poor conductance	Poor conductance	Poor conductance	Good conductance
6. Examples	NaCl (salt) $CaCO_3$ (lime stone) $MgSO_4$ (epsom)	C (diamond), SiC, SiO_2	CO_2 (dry ice), H_2O (ice),NH_3 HF, SO_2	Na, Cu, Fe, Pt, Ni, etc.

5-8 DEFECTS IN CRYSTALS

A crystal lattice is an idealization never achieved cent percent by a structure. Normally impurities and loose electrons generated by the absorption of heat or light cause these imperfections. Defects arise in all types of crystal lattices but the present discussion is restricted to ionic crystals of the type AB (e.g. NaCl, FeO, etc.).

There are three ways in which ionic crystals are defective or imperfect. (A) Defects in stoichiometric crystals, (B) Defects in non-stoichiometric crystals and (C) Mechanical defects. Defects in stoichiometric and non-stoichiometric crystals (A&B) arise either due to missing of an atom or ion from the lattice site (Vacancy defect) or due to a foreign atom or ion entering the lattice site (Impurity defect). As these defects are restricted to few lattice points they are also called *Point defects.* The mechanical defects arise due to displacement of a complete lattice plane and hence are called *Line defects.*

(A) Stoichiometric crystals: *The stoichiometric crystals are one in which the positive and negative ions are in the ratio represented by the chemical formulae of the compound.* The stoichiometric compounds exhibit both (i)Frenkel and (ii) Schottky defects:

(i) ***Frenkel defect**: A Frenkel defect is a misplaced ion, generally a cation, held interstitially in a region that would otherwise be perfect lattice. For every interstitial ion of this kind there is also a vacant ion site.* This defect arises due to vacant lattice point developed by the movement of one of the ion to another position, generally in between the lattice planes.

In Figure 5-26 the cation A^+ moved between the two lattice planes, vacating its original positions. In stoichiometric crystals,Frenkel defect occurs;

(a) When the cation is smaller than the anion, as in AgBr, thus allowing the cation to slip into a interstitial position.

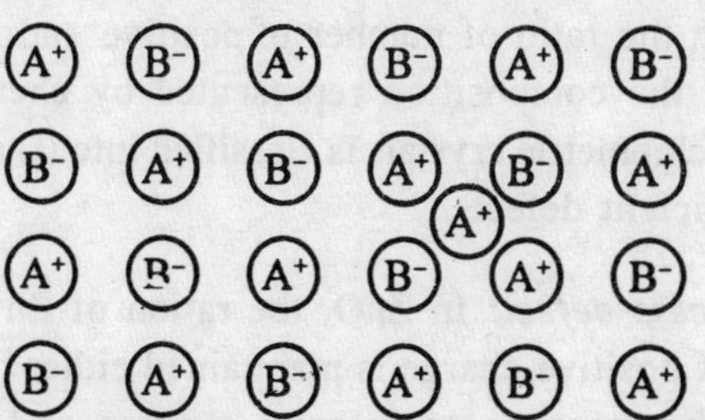

Fig.5-26 Frenkel defect in a stoichiometric crystal

(b) When the cation and anion have low oxidation numbers and larger sizes,resulting in weak electrostatic attractive forces (e.g. AgBr) *(ii)* ***Schottky defect:*** A Schottky defect is a vacant ion site. Both anions and cations are missing from what might have been a perfect structure. The ions from the vacated sites are considered to have moved to the surface of the crystal. In crystals like NaCl, a cation and an anion completely miss from their lattice points causing Schottky defect.

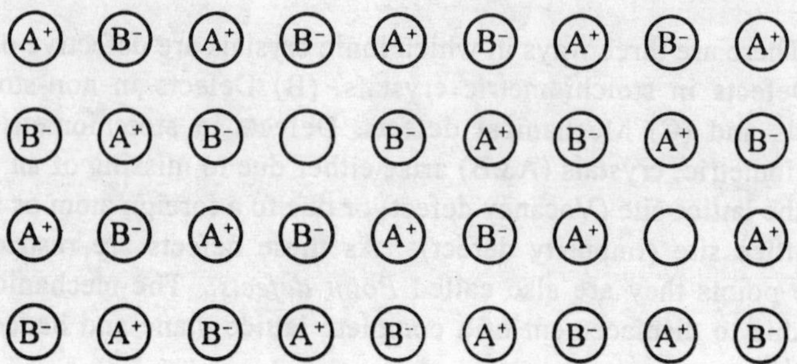

Fig.5-27 Schottky defect in a stoichiometric crystal

In Figure 5-27, a cation moved away along with an anion from the lattice points, maintaining electrical neutrality. This defect is common among the stoichiometric crystals having ;

(a) Ions of equal size and

(b) Ions with high oxidation numbers.

Schottky defects are generally one hundred times more probable than Frenkel defects. The energy requirements of Schottky defect site is about 25 kcal $mole^{-1}$ of defect sites. Defects of both kinds may be present at one in fifty sites near the melting point, where the ions are more free to move, but at ordinary temperatures it is only one in one million possible sites.

(B) Defects in non-stoichiometric crystals: Non-stoichiometric crystals are those in which the ratio of number of positive andnegative charges do not coincide with the composition represented by chemical formulae. The defects in non-stoichiometric crystals is classified into (i) metal excess defects and (ii) metal deficient defects.

(i) Metal excess defect: In ZnO, the ration of Zn^{2+} and O^{2-} is 0.95:1 and the balance of positive charge is maintained either by the higher valent metal ion or by the presence of an extra electron and an extra metal ion trapped in the interstitial planes. In either case there is an excess of metal ions in the crystal and hence called metal excess defect. The two ways of defects cited above are illustrated in Figure 5-28.

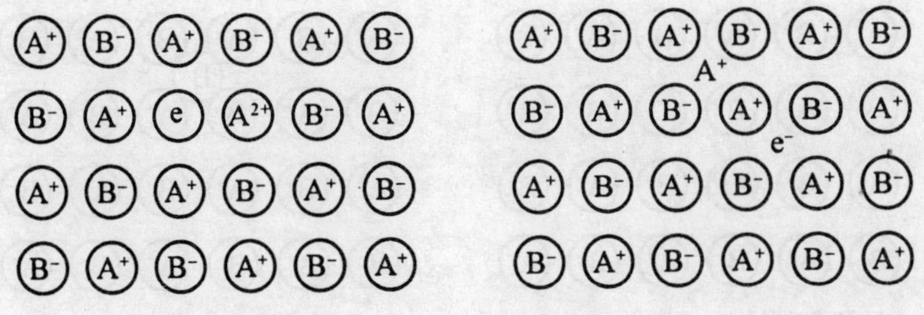

a) Defect due to missing anion b) Defect due to interstitial cation

Fig.5-28 Metal excess defects

(a) *Defect due to missing anion:* The illustration, Figure 5-28a, explains the defect due to missing anion. In this the loss of an anion from a lattice point creates an anion void and the charge balance is achieved by filling the void by an electron donated by a neighbouring cation which raises itself to a higher valence state.The missing anion leaves the crystal surface in the form of vapour or by combination with surrounding metal vapour. This defect arises in crystals susceptible for Schottky defects. When excess sodium metal is heated with chlorine gas and slowly cooled, yellow sodium chloride crystals are formed because of development of defects due to missing anion.

(b) *Defect due to interstitial cation:* In this a cation is trapped in interstitial sites along with the charge balancing electron as represented by Figure 5-28b. This defect is common with the crystals susceptible to Frenkel defects, e.g., ZnO.

(ii) Metal deficiency defect: Metal deficiency defects are just the converse of metal excess defects. Here cations are either completely removed or shifted to interstitial sites amounting to excess anions or deficient of metal ions in the crystal. We have two types of metal deficiency defects

(a) Defects due to missing cation and (b) Defects due to interstitial anion.

(a) *Defects due to missing cation:* In the lattice plane illustrated by figure 5-29a, a cation is missing from the lattice point. To compensate this loss and maintain the charge neutrality of the crystal, one of the cation reaches a higher oxidation state. Thus on the whole there is an excess of anion and the lattice remains deficient of cations. Considering a ferrous oxide, FeO, crystal, we notice some ferric, Fe^{3+}, ions in the defective crystals in addition to ferrous, Fe^{2+}, ions. The other compounds that show this defect are FeS, TiO, NiO, etc.

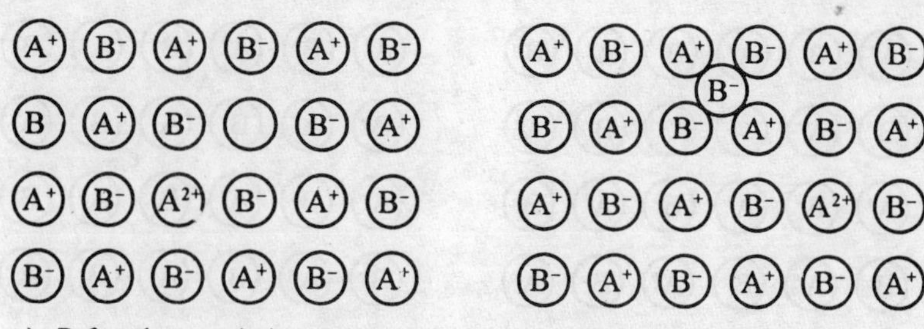

a) Defect due to missing cation b) Defect due to interstitial anion

Fig.5-29 Metal deficiency defects

(b) Defect due to interstitial anion: In this type of defect an anion is trapped in between the lattice planes of a crystal. The extra negative charge associated with the anion occupying interstitial site is compensated by an extra charge present in one of the cation, Figure 5-29b. In the crystals possessing this defect the anions are relatively small when compared to the size of cation, thus allowing the anion to find a place in the interstitial site.

(C) Mechanical defects: When a stress or strain is applied on a localized plane, there is a deformity of crystal lattice. They have a large impact on the mechanical properties of the material and hence termed as mechanical defects. Suppose that a path is followed round the crystal from atom to atom or from ion to ion. If this path returns itself it is called a **Burgers circuit.** If the **Burgers circuit** returns exactly to the starting lattice point, the circuit is said to enclose no dislocations or defects. Instead if the Burgers circuit did not end on the starting atom or ion it is said to enclose one or more dislocations. There are two types of dislocations (i) Edge dislocation (ii) Screw dislocation.

(i) ***Edge dislocation:*** When a strain because of compression or expansion is felt along a localized plane, part of the lattice gains extra half plane of atoms or ions. this dislocation is called edge dislocation. The deformity is produced in a plane perpendicular to the direction of applied strain.

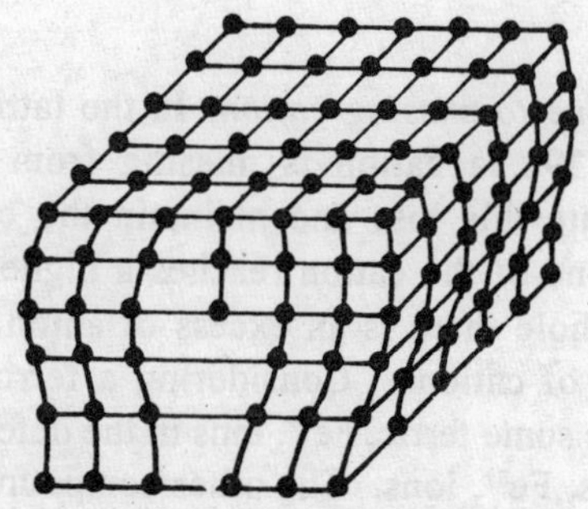

Fig.5-30 Edge dislocation

(ii) ***Screw dislocation:*** In screw dislocation, few lattice planes are dislocated by one or more lattice units relative to the other lattice planes by the applied stress, Figure 5-31.

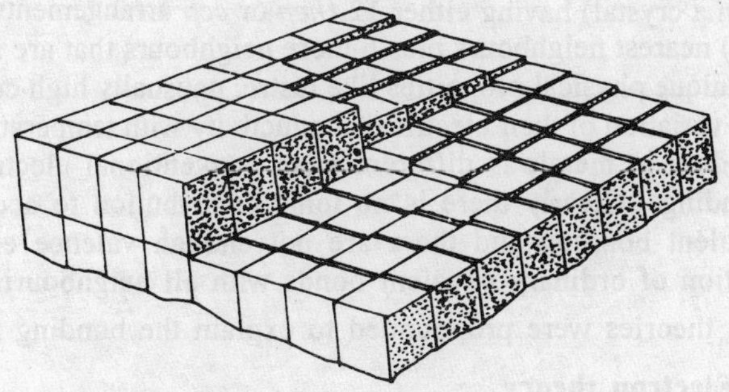

Fig.5-31 Screw dislocation

5-8-1 Better side of crystal defects

The free electrons available in imperfect lattice planes give special properties to the crystals like colour and conduction properties.

(a) *Colour Properties:* Depending on the extent of imperfection or in other words free electrons, crystals assume different colours. This is due to absorption of light by the free electrons and their excitation to higher energy levels. Thus we can prepare sodium chloride crystals that are yellow and blue. Similarly zinc oxide crystals turn yellow when hot and white when cold.

(b) *Semi-conductivity: Pure covalent and ionic crystals are non-conductors of electricity. Defects in crystals, like metal excess defect, put free electrons in their lattice planes and give them partially the capacity to conduct electricity known as semiconductivity and then the crystals are called semiconductors.* Generally quartz (SiO_2), TiO, etc., function as semiconductors and are used in the solar cells for the conversion of light energy into electrical energy.

5-9 METALLIC BONDING

As discussed earlier, metals are characterized by characteristic lustre, high electrical and thermal conductivities, malleability, etc. They are classified into two groups on the basis of the variation of their electrical conductivity with temperature.

1. **Metallic conductor :** *It is a substance with an electrical conductivity that decreases as the temperature is raised.*

2. **Semiconductor:** *It is a metallic substance with an electrical conductivity that increases as the temperature is raised.*

The semiconductors in general have low conductivity in comparison to pure metals and conventionally the semiconductors with extremely low electrical conductivities are called **Insulators**. Further, metals are also characterized by their high coordination numbers (number of nearest neighbours in a crystal) having either 12 (*hcp* or *ccp* arrangements) or 8 (*bcc* arrangement) nearest neighbours plus 6 more neighbours that are sufficiently close. The unique physical properties like lustre, unusually high coordination numbers and variation of their electrical conductivity with temperature suggest that the bonding in metals is different from conventional electrovalent or covalent bonding. Clearly there is no ionic contribution to accommodate the electrovalent bonding and there are not enough valence electrons to allow formation of ordinary covalent bonds with all neighbouring atoms.

Several theories were propounded to explain the bonding in metals.

5-9-1 Free Electron theory

In 1900 Drude propounded this theory. He regarded metal as a lattice through which electrons flow freely with the same swiftnes as gas molecules in air. This idea was refined by Lorenty in 1923 who described metal lattice points as made of positive and electrons moving freely. This assumption successfully explained many metallic properties but failed to provide any quantitative study as to the said properties.

5-9-2 Valence bond theory

Valence bond theroy, in its original form proposed by Linus Pauling, considers a chemical bond as due to sharing of electrons by two atoms. This was extended to explain metallic properties. Consider a simple metal like lithium, which has a body centered cubic structure, and therefore has eight nearest neighbours. Any atoms could share its electrons equally well with all eight neighbours, thus forming different arrangements. Fig. 5-32 a & b. The same Li atom could form two bonds if it ionizes and give structures like fig. 5-32 c & d.

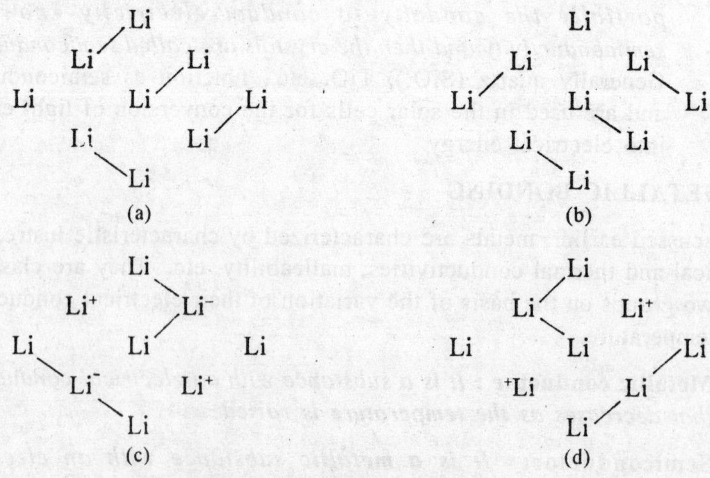

Fig.5-32 Valence band structures of Li.

According to Pauling the true structure in the metallic lattice is a mixture of different posssible structures. Electrical conduction is the result of the presence of number of ions. This theory failed to explain many metallic properties like the conduction of heat in solids, the lustre, etc.

5-9-3 The band theory or Molecular orbital theory of solids

The band theory is an extension of conventional Molecular orbital theory. The metal crystal is though as a three dimensional array in which the atoms are placed side by side. To make the case simple let us once again consider a one dimensional arrangement in which *1s* Li atoms are placed side by side along *x axis*. A single atom has a certain energy corresponding to one *1s* orbital electron. When another atom is brought into its vicinity the combination of the two s orbitals result in the formation of a bonding and antibonding orbital. Similarly a third atom in their neighbourhood lead to three molecular orbitals. Extending this argument to N number of atoms at a neighbourhood gives rise to N molecular orbitals covering a band of finite width, Fig.5-33.

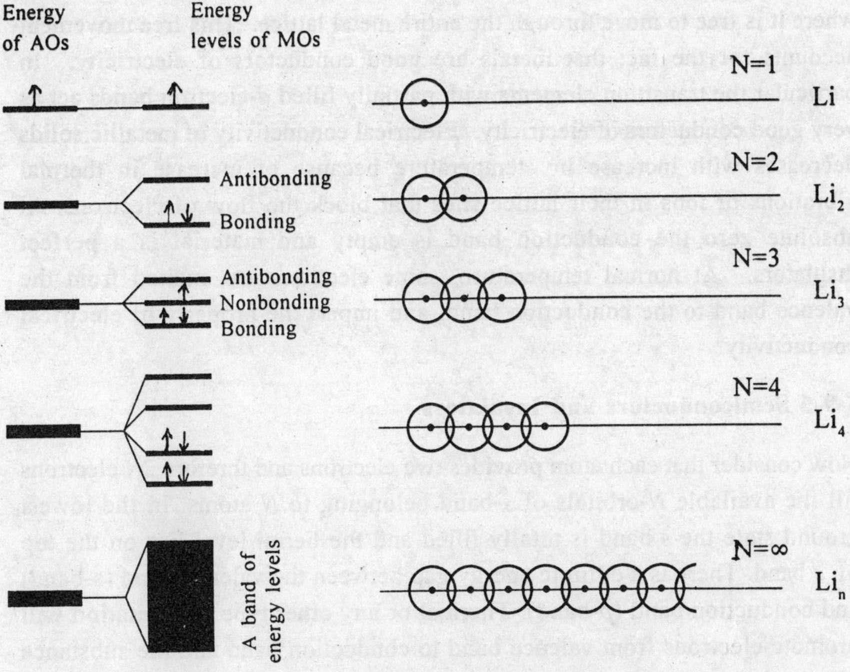

Fig.5-33 Different MO combinations

When N is infinitely large, there will be infinitely large number of molecular orbitals, the difference between the neighbouring energy levels being infinitely small to assume a continuum. This band formed from overlap of many *s* orbitals is called *s-band*. Similarly it is possible to have molecular orbitals formed by the overlap of *p* orbitals and is called *p-band.*

Since the atomic p orbitals in general lie, higher in energy than *s* orbitals, the *p* band lies at a higher energy than s band and the energy difference is referred as **band gap.**

5-9-4 The Valence band and Conduction band

Consider N one electron atoms in the neighbourhood. As discussed all the N electrons occupy a band formed by the closely spaced N orbitals. If this solid is cooled to absolute zero, the electrons occupy the lowest possible energy level and are distributed in N/2 bonding molecular orbitals leaving the antibonding orbitals unoccupied. The **H**ighest **O**ccupied **M**olecular **O**rbital (HOMO) is called the *Fermi level*, Fig.5-34. The set of closely spaced, occupied energy levels lying below Fermi level is called the *valence band* and the other half of unoccupied energy levels corresponding to empty antibonding molecular orbitals is called *conduction band.* In case of metals, the valence band and conduction band are adjacent with negligible energy gap allowing a facile promotion of valence electrons into conduction band, where it is free to move through the entire metal lattice. This free movement accounts for the fact that metals are good conductors of electricity. In particular the transition elements with partially filled *d*-electron bands act as very good conductors of electricity. Electrical conductivity of metallic solids decreases with increase in temperature because of increase in thermal vibrations of ions in their lattice sites that block the flow of electrons. At absolute zero the conduction band is empty and material is a perfect insulators. At normal temperature, some electrons are excited from the valence band to the conduction band, and impart the property of electrical conductivity.

5-9-5 Semiconductors and Insulators

Now consider that each atom provides two electrons and threfore *2N* electrons fill the available *N*-orbitals of *s*-band belonging to *N* atoms. In the lowest ground state the s band is totally filled and the Fermi level lies on the top of *s* band. There is a definite energy gap between the valence band (s-band) and conduction band (*p*-band). Thermal or any other type of excitation will promote electrons from valence band to conduction band and the substance gains conduction property because of these newly created partially filled bands. In a semiconductor this energy gap is small enough to allow the excitation while in an insulator the energy gap is large to inhibit the promotion of an electron from the valence band to conduction band. Lacking this energy, free motion of electrons is not possible and the solid is then called an insulator.

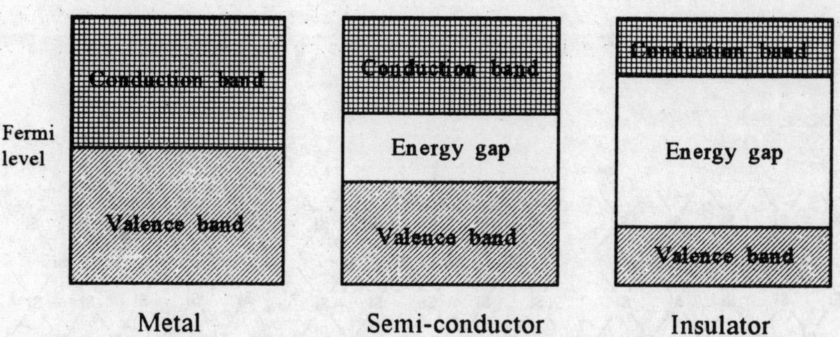

Fig.5-34 Comparison of energy gaps between metal, semiconductor and Insulator.

The elements like Si and Ge in group IV are the examples for semiconductors. The energy gap between the valence and conduction bands of these elements is much smaller than for insulators and their electrical conductance increases with increase in temperature. Other than thermal excitation the semiconductivity of substance can be enhanced by adding a certain impurity to the pure element. *This process of implanting a foreign substance into a pure element to enhance the ability of a semiconductor is called doping and the impurity is called* ***dopant***. There are two types of semiconductivity depending on whether a dopant is an acceptor or donor. A donor (eg., P) develops a negative centre in the crystal lattice and an acceptor (eg., B) develops a positive centre and correspondig semi-conductivities are referred to as *n*-type and *p*-type respectively.

n-TYPE SEMI-CONDUCTIVITY

The structure of solid silicon is similar to that of diamond, each atom of Si covalently bonded to four other Si atoms, fig.5-35a.. If phosphorous ($3s^2$ $3p^3$), having one valence electron more than the Si atoms [(Ne)$3s^23p^2$], is doped in the Si lattice, there is a valence electron left over unbonded after formation of covalent bonds with Si atoms (fig.5-35b) This free electron can occupy the conduction band and move in the Si lattice exhibiting n-type semiconductivity. This type of impurity is called a donor impurity. The letter *n* stands for negative indicating the formation of a negative centre.

p-TYPE SEMICONDUCTIVITY

If boron impurity is implanted in the Si lattice instead of phosphorous a positive centre develops in the crystal. This is because boron (He) $2p^1$ has one valence electron short of the host Si atoms (Ne)$3s^23p^2$ (fig.5-35c). The positive hole created at B attract the neighbouring electrons and thereby shifts its location to the corresponding atom. This movement of positive hole accounts for the electrical conductivity.

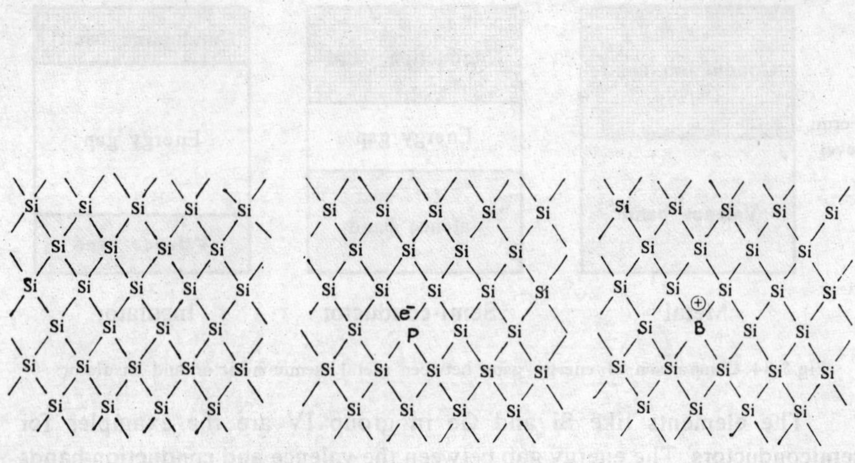

Fig.5-35 *n*-type & *p*-type semiconductors

In both *p*-type and *n*-type semiconductors the energy gap between valence and conduction bands is effectively reduced allowing a facile excitation of electrons. In a typical semiconductor the electrical conductivity is increased by a factor of 10^5 over the pure host metal.

5-10 PHOTOVOLTAIC CELL

An electrochemical cell is a device used for the conversion of electrical energy into chemical energy and vice-versa. It often consists of two electrodes dipped into same solution or two solutions separated by a partition and at these electrodes two half reactions of a chemical reaction occur. There are two types of electrochemical cells. A **galvanic cell** converts the energy released in a spontaneous chemical reaction into electrical energy. On the contrary in an **electrolytic cell**, the electrical energy forces a nonspontaneous chemical reaction to occur which is often referred to as **electrolysis**.

Harnessing of thermal component of solar energy is well known and finds application in solar cookers, water heaters etc. The light component of solar energy is converted into chemical and electrical energies by the usage of semiconductors. The device used for conversion of light energy into electrical energy is called **photovoltaic cell** and into chemical energy is called **photo-electrochemical cell**. The arrangement involves growth of a semiconductor crystal in such a way that one region is *n*-type and adjacent region p-type. The bonding between these two types is called *p-n* junction, fig. 5-36(a). In the absence of any external interference, the positive hole developed in the *p*-type side and the conducting electrons of the *n*-type side of the crystal are totally isolated and are at some equilibrium. Across the *p-n* junction charge is positive in *p*-type region and negative charge in *n*-type region, Fig. 5-36a.

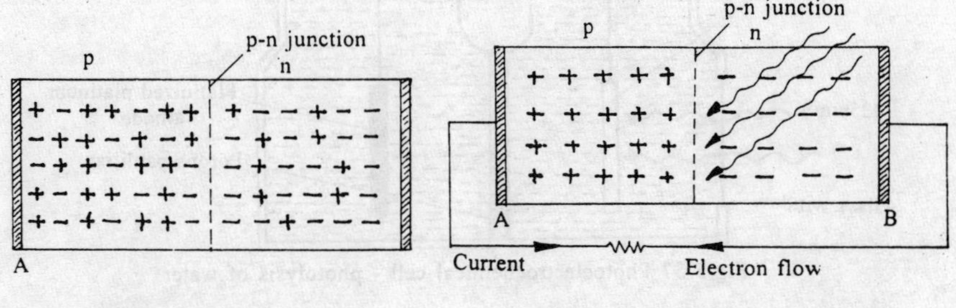

(a) Isolated semi-conductors (b) When light falls on *p-n* junction

Fig. 5-36 Photo-voltaic cell

When light shines on a *p-n* junction, if the energy of the light photon exceeds the band gap, then some bonds are borken generating electrons and postive holes. These electrons are then promoted from the valence bond to conduction band. The extra electrons in the conduction band make the *n*-type region of the semiconductor more negative, whilst in the *p*-type region these electrons are trapped by existing positive holes. This disturbance of equilibrium existing prior to irradiation of *p-n* junction allows the electrons to flow from *n*-type region to *p*-type region when these two regions are connected in an external circuit, the current flows from the *p*-type to *n*-type region, Fig. 5-36b. This set up acts as a battery that can generate electricity from light that can be directly used as in solar powered calculators and in powering satellites or could be stored in an accumulator like lead storage cell.

5-10-1 Photoelectrochemical cell

Photoelectrochemical cells convert the energy present in light photons into electrical energy which in turn used to convert a non-spontaneous reaction into a spontaneous process. The concept of semiconductors developing positive holes or electron rich centres on exposure to suitable light is employed to harness solar energy and to produce hydrogen, a clean and environment friendly fuel, from abundantly available water. The electrochemical photolysis of water was first reported by Honda and co-workers. Water is transparent to visible and near U.V.radiation. Direct photolysis of water requires radiation of wavelengh 190nm. But the electrochemical decomposition of water requires a potential of 1.23V which corresponds to radiation of wavelength 1000nm. Hence if the energy of visible photon is converted into electrical energy it is possible to split H_2O molecules. For this a cell comprising of *n*-type TiO_2 working as anode could be coupled with inert platinum electrode working as cathode, Fig.5-37.

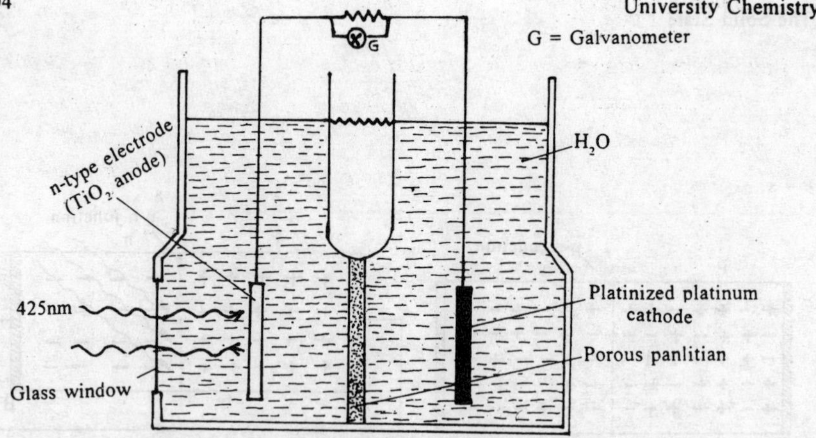

Fig.5-37 Photoelectrochemical cell - photolysis of water

Light of wave length <415nm is allowed to fall on a TiO_2 semiconductor electrode (*n*-type) working as an anode. This excites the valence electrons to conduction band, the TiO_2 being of *n*-type, turns more electronegative. A current flows from Pt electrode to the TiO_2 electrode through the external circuit whilst the electron flow occurs in the opposite derection.

At anode :

$$TiO_2 + 2h\nu \longrightarrow 2\bar{e} + 2p^+$$

(Positive holes)

$$2p^+ + H_2O \longrightarrow \tfrac{1}{2} O_2 + 2H^+$$

At cathode :

$$2\bar{e} + 2H^+ \longrightarrow H_2$$

Cell reaction

$$H_2O + 2h\nu \longrightarrow H_2 + \tfrac{1}{2} O_2$$

The quantum yield for this process was estimated to be 0.1. A better yield could be produced by replacing Pt cathode by a *p*-type electrode owing to increase in potential difference between the electrodes.

5-11 SOLVED PROBLEMS

5-11-1 Designate (200) and (110) Miller planes of a body centered cube

SOLUTION :

For (200) Miller plane the Weiss indices are $^1/_2$, $^1/_\infty$, $^1/_\infty$ Hence the plane is parallel to y and z axes and cuts the unit cell into two equal halves. Therefore the planes are indicated as in the following figure.

For (110) Miller planes the Weiss indices are 1, 1, $^1/_\infty$. Hence the plane is parallel to z-axis and intersects at the cell length.

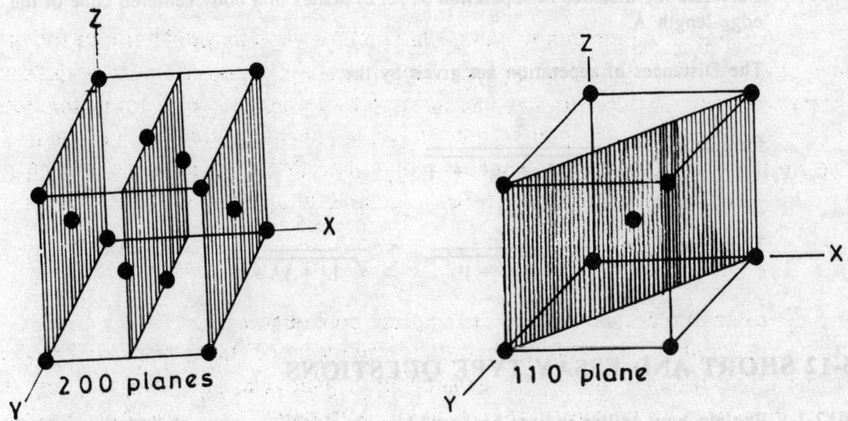

5-11-2 Illustrate (200) and (220) planes of a face centered cube.

SOLUTION :

For (200) planes, Weiss indices are $^1/_2, ^1/_\infty, ^1/_\infty$. Hence the plane is parallel to y and z axes and divides the unit cell on x-axis into two halves.

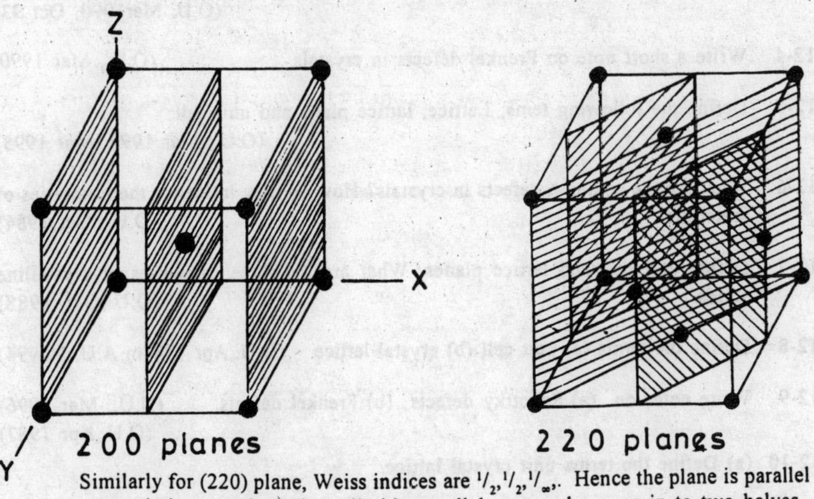

Similarly for (220) plane, Weiss indices are $^1/_2, ^1/_2, ^1/_\infty$. Hence the plane is parallel to z-axis but cut the units cell side parallel to x and y axes in to two halves.

5-11-3 Calculate the Miller indices of crysstal planes which cut the crystal sides at 2a,b,3c.

SOLUTION :

	x	y	z
The planes cut the axes at	2a	b	3c
Coefficients	2	1	3
Therefore Weiss indices are	2	1	3
Inverse of Weiss iindices are	1/2	1	1/3
Multiplying with 6 the LCM	3	6	2

Therefore Miller indices of the plane are (3 6 2)

5-11-4 Calculate the distance of seperation of (222) planes of a body centered cube of the edge length Å.

The Distances of seperation are given by the

formula $d_{(hkl)} = \dfrac{a}{\sqrt{h^2 + k^2 + l^2}}$

Therefore $d_{222} = \dfrac{5}{\sqrt{h^2 + k^2 + l^2}} = \dfrac{5}{\sqrt{h^2 + k^2 + l^2}} = 1.44A°$

5-12 SHORT AND ESSAY TYPE QUESTIONS

5-12-1 Explain how Miller indices are useful in the identificcation of different crystal planes in a crystal.

5-12-2 What are anisotropic and isotropic substances?

5-12-3 State the law of constancy of interfacial angles of a crystal.
(O.U. Mar1990, Oct 93)

5-12-4 Write a short note on Frenkel defects in crystals. (O.U., Mar 1990)

5-12-5 Define the following tems; Lattice, lattice plane and unit cell.
(O.U., Mar 1990, Mar 1995)

5-12-6 What are the common defects in crystals? How do they influence the properties of materials?
(O.U., Oct 1984)

5-12-7 Define unit cell and lattice planes. What are the three categories of crystalline solids.
(O.U.,Oct 1988)

5-12-8 Define the terms (a) unit cell (b) crystal lattice. (O.U.,Apr 1988m A.U. Nov94)

5-12-9 Write notes on (a) Schottky defects (b) Frenkel defects. (N.U., Mar 1996)
(O.U.,Apr 1987)

5-12-10 (a) Define the terms unit crystal lattice.

(b) Mention the types of defects that occur in crystals. (N.U., Mar 1996)
(O.U., Nov 1986)

5-12-11 State laws of crystallography. (O.U., Mar 1989, A.U. Sep 1993)

5-12-12 What is crystalline solid and amorphous solid? (K.U., Apr 1989)

5-12-13 Explain the point defects in crystals.
(A.U., Apr 1989), (N.U., Apr 1989) (O.U., Oct 1992)

5-12-14 Write a note on the defects in crystals (SKU., Sep 1991) (A.U., Mar 1993)

5-12-15 Explain the terms plane of symmetry, centre of symmetry with suitable examples.
(N.U., Mar 1996), (A.U., Mar 1990,), (N.U., Apr 1991)
(A.U., Sep 1993), (A.U., Jan 1990) & (N.U., Nov 1990)

5-12-16 Write two characteristics of crystal solids. (SKU., Apr 1990)

5-12-17 Write a note on Miller indices. (N.U., Mar 1996), (A.U., Dec 1989), (A.U., Mar 1993)

5-12-18 Derive Bragg's equation (N.U., Apr 1990)

5-12-19 Explain Plane of symmetry and centre of symmetry. Discuss the defects in crystals. (A.U., Nov 1994, Mar 94)

5-12-20 Explain the terms wein indices and Miller indices. Give suitable examples is them. (N.U., Apr 1992)

5-12-21 What are the symmetry elements in a single cubic system? (A.U., Mar 1993)

5-12-22 Explain the short key defects in crystals. (A.U., Sep 1993)

5-12-23 What is plane of symmetry and how many planes of symmetry are possible for a cubic. (O.U., Mar 1994)

5-12-24 Discuss briefly "Valence Bond theory (VBT) of metals" and explain the high electrical conductivity and thermal conductivity of metals on the basis of this theory. (O.U., Mar 1996)

5-12-25 Draw lattice panes of a simple cubic crystal system and mention their respective miller indices. (O.U., Mar 1996)

5-12-26 Explain metallic band theory on the basis of free electron theory and valency band theory. (N.U., Sept 1994), (N.U., Mar 1995), (N.U., Sep 1996)

6

PHASE RULE

6-1 HOMOGENEITY AND INHOMOGENEITY OF SYSTEMS

A system which exhibits identical physical and chemical properties throughout and possessing physical indistinguishability is called a **homogeneous system.** Thus we say that copper sulphate solution is homogeneous as it exhibits identical physical properties like uniform colour, physical indistinguishability of copper sulphate even under ultra microscope, etc. Though milk possesses uniform white colour, we classify it under heterogeneous solution, as milk is a colloidal solution and the presence of fat particles can be felt under ultra microscope. *The particle of a solute in a true solution which generally has a diameter of few angstroms, could be detected employing X-rays as they interact with the solute particles of true solution and give characteristic diffraction pattern.* To solve this homogeneity and inhomogeneity of the true solutions, we have to restrict out physical indistinguishability to visible light, wherein the colloidal particles are large enough for detection. Keeping this boundary condition in view, we can comfortably identify all the constituents of heterogeneous systems, which are generaly termed as **phases.**

6-2 PHASE RULE

Fusion, vaporization, sublimation, dissolution of a salt, liquefaction of a gas etc., all involve changes in the phases of the constituents. These phase changes are completely reversible and are said to make a **heterogeneous equilibrium.** A study of variations in the physical parameters like pressure, temperature, volume, concentration etc., of constituents on the heterogenous equilibria were discussed quantitatively by Henry's law, Clapeyron-Clasius equation, Raoult's law, Nernst distribution law, Le Chatelier principle, law of mass action etc. *Phase rule is a generalization of the above mentioned laws and gives a unified approach as regards to fixing the physical variables to decide the existence of the phases present in heterogenous equilibria. It is not meant for either to invalidate or supersede the existing laws, which mainly describe the quantitative relation among the variables.*

The **Phase rule** was deduced theoretically by J. Willard Gibbs (1878), but its application to physical chemistry was made by H.W.B. Roozeboom (1884). It is described by mathematical expression,

$$F = C - P + 2 \quad (6\text{-}1)$$

where F is called **number of degrees of freedom** of variables

C is the number of **components** of the system

P is the number of **Phases** of the system

The phase rule is defined as, *if an equilibrium is dependent only on temperature, pressure and concentration variations and not influenced by gravity, surface forces, electrical or magnetic forces, then **the number of degrees of freedom,** (F), exceeds the difference of number of **components,** (C), and the number of **phases,** (P) by 2.*

6-2-1 Phase

A phase is defined as any homogeneous, physically distinct with well defined boundaries and mechanically separable part of a system. Boiling, freezing, sublimation and conversion of one allotropic form to other all involve changes of phase without change of chemical composition. Ice below 0°C and normal pressure conditions exists in mere solid phase. As we raise the temperature, it melts to give liquid water and water vapour, thus forming a system of two or three phases that are homogeneous, physically distinct and mechanically separable. As we cool they once again form solid ice. Hence we refer that these three phases of water are at a heterogeneous equilibrium, Figure 6-1. Addition of few ml of alcohol to this three phase equilibria, will not generate another phase, although, the liquid phase is now referred to as a solution of alcohol in water and the vapour phase is a homogenous mixture of alcohol vapour and water vapour. But carbon tetrachloride being immiscible with water, its addition to above three phase equilibria genearate a fourth liquid phase, Figure 6-2.

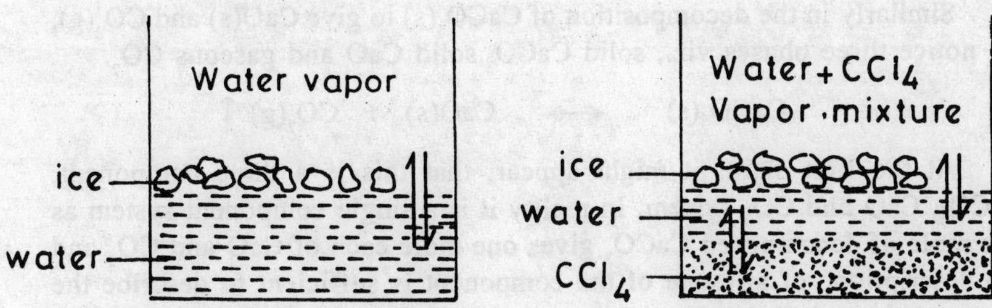

Fig. 6-1: Three phase equilibria

Fig.6-2: Four phase equilibria

Similarly addition of every solid constituent generates a separate phase as they are once again physically distinct, defined by separate boundary conditions. Thus addition of naphthalene balls to the three phase water equilibria generates a fourth phase of solid naphthalene which then involves in an exchange of naphthalene with vapour and liquid water. On the other hand addition of sand to the three phase equilibrium under consideration is not considered as a creation of fourth phase as it lacks any exchange of chemical species. For the same reason, the breaker in which the constituents are present is also not a phase of the system.

Similar to liquid solution, a solid solution (e.g.an alloy) is also considered as a single phase, no matter how many individual substances are present in it.

A gas or mixture of gases always constitute a single phase as they are physically indistinguishable and always form a true solution.

6-2-2 Components

The number of components in a system is the minimum number of independently varaible chemical species necessary to describe the composition of each phase. Thus at 100°C temperature when we take a mixture of $H_2(g)$, $O_2(g)$ and $H_2O(g)$, it forms a threee component system. At arc temperatures, if this mixture is allowed to equilibrate to give water

$$2H_2\,(g) + O_2\,(g) \longleftrightarrow 2H_2O\,(g) \qquad K_{eq} = \frac{[H_2O]^2}{[H_2]^2\,[O_2]}$$

vapour, it then becomes a two component system as the description of any two components of the system, and a knowledge of equilibrium constant allows one to calculate the third component. Instead, if we start with a known amount of water vapour and allow it to dissociate to give O_2 and H_2, it becomes a one component system as the knowledge of dissociation constant of water and concentration of any one of the component fully describes all the three constituents of the vapour phase.

Similarly in the decomposition of $CaCO_3(s)$ to give $CaO(s)$ and $CO_2(g)$, we notice three phases viz., solid $CaCO_3$,solid CaO and gaseous CO_2.

$$CaCO_3(s) \longleftrightarrow CaO(s) + CO_2(g)\uparrow$$

At the first sight, it might appear, that this is a three component, $CaCO_3$, CaO and CO_2 system. In reality it is a single component system as one mole of decomposed $CaCO_3$ gives one mole each of CaO and CO_2 and the description of any one of the component is sufficient to describe the concentration of all the three phases present in the system. In the formation

$$CO_2 + CaO(s) \longleftrightarrow CaCO_3$$

of $CaCO_3(s)$ by the combination of $CaO(s)$ and $CO_2(g)$, we have an open option of taking different amounts of CaO and CO_2. Hence we have to know at least, the concentration of two of the components, to fully describe the equilibrium under investigation, making it a two component system.

Another interesting example is the sodium sulphate-water system. Na_2SO_4 and H_2O combine in different proportions to give, compounds ranging from anhydrous Na_2SO_4 to aqueous solution of Na_2SO_4. Some of the equilibria are,

$$Na_2SO_4(s) + 0\ H_2O(l) \longleftrightarrow Na_2SO_4(s)$$
$$Na_2SO_4\ (s) + 7\ H_2O(l) \longleftrightarrow Na_2SO_4\ .\ 7\ H_2O(s)$$
$$Na_2SO_4(s) + 10\ H_2O\ (l) \longleftrightarrow Na_2SO_4\ .\ 10\ H_2O\ (s)$$
$$Na_2SO_4(s) + x\ H_2O(l) \longleftrightarrow Na_2SO_4(aq)\ (l)$$
$$0\ Na_2SO_4(s) + \underset{(ice)}{H_2O(s)} \longleftrightarrow H_2O\ (s)\ or\ (l)\ or\ (g)$$
$$0\ Na_2SO_4(s) + H_2O(s) \longleftrightarrow H_2O\ (s)\ or\ (l)\ or\ (g)$$
$$0\ Na_2SO_4(s) + H_2O(g) \longleftrightarrow H_2O\ (s)\ or\ (l)\ or\ (g)$$

Although compostion of certain phases (last three equilibria) may be stated in terms of only one of the constituent, certain others require the knowledge of the amounts of any two constituents to specify unambiguously the composition of each phase of the system. The smallest number by which the compostions of all the phases can be described being two, sodium sulphate water system is called a two component system.

Liquid water contains presumable enormous number of associated species; $H_2O, (H_2O)_2, (H_2O)_3 ... (H_2O)_n$

$$H_2O + H_2O \longleftrightarrow (H_2O)_2$$
$$H_2O + (H_2O)_2 \longleftrightarrow (H_2O)_3$$
$$\vdots \qquad\qquad \vdots$$
$$\vdots \qquad\qquad \vdots$$
$$\vdots \qquad\qquad \vdots$$
$$\vdots \qquad\qquad \vdots$$
$$H_2O + (H_2O)n\text{-}1 \longleftrightarrow (H_2O)n$$

The equilibria between various polymers of water shift very rapidly, within 10^{-11} seconds at the most. Ordinary measurements require much longer time. Of the n species, representing (n-1) equilibria, only one species is chemically independent. Hence the system is effectively one component system.

From the above discussion it is clear that an accurate assignment of the number of components in a system pre-supposes some experimental knowledge of various equilibria involved in the system.

6-2-3 Degrees of freedom

The number of degrees of freedom or variance (F) is the minimum number of ***intensive variables*** *(mass independent properties such as T, P, and concentration) that must be specified in order to fix the vaalues of all remaining intensive variables.* The mass dependent properties like weight, volume etc., are called **extensive properties** and they do not determine the equilibrium state to the phase. Physical properties like temperature, pressure, density, refractive index, viscosity,cocentration (moles per unit volume), etc., are mass independent variables, and they only define the phases of a system.

Considering a system of pure water, to describe the phase and know whether water is present as solid,liquid or vapour it is just sifficient to know the pressure and temperature conditions as the other intensive varaibles (refractive index,viscosity etc) are automatically defined. Thus we have a freedom choice of only two variables, viz., the pressure and the temperature, of the various permitted intesive varaibles, and the water present in a single phase thus has only two degrees of freedom. Any change in either of these two variables could change the phase of the water. Thus water,existing in the vapour phase at 100°C and 1 atm pressure, converts into liquid water either by increasing the pressure or decreasing the temperature.

Systems with $F = 0$ are known as **invariant** or as having **no degrees of freedom**; with $F = 1$ they are called **univariant** (one degree of freedom), with $F = 2$ they are called **divariant** (two degrees of freedom); and so on. These are demonstrated in the following examples:

(i) **A pure gas**: A pure gas is completely described by just knowing the temperature and pressure. Hence it is divariant system or has two degrees of freedom. This conslusion could be arrived from the phase rule expression (6.1) as a pure gas is a one component, ($C = 1$), one phase ($P = 1$) system.

Therefore;

Degrees of freedom, $F = 1 - 1 + 2 = 2$ (Divariant)

(ii) **A gaseous system consisting of two gases:** To describe the system fully, we need to define the composition of the mixture, temperature and pressure of the system. Hence it becomes a trivariant system. The same conclusion could be arrived as the gas mixture is a two component and one phase system.

Degrees of freedom, F = 2 - 1 + 2 = 3 (Trivariant)

(iii) **A gaseous system consisting of three gases:** Here the system could be fully described, if the composition of three of the constitutents, temperature and pressure are known. Now substituting the number of components 3 and phases 1 we have from the phase rule expression.

Degrees of freedom, F = 3 - 1 + 2 = 4

(iv) An undissolved solute in contact with its solution: When CO_2 is filled in a jar containing water, the water gets saturated with CO_2 and the dissolved CO_2 is in equilibrium with free CO_2 gas. From Raoult's law, we know that the partial vapour pressure of a component at a given temperature is directly proportional to the mole fraction of that component. Just the description of temperature defines the solubility and the vapour pressure of the gas. Hence it is a univariant system. The sàme conclusion can be drawn from phase rule as the number of components is one and phases are two.

Degrees of freedom , F = 1 - 2 + 2 = 1

(v) Ice-Water-Vapour System: All the three phases of water, viz., ice, water and vapour, are at equilibrium at a temperature of 0.0098°C, pressure 4.58 *mm* Hg. Any change in temperature or pressure disturbs the three phase equilibrium and we then get either one or two phases. Hence such a system has no degrees of freedom or invariant. As the number of components = 1, phases = 3; it could be seen from the phase rule equation that;

Degrees of freedom F = 1 - 3 + 2 = 0

6-3 PHASE DIAGRAM

A diagram which illustrates the conditions of equilibrium between various phases of a substance is called a ***phase diagram or equilibrium diagram.*** A polyphasic equilibrium which can be studied with the help of any two pairs of internsic properties, say temperature and pressure or temperature and composition or pressure and composition, are represented by a two dimensional phase diagram. A trivariant system requires a three dimensional phase diagram. The importance of the phase diagrams could be understood by considering the following one component system.

6-4 ONE COMPONENT SYSTEM

Pure water forms an example of one component system. Similarly the physical equilibria like melting, sublimation, evaporation etc., of pure substances are all example of one component heterogenous systems.

Solid ⟷ Solid (Allotrophic transition)
Solid ⟷ Liquid (Fusion)
Liquid ⟷ Vapour (Evaporation)
Solid ⟷ Vapour (Sublimation)

A one component system, should have at least one phase. Thus a one component system possesses a maximum variance of two as seen by the phase rule expresion.

Degrees of freedom $F = 1 - 1 + 2 = 2$

Hence a one component system is defined by at least two physical parameters viz., temperature and pressure. As we increase the number of phases of the one component system we can achieve a variance of 1 and zero.

(a) When two phases of the system are at equilibrium,degrees of freedom,

$F = 1 - 2 + 2 = 1$

Hence it becomes a univariant system, allowing a variation of either temperature or pressure. This equilibrium vanishes by the variation of both the phyiscal variables.

(b) when any three phases of the system are at equilibrium; degrees of freedom = 1 - 3 + 2 = 0

Hence it becomes an invariant system, allowing no variation of any of the intensive properties.

These results infer that it is not possible to have any four or higher polyphasics equilibria for a one component system.

6-4-1 Water system

Water system forms a typical example of one component system. Argument in support of water being a one component system has already been provided in section (6-2-2). Water system comprises of three phases, viz., ice,liquid water and water vapour, which show the following equilibria.

Ice $\longleftrightarrow$ Liquid water

Ice $\longleftrightarrow$ Water vapour

Liquid water $\longleftrightarrow$ Water vapor

Ice $\longleftrightarrow$ Liquid water $\longleftrightarrow$ Water vapour

similar to other one component systems, water can have 2, 1 and 0 degrees of freedom. Because the maximum variance is 2 we can build the phase diagram of water system by studying the affects of pressure and temperature on various phase equilibria, Figure 6-3.

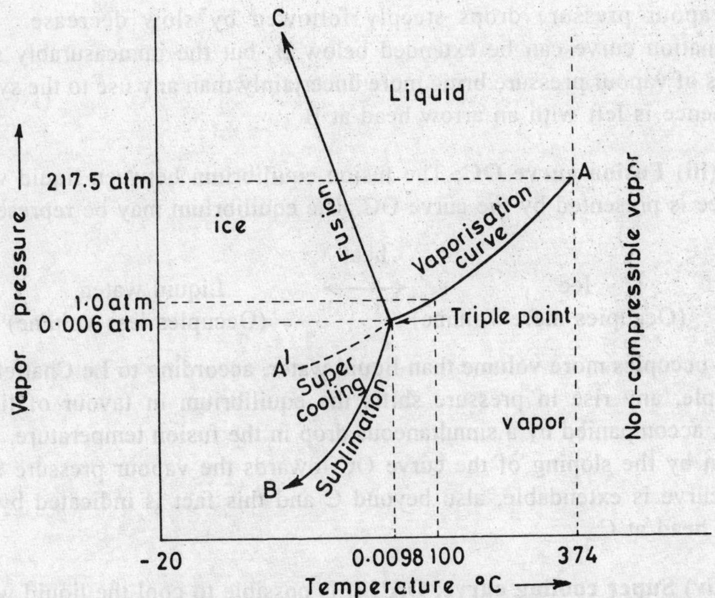

Fig 6-3: Phase diagram of water system

(i) Vaporization curve, OA: Liquid water at any given temperature has a precise vapour pressure. With the increase in temperature we notice an increase in the vapour pressure, but the rate of increase at higher temperatures is higher than at lower temperatures making the vaporization curve, OA, non-linear and slant upwards away from the temperture axis. This increasing trend continues till the system reaches a critical temperature of 374°C and a critical vapour pressure of 217.5 atm. Any attempt to pass beyond **A** destroys the liquid vapour equilibrium leaving the water totally in the incompressible vapour phase. It is not posssible to liquify water vapour beyond 374°C (critical temperature) by the application of any amount of pressure.

Point **O** on the lower part of vaporization curve represents the freezing point of water, 0.0098°C, under its own vapour pressure, 4.58 atm of Hg or 0.008 atm. The lower value of normal freezing point of water, (0°C), measured under laboratory conditions is due to

(a) depression of freezing point by the dissolved air,

(b) the external pressure being 1 atm. far higher than the critical vapour presssure of water vapour at the freezing point.

(ii) Sublimation curve, OB: Below the freezing point the liquid phase totally converts into solid ice and then we observe only one equilibrium between solid and vapour phases and the curve representing this equilibrium is called **sublimation curve OB**. With the decrease in temperature initially the vapour pressure drops steeply followed by slow decrease. This sublimation curve can be extended below B, but the immeasurably small values of vapour pressure bring more uncertainly than any use to the system and hence is left with an arrow head at B.

(iii) Fusion curve OC: The fusion equilibrium between liquid water and ice is presented by the curve OC. The equilibrium may be represented

$$\underset{\text{(Occupies more volume)}}{\text{Ice}} \xleftrightarrow{\text{heat}} \underset{\text{(Occupies less volume)}}{\text{Liquid water}}$$

as ice occupies more volume than liquid water, according to Le Chattelier's principle, any rise in pressure shifts the equilibrium in favour of liquid water, accompanied by a simultaneous drop in the fusion temperature. This is seen by the sloping of the curve OC towards the vapour pressure axis. This curve is extendable, also beyond C and this fact is indicated by the arrow head at C.

(iv) Super cooling curve, OA': It is possible to cool the liquid water at 0,by gradually lowering the temperature without making any mechanical movements like stirring, scratching etc., and avoiding seeding by the dust particles or a tiny crystal of ice. Any mechanical shock, or the presence of a tiny crystal of ice result in shooting up of temperature to 0°C. This cooling below the freezing point indicated by the broken line OA' at any temperature is greater than the vapour pressure of stable phase (curve OB). *In the super cooled liquid, the solid and super cooled liquid are imagined to be separated by a thin vapour membrane and any slightest disturbance removes this insulatory vapour membrane leading to freezing of water and shooting of temperature to 0°C.*

(v) Super heating: Similar to super cooling curve, attempts to obtain super heating of a solid beyond the freezing point O were found futile. Hence no super heating curve is shown in the phase diagram.

After obtaining the phase we can make an attempt to interpret the results obtained, in terms of number of degrees of freedom or variance of water system. The phase diagram contains four curves, a point of intersection of these four curves and areas separated by these curves.

Curves: Out of the four curves OA,OB,OC and OA' present in the phase diagram of water system,the former three curves (OA,OB and OC) represent phase equilibria of any two of the stable phases. Applying the phase rule equation for this one component ,two phase equilibria represented by these curves;

the number of degrees of freedom $F = 1 - 2 + 2 = 1$

Thus the curves represent a univariant system , i.e., along the curve on the phase diagram fixing any one of the intensive properties (either pressure or temperature), automaticaly fixes the other variable. The broken curve OA'represents a metastable equilibrium of vapour and super cooled liquid and is also a univariant system.

Areas: Areas in a one component phase diagram indicate the existence of single phase. Applying the phase rule equation for this one component, one phase equilibria represented by the areas AOB, BOC and COA;

the number of degrees of freedom, $F = 1 - 1 + 2 = 2$

Thus the areas represent a bivariant system, suggesting that in these areas for locating any point both temperature and pressure have to be fixed.

In these areas:

(a) area covered by AOB represents only vapour phase

(b) area covered by BOC represents only solid ice phase, and

(C) area covered by COA represents only liquid water phase.

Point of intersection of curves or Triple point: All the four curves intersect at O and this point of intersection is called the **triple point**. At the triple point all the three phases coexist and are at equilibrium. Applying the phase rule equation for the *one component, three phase equilibrium represented by point O;*

the number of degrees of freedom, $F = 1 - 3 + 2 = 0$

Thus the triple point represents an invariant system, i.e., any variation in even one of the variables (either pressure or temperature) breaks the three phase equilibrium and transforms into a two phase equilibria represented by the curves. The temperature and pressure of the triple point are fixed at 0.0098°C and 4.58 mm Hg respectively.

6-5 TWO COMPONENT SYSTEM

In a two component system the composition of all the individual phases available in various phase equilibria are expressed with the help of two components. Systems involving fusion of any two metals, formation of freezing mixtures and salt hydrates are some of the typical examples of two component systems. In a two component system, minimum number of possible phases could be taken as unity and this corresponds to three degrees of freedom (F = 2 – 1 + 2 = 3). Hence to illustrate fully a two component system we require to define at least three physical variables, viz., temperature, pressure and composition. As explained earlier the three possible variables could only be illustrated on a three dimensional phase diagram. To simplify the situation and bring it down to a two dimensional illustration, normally one of the variables is kept constant throughout and only two of the intesnive properties are varied and corresponding curves are plotted. We now illustrate the phase diagrams of silver-lead system and KI-water system and explain the finer details of these phase equilibria.

6-5-1 Silver-Lead system

The four possible phases of the silver-lead system are (a) solid silver (b) solid lead (c) solution lof silver and lead in molten state and (d) vapour. Silver and lead being very high temperature boiling metals, the vapour is practically absent, therefore pressure changes have a negligible impact on the phase diagram. Negligible pressure effects on the phase equilibria led to a practice of plotting the temperature against percentage composition on one of the metals. As discussed already, this fact helped in illustrating the system by a two dimensional phase diagram, Figure 6-4.

In plotting the phase diagram, because the measurements are made at a constant pressure of 1 atmosphere, the variance of the system reduces by a factor of one and corresponding phase rule eqution, commonly called **reduced or condensed phase rule equation** may be written as ;

$$F' = C - P + 1$$

The freezing point curve OA: The point A represents the melting point of 100% pure silver (M.P. = 961°C) where the solid silver and silver melt (liquid) coexist. The melting point of silver falls gradually by the addition of lead (depression of freezing point) till the silver melt gets saturated with lead as represented by the curve AO. At O the Ag solution is fully saturated with Pb and any further addition of lead is not accepted

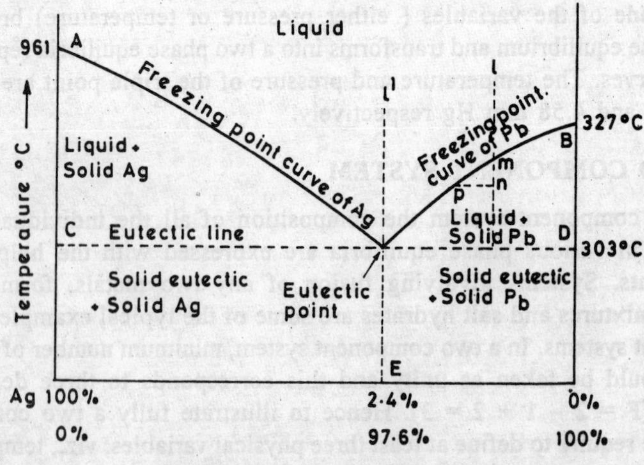

Fig 6-4 : Phase diagram of silver-lead system

and it simply separates out as solid Pb. This saturation limit represented by O is called the **eutectic point** (Greek; eutectos = easy melting).

All along the curve AO, we notice two phases, solid silver and silver melt in equilibrium. Applying the reduced phase rule equation for this *two component, two phase equilibrium*

the number of degrees of freedom $F' = 2 - 2 + 1 = 1$.

Hence the curve AO represents a univariant system,where fixing any one of the vairables automatically fixes the other two physical variables and even a slight variation of one of the variables leads to the total disappearance of one of the three phases.

Freezing point curve of lead OB: In the freezing point curve OB, the point B represents the melting point (M.P. = 327°C) of 100% pure lead where the solid lead and lead melt coexist. By the gradual addition of silver, the melting point of lead falls till the lead melt gets saturated by silver, as represented by the curve OB. Any further addition of lead beyond O, simply separate out as solid lead.

Similar to the freezing point curve of silver, OB also represents a phase equilibrium between solid lead and lead melt. Applying the reduced phase rule equation for this two component, phase equilibrium;

the number of degrees of freedom, $F' = 2 - 2 + 1 = 1$

Hence the curve BO also represents a univariant system where defining one vairable amounts to fixing of the other varaibles automatically.

Eutectic Point O: At this point all the three phases, viz.,solid Ag,solid Pb and liquid solution coexist. Applying the reduced phase rule equation;

The degree of freedom, $F = 2 - 3 + 1 = 0$

Hence point O represents an invariant system, where any minor variation in one of the variables shatters the three phase equilibrium in a three phase equilibrium. The temperature and composition of the eutectic mixture corresponding to eutectic point O at 1 atmospheric pressure are 303°C and 2.4% Ag + 97.6% Pb respectively.

The salient features of eutectic point are:

(i) Eutectic point refers to lowest freezing point of a liquid mixture of any two metals at a given external pressure (conventionally fixed at 1 atm).

(ii) Freezing any unsaturated liquid solution described by areas AOC and BOD leaves an eutectic mixture and pure solid metal corresponding to that area.

(iii) Similar to a molten metal, eutectic mixure freezes at a constant temperature. The solid eutectic so formed as a compound has a fixed composition and is called an alloy.

(iv) In spite of having a definite composition and fixed freezing point,eutectic mixture is never considered as a compound as the crystals of components are clearly observable under microscope.

The Areas: Drawing a horizontal line CD passing through the eutectic point, we notice three areas above the eutectic line, CD, and two areas below the eutectic line.

The Area AOB: The area AOB represents a liquid solution of Ag and Pb in equilibrium with its vapour. As the pressure is arbitararily fixed at 1 atm;

the number of degrees of freedom in the area AOB are $F' = 2 - 1 + 1 = 2$.

Hence AOB is a bivariant region. On cooling the liquid at any point ***l'***, vertically above the eutectic point, eutectic solid is separated only after touching the eutectic point O. The liquid at any other point, say, ***l*** when cooled, remain unchanged until it reaches a temperature corresponding to ***m*** on the freezing curve. At ***m***, solid Pb will commence to deposit keeping the liquid composition at ***m***. On further cooling it enters a nuivariant region BOD. At any point in this region, say ***n***, the composition of the liquid is given by the point ***p***, corresponding to the intersection of the tie line nad BO curve. Further cooling to eutectic temperatue, the point ***q*** on teh eutectic line corresponds to solid eutectic and the solid *Pb*.

The Area AOC: The area AOC has solid silver and liquid melt in equilibrium. The composition of the liquid melt directly corresponds to the tie line intersection point on the AO curve. This area is a univariant region ($F' = 2 - 2 + 1 = 1$).

The Area BOD: The area BOD has solid lead and liquid melt in equilibrium. The composition of the liquid melt directly corresponds to the tie line inter-section point on the BO curve. This is also a univariant region.

The Area COE: The area COE has solid eutectic and solid Ag in equilibrium. Applying reduced phase rule equation we find that the region is univariant ($F' = 2 - 2 + 1 = 1$).

The Area DOE: The area DOE has solid eutectic and solid Pb in equilibrium. This is also a univariant system just like area COE.

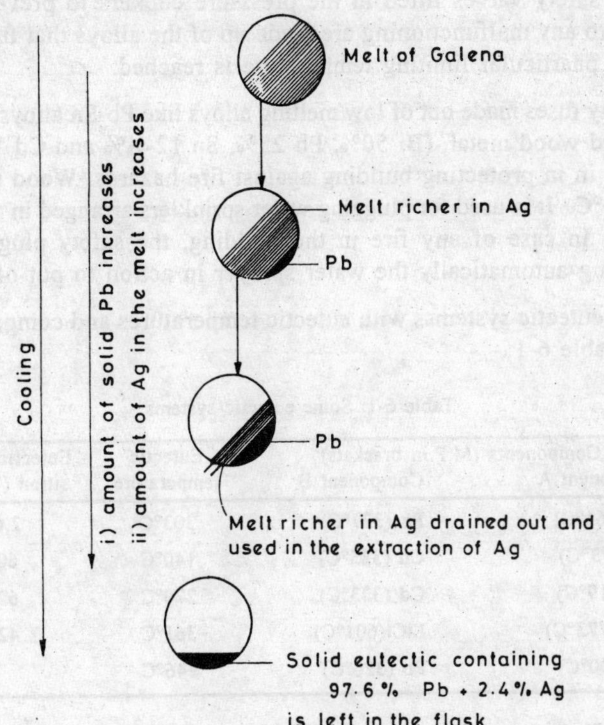

Fig 6-5: Pattinson's process

6-5-2 Pattinson's process for the desilverzation of lead

Silver is often recovered from the work up of lead and copper ores. Gaalena is a lead ore containing 0.1% silver in it. The enrichment of silver in this ore to 2.4% was achieved from a knowledge of Ag-Pb phase diagram. This proces is called **Pattinson's process or desilverization process.** It was already discussed that under area AOB, a molten liquid ***l***, which is rich in lead, when cooled traces a path ***l m n q***, with a separation of lead metal. The concentration of the liquid at points ***m, n***, and ***q***, correspond to the intersection point of BO curve with the tie lines passing through these points. Hence, as the liquid cools to q, the composition of the melt aproaches the composition of the eutectic mixture (2.4% Ag and 97.6% Pb). This solid eutectic rich in Ag is profitably then used for the subsequent recovery of silver employing electrochemical techniques. The relative increase in the amount of lead in the solids and increase of silver in the melt on cooling argentiferrous lead is shown in Figure 6-5.

6-5-3 Other applications of eutectic mixtures

(i) In the soldering of two metal pieces, often an alloy (solid eutectic) that has much lower melting point than the individual metal pieces to be joined is used and the processs is called soldering.

(ii) The safety valves fitted in the presssure cookers to prevent bursting due to any malfunctioning are made up of the alloys that fusse as soon as a paarticular limiting temperature is reached.

(iii) Safety fuses made out of low melting alloys like Pb-Sn alloys, commonly called wood metal, (Bi 50%, Pb 25%, Sn 12-½% and Cd 12-½%) are used in in protecting building against fire hazards. Wood metal melts at 65°C. It is used for plugging water sprinklers arranged in the building roof. In case of any fire in the building, the safety plug melts off, putting automatically the water sprayer in action to put off the fire.

Few eutectic systemss with eutectic temperatures and compositions are given in table 6-1.

Table 6-1: Some eutectic systems.

Components (M.P.in brackets) component A	Component B	Eutectic temperature	Eutectic composition (% of A)
Ag (961°C)	Pb (327°C)	303°C	2.6%
Bi (273°C)	Cd (323°C)	140°C	60%
Zn (419°C)	Cd (323°C)	270°C	67%
KCl (773°C)	LiCl(601°C)	361°C	42%
Sb (630°C)	Pb (326°C)	246°C	-

6-5-4 KI-Water system

This is another example of a two component system and is a representative of freezing mixtures. Since *Kl* does not form hydrates, the four possible phases of *Kl*-water system are ice, solid, *Kl*,solution of *Kl* in water and vapour. As in Ag-Pb system here also the pressure is maintained at 1 atm and the phase diagram is illustrated between temperature and composition of *Kl*. Most of the treatment of this system is similar to Ag-Pb system. The phase diagram is given in figure 6-6.

Freezing point curve AO: The point A is the freezing point of water under normal conditions. Here the phase equilibrium exists between solid ice, liquid water and vapour. This freezing point is lowered by the addition of KI as represented by the freezing point curve AO. Point O represents lowest freezing point attainable by the addition of KI to ice-water equilibrium (freezing) and is called the eutectic temperature. The eutectic point also represents the maximum solubility of KI in water. Any further addition to the eutectic mixture only leads to simple separation of solid KI.

Along the curve we observe a two phase equilibrium between ice and KI solution. Applying the reduced phase rule equation for this two component, two phase equilibrium curve AO;

the number of degrees of freedom F' = 2 - 2 + 1 = 1

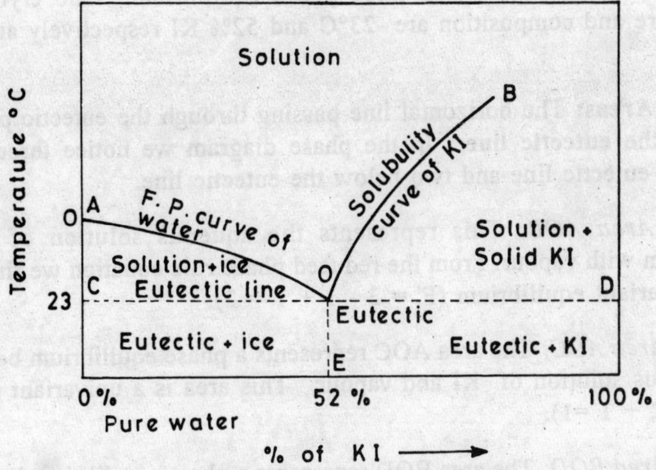

Fig 6-6: Phase diagram of KI-water system

Thus curve AO represents a univaaariant system, where fixing any oneof the variable automatically defines the other varaible and even slight variation of the variable leads to the total disappearance of one of the two phases.

Solubility curve of KI, OB: The curve OB represents the effect of increase in solubility with increase in temperature. Along OB we find a two phase equilibrium between solid KI and saturated KI solution. This solubility curve can not be traced to 100% KI because the saturation of KI puts a limit at B on the boiling point. Thus the point B, owing to its constant temperature, pressure and composition conditions becomes an invariant system.

Applying the reduced phase rule equation for this two component, two phase equilibrium curve BO;

the number of degrees of freedom $F' = 2 - 2 + 1 = 1$

Thus curve BO represents a univariant system, where fixing any one of the variable automatically defines the other variable and even a slight variation of one of the variables leads to the total disappearance of one of the two phases.

Eutectic point or Cryohydric point: The melting point O, of the curves AO and OB is as usual called an eutectic point and in these freezing mixture equilibrium in specific it is termed as **cryohydric point**. At this point all the four phases, viz., solid KI, ice, solution of KI and vaapour coexist. Applying the reduced phase equation;

the number of degrees of freedom $F' = 2 - 3 + 1 = 0$

Hence cryohydric point, O, represents an invariant system, where any minor variation in one of the vaariables shatters the four phase equilibrium, resulting in the creation of a three phase equilibrium. The cryohydric temperature and composition are -23°C and 52% KI respectively at 1 atm pressure.

The Areas: The horizontal line passing through the eutectic point O is called the **eutectic line**. In the phase diagram we notice three areas above the eutectic line and two below the eutectic line.

The Area AOB: This represents the aqueous solution of KI in equilibrium with vapour. From the reduced phase rule equation we find that it is a bivariant equilibrium (F' = 2 - 1 + 1 = 2).

The Area AOC: The area AOC represents a phase equilibrium between ice, aqueous solution of KI and vapour. This area is a univariant region (F' = 2 - 2 + 1 =1).

The Area BOD: The area BOD represents a phase equilibrium between solid KI, aqueous solution of KI and vapour. This is also a univariant region.

The Area COE: The area COE represents a phase equilibrium between eutectic, ice and vapour. This is also an univariant region.

The Area DOE: The area DOE represents the equilibrium between the eutectic, ice and vapour. This is univariant region.

6-5-5 Applications of freezing mixtures

From the phase diagram of KI-water system it is clear that when we add salt to ice, the freezing point falls. Initially this fall in the freezing point is directly proportional to the amount of salt added and addition of excess of salt drops the temperature to touch the cryohydric temperature (eutectic temperature) and no further drop occurs. In the laboratory the freezing mixtures are used conveniently to bring down the reaction temperature to sub zero levels. Some of the commonly employed cryohydric mixtures and their temperatues are given in Table 6-2.

Table 6-2: Some cryohydric temperatures

Cryohydric mixture	Composition at cryohydric point	Cryohydric temperature
KI + ICe	52%	-23°C
NaCl + Ice	24%	-22°C
$CaCl_2$ + ICe	15.2%	-55.9°C
NH_4Cl + ICe	20%	-17°C
KCl + ICe	--	-11.4°C
KNO_3 + ICe	--	-2.6°C
Alcohol + Solid CO_2	--	-72°C
Chloroform + Solid CO_2	--	-77°C

6-6 SHORT AND ESSAY TYPE QUESIONS

6-6-1 Draw the phase diagarm for water and mention its parts.
(N.U., Sept. 1995), (K.U., Apr.1989), (O.U.,Oct.1990), (O.U.,Oct.1993)

6-6-2 Construct the phase diagram of silver-lead system and explain the importance of eutectic point. (A.U., Apr.1989)

6-6-3 State phase rule and explain the terms involved. Apply the phase rule to silver-lead system. (A.U.,Jun. 1989), (SKU., Sep.1991), (N.U., Mar.1996)

6-6-4 What is a phase diagram? Explain. (A.U.,Dec.1989) & (O.U.,Oct.1987)

6-6-5 What is an eutectic? Give one example.
(A.U.,Jun.1990), (N.U., Sept.1994), (N.U., Sept.1995)

6-6-6 Explain with suitable examples the following terms used in the phase rule study of heterogenous equilibrium

(i) Triple point

(ii) Eutectic point

(iii) Univariant system (K.U., Jun.1990)

6-6-7 Discuss the application of phase rule to potassium iodide-water system.
(O.U.,Apr.1989)

6-6-8 (a) Write the phase rule equation. Discuss the application of the phase rule to a one component system.

(b) Define the terms "Eutectic point" and triple point.
(O.U., Mar. 1989) & (O.U., Oct. 1991)

6-6-9 Define the tems in the phase rule. Explain the tems in the following system is.

(i) A gas in equilibrium with its solution in a liquid

(ii) Ammonia gas

(iii) $I_2(s) \longleftrightarrow I_2(g)$ (O.U., Mar. 1989)

6-6-10 (a) Define the terms (i) Component (ii) Phase of a system

(b) Predict the degrees of freedom for the following

(i) A one component system having two phases in equilibrium

(ii) A one component system having a single phase

(iii) A one component system having three phases in equilibrium
(O.U., Nov. 1986)

6-6-11 Explain the term component. How many components are present in the following systems?

(a) water $\longleftrightarrow$ water vapour

(b) KCl + Water $\longleftrightarrow$ KCl hydrate. (O.U., Mar. 1986)

6-6-12 Discuss a method to obtain the phase diagram of Ag-Pb system and give details of the number of degrees of freedom of the curves and areas in it. (O.U., Oct. 1992)

6-6-13 (a) Draw the phase diagram of KI-H_20 systems

(b) Answer the following questions pertaining to the diagrams.

(i) How many components are there in the system.

(ii) How many phases are in equilibrium along any one line.

(iii) What is the number of degrees of freedom along any line.

(iv) Name the phases which are in equilibrium at the eutectic point.

(v) What is the number of degrees of freedom at the eutectic point.

6-6-14 Find the degrees of freedom aand number of components in the following systems.

(a) $CaCO_3(s) \longleftrightarrow CaO(s) + CO_2(g)$

(b) A two phase system made by mixing arbitrary amounts of C, O_2 and CO_2 at a temperature at which the reversible equilibrium establishes

$$C(s) + O_2(g) \leftrightarrow CO_2(g)$$

6-6-15 Explain why the freezing point curve in the phase diagram of water is inclined towards the vapour pressure axis.

6-6-16 Plot the phase diagram of a one component system and identify all the lines,areas and phases.

6-6-17 State phase rule and apply it to water system.
(N.U., Apr.1991) (A.U., Mar.1990), (N.U., Sept.1995)

6-6-18 Explain the phase diagram of salt-water system.
(N.U., Nov.1990), (SKU., Apr.1990)

6-6-19 What are the differences between triple point and eutectic point and what is meant by invariant system. (O.U., Mar.1992), (O.U., Mar.1995)

6-6-20 State phase rule. Explain the terms phase, component and degrees of freedom. Discuss the application of phase rule of a one component system.
(A.U., Mar.1993), (N.U., Apr.1992), (O.U., Mar.1993), (O.U., Mar.1994), (O.U., Mar.1995), (N.U., Mar.1995)

6-6-21 Discuss the significance of triple point in the phase diagram of water.
(A.U., Sep.1993), (N.U., Mar.1995), (N.U., Mar.1996), (O.U., Mar.1996)

7
COLLOIDS AND ADSORPTION

COLLOIDS

7-1 INTRODUCTION

All the macroscopic and microscopic particulate matter may be grouped into two categories (i) pure substances (ii) Mixtures.

(i) Pure substances: All the elements and compounds fall in this category.They have a fixed composition and generally have sharp melting and boiling points.

(ii) Mixtures: A mixture may be defined as a substance containing two or more pure substances whose particles are in intimate contact but have not lost their individuality. These substances can be separated by simple sepa- ration techniques and the properties of mixtures vary according to percentage of components present. They usually do not have sharp melting and boiling points. We find a variety of mixtures in this universe. They can be further grouped into three general categories depending on the size of particles that are mixed, *viz.,* (A) Solutions (B) Suspensions and (C) Colloids.

(A) Solutions: On adding a spoon of sugar to a glass of water and stirring, we find that it vanishes, but leaves a sweet taste to water. The sugar is distributed uniformly and can be regained by evaporating water. Varying amounts of sugar can be accommodate by the water. This mixture is known as a solution of sugar in water. As long as the composition of sugar is less than that of water, sugar is called the **solute** and water the **solvent**. Properties of solution are different from the properties of solvent. Though the presence of sugar in water could be realized from its taste, attempts to detect the sugar molecules in a sugar solution using a microscope did not yield fruitful results because they assume a size of few angstroms and for the same reason they readily diffuse through parchment membrane. As these substances are generally in crystalline form Graham called them **crystalloids.**

(B) Suspensions: On the other hand, by adding a spoon of sand to a glass of water, we find the clarity of water is lost and it becomes turbid. If we keep this mixture aside for a while we find that the sand settles at the bottom under the influence of gravity and the clarity of solvent is restored. Clear water could also be obtained quickly by filtering it. This type of mixtures are called a **suspensions**. The particles in a suspension are large enough to be seen either with the naked eye or through a microscope. The suspended particles or **suspensoids** are not distributed uniformly through the medium (here water) and they normally do not have any significant impact on the properties of the medium.

In the laboratory one can prepare a suspension by adding few drops of $AgNO_3$ solution to a solution of NaCl. The white cloud, appeared due to formation of water insoluble AgCl, is an example of suspension. Either by filtration or centrifuging the mixture it is possible to separate the AgCl and get a clear solution of $NaNO_3$

(C) **Colloids:** This category of mixtures fall between true solutions and suspensions. The term colloid is derived from the Greek Word **'kolla'** meaning glue. The term **Sol** is often used to represent a colloid. The natural glues are colloidal dispersions of glue in water. Starch solution is a typical example of a colloid. The particles are invisible but their presence could be detected by their scattering effect. The main properties that differentiate colloids from suspensions are their uniform distribution throughout medium or stability towards settling of particles under gravity and passage of particles through the filter paper without getting separated. These two properties resemble that of a true solution. The main property that distinguishes a colloidal solution from a true solution is the detection of particles under ultra microscope and their precipitation by the addition of an electrolyte. The terms solute and solvent used in connection with a true solution are replaced by **disperse phase** and **dispersion medium** in case of colloidal solutions.

Colloidal particles range from 1 nm, (10Å), to 1000nm and they consist of neutral molecules and charged particles like ions. They are not single phase solutions like true solutions but generally contain two phases. This fact could be demonstrated by studying the optical properties of a true solution, a suspension and a colloidal solution. For this take the suspension of soil in water in a test tube say 'a', the true solution of sodium chloride in water in test tube 'b' and a colloidal starch solution in test tube 'c'. Just by looking at the test tubes with our naked eye, we can identify that test tube 'a' is a suspension but test tubes 'b' and 'c' both look fully transparent.

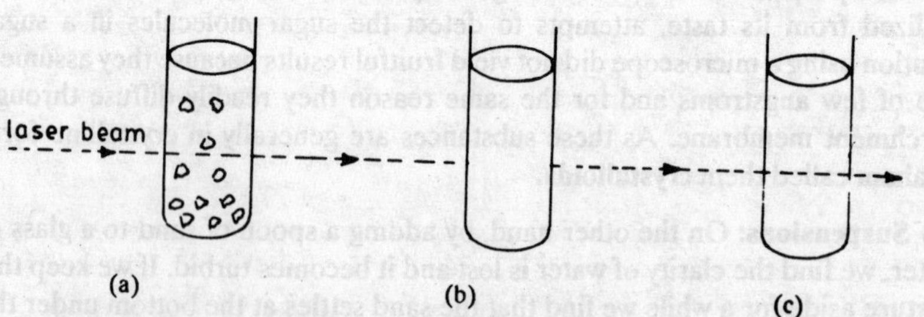

Now pass an intense beam of light through these test tubes. We will be able to trace the path in test tubes 'a' and 'c' while we can not locate any ray in the test tube 'b'. This is due to scattering of the light by the suspension particles and colloidal particles. The scattering of light by the colloidal particles was first noticed by Tyndall and is known as Tyndall effect. The distinguishing features of colloids are given in Table 7-1, although a detailed account of properties of colloids will be given later in **Section 7-6**

Table 7-1 Comparative statement of properties of true solutions, colloids and suspensions

Property	*True solutions*	*Colloids*	*Suspensions*
1. State	Monophasic (Homogeneous)	Biphasic (Heterogeneous)	Biphasic (Heterogeneous)
2. Particle size	< 1 nm	1-1000 nm	> 1000nm
3. Visibility of particles	Particle are not visible	Particles are invisible but are discernible under ultra microscope.	Particles are either visible to naked eye or under microscope
4. Settling	Does not settle under gravity	Does not settle under gravity	Settles under gravity
5. Appearance	Clear	Clear	Opaque
6. Filtration	Particles can not be separated by filtration through filter paper	Particles can not be separated by filtration through filter paper	Particles separated out by passing through filter paper
7. Dialysis	Easily pass through parchment membrane	Passes through parchment membrane but slowly	Does not pass through parchment membrane.
8. Brownian motion	The solute particles have constant kinetic motion	The dispersion phase shows Brownian motion	The motion of the suspension particles is negligible.
9. Diffusion	Diffuse very fast	Diffuse slowly	No diffusion
10. Tyndall effect	Do not scatter light. Tyndall effect not observed	Shows scattering of light. Tyndall effect observed	Shows scattering of light. Tyndall effect observed.

7-2 TYPES OF COLLOIDAL DISPERSION

We come across various colloids daily. The milk we drink, the blood flowing through our veins, the protein solutions, the butter, cheese,jelly etc., that we eat, the clouds, the fog, the smoke are all different forms of **colloidal dispersions or colloidal suspensions.**

The colloidal phase and colloidal medium which form the colloidal dispersions can take all the possible combinations of three states of matter *viz,* solid, liquid and gas except for a colloid formed by a gas dispersed in a gas, as a gas mixes in another gas in all proportions, uniformly at molecular level, this combination only results in a solution. Various colloidal dispersions with suitable example are listed in Table 7-2.

Table 7-2 Types of colloidal dispersions

Disperse phase	*Dispersion medium*	*Colloidal type*	*Examples*
1. Solid	Solid	Solid sol	Colored glasses, pearls many precious stones
2. Solid	Liquid	Sol, gel	Gold sol, $Fe(OH)_3$ sol, starch in water, paint
3. Solid	Gas	Solid aerosol	Smoke (carbon particles in air), Haze (dust in air)
4. Liquid	Solid	Solid emulsion	Cheese, butter (water in fat), jellies (water in fruit pulp)
5. Liquid	Liquid	Emulsion	Milk(fat in water), cream, mayonnaise
6. Liquid	Gas	Liquid aerosol	Fog, cloud (water in air), mist, fog
7. Gas	Solid	Solid foam	Pumice stone (air in silicates), marshmallow
8. Gas	Liquid	Foam	Foam (water in air), froth, shaving cream, whipped cream
9. Gas	Gas	**Always produces a solution**	

7-3 CLASSIFICATION OF COLLOIDS

A colloidal suspension has two phases, the disperse phase and dispersion medium. Based on the affinity between these two phases colloids are classified into **lyophobic sol and lyophilic sol**. When the dispersion medium is water these are said to be **hydrophobic** and **hydrophilic sols** respectively.

Lyophobic sols: These are also termed as irreversible sols or suspension. This term has its origin in greek and means solvent hating or solvent averting. *In the lyophobic colloids the disperse phase has very little affinity towards the dispersion medium.* This lack of affinity accounts for the difficulties encountered in the preparation and their easy precipitation. As the colloids are readily precipitated giving larger visible particles, they are called **suspensoids** and this process of precipitation is known as **coagulation,** irreversibility of this process brought the name **irreversible colloids**. Colloidal dispersions of metals like gold, silver, iron etc., and sparingly soluble salts like As_2S_3, $Fe(OH)_3$ etc., are the examples for this class. They coagulate readily by the addition of even a trace amount of electrolyte. Normally stabilizing agents are added to prevent coagulation.

Lyophilic sols: Lyophilic means solvent loving or solvent liking. These colloids are formed readily by employing simple techniques like heating, shaking a colloidal particle phase in a suitable dispersion medium and called **emulsoids** or **reversible colloids**. In the house we notice the preparation of starch solution, just by adding starch powder to the boiling water. Protein solution, glue (gum), etc. are some typical examples of lyophilic sols. These sols obtained are of reversible type. A coagulated disperse phase could be brought back to the colloidal state just by diluting and warming it again. These sols are self stabilized and require no stabilizer and small amounts of electrolyte have no effect.

The characteristics of lyophobic and lyophilic sols are given in Table 7-3.

Table 7-3 Characteristics of Lyophobic and Lyophilic sols

Property	*Lyophobic sol*	*Lyophilic sol*
1. Visibility	The particles are detectable with an ultra microscope	The particle can not be detected
2. Stability	Less stable	More stable
3. Reversibility	Irreversible	Reversible
4. Surface tension	The surface tension of colloidal dispersion is almost of the same order as colloidal medium	The surface tension is less than that of colloidal medium
5. Viscosity	Viscosity is almost of the same order as dispersing medium	Viscosity is higher than dispersing medium.
6. Effect of electrolytes	Small quantities of electrolyte is sufficient to cause coagulation	Large quantities are needed for coagulation
7. Hydration	Particles are usually hydrated poorly	Particles are heavily hydrated
8. Charge	Particles carry a characteristic charge	The charge of the particles depends on the pH, may carry negative, positive or even no charge.
9. Electrophoresis	They show electrophoresis	Depending upon the charge on the particle, it may or may not show electrophoresis
10. Concentration	Can have only very low concentration	Higher concentrations are possible.

7-4 METHODS OF PREPARATION OF COLLOIDAL SOLUTIONS

It is possible to prepare colloidal solutions of all substances by selecting suitable dispersion medium. The particle size of colloidal dispersion is intermediate to true solutions and suspension. So we can have two approaches in preparing these dispersions. One approach involves breaking of macroscopic suspension particles into particles of the size of colloidal phase (1nm-1000 nm), employing mechanical methods or electrical methods or by creating conditions for the reversal of coagulation. The methods employed are called **dispersion methods**. The second approach involves bringing the microscopic atoms or molecules of true solutions together forming colloidal particles. These methods are called **condensation methods**. These two approaches can be represented by the following flow chart.

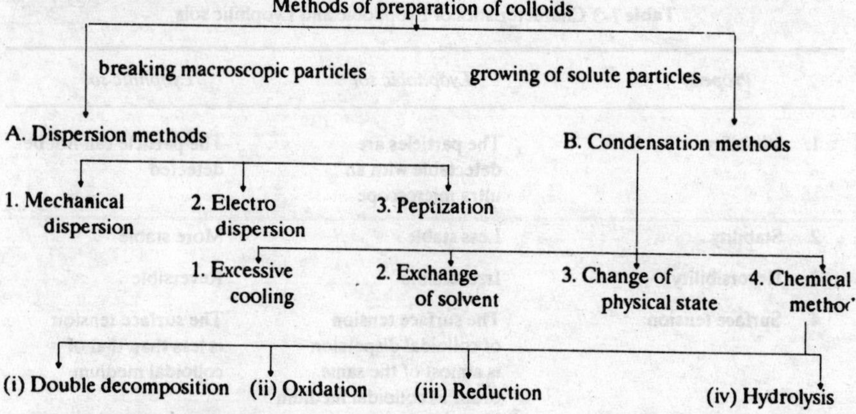

A. Dispersion methods

1. Mechanical dispersion: Coarser particles of a substance are broken into colloidal size by grinding them between two steel discs having a narrow clearance. For this a suspension along with a suitable stabilizer is charged through a hollow shaft of a colloidal mill (fig 7-1), where the steel discs are rotating at a high speed in opposite direction, and the colloidal dispersion is collected.

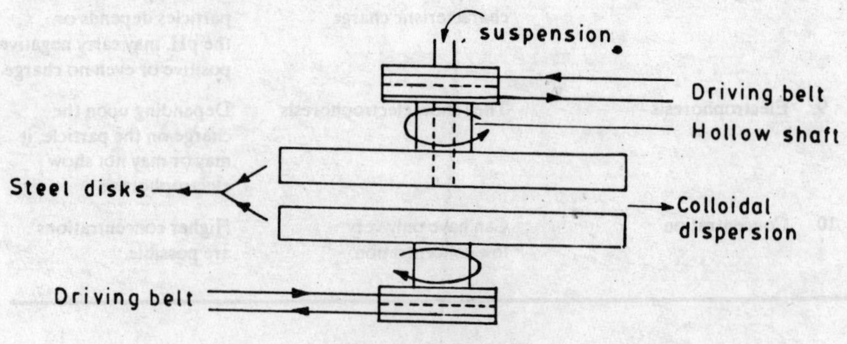

Fig 7-1 Colloidal mill

This way Indian ink is prepared by grinding lamp black in the colloid mill and stabilizing the dispersion with gum arabic.

2. Electro-dispersion Bredig's arc method: The colloids dispersion of metals like gold, silver, copper etc., in water are prepared by this method. The metal rods are dipped in the solvent and an arc is set by passing current through the rods. This arc liberates metal particles of colloidal size which are readily absorbed by the medium giving a colloidal suspension. To prevent evaporation of the medium due to heat liberated by the arc the vessel is surrounded by ice and kept cool.

This method is not useful when we take organic solvents as dispersion medium, as the heat generated at the arc chars the medium.

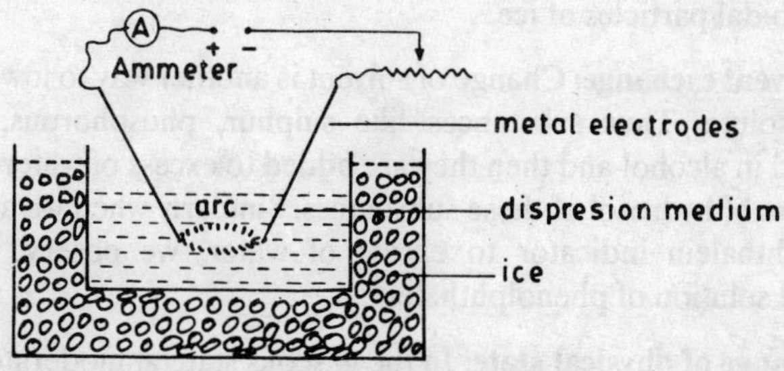

Fig 7-2 Bredig's setup for electro dispersion

3. Peptization: Peptization is inverse of coagulation. As discussed earlier coagulation is the result of lack of affinity between dispersed phase and dispersing medium. An affinity is created by the addition of a stabilizing agent, known as peptizing agent, which prompts the coagulated particles back to give colloidal solution and the process is called peptization.

A freshly prepared precipitate of ferric hydroxide, $Fe(OH)_3$, is peptized, giving a colloidal dispersion, by the addition of some ferric chloride, $FeCl_3$, solution. This is due to masking of uncharged colloidal particle of $Fe(OH)_3$ by the aquated ferric ions Fe^{3+} aq, and creating a net positive charge on the colloidal particle and thus bridging the affinity gap between hydrophobic $Fe(OH)_3$ particle and dispersion medium.

$Fe(OH)_3$

Chemical dispersion

Cl^- Cl^- Cl^- Cl^- Fe^{3+} Fe^{3+} Fe^{3+} Fe^{3+} Cl^- Fe^{3+} Fe^{3+} Fe^{3+} Cl^- Fe^{3+} Cl^- Cl^- Cl^- Fe^{3+} Cl^- Fe^{3+} Cl^-

Colloidal particle

Precipitate particle

Colloidal particles

As these colloidal particles are positively charged, addition of a negatively charged PO_4^{3-} ion to the colloidal solution neutralizes the change causing destabilization of colloid and precipitation of disperse phase.

We come across many such systems in our day to day life. Milk is a colloidal dispersion of oil in water. As oil is a hydrophobic substance, casein in milk serves the purpose of peptizing agent. Milk coagulates by the addition of common salt. In the formation of indian ink, lamp black is peptized by gum arabic.

B. Condensation Methods

1. By excessive cooling: Colloidal solution of ice in organic solvents like chloroform, ether are formed by cooling below 0°C, a solution of water in these solvents. Water is only sparingly soluble in these solvents and on excessive cooling the solubility, further decreases throwing dissolved water out to form fine colloidal particles of ice.

2. By solvent exchange: Change of solvent is another way to lower the solubility of the solute. Thus substances like sulphur, phosphorous, etc., are first dissolved in alcohol and then they are added to excess of water resulting in the formation of hydrosol of these substances. Similarly when we add an alcoholic phenolphthalein indicator to excess of water, we observe a white milky colloidal solution of phenolphthalein.

3. By change of physical state: In the gaseous state at moderate pressures and temperature, many substances attain a macromolecular size. Hence by passing the vapour, *e.g.*, sulphur and phosphorous, through the dispersion medium containing stabilizer yields a colloid.

4. Chemical methods: Chemical reactions that produce partially soluble and insoluble substances lead to the formation of colloids. Depending on the nature of chemical reaction. They can be classified into (i) double decomposition (ii) oxidation (iii) reduction and (iv) hydrolysis.

(i) Double decomposition: On passing H_2S through a dilute solution of arsenious oxide we get a colloidal solution of arsenious sulphide.

$$As_2O_3 + 3H_2S \longrightarrow As_2S_3 + 3H_2O$$

Concentrated solutions yield only suspensions. The excess of hydrogen sulphide is removed by boiling the solution which may otherwise lead to precipitation. Similarly on adding a few drops of ferric chloride to a dilute solution of potassium ferrocyanide in the presence of a peptizing agent we get a sol of ferri cyanide.

$$K_4Fe(CN)_6 + FeCl_3 \longrightarrow \underset{\text{Prussian blue colloid}}{KFe[Fe(CN)_6]} + 3\,KCl$$

(ii) Oxidation: H_2S on oxidation gives a colloidal solution of elemental sulphur

$$2H_2S + O_2 \longrightarrow 2H_2O + \underset{\text{Sulphur colloid}}{2S}$$

A moderate oxidizing agent as bromine or nitric acid could also be used for this purpose.

$$H_2S + Br_2 \longrightarrow 2HBr + S$$

(iii) Reduction: The colloids of metals such as gold, silver, lead and platinum could be conveniently prepared by reducing the metal salts with reducing agents such as formaldehyde, tannic, acid, stannous chloride, etc.

$$2AuCl_3 + 3SnCl_2 \longrightarrow 3SnCl_4 + \underset{\text{gold sol}}{2\,Au}$$

Zigmondy prepared a bright red gold sol by reducing gold chloride ($HAuCl_4.3H_2O$), and phosphorous colloid by reducing a dilute etherial layer of phosphorous.

(iv) Hydrolysis: Colloidal dispersions are obtained by the hydrolysis of metal hydroxides, *viz,* $Fe(OH)_3$ or $Al(OH_3)$ and stabilizing the colloids by the addition of little excess of solutions of corresponding metal chlorides.

$$FeCl_3 + 3H_2O \longrightarrow Fe(OH)_3 + 3HCl$$

For achieving stability of the colloid it is essential to remove HCl by dialysis.

7-5 PURIFICATION OF COLLOIDAL DISPERSION

The general impurities present in a colloidal dispersion are mainly the suspended particles and the electrolytes. The suspended particles could be easily separated by simple filtration. The following methods are employed for freeing a colloidal solution from electrolytes.

1. Dialysis: The process of separating a colloidal solution from a crystalloid (true solution) is known as dialysis. The property of passing through an animal membrane by the solute and solvent molecules of a crystalloid and inability of colloidal particles was exploited by Graham in designing the apparatus used for purifications a colloid. This apparatus of known as dialyzer. The colloidal solution to be dialyzed is taken in a parchment bag and suspended in a running water trough. The soluble impurities pass through the membrane into the running water leaving pure colloidal suspension in the parchment bag. Completion of dialysis could be ensured by monitoring the conductivity of water in the trough. Presence of ions increases the conductivity of pure water.

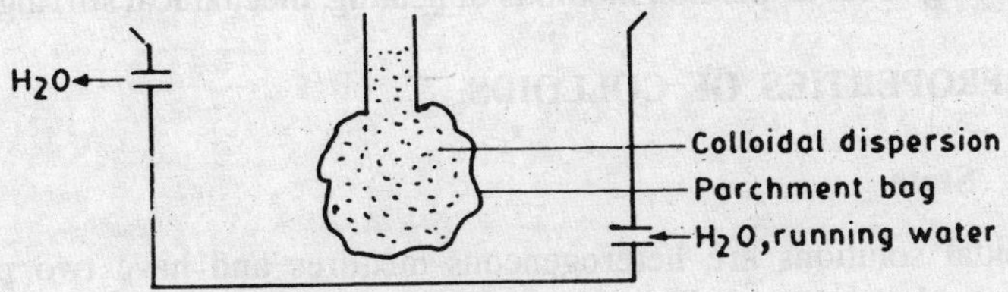

Fig 7-3 Graham dialyzer

2. Electro–dialysis: The principle behind this technique is same as simple dialysis. Normal dialysis is a slow process and usually takes several hours or sometimes days for completion. Colloidal particles normally have a charge due to adsorption or resting of ions over the surface. Hence diffusing of electrolyte through the membrane is fastened by carrying out the dialysis in the presence of electric field and is known as **electro-dialysis.**

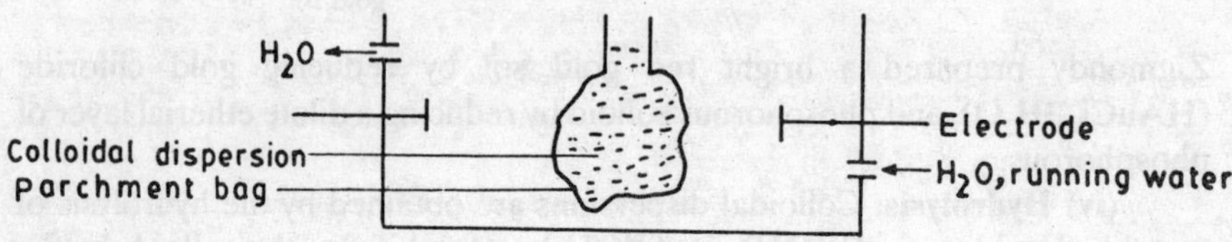

Fig 7-4 Electro-dialysis

Electrodialysis is obviously not meant for removing uncharged impurities such as sugar, urea etc.

3. Ultra-filtration: The pore size of ordinary filter paper is > 1000 nm allowing the sol particles, which have a diameter of 10nm-1000 nm, to pass through them. In the ultra-filtration, the filter paper is soaked in a colloidal solution such as gelatin or collodion and dried. This reduces the pore size considerably and the impregnated filter paper, known as ultrafilter, is used in the purification of colloid, similar to a dialysis membrane. By selectively choosing impregnation suspensions of different sizes and preparing corresponding ultra filters, it is also possible to separate a mixture of colloidal solutions having particles of different sizes .

4. Ultra-centrifuging: Colloidal particles and crystalloid, both, do not settle under gravitation and exhibit Brownian (zig-zag) motion. The continuous zig zag motion of the colloidal particles prevents the colloidal particles from settling under gravity. But if one considers the velocity of this motion colloids move rather slow in comparison to crystalloids. Hence when a sol is placed in a ultra-centrifuge and rotated at a speed of 15000 or more revolutions per minute, the colloids settle at the bottom of centrifuging tube leaving the electrolyte free. This suspension can be reconverted into a colloidal solution, employing usual dispersion methods of heating, mechanical stirring etc.

7-6 PROPERTIES OF COLLOIDS

7-6-1 State

Colloidal solutions are heterogeneous mixtures and have two phases, the disperse phase and the dispersion medium.

7-6-2 Particle Size

They have a particle size of 1-1000 nm and can be seen only under a ultra microscope. They are invisible to naked eye and to even for powerful microscopes.

7-6-3 Filtration

Colloidal particles pass through ordinary filter paper. As mentioned in the purification methods, specially prepared ultra filters retain the colloidal particles. They also can not pass through animal membranes.

7-6-4 Colour

The colour of a colloidal dispersion depends on

(a) The size and the shape of the particle.
(b) The characteristic absorption, the disperse phase and medium.
(c) Whether the observer is receiving the transmitted light or reflected light.
(d) The wave length of the light falling.

For example, with the increase in the particle size gold sol changes its colour gradually from red to blue. When the particles are small (say 200nm) the blue component of the white light is reflected and when we observe in the direction of incidence only red colour is observed. Similarly with the increase in particle size the gold particles start reflecting even the light of other wave lengths there by start looking blue.

Dilute milk appears blue when observed at right angle to the incident light (reflected rays) and red, when observed in the direction of incident light (transmitted light).

A beautiful demonstration of the colour properties of colloidal particles is the chemical sunset scene produced by the transmitted light by the addition of hydrochloric acid to sodium thio sulphate. Here a sulphur colloid is produced initially, which gradually coagulates to give suspension particles.

7-6-5 Colligative Properties

The colligative properties *viz.,* osmotic pressure, relative lowering of boiling point depend on the number of solute particles in a given weight of solvent. For a given concentration true solutions have far larger number of solute particles than the colloidal particles. This is due to aggregation of solute particles (about 1000) to give new bigger size colloidal particles. Of the above mentioned colligative properties only osmotic pressure has a measurable value and changes in other properties are insignificant. The change in osmotic pressure of a colloid is a highly useful data in determining the average particle mass of starch particle which was found to be about 32,000 atomic mass units.

7-6-6 Optical Property – Tyndall Effect

As discussed already, when light passing through a true solution is observed at right angle to the path of incident ray, it looks completely dark. The same experiment if we repeat with a colloidal solution, we notice a luminescent path of the ray passing through the solutions. The path of the ray is made visible by the scattering of light by the colloidal particles. The particles of true solution can not scatter light because the wave length of light is larger than the size of the particles (< 10nm).

This effect was first noticed by Faraday but detailed studies were made by Tyndall giving it the name **Tyndall effect**. The appearance of bright path of the projection lamp in a theatre is due to scattering of light by the dust particles.

The two conditions that must be satisfied to show Tyndall effect are:

(i) The wave length of light used must not be much larger than the diameter or size of the particle.

(ii) The dispersed phase and dispersed medium should differ greatly in the magnitude of refractive indices.

Condition (i) was already explained and condition (ii) could be understood if we consider the colloidal suspension of freshly prepared albumin. Here the disperse phase, albumin and the medium have a very small difference of refractive indices and the Tyndall effect is not observable. On standing, the particle size increases, bringing a change in the refractive index thus showing Tyndall effect.

Tyndall effect is strong for hydrophobic sols due to large difference in the refractive indices of disperse phase and dispersion medium. In case of lyophilic sols the difference in refractive indices is small, resulting in a weak Tyndall effect.

Siedentopf and Zsigmondy made successful attempt to see the colloid particles with the help of a microscope known as **ultra-microscope**. Here a bright light beam from an arc is converged with a convex lens and allowed to fall on the colloidal solutions. We find two self-luminous bluish colored cones, called **Tyndall cones**, meeting at the focus of the lens. When a microscope is placed at right angles to the incident light, presence of colloidal particles can be detected as bright moving spots. This is due to initial absorption of light by the colloidal particle before it is scattered.

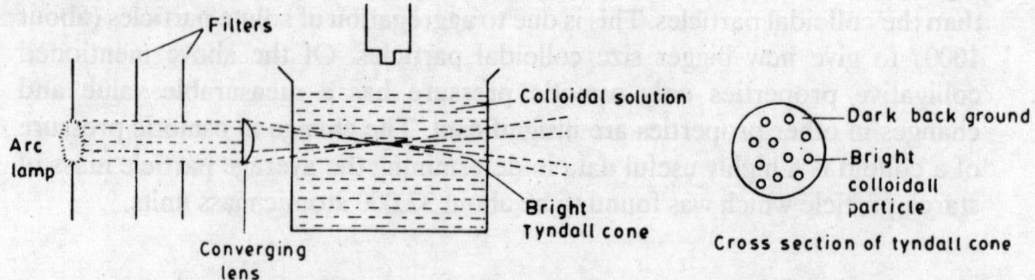

Fig 7-5 Tyndall effect

7-6-7 Kinetic Property – Brownian Movement

The invention of ultra-microscope by Siedentopf and Zsigmondy led to another invention by Robert Brown in 1827. Just like the constant directionless movement of pollen grains in an aqueous suspension, he noticed under microscope that the colloidal particles also move continuously in a zig-zag path without any destiny and this type of motion is called **Brownian movement.**

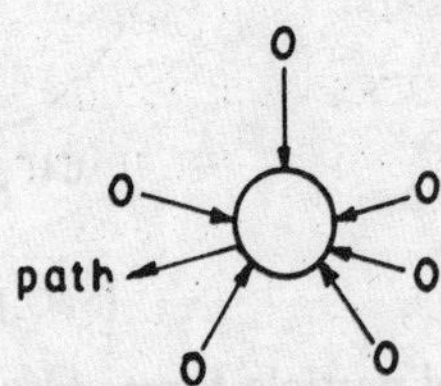

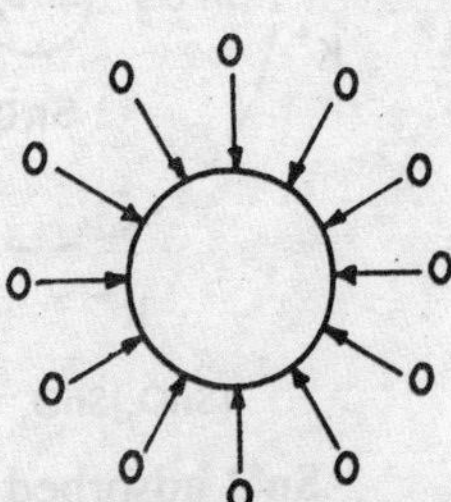

Zig-Zag motion

Small colloidal particle (large probability of motion of small particles)

Large colloidal particle (no resultant; less probability of movement of large particles)

Fig 7-6 **Brownian movement**

This Brownian movement arises due to constant collisions or striking of the molecules of dispersion medium and colloidal particles, Fig 3-6. Owing to the smaller size of the colloidal particles, the striking are limited and the colloidal particles always move in a resultant direction.

A large particle (e.g: a suspension particle) has a higher inertia (inability of a body to move) due to its heavy mass and in addition to numerous collisions in all the directions and cancellation of opposing forces the resultant force becomes extremely weak to cause any significant impact. Hence the particles of a suspension show no Brownian movement and settle under gravity.

7-6-8 Electrical Charge

Colloids often carry a charge on their surface. For example, ferric hydroxide sol particles are positively charged while the arsenous sulphide sols are negatively charged. The electrical neutrality of the sol is maintained by the dispersion medium taking equal and opposite charge of the colloidal particle.

A colloidal particle can take both positive and negative charges depending on the conditions employed in the preparation of the colloidal dispersions. The following two examples demonstrate this fact.

Example 1: If a freshly prepared stannic oxide precipitate is peptized by a few drops of KOH solution the sol assumes a negative charge.

This is due to the adsorption of stannate ions,SnO_3^{2-} formed by the reaction between SnO_2 and KOH.

$$SnO_2 + 2KOH \longrightarrow SnO^{2-}{}_3 + 2K^+ + H_2O$$

This colloidal particle is represented as $[SnO_2]\ SnO_3^{2-}\ 2K^+$

When the SnO_2 precipitate is peptized with the addition of few drops of HCl, the sol takes a positive charge.

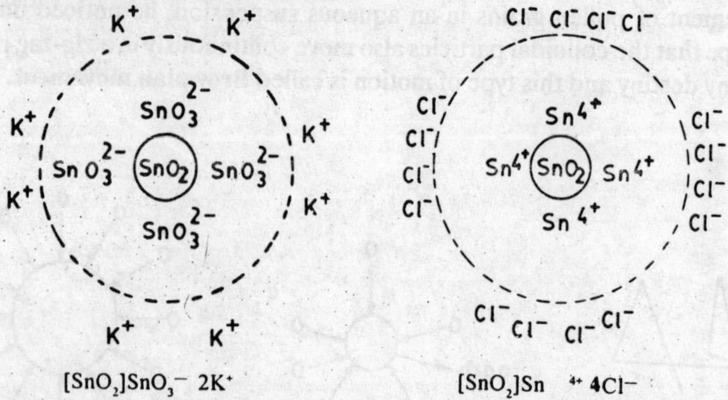

Sn^{4+} adsorbed on the sol particle formed by the reaction between SnO_2 and HCl

$$SnO_2 + 4HCl \longrightarrow Sn^{4+} + 4Cl^- + 2H_2O$$

this is represented as $[SnO_2]\ Sn^{4+}4Cl^-$

Example 2: If a dilute solution of potassium iodide is added to a concentrated solution of $AgNO_3$ a positively charged colloid is formed due to adsorption of free Ag^+ ion on the AgI surface and is represented as $[Ag\ I]\ Ag^+NO^-_3$. Similarly a negatively charged colloid is produced when dilute silver nitrate is added to a concentrated solution of KI and is represented as $[Ag\ I]\ I^-\ K^+$.

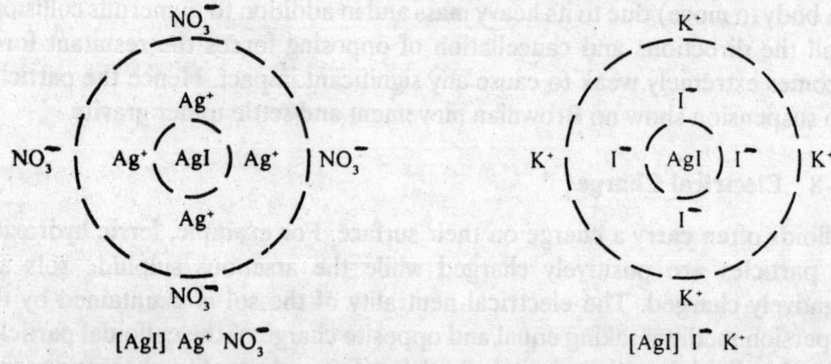

Inspite of co-existence of oppositely charged disperse phase and dispersion medium, the reason for no charge (neutralization), and unexpected stability is explained by the **electrical double layer theory.**

7-6-9 Electrical Double Layer Theory

This theory was propounded by Helmholtz. Consider a positively charged fixed layer in contact with a **mobile, movable or diffused medium.** The negatively charged ions of the movable medium try to approach the fixed layer to neutralize the charge. But the constant thermal agitation or kinetic motion of the ions of the mobile layer create a varying electrical environment for the positively charged fixed layer. *This constant shifting environment results in a charge difference between the fixed layer and the mobile layer is known as* **Zeta potential or electro kinetic potential.**

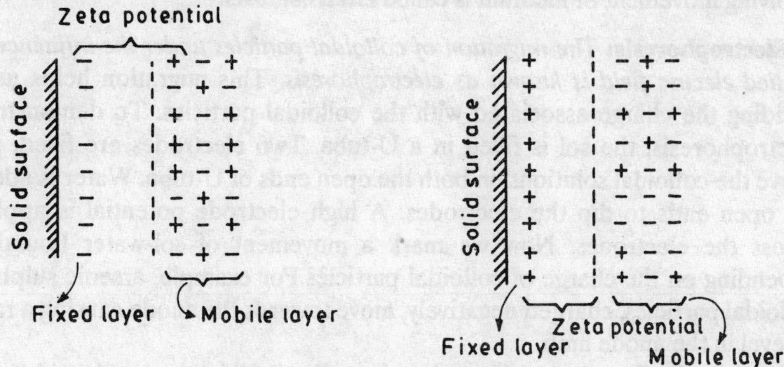

Fig 7-7 The electrical double layer.

The combination of the fixed layer and mobile layer forms an electrical double layer system. *The stability of a colloidal dispersion,according to electrical double layer theory, is due to formation of a zeta potential screen between the dispersed particles (fixed layer) and dispersion medium (mobile layer) that restricts the neutralization of electrical charge on the fixed layer.* This screen of electrical double layer is broken down when we add multivalent ions carrying the charge of dispersion medium, leading to coagulation of the sol. Thus positively charged ferric hydroxide, $Fe(OH)_3$, sol is coagulated by the addition of trivalently charged PO^{3-} ions as shown in fig 3-8.

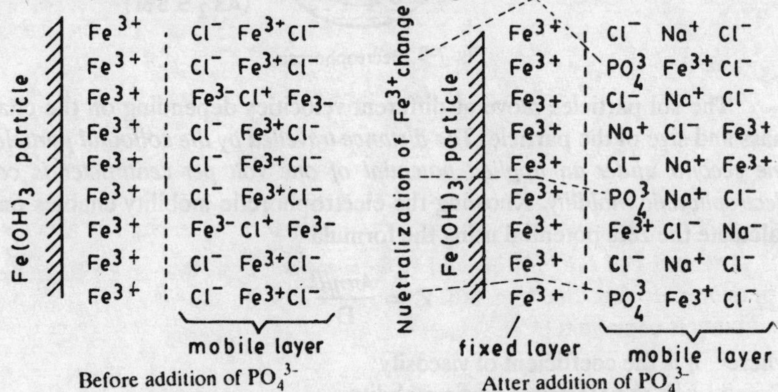

Before addition of PO_4^{3-} Atter addition of PO_4^{3-}

Fig 7-8 Electrical double layer of $Fe(OH)_3$ sol

The phosphate anion, where ever it gets in contact with the colloidal particles, the Zeta potential approaches zero resulting in the coagulation of the sol particle.

7-6-10 Passage of Current

Passage of current through a colloidal solution results in the movement of colloidal particles, as they are charged, towards oppositely charged electrodes. We can design experiments either to allow the movement of the colloidal particles or restrict their motion. The former process involving movement of particles is known as **electrophoresis** or **cataphoresis** and the later process involving movement of medium is called **electrosmosis.**

A. Electrophoresis: *The migration of colloidal particles under the influence of applied electric field is known as electrophoresis.* This migration helps us in deciding the charge associated with the colloidal particles. To demonstrate electrophoresis, the sol is filled in a U-tube. Two electrodes are fixed, just above the colloidal solutions, in both the open ends of U-tube. Water is added the open ends to dip the electrodes. A high electrode potential is applied across the electrodes. Now we mark a movement of sol-water boundary depending on the charge of colloidal particles.For example, arsenic sulphide colloidal particles, charged negatively, move towards the anode causing a raise of level in the anode limb.

As in the first observation, migration of boundaries was made with ferric hydroxide colloid, which migrated towards cathode, the phenomena was given a misnomer as **cataphoresis.**

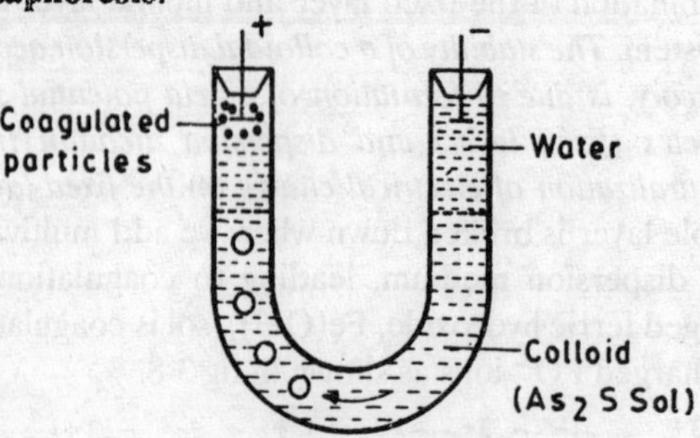

Fig 7-9 **Electrophoresis**

The sol particles move at different velocities depending on the charge, mass and size of the particle. *The distance travelled by the colloidal particles in one second under an applied potential of one volt per centimeter is called electrophoretic mobility.* Knowing the electrophoretic mobility enables one to calculate the zeta potential using the formula

$$Z = \frac{4\pi\eta\mu}{D}$$

where η is the coefficient of viscosity

μ is the electrophoretic mobility

and D is the dielectric constant of the dispersion medium

Knowledge of electrophoretic mobility also helps in separating the sols from a colloidal mixture. Thus from a mixture of metal sol and metal sulphide sol, the lighter colloidal metal sulphide particles are recovered first, followed by the metal sol particles. Biochemists use this technique is fractionating proteins,polynucleic acids and polysaccharides from their colloidal mixtures.

Depending on the charge carried by colloidal particles they are grouped into positively charged sols and negatively charged sols. Few common sols are listed below:

Table 7-4

Positively charged sols	*Negatively charged sols*
1. Metal hydroxides: $Fe(OH)_3$, $Cr(OH)_3$, $Al(OH)_3$ etc.	1. Metals: Au, Ag, Cu, Fe,etc
2. Metal oxides: TiO_2 etc.	2. Metal sulphides: As_2S_3, CdS, Sb_2S_3 etc.
3. Basic dye stuffs: Bismark brown, methylene blue etc.	3. Acid dye stuffs: Eosin, congo red etc.
	4. Organic colloids: Starch, gums, proteins etc.

Applications of electrophoresis: Electrophoresis has many industrial applications:

(i) Carbon from the colloidal smoke going out of an industrial chimney is coagulated and removed on a cathode.

(ii) Suspended impurities from the dust sensitive assembly units like computer and watch are removed taking advantage of this phenomena.

(iii) Negatively charged latex rubber particles are electroplated on the surface of metal, wood or cloth.

B. Electroosmosis: *The movement of the dispersion medium, when the colloidal particles are restricted from their motion, under the influence of electric field is known as electroosmosis.* This could be demonstrated by constructing a simple apparatus shown in fig 7-10.

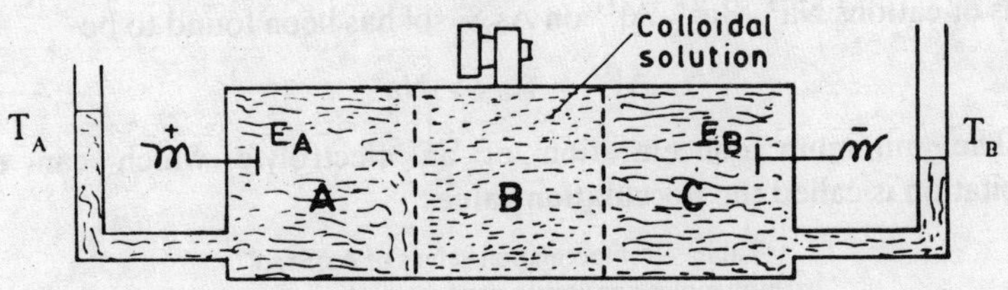

Fig 7-10 Electroosmosis apparatus

The apparatus consists of three compartments A,B and C separated by two dialyzing membranes M_A and M_B. The central compartment 'C' is filled with the colloidal solution, and the side compartments 'A' and 'B' are fitted with side tubes T_A and T_B respectively.

The side compartments are fitted with platinum electrodes E_A and E_B. An electrode potential (D.C current) is applied across E_A and E_B. The dispersed medium and disperse phase start moving towards oppositely charged electrodes. For example if we take arsenic sulphide sol, the negatively charged As_2S_3 particles start moving towards E_A, the anode and the dispersion medium towards E_A, the cathode. But the sol particles are not allowed to get into the anodic compartment'A' by the dialyzing membrane M_A. The dispersing medium passes into the cathodic compartment 'B' without any hindrance by the membrane M_B. This results in a raise in the level of T_B with a corresponding fall in the level of T_A. In case of positively charged sol particle the raise is in the anodic compartment.

7-6-11 Coagulation

Coagulation is a process of growing the colloidal particle size resulting in the formation of suspension and ultimately leading to the precipitation of the particles of the dispersed medium. It may be spontaneous or accomplished either by the addition or complete removal of electrolyte, or by mixing two oppositely charged colloidal solutions.

(i) Spontaneous coagulation: When some sols, on long standing, the particles tend to come closer and grow in their size leading to precipitation. This spontaneous coagulation is called **ageing.**

(ii) By the addition of electrolyte: The colloidal particles are stabilized by assuming a charge cloud on its surface and by forming an electrical double layer or having a zeta potential. Thus as we already discussed As_2S_3 sol particles are negatively charged with positive dispersion medium around them. Addition of an electrolyte brings a change in the zeta potential and the colloidal particle is robbed away of its charge leading to coagulation.

Since it is the charge on the ions of added electrolyte which is responsible for the coagulation, it is rational to assume that the valency of the ions is the root cause of precipitation. Hardy and Schulze after rigorous experiments found that higher the valency of the active ion the greater is its precipitating action and is known after them as **Hardy-Schulze law**. The precipitating actions of cations Na^+, Ba^{2+}, Al^{3+} on As_2S_3 sol has been found to be

$$Al^{3+} > Ba^{2+} > Na^+$$

The minimum concentration of an electrolyte which can cause precipitation is called the **flocculation value.**

Table 7-5 Flocculation values of As_2S_3 sol
(milli moles per litre to produce precipitation)

NaCl	51.0	$MgSO_4$	0.81	$Al(NO_3)_3$	0.095
KCl	49.5	$BaCl_2$	0.69	$AlCl_3$	0.093
		$CaCl_2$	0.65		

Similarly for the positively charged colloidal particles (*e.g.*, $Fe(OH)_3$ sol) we have to take the anion as the active ion. There the precipitating action is of the order

$$PO_4^{3-} > SO_4^{2-} > NO_3^{+}$$

Table 7-6 Flocculation value of $Fe(OH)_3$ sol
(milli mole per litre required to produce precipitation)

KCl	9.5	K_2SO_4	0.20
NaCl	9.3	$MgSO_4$	0.22
$BaCl_2$	9.6		

When we compare the relative stability of hydrophobic and hydrophilic sols, the hydrophobic sols readily coagulate by the addition of electrolytes. In case of hydrophilic sols even after stripping away the charge present on the colloidal particle by the added electrolyte, the sol is not coagulated. This is due to a shield of water molecules formed around the colloidal particle as a result of aquation. Addition of a few drops of dehydrating agent like alcohol or acetone breaks open the shield and coagulation sets in. This is represented in Fig 3-11.

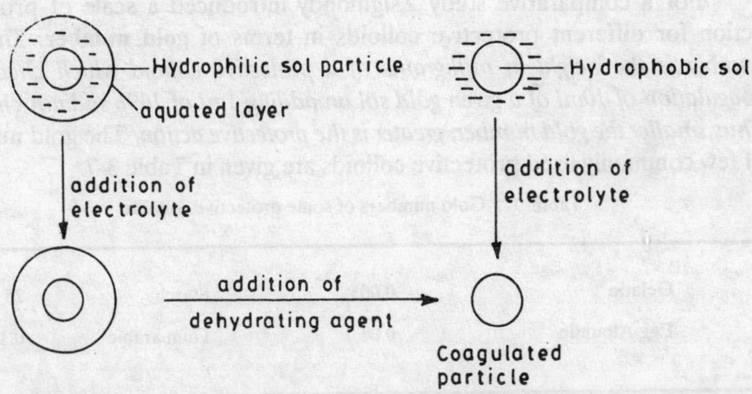

Fig 7-11 Coagulation of hydrophobic and hydrophilic sols.

(iii) By the complete removal of electrolyte: Traces of electrolytes are needed for the stability of a colloid. Prolonged dialysis removes these traces of electrolytes causing precipitation of colloidal dispersion.

(iv) Mutual coagulation: When oppositely charged colloidal dispersions are mixed they neutralize their charges mutually causing precipitation known as mutual coagulation. Thus, when As_2S_3 sol and $Fe(OH)_3$ sol are mixed, both the colloidal particles coagulate.

7-6-12 Protection

As mentioned above the lyophilic colloids achieve extraordinary stability due to solvation of colloidal particles. When lyophilic sols are added to lyophobic sols, depending on their sizes, either lyophobic sol is adsorbed in the surface of lyophilic sol or lyophilic sol is adsorbed on the surface of lyophobic sol as represented in fig 3-12. In both the cases the hydrophobic sols achieve additional stability. *The process of addition of lyophilic sols by which the lyophobic sols are protected from precipitation on adding an electrolyte is known as protection, and the lyophilic colloid added as the protective colloid.*

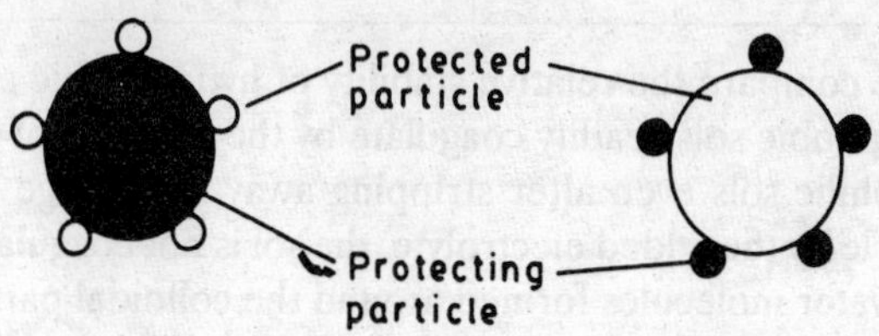

Fig 7-12 Protection of a colloid

Thus hydrophilic colloids such as gelatin, starch etc., act as protective colloids for the gold sol and prevent their precipitation by the addition of limited amounts of NaCl.

For a comparative study Zsigmondy introduced a scale of protective action for different protective colloids in terms of **gold number**. *The gold number is the weight in milligrams of a protective colloid which checks the coagulation of 10ml of a given gold sol on adding 1ml of 10% sodium chloride. Thus smaller the gold number, greater is the protective action.* The gold numbers of few commonly used protective colloids are given in Table 3-7.

Table 7-7 Gold numbers of some protective colloids

Gelatin	0.005	Starch	25
Egg Albumin	0.08	Gum arabic	0.10

The data in Table 7-7. suggests that gelatin is a very effective protective colloid.

7-6-13 The Donnan Membrane Equilibrium

The Donnan membrane equilibrium is a special case where a true solution of an electrolyte is separated from a salt which has a non-diffusible ion by a membrane, and at equilibrium the concentration of diffusible electrolyte on both the sides of membrane was found to be different. In case of colloids the nondiffusible ion is the charged colloidal particle.

Thus when we take C_1 moles of sodium chloride on one side of the membrane compartment 'A' and sodium salt of non-diffusible colloidal ion, R^-, on the other side of the membrane (compartment B), after allowing the system to attain an equilibrium, one finds an increase in the concentration of diffusible electrolyte in the compartment 'B'. If x moles/lit of sodium chloride diffused into the compartment 'B' dropping the concentration of electrolyte in compartment 'A' from an initial value of C_1 to $C_1 - X$, then the equilibrium may be represented as

	Compartment 'A'		Compartment 'B'		
Ions	Na^+	Cl^-	Na^+	R^-	Cl^-
Initial conc. (moles lit^{-1})	C_1	C_1	C_2	C_2	0
At equilibrium (moles lit^{-1})	$C_1 - X$	$C_1 - X$	$C_2 + X$	C_2	X

Donnan (1911) gave explanation for the migration of diffusible ions from one compartment to another compartment separated by a membrane that restricts the movement of non diffusible ions. If no non-diffusible ion is present in the system the electrolyte is present equally in both the compartments. The driving force behind this diffusion is known as **Chemical potential** and for an ideal electrolyte it is equal to the product of concentration of the ions in the electrolyte. At equilibrium the chemical potential in both the compartments is equal.

When non-diffusible electrolyte of concentration C_2 is separated from diffusible electrolyte of concentration C_1, as represented above, the chemical potential of diffusible ions at equilibrium are equal.

Therefore $\quad (C_1 - X)(C_1 - X) = (C_2 + X)X$

or $\quad X/C_1 = C_1/(C_2 + 2C_1)$

X/C_1 is the fraction of sodium chloride (diffusible electrolyte) that has diffused through the membrane. Higher the concentration of non-diffusible electrolyte present, lower is the fraction of electrolyte diffused. From the above expression it is also clear that in the absence of non-diffusible electrolyte, half of the NaCl migrates from Compartment 'A' to Compartment 'B'.

Total concentration of ions at equilibrium in compartment 'A' is $2(C_1 - X)$ and in compartment 'B' it is $2(C_2 + X)$. Hence the concentration difference across the membrane is

$$= 2(C_2 + X) - 2(C_1 - X)$$

$$= 2(C_2 - C_1 + 2X)$$

As this system is comparable to an osmometer, the osmotic pressure 'π' of the solution is given by

$$\pi = 2(C_2 - C_1 + 2X)\,RT$$

but $\quad X = C_1^2/(C_2 + 2C_1)$

Therefore $\quad \pi = 2\{C_2 - C_1 + 2C_1^2/(C_2 + 2C_1)\}\,RT$

By rearranging we have

$$\pi = 2\,RT\,C_2\,(C_1 + C_2)/(2C_1 + C_2)$$

When $C_1 > C_2$ the above expression reduces to

$$\pi = RTC_2$$

Thus a measurement of osmotic pressure allows us to calculate the concentration of a non-diffusible electrolyte or a charged colloidal particle, there by its molecular weight.

7-7 EMULSION

When liquids form both the disperse phase and dispersion medium of a colloid, then that colloid is said be an emulsion. Any two partially miscible liquids can yield an emulsion. Either liquid can be dispersed in the other, fig 3-13.

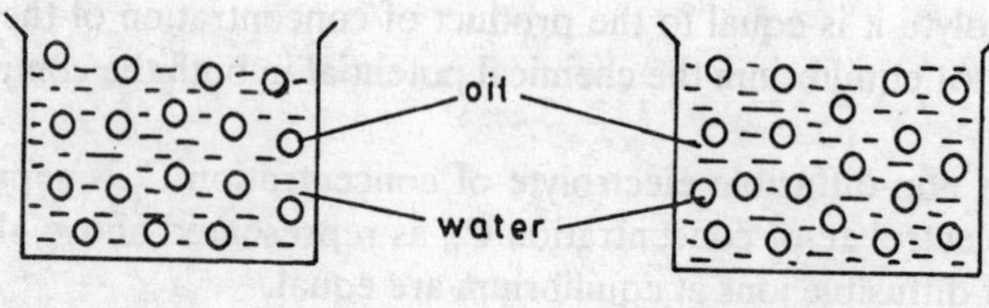

oil
water

emulsion of oil in water

emulsion of water in oil

Fig 7-13 Emulsion of oil in water and water in oil

Thus we can have an emulsion of oil in water by shaking vigorously few drops of oil in excess water. Similarly an emulsion of water in oil can be obtained by taking few drops of water in oil. These are illustrated in fig 7-13.

The emulsions formed entirely of oil and water are quite unstable, as the molecules of the dispersed phase collide with one another and separate out into two layers on standing. Stability of an emulsion is achieved by the addition of a third substance known as **emulsifying agent or emulsifier.** The commonly used emulsifiers are gelatin, soap and gum arabic. The role of emulsifier, according to Brancraft, is to surround the particles of dispersed phase reducing the surface tension on one side of the water oil interphase than the other. As the spherical surface has always a minimum area the emulsion particle forms a spherical envelope that inhibits the colloidal particles to come closer, fig 7-14.

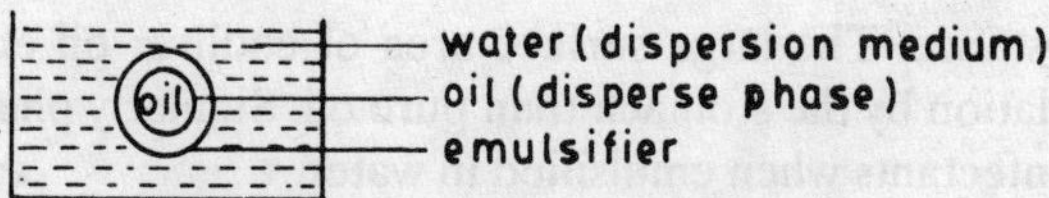

Fig 7-14 Oil particle dispersed in water

Certain finely divided solid particles can also act as emulsifiers. They are wetted by both the liquids. The liquid that wets more forms the dispersion medium and the liquid that wets less is protected by the solid particles, fig 3-15.

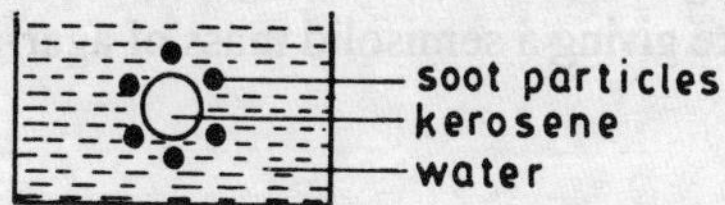

Fig 7-15 Protective action of soot particles

We get an emulsions of water in kerosene on shaking them along with some soot particles.

1. Preparation of emulsions: Vigorous shaking or stirring of few drops of the disperse phase with dispersion medium in the presence of a liquid emulsifier or finely divided solid particles is sufficient to give an emulsion.

Emulsions can be obtained by adding a solution in excess to another solvent, which readily dissolves the solvent of the added solution but keeps liquid solute undissolved. Thus an emulsion of oil in water is obtained by adding a dilute solution of oil in alcohol to excess of water.

Passing a vapour of oil into water containing emulsifying agent also gives an emulsion.

2. Properties of emulsions: The properties of emulsions are same as that of colloids. To avoid repetition of electrical,optical, brownian movement etc., properties, only few selected properties are discussed here.

A. The Stability: The stability of an emulsion is determined by two factors (i) the strength of the protective layer of the emulsifying agent, and

(ii) the charge on the emulsoid (dispersed particle).

B. Coagulation: Coagulation or demulsification takes place by adding a substance that uncovers the emulsifier film.This can be affected

(i) by adding an electrolyte that neutralizes the charge on emulsoid.

(ii) by excessive heating, cooling and centrifuging.

(iii) by passing current.

(iv) by adding dehydrating agent to an emulsion of water in oil.

C. Vision: When the liquids participating in the formation of emulsion have refractive indices of the same order the emulsion looks transparent and when they are different the emulsion is opaque or turbid.

3. Uses of emulsions: Emulsions of a substance have more surface area than the pure substance. The large surface area of codliver oil emulsions help in better assimilation by the stomach than pure oil. Similarly phenyl and lysol act as better disinfectants when emulsified in water.

7-8 GELS

Colloidal systems of liquids dispersed in solids are called **gels**. They are obtained by coagulation of the colloidal solutions yielding a semisolid mass. Thus on heating agar-agar with water, first we get a colloid and on further boiling this solution agar-agar molecules come closer trapping water in a honey combed like structure giving a semisolid mass of agar-agar gel, fig 3-16.

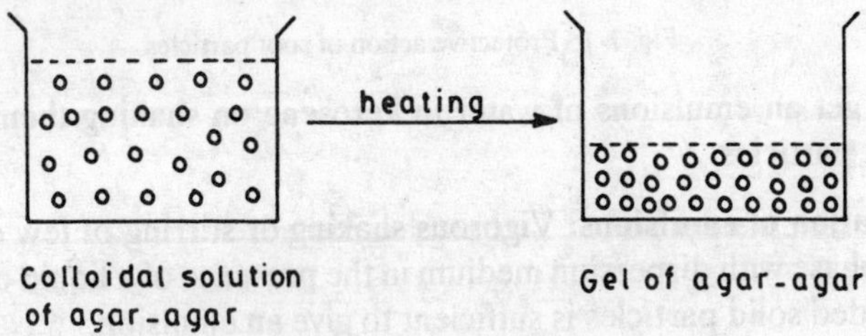

Fig 7-16 Formation of a gel

Similarly adding alcohol to a saturated solution of calcium carbonate in water initially produces a colloidal solution which coagulates to give a semisolid mass of $CaCO_3$ in which alcohol is trapped. This gel of alcohol in calcium carbonate, called solid alcohol, is used as solid fuel in picnic stoves, military camps etc.

1. Properties of gels: On standing, gels loose the medium (*e.g.,* water) and shrink. This shrinking of gel with simultaneous exudation (sweating) of the gel is called **synergisis** or weeping of the gel. Gels are classified into three categories (i) Elastic gels (ii) Rigid gels (iii) Thixotropic gels.

(i) Elastic gels: These gels are defused by the application of force but regains its original shape on the removal of applied force. *e.g.,* Starch, gelatin, fruit jam and jellies.

(ii) Rigid gels: There is no deformation on applying force. The strong covalent bonds between molecules give the rigidity of shape, *e.g.,* silicic acid or commonly known as silica gel.

(iii) Thixotropic gels: These become semisolids when kept aside but become liquids when agitated. Kiesel gel forms a thin paste on standing but readily flows on shaking. Solutions of high polymers exhibit these properties.

3. Uses of gel: Commercially many gels are available in the market. The commonly used boot polishes, polishing waxes, soaps, some brands of tooth pastes and various foodstuffs as petha are all examples of gels.

7-9 APPLICATIONS OF COLLOID CHEMISTRY

Applications of a class of colloids namely emulsions and gels have already been discussed in subheads 7-7 and 7-8. Application of other forms is given below.

1. Colloids in medicine: Medicines in the form of colloids are easily absorbed by the body tissues. Some of the essential metals like silver, gold which can not be absorbed in the metallic form are administered as colloids.

(i) Argyrol and protargal are protected silver colloids used in the cure of granulations.

(ii) Colloids of gold, calcium, manganese, etc., are administered as intra-muscular injections to enhance the vitality and cure tuberculosis and rickets.

2. Colloids as pesticides: Sulphur colloids are sprayed on plants to kill various germs.

3. Smoke precipitation: Smoke in the industrial exhausts contain particulate carbon and dust in the colloidal form and these are removed by the application of a high voltage of 30,000 volts and then allowed to escape into the atmosphere through the chimney.

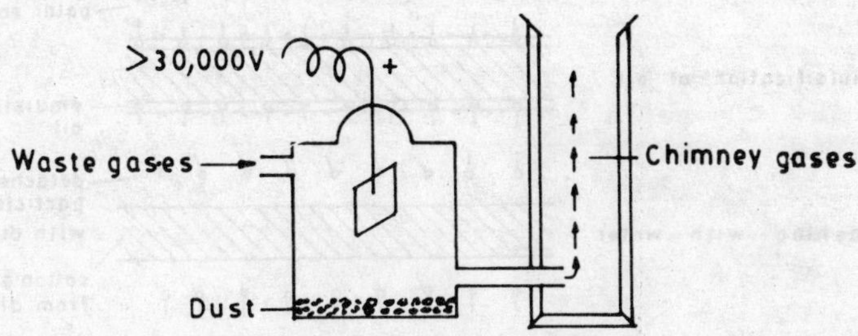

Fig 7-17 Cotrel smoke precipitator

4. Dialysis: Kidney removes urea and other unwanted electrolytes from the impure blood by passing it through a dialyzing membrane which holds back the colloidal blood. A kidney failure sends these toxic wastes into the blood. Patients having kidney problem are dialyzed externally.

5. Purification of water: The clay particles present in the muddy water are removed by the addition of alum (contains Al^{3+} ions) which coagulates them leaving a clear water.

6. Sewage disposal: The dirt particles present in the sewage water are in the colloidal form. These are removed in the presence of a high voltage and the clear water free from sewage sludge is let out into the irrigation fields. The sludge is also used as a manure.

7. Cleaning action of soap: A cloth gets dirty by the sticking of grease and oil particles which in turn adsorb the dust from the air. Soap is a sodium salt of fatty acid.

R ~~~~~~~~~~ $^{-}Na^{+}$
Polar end
Fatty acid (non polar end)

It has a polar end and a non-polar fatty acid chain. Polar end is water soluble (hydrophilic) while non polar end is oil soluble (hydrophobic). When soap is applied to the cloth the oil soluble end of soap emulsifies with the grease along with the dirt particles sticking to it. Polar ends are present away from the cloth and are water soluble. Hence after the application of soap to the cloth, washing with excess water removes the emulsified grease, dirt particles by solubilizing the polar end of the emulsified soap, fig 7-18.

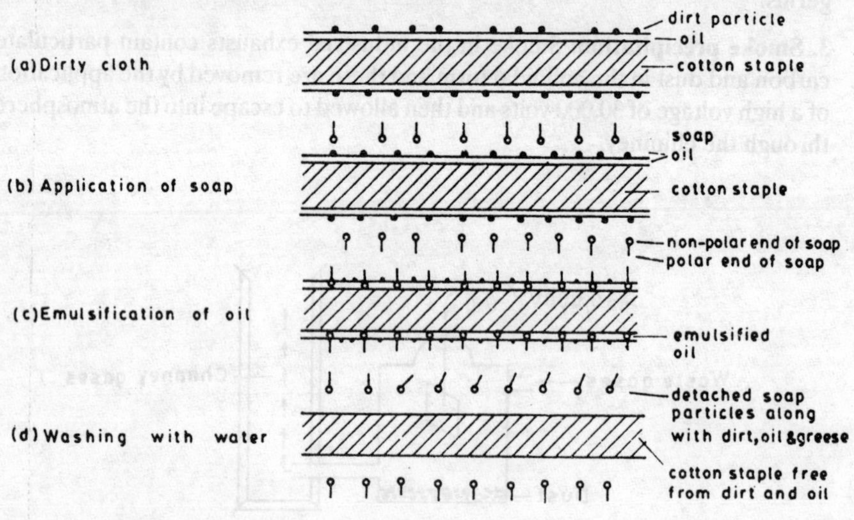

Fig 7-18

8. Other Miscellaneous uses

(a) Photographic plates: Silver bromide emulsions are coated on the photographic plates, films etc.,

(b) Formation of deltas: Rivers carry silt and other mud particles with them which are negatively charged, when the river water meets sea, the electrolytes in the sea water coagulate the silt forming fertile delta lands.

(c) Industrial products: Various industrial products as inks, boot polish, creams, paints, rubber lubricants are all colloidal particles.

(d) Rubber industry: Latex is a negatively charged colloidal solution. Coagulation of latex yields rubber.

(e) Dairy industry: Dairy products as creams, butter, cheese, milk are all colloidal solutions.

(f) Leather Industry: Skins and hides have proteins in them in the colloidal state. Skins are first soaked and put in a colloidal solution of tannin, a negatively charged colloid. Tannin enters the skins and mutual coagulation of tannin and protein takes place giving hardness to the leather.

(g) Stopping bleeding: Blood is a colloid. So shaving cuts and other bleedings are stopped using alum which coagulates the blood forming a clot.

9. Explanation of some interesting observations

(a) Blue colour of sky: Sun light falling on the earth is partially reflected back by the earth's surface. On its way back this light hits the dust particles present in the atmosphere. The colloidal particles of dust are quite small and are capable of reflecting only blue radiation while the radiations of other colours escape through the sky. This given the sky a blue colour.

(b) Tail of comets: Comets are celestial bodies moving at tremendous speeds. During their motion they leave behind tiny solid particles suspended in air. The light falling on them is scattered leaving a Tyndall cone like structure which we call the tail of comet.

(c) Artificial rains: Clouds are colloids of tiny water droplets suspended in air. Artificial rains are caused by spraying a sol or an electrolyte that coagulates these water droplets into rain and this process is called seedling of clouds.

ADSORPTION

7-10 INTRODUCTION

The modern theories of atomic structure describe an atom or a molecule not as any rigid structure with a central nucleus or nuclei, around which the electrons are revolving in a fixed orbit but more as a mathematical model with electrons possessing a dual nature of wave and matter and the probability of finding an electron described by a wave function. As any movement of charged particles builds a field, it is rational to assume forces of attraction between the atoms and molecules. If we consider molecules in the bulk of the medium and on the surface, the former experience equal and opposite pulls (forces of attraction) in all directions while the molecules on a surface experience unbalanced forces, Fig 7-19.

These intermolecular forces of attraction are very strong in solids, weak in gases and intermediate in liquids. In all the three phases the molecules at the surface experience a resultant pull directed inwards.

Thus considering a solid-liquid or solid-gas interface both the phases undergo an inward pull but the attractive forces of solid are stronger than less dense liquid or gas molecules. The consequence of this behaviour is accumulation of gas or liquid molecules on the solid surface, Fig 7-20. *The existence of a higher concentration of any particular liquid or gas at the surface of a solid/liquid than is present in the bulk of the medium is called adsorption.* The solid surface where adsorption takes place is called **adsorbate.**

Adsorption is purely a **surface phenomenon**. In some solids like a filter paper or a chalk the adsorbate, say ink, also penetrates into the solid surface

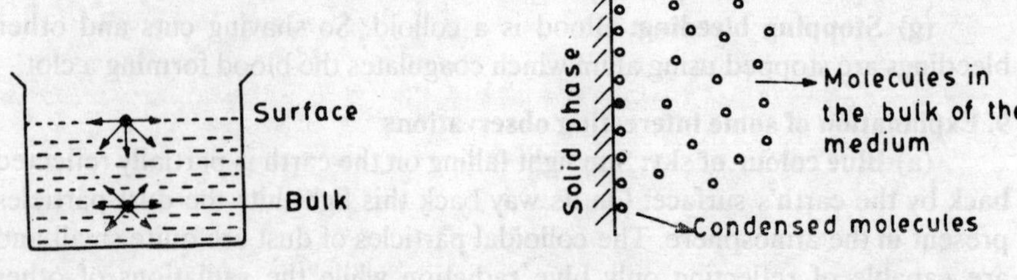

Fig 7-19

Fig 7-20

and the phenomenon is called absorption. In absorption the concentration of the absorbing substance is uniform through out the solid phase.

The primary requirement for effective adsorption is the availability of a large surface area. The extent and the strength of adsorption depends on the inherent characteristics of adsorbate and adsorbent. A given metal may adsorb a particular gas and may not adsorb the other. Thus Ni adsorbs hydrogen gas readily but is a poor adsorber of nitrogen. Animal charcoal is a less selective adsorbent. It is used for the removal of colouring matter present in traces as impurity. It has long been known that on boiling jaggery syrup with animal charcoal removes the colouring matter giving a colourless sugar syrup. Another example that could be easily demonstrated in a class room is the adsorption of bromine vapour by animal charcoal. On taking a drop of bromine in a test tube or just by inverting the test tube over a bottle containing bromine we get a test tube filled with brown coloured bromine vapour. Just by the addition of a pinch of charcoal into the test tube the colour totally disappears indicating its adsorption by the charcoal. The volume of gases adsorbed by 1g of charcoal are given in the Table 3-8.

Table 7-8 Adsorption of gases by 1g charcoal at 15° C

Gas	*Volume of gas in c.c*	*Critical temperature (K)*
Hydrogen	**4.7**	**33**
Nitrogen	**8.0**	**126**
Methane	**16.2**	**190**
Carbon dioxide	**48.0**	**304**
Ammonia	**181.0**	**406**
Chlorine	**235.0**	**417**
Sulphur dioxide	**380.0**	**430**

In the adsorption a true dynamic equilibrium is set up between the gas in contact with the surface and the gas present in the bulk of the medium. Hence amount of gas adsorbed is dependent on the pressure, temperature and concentration of adsorbate molecules just as in a chemical equilibrium and in

addition also on the active surface area of the adsorbent. An increase in pressure and concentration of adsorbate molecules or a decrease in the temperature increases the amount adsorbed. Adsorption is invariably an exothermic process. Thus, at 600mm pressure a gram of charcoal adsorbs about 10 c.c. of nitrogen at O°C, 20 c.c. at −29°C and 45 c.c at −78°C. The heat liberated during adsorption is called **heat of adsorption.**

7-11 TYPES OF ADSORPTION

Studies of adsorption of various gases on solid surfaces revealed that the forces that operate in adsorption are not same. The heats of adsorption were found to have a wide range from 0-100 kcal. Normally chemical forces were found to accompany a heat change (20 kcal) and an energy (less than 10 kcal) indicates weak forces of attraction called Van der Waals forces. Thus the adsorption is classified into **physical or Van der waals adsorption,** accompanied by heat of adsorption (10 kcal $mole^{-1}$) and **chemical or activated adsorption** or **chemisorption** accompanied by a heat change greater than 20 kcal $mole^{-1}$.

(A) PHYSICAL ADSORPTION: Physical adsorption is a phenomenon where the adsorbate (gas or liquid) molecules are bound to the molecules presents on the surface of adsorbent (solid) by weak Van der Waals forces with a heat of adsorption lying below 10 kcal mol^{-1}. The characteristics of physical adsorption are:

1. Lack of specificity: Van der Waals forces are same as universal weak inter molecular forces. They are mainly responsible for the molecules to come closer, occupying a space and taking the shapes of any container. Hence they lack specificity and adsorption takes place between any pair of adsorbent-adsorbate system.

2. Nature of gas: Physical adsorption depends upon the critical temperature of the gas. Gases having higher critical temperature, liquified easily and hence adsorbed more readily (Table 3-9).

3. Effect of temperature: Van der Waals forces are stronger in liquid state than in gaseous state. Hence physical adsorption is significant at and below the critical temperature. On heating the process of adsorption is reversed and gas is released into the medium from the surface and is called desorption. With the increase in temperature, due to supply of excess energy, it is also possible for some systems to change the forces of attraction from physical to chemical in nature. Thus nitrogen at 83° K undergoes physical adsorption on iron but at 775° K it is chemisorbed on iron forming iron nitride on the surface.

4. Effect of pressure: Increase in pressure increases the condensation of gas and hence the adsorption. As the gas approaches critical pressure the adsorbate starts forming multilayers on the adsorbent with a rapid increase in the rate of adsorption.

5. Reversibility: Physical adsorption is completely reversible as suggested earlier. Adsorption increases with increase in pressure, concentration and decrease in temperature and the process reverses on reversing these physical parameters.

6. Surface area of adsorbent: With the increase in the surface area of adsorbent the physical adsorption also increases.

7. Heat of adsorption: As suggested in the classification the heat of adsorption is very low and ranges between 1-10kcal mol^{-1}.

(B) CHEMISORPTION: Chemisorption is a phenomenon where the adsorbate molecules are bound to the molecules present on the surface of adsorbent by strong chemical forces with a heat of adsorption lying above 20 kcal. Some of their characteristics are:

1. Specificity: Chemisorption is highly specific unlike physical adsorption because of a chemical bond that is formed between adsorbate and adsorbent molecules. For example, hydrogen is adsorbed on iron at fairly high temperatures forming metal hydrides. This involves a chemical interaction between the d–orbital electrons of iron and s-orbital of hydrogen.

2. Nature of gas: The primary requirement for chemisorption is the formation of a chemical bond between the adsorbate and adsorbent. Hence only the adsorbates that have the capability for the formation of a chemical bond with adsorbent are chemisorbed.

3. Effect of temperature: Temperatures around critical temperature result in the physical adsorption while chemisorption is favoured at on high temperatures above critical temperature. They require a high activation energy and hence higher temperatures facilitate chemisorption. Thus as cited already under physical adsorption hydrogen is bound by Van der Waals forces at low temperatures but chemisorbed at 775°K on an iron surface. The amount of gas adsorbed increases with increase in temperature due to increase in collisions with the surface but after a particular temperature if decreases due to exothermicity of the reaction.

4. Effect of pressure: Increase in pressure makes more adsorbate molecules available to the adsorbent surface and hence increases chemisorption.

5. Irreversibility: Unlike physical adsorption chemisorption is irreversible because of formation of a chemical bond.

6. Surface area: Similar to physical adsorption increase in the surface area increases chemisorption.

7. Heat of adsorption: The heat of adsorption is comparable to that of heat of reaction and is above 20 kcal $mole^{-1}$. It is always exothermic like physical adsorption.

The differences between physical adsorption and chemisorption is summarized in the Table 7-9.

Table 3-9 A comparison between physical adsorption and chemisorption

Physical adsorption	Chemisorption
1. Van der Waals forces operative	1. Chemical forces are operative.
2. Non-selective	2. Highly selective
3. Multi layer adsorption possible	3. Unimolecular layer is formed.
4. Does not depend on the nature of gas	4. Depends on the nature of gas.
5. Decreases with the increase in temperature.	5. Increases with increase in temperature to a particular limit but later it falls.
6. Facilitated near and below critical temperature	6. Facilitated far above the critical temperature.
7. No appreciable activation energy involved	7. Involves high activation energy.
8. Heat of adsorption below 10 kcal mol^{-1}	8. Heat of adsorption above 20 kcal mol^{-1}.
9. Reversible in nature	9. Irreversible in nature
10. The rate of adsorption increases with increase in pressure	10. Here also rate increases with increase in pressure.

7-12 ADSORPTION ISOTHERMS

The amount of gas adsorbed depends on the temperature and pressure of a gas. *The relation between the amount of substance adsorbed by an adsorbent and the equilibrium pressure or concentration at constant temperature is called an adsorption isotherm.* Five general types of adsorption isotherms were observed in the adsorption of gases on solids, fig 7-21.

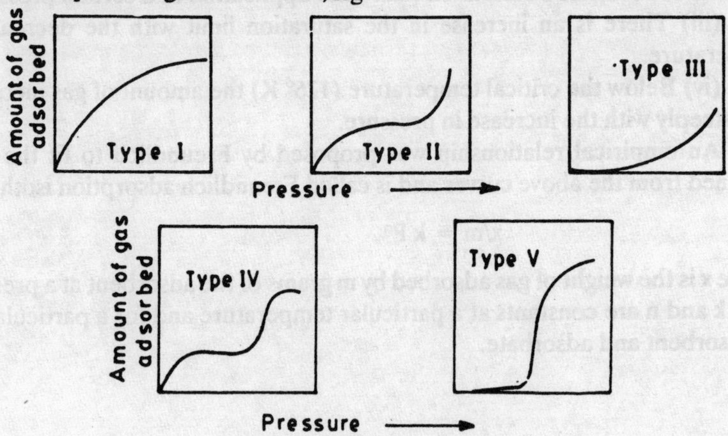

Fig 7-21 Adsorption isotherms

In chemisorption only curves of type I are observed.

Attempts were made by various groups to explain the observed curves and fit the curve in a mathematical relationship. In these attempts the work of Freundlich and Langmuir requires a special mention and the relationship propounded by them are called as Freundlich adsorption isotherm and Langmuir adsorption isotherm.

(A) Freundlich adsorption isotherm: Freundlich studied a number of systems like adsorption of nitrogen on charcoal, and plotted volume of gas adsorbed vs pressure curves at different temperatures fig 7-22.

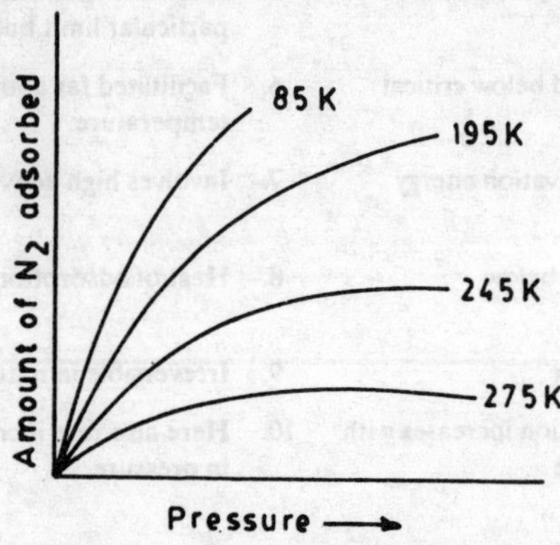

Fig 7-22 **Adsorption isotherms for the adsorption of N_2 on charcoal.**

All these curves belong to type I adsorption isotherms. The following observations could be made from the adsorption isotherms given in the Fig 7-22.

(i) Amount of N_2 adsorbed increases with decrease in temperature.

(ii) Amount of N_2 adsorbed increases initially with the increase in pressure but reaches a saturation limit after application of a certain pressure.

(iii) There is an increase in the saturation limit with the decrease in temperature.

(iv) Below the critical temperature (126° K) the amount of gas increases very steeply with the increase in pressure.

An empirical relationship was proposed by Freundlich to fit the data obtained from the above curves and is called Freundlich adsorption isotherm.

$$x/m = k P^n \tag{7.1}$$

where x is the weight of gas adsorbed by m grams of the adsorbent at a pressure of P. k and n are constants at a particular temperature and for a particular set of adsorbent and adsorbate.

In the above equation x/m gives the amount of gas adsorbed by 1 gram of the adsorbent.

1. Testing the validity of Freundlich adsorption isotherm: For testing this equation it is convenient if we convert it into a linear equation. Hence taking logarithm of the above equation we have

$$\log (x/m) = \log k + n \log P \tag{7.2}$$

or

$$\log (x/m) = n \log P + \log k \tag{7.3}$$

This equation is similar to an equation of a straight line

$$Y = m X + C \tag{7.4}$$

where Y is the value on y-axis, X is its corresponding data value on x–axis, m the slope and C the intercept of the straight line.

In the equation(3) Y = log x/m, X = log P and slope m = n and intercept C = log k, fig 7-23.

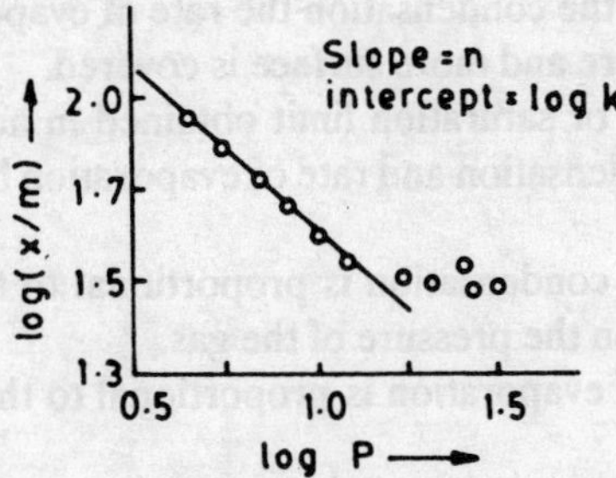

Fig 7-23 Application of Freundlich equation to adsorption of N_2 on mica at 90° K.

A reasonably straight line plot, as predicted by the Freundlich isotherm, was obtained but deviations were noticed at high pressures. This tells the success of Freundlich equation with some limitation at higher pressures. The value of slope 'n' was usually found to be less than unity.

This equation was found equally applicable to solid-liquid adsorption equilibria. Here the pressure term is replaced by concentration of adsorbate. For example, when we consider the adsorption of acetic acid on charcoal a linear plot was obtained by plotting log x/m vs log c. (fig 7-24)

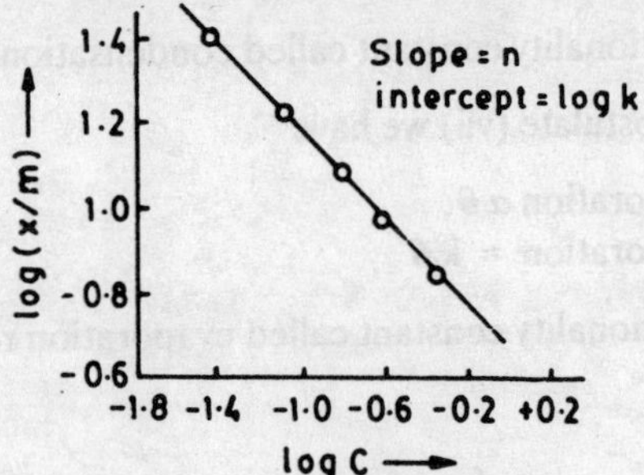

Fig 7-24 Application of Freundlich equation to adsorption of acetic acid on charcoal at 300°K

(B) Langmuir adsorption isotherm: Freundlich adsorption isotherm is only empirical without any theoretical explanation and even that failed at high pressure. Langmuir suggested a much better equation for type I isotherms and having theoretical validity.

1. Langmuir's postulates

(i) The gases being adsorbed always form only a unimolecular layer or a layer of single molecule depth.

(ii) Adsorption process consists of two apposing actions, (a) **The condensation** of gas molecules from the bulk of the medium on to the adsorbent surface and (b) the **evaporation** or **desorption** of gas molecules from the surface back into the bulk of the medium.

(iii) Initially all the gas molecules striking the bare adsorbent surface are adsorbed but later only the molecules striking the uncovered patches are adsorbed. Consequently the initial rate of condensation is the highest which slowly falls as the surface is covered by more and more molecules.

(iv) Contrary to the condensation the rate of evaporation is zero initially which increases as more and more surface is covered.

(v) The plateau or saturation limit obtained in adsorption isotherms is due to the rate of condensation and rate of evaporation becoming equal after a particular pressure.

(vi) The rate of condensation is proportional to the uncovered surface area and depends upon the pressure of the gas.

(vii) The rate of evaporation is proportional to the area covered by the adsorbate.

2. Derivation of langmuir adsorption isotherm

Let the total area available for adsorption be unity '1'

The total area covered at any instant be $= \theta$

Therefore the fraction of the bare surface (uncovered area) $= 1-\theta$

Pressure of the gas $=$ P

From the (vi) Postulate the rate of condensation is directly proportional to the bare surface area and the pressure of gas.

Therefore rate of condensation $\alpha\ (1-\theta)\text{P}$

or rate of condensation $= k_1\ (1-\theta)\ \text{P}$ (7.5)

where k_1 is a proportionality constant called condensation rate constant.

Similarly from the postulate (vii) we have

rate of evaporation $\alpha\ \theta$

or rate of evaporation $= k_2\theta$ (7.6)

where k_2 is a proportionality constant called evaporation rate constant.

At equilibrium

rate of condensation = rate of evaporation

Therefore $k_1(1-\theta)P = k_2\theta$

$$k_1P - k_1\theta P = k_2\theta$$

taking all the terms involving θ on one side

we have $k_2\theta + k_1\theta \;\; P = k_1 P$

or $\theta(k_2 + k_1 P) = k_1 P$

$$\theta = k_1 P / (k_2 + k_1 P) \tag{7.7}$$

Dividing numerator and denominator with k_2

therefore $$\theta = \frac{k_1/k_2 P}{1+k_1/k_2 P}$$

Or $$\theta = \frac{aP}{(a+aP)} \tag{7.8}$$

where $a = k_1/k_2$ is another constant.

If x grams of gas adsorbed on m grams of the adsorbent, then x/m gives the amount of gas adsorbed per gram of adsorbent. x/m is also proportional to θ the fraction of area covered at any instant.

Therefore $x/m \; \alpha \; \theta$

or $x/m = k_3\theta$

where k_3 is a proportionality constant. Substitute the value of θ from eq (3.8)

we have $$x/m = k_3 a_p/(1+a_p) \tag{7.9}$$

replacing k_3a with another constant b we have

$$x/m = b_p/(1+a_p) \tag{7.10}$$

taking the inverse $1/(x/m) = (1+a_p)/b_p$

$$P/(x/m) = (1+a_p)/b$$

$$P/(x/m) = (1/b) + (a/b)P \tag{7.11}$$

This equation is known as **Langmuir adsorption isotherm.** This is once again similar to the equation of a straight line eq (7.4).

Hence on plotting P/(x/m) vs P we expect a straight line with a positive slope equal to a/b and intercept on Y axis equal to (1/b). The data of Freundlich for the adsorption of nitrogen on mica at 90°K was re-plotted using Langmuir adsorption isotherm (Fig 7-25).

3. Advantages of Langmuir adsorption isotherm

1. It is not empirical and has theoretical back ground.
2. It gives a better fit of experimental data than Freundlich adsorption isotherm.

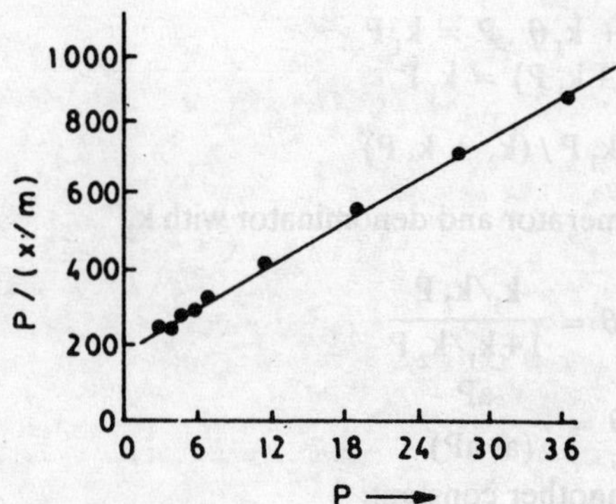

Fig 7-25 Application of Langmuir equation to adsorption of N_2 on mica at 90°K.

3. There is no deviation at moderately high pressure unlike Freundlich adsorption isotherms.

4. From the slope (a/b) and intercept (1/b) it is possible to calculate 'a' and 'b' values and there by the constants involved in the equation.

4. Limitations of Langmuir adsorption isotherm

(i) It explains fairly well the type I curves but fails to explain the other curves because they correspond to multilayer formation.

(ii) It is applicable to comparatively low pressure and high temperatures. As the temperature is lowered and approaches critical temperature it leads to multilayer formation. Similarly at very high pressures also multilayer formation is detected.

7-13 APPLICATIONS OF ADSORPTION

Adsorption has many applications in the research laboratory and in industry. Some of them are given below:

1. In the Dewar Flask: These flasks are used for the storage of liquid air, liquid hydrogen etc. Normal vacuum flasks are not sufficient for storing these liquid gases as they can diffuse through the imperfection of the glass vacuum jacket and may build up pressure. This problem is solved by placing animal charcoal between vacuum jacket and outer jacket which adsorbs these diffused gases.

2. Refining of vegetable oils and sugar solutions: Normally vegetable oils and sugar solutions have colouring material in them. These are removed by adsorption of these substances either with fullers earth or animal charcoal.

3. In gas marks: One important usage of various adsorbents is their usage in gas masks. Gas masks are used by the scientists, in the chemical industries and research laboratories, working with toxic gases like chlorine etc. These toxic gases are adsorbed by adsorbents present in the mask without hindering the flow of oxygen.

4. Chromatography: The fact that all the organic and inorganic substances are adsorbed to different extent by an adsorbent is used in developing chromatography in the separation and identification of the substances present in mixtures of these substances. Various techniques such as thin layer chromatography (TLC), gas chromatography (GC), high pressure liquid chromatography (HPLC), paper chromatography etc., all based on the principle of preferential adsorption of substances by various adsorbents are widely used and is a boon to industry and research work.

5. Surface active agents: They find extensive application in paints, lubricants, detergents and many other fields.

6. Softening hard water: The softening of hard water is done by passing the hard water through an ion exchange column which adsorbs heavy metal ions like Ca^{2+}, Mg^{2+} etc., responsible for hardness. Demineralized water is added to the car batteries.

7. Dyeing: To help the penetration of dye into the fabric it is first adsorbed on the mordant and then spread onto the fabric with a suitable surface active agent.

8. In curing diseases: A number of drugs are adsorbed on the germs to destroy them.

9. Heterogeneous Catalysis: The first step in the action of a heterogeneous catalyst like Ni, Palladium black, Platinum etc., is the adsorption of substances on selective active sites. By blocking these sites with a suitable catalytic poison the activity of the catalyst is lost.

10. Adsorption indicators: We use a number of adsorption indicators in titrimetry. For example starch solution is used as an adsorption indicator in the iodometric titrations.

11. Separation of rare gases: Rare gases are separated employing the adsorption techniques.

7-14 WORKED OUT NUMERICAL PROBLEMS

7-14-1 6.23g of a fatty acid of molecular weight 280 is spread on the surface of water. It could be compressed in an area of 300 sq. cm.before resisting force increased sharply. Calculate the area occupied by a single molecule in the closely packed layer.

(Hint: Assume a unimolecular adsorbed layer formation)

Given data:

Weight of fatty acid = 6.23g

Molecular weight of fatty acid = 280

Area occupied by = 300 sq cm

therefore no. of moles of fatty acid = 6.23/280 = 2.225×10^{-2}

According to Avogadro hypothesis a mole of any substance comprises of 6×10^{23} molecules.

Therefore no. of molecules present in 2.225×10^{-2} moles of fatty acid

$= 2.225 \times 10^{-2} \times 6 \times 10^{23}$

$= 13.35 \times 10^{21}$

But 13.35×10^{21} molecules occupy an area of 300 sq cm.

Hence area occupied by a single molecule $= 300/(13.35 \times 10^{21})$

$= 2.25 \times 10^{-20}\ cm^2$

7-14-2 When 0.2 ml of 0.01 M solution of palmitic acid was placed on a water surface, the resulting monolayer spreads to an area of $2.5 \times 10^3\ cm^2$. Find the cross sectional area per hydrocarbon chain.

Volume of palmitic acid taken = 0.2 ml

Concentration of palmitic acid taken = 0.01 M

i.e 0.01 moles of palmitic acid are present in 1 litre (1000 ml) of solvent.

No. of moles of palmitic acid present in 2ml

$= (2/1000) \times 0.01 = 2 \times 10^{-5}$

No. of molecules of palmitic acid present in 2×10^{-5} moles

$= 2 \times 10^{-5} \times 6 \times 10^{23}$

$= 12 \times 10^{18}$

Area occupied by the acid $= 2.5 \times 10^3\ cm^2$

Therefore area occupied by 1 molecule of the acid

$$= \frac{2.5 \times 10^3}{12 \times 10^{18}} = 2.08 \times 10^{-16}\ cm^2$$

7-15 SHORT AND ESSAY TYPE QUESTIONS

7-15-1 (a) What are the distinguishing features of the colloidal state? Describe the optical properties of colloids.

(b) Differentiate clearly between physical adsorption and chemisorption. Mention the factors that effect adsorption.

(O.U., Mar. 1990)

7-15-2 Write a brief notes on

(i) Protective colloids (ii) Zeta potential

(O.U., Mar. 1990) & (K.U., May. 1988)

7-15-3 Write a brief notes on

(i) Emulsion and (ii) gels. Give two examples each?

(O.U., Oct. 1989)

7-15-4 Derive the Langmuir adsorption isotherm and mention its significance.

(O.U., Oct. 1989) & (S.K.U., Apr. 1988)

7-15-5 (a) Discuss Langmuir theory of adsorption and derive the expression for Langmuir's adsorption isotherm?

(b) What do you understand by the term protection of colloids? Define the term 'Gold Number'. Three organic substances M, N, P have their gold numbers as 0.5, 0.05 and 0.005 respectively. Which one of these substances has the greatest protective action?

(O.U., Mar. 1989)

7-15-6 Write short notes on

(a) Tyndall effect and

(b) Cataphoresis

(O.U., Mar. 1989) & (K.U., Oct. 1988)

7-15-7 Briefly explain the following:

(a) Zeta potential

(b) Hardy -Schultz rule

(O.U., Oct. 1988)

7-15-8 What do you understand by physical adsorption and chemisorption? Differentiate clearly between the two types?

(O.U., Oct. 1988) & (K.U., Oct. 1988)

7-15-9 Explain fully what happens when a colloidal solution of gold is treated as follows.

(a) an electrolyte is added to it.

(b) a beam of light is passed through it.

(c) it is brought under the influence of an electric field.

(O.U., Oct. 1988)

7-15-10 Distinguish between a true solution and a colloidal solution. Describe briefly any two methods of preparing colloidal solutions. Explain the optical and electrical properties of colloids.

7-15-11 What do you understand by physical adsorption and chemisorption?

(A.U., Jun. 1988)

7-15-12 Derive Langmuir's adsorption isotherm and explain its importance?

(K.U., May. 1988) & (A.U., May. 1989)

7-15-13 What is a protective colloid? Describe its working?

7-15-14 Discuss the origin of the charge on colloidal particle. How would you measure the charge and how is it influenced by addition of electrolyte?

7-15-15 Give an account of methods of preparation and purification of colloidal solutions?

7-15-16 Write short notes on
(a) Adsorption and absorption
(b) Ultra centrifuge
(c) Charge of a colloidal particle and its stability
(d) Bredig's arc method for the preparation of a colloid
(e) Ultra microscope.

7-15-17 Explain how the phenomenon of adsorption is applied in chromatography and the ion exchange separation?

7-15-18 Give general characteristics of chemisorption and show how chemisorption on metals differ from that of oxides?

7-15-19 Explain the following;
(a) A sulphur sol is coagulated by adding a little electrolyte, where as a gelatin sol is apparently unaffected?
(b) A colloidal solution is stabilized by addition of a gel.

7-15-20 Give a detailed account on the applications of colloidal chemistry?

7-15-21 Derive Langmuir adsorption isotherm and discuss its applications in the determination of surface area of solids.
(A.U., Apr. 1989) & (N.U., Apr. 1989)

7-15-22 Give an account for the stability of colloids in terms of electric double layer.
(A.U., Apr. 1989) & (N.U., Oct. 1989)

7-15-23 Discuss various factors which effect the adsorption of a gas on solid adsorbent.
(A.U., Sept. 1988)

7-15-24 How is colloidal sulphur prepared? Write a note on the optical & kinetic properties of colloids.
(S.K.U., Apr. 1988)

7-15-25 Write & explain Frendlich isotherm.
(N.U., Oct. 1989)

7-15-26 What are lyophobic and lyophilic colloids.
(N.U., Oct. 1989 & K.U., May 1988)

❋

8

MACROMOLECULES

8-1 INTRODUCTION

Although the large or macromolecules, containing few hundred to few thousand atoms all linked to one another firmly, have evolved along with the universe, their science remained obscure and emerged suddenly into lime light, to name this technology era as **polymer or plastic age**. Their actual importance may be realized from the statistics that half the American chemists and chemical engineers are employed in industries concerned with polymeric materials. A common man can readily realize their importance by attempting to recognize the polymeric materials around him and within him. The naturally occurring macromolecules, viz., cellulose, wool and rubber, are used to drape ourselves, to live, to walk and to roll on them. Majority of the food items contain polysacharides, proteinous matter and fibers and the very existence of life and its replication is the result of bio-polymers called DNA and RNA. Similarly most of the synthetic materials existing around us in the form of plastics, fibers, gums, laminates, paint, gadgets, etc., are all different forms of macromolecular materials. Some of the synthetic polymers have gone to the extent of replacing ailing hearts and other very sensitive organs.

The word "Polymer" is derived from two greek words - Poly (= many) and meros (= parts or units). A polymer (often called as macromolecule), is a very large molecule formed by linking together small, simple chemical units called **monomers**. The resulting structure comprises of repetition of relatively small molecular fragments usually referred to as **repeating units**. For example, polyethylene is a polymer made by inter linking a number of ethylene monomer molecules. In this process the double bonds of ethylene molecules open up and bind with the adjacent C_2H_4 unit. Here - C_2H_4 - is the repeating unit.

$$n\ \begin{matrix}H\\|\\C\\|\\H\end{matrix} = \begin{matrix}H\\|\\C\\|\\H\end{matrix} \longrightarrow \cdots - \begin{matrix}H\\|\\C\\|\\H\end{matrix} - \begin{matrix}H\\|\\C\\|\\H\end{matrix} - \begin{matrix}H\\|\\C\\|\\H\end{matrix} - \begin{matrix}H\\|\\C\\|\\H\end{matrix} - \cdots \text{ or } \left(\begin{matrix}H\\|\\C\\|\\H\end{matrix} - \begin{matrix}H\\|\\C\\|\\H\end{matrix} \right)_n$$

Ethylene
A monomer

Polyethylene
A polymer

Repeating unit

A macromolecule may consist of monomers of identitical or different chemical structure and accordingly they are called homopolymers and copolymers (or mixed polymers) respectively.

Homopolymer --- A–A–A–A–A---

Copolymer --- A–B–A–B–B---

A copolymer could possess a regular or random arrangement of monomeric units.

Regular arrangement --- A–B--A–B–A–B ---

Random arrangement --- A–B–B–A–B–A ---.

Further, both the homo and copolymers could be arranged to give linear, branched, cross linked or graft structures.

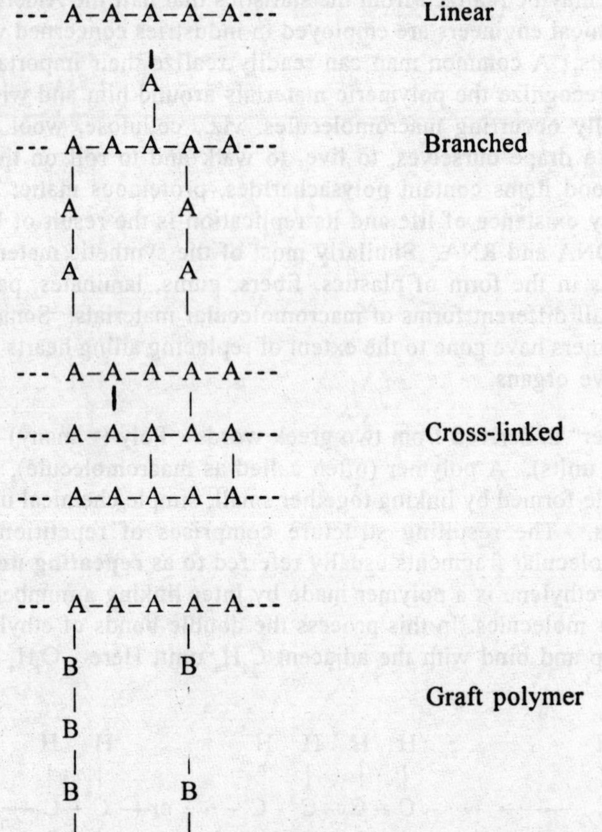

In linear polymer the repeating units are joined together to give fairly long continuous sequences. In branched polymers side chains of the repeating units are attached to main polymer chain.

A branched copolymer with one kind of monomers in their main chain and another kind of monomers in their side chains is called a **graft polymer**.

Block copolymers are linear copolymers in which the like monomer units occur in relatively long sequences.

--- A–A–A–A–A–A–A–B–B–B–B–A–A–A–A–A–A–A ---

The physical and chemical properties of the polymer are much different from the monomer. In fact polyethylene behaves much like saturated hydrocarbons with their inactivity towards chemical reactions characteristic of double bond like facile bromination. A list of some homo polymers with corresponding monomers and repeating units is given in Table 8-1. Another unique feature of the polymers is the difference between expected colligative properties on the basis of number of monomer units and the observed values.

Table 8-1 Some homopolymers, Their monomers, Repeating units and Uses.

Polymer	*Manomer*	*Repeat unit*	*Uses*
1. Polyethlene	$CH_2 = CH_2$	$\text{-(}CH_2\text{-}CH_2\text{)-}$	Plastic piping, bottles, insulation cables, moisture proofing.
2. Polyvinyl chloride (PVC)	$CH_2 = CH(Cl)$	$\text{-(}H_2C\text{-}CHCl\text{)-}$	Piping, floor tiles, fabric coating, food wrap.
3. Polyiso-butylene	$CH_2 = C(CH_3)(CH_3)$	$\text{-(}CH\text{-}C(CH_3)(CH_3)H_2\text{)-}$	Butyl rubber, bullet proof tanks, cable coatings.
4. Polypropylene	$CH_2 = C(H)(CH_3)$	$\text{-(}CH\text{-}C(CH_3)H_2\text{)-}$	Packaging, carpets and soft drink bottles.
5. Polystyrene	$CH_2 = C(H)(C_6H_5)$	$\text{-(}CH_2\text{-}CH(C_6H_5)\text{)-}$	Refrigerator cabinet components, insulation, lenses, panels.
6. Polyacrylo nitrile (PAN)	$CH_2 = C(H)(CN)$	$\text{-(}CH_2\text{-}C(H)H(CN)\text{)-}$	Lenses, Airplane covers and shields, costume jewellery.
7. Polytetrafluoro ethylene (Teflon)	$CF_2 = CF_2$	$\text{-(}CF_2\text{-}CF_2\text{)-}$	Non-stick ware, electrical insulation, bearings, gaskets.
8. Polybutadiene	$H_2C{=}C(H)\text{-}C(H){=}CH_2$	$\text{-(}CH_2\text{-}CH{=}CH\text{-}CH_2\text{)-}$	Tire tread, coating resins.
9. Polyisoprene (Natural rubber)	$H_2C{=}CH\text{-}C(CH_3){=}CH_2$	$\text{-(}CH_2CH{=}C(CH_3)\text{-}CH\text{)-}$	Flexible pipes, gloves.
10. Polychlorprene (Neoprene)	$H_2C{=}CH\text{-}C(Cl)\text{-}CH_2$	$\text{-(}CH_2CH{=}C(Cl)\text{-}CH_2\text{)-}$	Hose pipes, shoe soles, belts, gloves, cable jackets.

Although polymeric materials were discovered by early chemists, polymers science may be thought to have entered this new era largely because of untiring efforts by Staudinger through his macromolecular hypothesis presented in 1920, which earned him the 1953 Noble Prize. He proposed and substantiated the molecular formula of polystyrene, rubber and other polymers. This soon paved way for commercial exploitation of polymers like PVC, Nylon, Neoprene, Styrene-Butadiene synthetic rubber, etc. Ziegler and Natta (1955) developed new polymerization catalysts with stere o-regulating powers and obtained the second Nobel prize of polymer chemistry in 1963. Flory elucidated the mechanisms of polymerization processes and added one more Noble prize to the field in 1974. As a result of these discoveries the polymer industry currently produces around 30 billion pounds of polymer materials annually.

8-2 CLASSIFICATION

Two types of classification of polymeric materials exist basing on physical and chemical properties.

8-2-1 Physical Classification of Polymers

Polymers materials are classified on the basis of a physical verification of origin, elements involved in the backbone chains and physical properties.

(A) Natural and Synthetic Polymers: Depending on their origin, polymer can be grouped into **Natural and Synthetic polymers**. The natural polymers are either naturally occurring materials or their chemical modifications. In spite of discovery of synthetic routes, for example - natural rubber from isoprene, these materials, still fall under natural polymers. Cell constituents like DNA, proteins etc., also fall under this category. Some of the commercially used natural polymers are cotton, silk, leather, wool, rubber, cellulose acetate and cellophane.

The synthetic polymers are relatively of *lower molecular weight* when compared to their natural counter parts but find use in variety of applications. Classifications and uses of polymers of this class is provided in the next section.

(B) Organic and Inorganic Polymers : Organic polymers, as expected, have carbon back bone structures while inorganic polymers usually have back structures made out of non-metals other than carbon. Elements like Si, Ge, N, P, As, O, S, Se, etc., with limited **catenations** (inter linking) usually make the back bone of inorganic polymers. In either case the side valencies of the back bone elements are satisfied by elements like hydrogen, halogens, oxygen, nitrogen, etc., Almost all the polymers of biological origin and majority of the synthetic origin belong to the group of organic polymers. Therefore, when we mention simply **polymers**, we generally mean organic polymers. As we already stated, organic polymers have a variety of applications and does not fit in any simple frame work. However, inorganic polymers have some specific and unique applications, where organic polymers are not of much use. The typical and most useful polymers of this class are glass and silicones (-Si-O-Si-back bone). Depending on the size and cross linking silicones may be fluids, greases, resins or rubber. Silicone rubbers retain elasticity over a wide range of temperatures and silicone lubricants are mixed in mineral oils.

$$\left[CH_2-CH_2 \right]_n$$

Polyethylene-An organic polymer

```
    H    H    H    H
    |    |    |    |
--- Si—Si—Si—Si ---
    |    |    |    |
    H    H    H    H
```

Polysilane-An inorganic polymer

(C) Thermosetting And Thermoplastic Polymers: This classification depends on the differences in physical properties on heat treatment. Polymers that soften on heating and regain original physical properties on cooling are known as **Thermoplastics**. On the contrary, **Thermosetting plastics** are deformed irreversibly on heat treatment. Linear chain polymers fall under the first category of thermoplastics. The process of heating, reshaping and retaining the same on cooling can be repeated several times on this class of polymers. Polyethylene, PVC, Nylon, etc., are typical examples of thermoplastic polymers. In case of cross linked polymers, heat treatment brings charges in the cross linkages and converts them into infusible mass.

Table 8-2 : Thermoplastics and Thermosetting resins

Thermoplastics	Thermosetting series
1. They are formed by addition polymerization	1. They are formed by condensation polymerization.
2. They consist of long chain linear polymers with negligible cross links.	2. They have three dimentional network structure.
3. They soften on heating readily because the interlinking forces between individual chains are weak and break on heating	3. They retain their shape even on heating, however, prolonged heating could lead to charring of the polymer.
4. They are soft, weak and less brittle.	4. They are usually hard and brittle.
5. They can be reclaimed from their waste.	5. They can not be reclaimed from their waste.
6. They are usually soluble in organic solvents.	6. They are insoluble in organic solvents.

(D) Fibers, Elastomers, Plastics and Liquid Resins: This nomenclature is based on the ultimate form which in turn depends on dominant physical property. Fibers, elastomers, plastics and liquids-resins differ in the extent of the crystallinity and physical state and are referred as crystalline, semi-crystalline, amorphous and liquid respectively.

i) Fibers

Being highly crystalline, possess high tensile strength. Because of this property these materials are drawn into fibers used in weaving and making ropes, e.g.., nylon, terelene.

ii) Elastomers

These polymers possess only partial crystallinity and therefore can be streched to one and half times its original length, but regain original shape on releasing. Because of their rubbery characters they are also used as synthetic rubbers. They exhibit good strength and elongation, eg., silicone rubber, neoprene, natural rubbers.

iii) Plastics

Polymers that can be shaped into hard and tough utility articles by the applications of heat and pressure are called plastics, eg., PVC, teflon, polymethylmethacrylate, polystyrene, etc.

iv) Liquid resins

Polymers existing in liquid state are used as adhesives, lubricants, sealants and resins, eg., some silicones, polysulphide sealants, epoxy resins.

8-2-2 Chemical Classification of polymers

W.H. Carothers in 1929 suggested a very useful classification which banks on chemical process through which a polymer is formed from its monomer units, viz., **Addition polymers** and **Condensation polymers.**

(A) Condensation polymers: Condensation Polymers are those in which the repeat unit, the polymer, lacks certain atoms present in corresponding monomer(s). Two like or unlike monomers join each other by the elimination of small molecules, such as water, to give the polymer chain. Nylon is a best example of condensation polymers. It is a copolymer, formed by the combination of dicarboxylic acids with diamines. A general mechanistic scheme for condensation polymerization involves a simultaneous combination of any two monomers to give dimers, which in turn combine to give tetramers and so on and finally grow into a polymer in a step-wise manner. For this reasons condensation polymerization is called by H.Mark (1950) as **Step polymerization.**

$$\underset{\text{Hexamethylene diamine}}{H-\overset{H}{\overset{|}{N}}-(CH_2)_6-\overset{H}{\overset{|}{N}}-H} + \underset{\text{Adipic acid}}{\overset{O}{\overset{||}{C}}-(CH_2)_4-\overset{O}{\overset{||}{C}}-OH} \xrightarrow{-H_2O}$$

(with OH attached to the first C of adipic acid)

$$H-\overset{H}{\overset{|}{N}}-(CH_2)_6-\underbrace{\overset{H}{\overset{|}{N}}-\overset{O}{\overset{||}{C}}}_{\text{Amide bond}}-(CH_2)_4-\overset{O}{\overset{||}{C}}-OH$$

$$..\overset{H}{\overset{|}{N}}-(CH_2)_6-\underbrace{\overset{H}{\overset{|}{N}}-\overset{O}{\overset{||}{C}}}_{\text{Amide bond}}-(CH_2)_4-\underbrace{\overset{O}{\overset{||}{C}}-\overset{H}{\overset{|}{N}}}_{\text{Amide bond}}-(CH_2)_6-\underbrace{\overset{H}{\overset{|}{N}}-\overset{O}{\overset{||}{C}}}_{\text{Amide bond}}-(CH_2)_4-\underbrace{\overset{O}{\overset{||}{C}}-\overset{H}{\overset{|}{N}}}_{\text{Amide bond}}...\text{etc}$$

Poly (Hexamethelene adipamide) or NYLON - 66

Molecular weight of condensation polymers is determined by the concentration, time of reaction and number of reactive functional groups of the monomers. Monofunctional monomers give only low molecular weight condensation products. Bi and poly functional monomers and longer reaction times allow further condensation and thereby give high molecular weight (longer chain) condensation polymers. Other most useful condensation polymers are polyesters (trade name Dacron) and polyurethane (trade name PUF).

$$\underset{\text{Acid}}{HOOC-C_6H_4-COOH} + \underset{\text{Glycol}}{HO-(CH_2)_2-OH} \xrightarrow{-H_2O} \underset{\text{Polyester}}{\left(\overset{O}{\overset{||}{C}}-C_6H_4-\overset{O}{\overset{||}{C}}-O-(CH_2)_2-O\right)_n}$$

$$OCN-(CH_2)_2-NCO + HO-(CH_2)_2OH \xrightarrow{-CO_2} \left(NH-(CH_2)_2-NH-C-O-(CH_2)_2-O\right)_n$$

Polymers having more than one atom in their back bone structure are called heterochain polymers or ***heteropolymers****. Because condensation polymers have functional groups as a part of polymer back bone, they invariably have more than a single element in the backbone structure and therefore are always give heterochain polymers.*

There are many types of condensation polymers and characterized on the basis of linkage. Five typical types are discussed below;

1. Polyamides : Nylon is a typical example. These compounds contain –CO–NH– groups as interlinkages. Scheme for preparation of Nylon 66 is already given and we now give a general scheme for synthesis of an aromatic polyamide.

ClOC–C_6H_4–COCl (Acid Chloride) + H_2N–C_6H_4–NH_2 (Diamine) $\xrightarrow[\text{CaO}]{\text{dimethyl acetamide}}$ [–HN–C(=O)–C_6H_4–C(=O)–NH–C_6H_4–$]_n$ (Aromatic Polyamide)

2. Polyester : The reaction scheme involves formation of an ester linkage. The reaction between terepthalic acid and ethylene glycol to form ester inter linkage in given below.

H_3CO–C(=O)–C_6H_4–C(=O)–OCH_3 (Dimethyl terephthalate) + HO–CH_2–CH_2–OH (Ethylene glycol) $\longrightarrow$ [–C(=O)–C_6H_4–C(=O)–O–CH_2–CH_2–O–$]_n$ (Terylene)

3. Poly(urethane) : Polyurethanes are obtained by condensation of a dihydric alcohol and diisocyanate.

$$n\ \text{HO–R–OH} + \text{OCN–R'–NCO} \longrightarrow (\text{O–R–O–C–NH–R'})_n$$

Dihydric alcohol Diisocyanate Polyurethane

4 Phenol-Formaldehyde : The reaction between phenol and formaldehyde is catalysed by an acid or alkali.

Phenol (OH) + HCHO $\xrightarrow{\text{Acid}}$ [HO–C_6H_4]–CH_2–[C_6H_3(OH)]–CH_2–[C_6H_2(OH)(CH_2–)]–CH_2–

Phenol (OH) + HCHO $\xrightarrow{\text{Alkali}}$ [HO–C_6H_4]–CH_2–O–CH_2–[C_6H_3(OH)]–CH_2–[C_6H_3(OH)]–CH_2–

5 Silicones : These polymers, unlike organic polymers are flame resistant and stable to high temperatures.

$$\underset{\text{Dimethyl dichlorosilane}}{Cl{-}\overset{\overset{CH_3}{|}}{\underset{\underset{CH_3}{|}}{Si}}{-}Cl} + H_2O \longrightarrow HO{-}\overset{\overset{CH_3}{|}}{\underset{\underset{CH_3}{|}}{Si}}{-}OH \longrightarrow \underset{\text{Silicone Oil}}{\left[-\overset{\overset{CH_3}{|}}{\underset{\underset{CH_3}{|}}{Si}}{-}O- \right]_n}$$

Some of the typical condensation polymers are listed table 8-3.

Table 8-3 : Typical condensation polymers

Type	Inter Unit linkage	Monomers	Trade name	Uses.
Phenol–Formaldehyde	–Ar–CH_2–	Phenol + Formaldehyde	Bakelite, PF	Moulded objects laquers, Protective coatings
Urea–Formaldehyde	–NH–CH_2–	Urea + Formaldehyde	Beetle, Scarab	Textile coatings, adhesives
Malamine–Formaldehyde	–NH–CH_2–	Malamine + Formaldehyde	Malamine	Unbreakable utencils and moulded objects
Poly (amide)	–NH–CO–	Dibasicacid + Diamine or ε–Amino carboxylicacid	Nylon Perlon	Fibers & Moulded objects
Poly (ester)	–CO–O–	Polyhydroxy glycol + dicarboxylic acid	Mylar Alkyl	Surface coatings, films and fibres.
Poly (urethane)	–O–CO–NH–	Glycol + Diisocyanate	Terylene PUF	Fibers, regid & flexide foams used as insulating material in refrigerators and other applainces.
Poly (siloxane)	$-\underset{\mid}{\overset{\mid}{Si}}-O-$	$HO{-}\overset{\overset{CH_3}{\mid}}{\underset{\underset{CH_3}{\mid}}{Si}}{-}OH$	Silicone	Libricants, flame resistant polymers
Poly (ethers)	–R–O–	Epichlorohydrin + Bisphenol–A or other alcohols	Araldite Epoxy resins	Adhesives, surface coatings.
Poly (carbonates)	–O–CO–	Phosgene + Bisphenol–A	–	Electrical mouldings.

(B) Addition Polymers: *Addition polymers are those in which the repeat unit and monomers (s) have same composition. Additional polymers are formed by simply adding monomer units together without undergoing any loss of atoms or groups.* Addition polymerization is usually initiated by a radical or a cation or an anion. Most of the vinyl monomers undergo addition polymerization. For example ethylene polymerizes into plyethylene at high temperatures (100-200°C), high pressure (100-300 atm) and in the presence of benzoyl peroxide. The reaction is split into three parts.

Initiation: Benzoyl peroxide decomposes to give benzoyloxy free radicals.

Propagation: Benzoyloxy radical (denoted by R$^\bullet$) readily combines with ethylene, causing opening of double bond and yielding another highly reactive radical.

This combines further with another ethylene molecule and this process goes on till a very long chain of CH_2 groups in built.

$$R^\bullet + CH_2 = CH_2 \longrightarrow R{-}CH_2{-}{}^\bullet CH_2$$

$$R{-}CH_2{-}{}^\bullet CH_2 + CH_2 = CH_2 \longrightarrow R{-}CH_2{-}CH_2{-}CH_2{-}{}^\bullet CH_2$$

$$\longrightarrow R{-}(CH_2{-}CH_2)_x{-}CH_2{-}{}^\bullet CH_2$$

Termination : Eventually, the reaction is terminated by the recombination of final free radicals.

$$R{-}(CH_2{-}CH{-})_x\ CH_2{-}{}^\bullet CH_2 + R'{-}(CH_2{-}CH_2{-})_y\ CH_2{-}{}^\bullet CH_2$$

$$\longrightarrow R{-}(CH_2{-}CH_2)_n{-}R'$$

Here *n* gives the total monomer molecules combined to give the polymer.

The above mechanism suggests a gradual but fast disappearance of monomer to give a long polymer chain. Since the process is akin to interlinking of metal rings one by one to give a chain, addition polymerization is also called Chain polymerization. With some exceptions addition polymers often contain only carbon atom back bone structure allowing them to be called homochain polymers.

Some typical examples of thin class are

$$n\ CF_2{=}CF_2 \longrightarrow {-}(CF_2{-}CF_2){-}_n$$

Tetra fluoroethene → **Teflon**

$$n\ CH_2{=}C(CH_3){-}CH{=}CH_2 \longrightarrow {-}(CH_2{-}C(CH_3){=}CH{-}CH_2){-}_n$$

Isoprene → **Polyisoprene or natural rubber**

A comparative account of the distinguishing features of addition and condensation polymers is given in Table 8-4

Table 8-4 Distinguishing features of Addition and Condensation polymers

Condensation polymers	*Addition Polymers*
1. Repeat units and monomers are different.	1. Repeat units and monomers are same.
2. Polymerization is accompanied by loss of molecules/ atoms	2. Polymerization involves no loss of molecules or atoms.
3. Almost all the monomer disappears in the first 10% of the reaction	3. Monomer concentration decreases steadily
4. Molecular weight of the polymer rises throughout reaction.	4. All the polymer is formed at once and molecular weight of the polymer changes little throughout the reaction.
5. Low molecular weight polymers are formed. Longer reaction times are essential for obtaining high molecular weight polymers.	5. High molecular weight polymers are formed all at once. Longer reaction times have a little effect on molecular weight.
6. Mainly heterochain polymers often have functional groups atoms as a part of chain.	6. Mainly homochain polymers often have carbon chains.
7. Monomer needs to be bi- or poly-functional.	7. Monomer could be monofunctional.

8-3 PHYSICAL PROPERTIES OF MACROMOLECULES

Physical properties of macromolecules are determined by the extent of cross linking of chains, nature of bonds making the back bone structure, molecular weight, etc.,

i) Crystallinity

Majority of the macromolecules are amorphous or partially crystalline.

ii) Solubility

Mostly soluble in organic solvents and form lyophilic sols. The solutions are highly viscous.

iii) Similarity with Colloids

Like colloids, macromolecular solutions also scatter light, show Tyndall effect and undergo sedimentation on centrifuging.

iv) Dissimilarity with Colloids

Unlike colloids, macromolecular solutions are poor conductors of electricity and do not coagulate on passing electric change. They also have very high molecular weight in comparison to colloidal particles. Macromolecules can be precipitated and redissolved without any difficulty.

v) Thermal properties

Linear organic high polymers have no sharp melting point but soften and flow over a temperature range and are used as thermoplastics. The thermoset polymers under go thermal decomposition and undergo no appreciable softening or flow.

vi) Mechanical properties

Mechanical properties are decided by the crystalline nature, interlinking of chains and polar groups present in the chain.

vii) Deformation and Flow

Plastic materials, which are amorphous, not only undergo deformation but also flow if suitable stress is applied. The crystalline polymer on the contrary, are quite hard.

viii) Elasticity

Thermoplastic high polymers are characterized by high order of elastic extendibility. Multiplicity of bonds in a polymer chain accounts for the remarkable flexibility and extendibility.

ix) Tensile strength

The tensile strength of a high polymer is affected mainly by the polar nature of the molecule. Polymers like cellulose are much stronger than nonpolar materials like polybutylene.

x) Molecular weights of Macromolecules

Molecular weight of a polymer is an important value in deciding various physical quantities like viscosity, colligative properties, hardness, fluidity, flexibility, etc. Molecular weights of simple molecules like water, hydrogen or benzene are definite values and are equal to simple algebraic sum of constituting atoms. Thus benzene, irrespective of its source or synthetic route, always has a fixed molecular weight of 78. Although, the polymers, are expressed with a definite structural formula, their composition and their molecular weights were found to be quite uncertain. Thus DNA strands were found to have molecular weights ranging from few thousands to few billions. Similarly polyethylene molecules formed by the polymerization of ethylene (M.Wt.= 28) were also found to have varying molecular weights, depending on the conditions employed for the polymerization. Further, under any given experimental conditions also, the number of monomer units, *n*, usually called **degree of polymerization,** is not a fixed number, because the growing chains terminate at different levels. Assuming degree of polymerization of 500, 600, and 700, the structural formula $-(CH_2-CH_2)-_n$ gives different molecular weights.

$R -(CH_2-CH_2)-_{500} R$	Molecular Weight ≈ 14000
$R -(CH_2-CH_2)-_{600} R$	Molecular Weight ≈ 16800
$R -(CH_2-CH_2)-_{700} R$	Molecular Weight ≈ 19600

8-3-1 Number and Weight average molecular weights

Molecular weights of polymers are expressed in terms of average weights of all the possible chains. The two most practical averaging methods that give molecular weights useful in the calculation of colligative properties and other physical properties are i) Number average method ii) Weight average method.

(A) Number Average Method: To get a clear concept of this method, consider a fruit basket containing bananas, apples, oranges and guava.

Variety	*Number of Each variety(n)*	*Weight of each fruit (Mg)*	*Total weight of each kind (W =nM)*
Bananas	6	100 g	600 g
Apples	4	125 g	500 g
Oranges	5	130 g	650 g
Guava	10	140 g	1400 g
	Σ n = 25		Σ nM = 3150 g

Total number of fruits in the basket = 25

Fraction of Bananas present = 6/25

No.average of Bananas (contribution made by 6 bananas to the average weight of fruits in the basket) $= \frac{6}{25} \times 100 = 24$ g

Similarly

Number average of Apples $= \frac{4}{25} \times 125 = 20$ g

Number average of Oranges $= \frac{5}{25} \times 130 = 26$ g

Number average of Guava $= \frac{10}{25} \times 140 = 56$ g

Number average weight of all fruits (Total contribution to the average weight) = 24+20+26+56 = 126 g

$$\text{Average weight} = \frac{3150}{25} = 126 \text{ g}$$

To apply the above example in case of polymer let us assume that there are n number of molecules in the polymer with n_1 of them having a molecular weight of M_1, n_2 have molecular weight M_2 and so on.

Total number of molecules $= n = \Sigma n_i = n_1 + n_2 + ... + n_i$

Number fraction of n_1 molecules $= \frac{n_1}{n} = \frac{n_1}{\Sigma n_i}$

Number average weight of n_1 molecules $= \frac{n_1 M_1}{\Sigma n}$

Similarly

Number average weight of n_2 molecules $= \frac{n_2 M_2}{\Sigma n_i}$

$\vdots \qquad \vdots$

Number average weight of n_i molecules $= \frac{n_i M_i}{\Sigma n_i}$

Total number average weight of polymer ($\overline{M}_n$) $= \frac{n_1 M_1}{\Sigma n_i} + \frac{n_2 M_2}{\Sigma n_i} + ... + \frac{n_i M_i}{\Sigma n_i}$

or

$$\overline{M}_n = \frac{\Sigma n_i M_i}{\Sigma n_i} \qquad (8.1)$$

but $\Sigma n_i M_i$ = *Weight of polymer*

$$\overline{M}_n = \frac{\textit{Weight of polymer}}{\Sigma n_i} \qquad (8.2)$$

Therefore the number average weight ($\overline{M}_n$) may be defined as the total weight of polymer divided by the total number of molecules present in the sample.

Colligative properties depend exclusively on the number of particles present in solution. Hence the molecular weight determined by the measurement of any colligative property gives only Number average molecular weight.

(B) Weight Average Molecular Weight: Properties like viscosity, density, etc., depend on the masses of polymer fractions rather than their number. To determine this value, once again consider the earlier example of fruits basket.

Total weight of all fruits $= \Sigma n_i M_i = 3150$ g

Weight of 6 bananas $= 600$ g

Weight fraction of bananas $= \dfrac{600}{3150}$

Weight average of bananas (contribution of bananas to the average weight of fruits) $= \dfrac{600}{3150} \times 100 = 19.05$ g

Similarly

Weight average of 4 Apples $= \dfrac{500}{3150} \times 125 = 19.84$ g

Weight average of 5 Oranges $= \dfrac{650}{3150} \times 130 = 26.82$ g

Weight average of 10 Guava $= \dfrac{1400}{3150} \times 140 = 62.62$ g

Total weight average $= 19.05 + 19.84 + 26.82 + 62.62$

$= 128$

To apply the above example in case of polymers, let M_w be the weight average of polymers.

Total weight of the polymer $= W = \Sigma n_i M_i$

Weight of 1st fraction $= n_1 M_1$

Weight ratio of 1st fraction $= \dfrac{n_1 M_1}{\Sigma n_i M_i}$

Weight average of 1st fraction $= \dfrac{n_1 M_1}{\Sigma n_i M_i} \times M_1$

$$= \frac{n_1 M_1^2}{\Sigma\, n_i\, M_i}$$

Similarly,

$$\text{Weight average of 2nd fraction} = \frac{n_2 M_2}{\Sigma\, n_i\, M_i} \times M_2 = \frac{n_2 M_2^2}{\Sigma\, n_i M_i}$$

$$\text{Total average weight contribution } \overline{M}_w = \frac{n_1 M_2}{\Sigma\, n_i\, M_i} + \frac{n_2 M_2^2}{\Sigma\, n_i\, M_i} + ...$$

$$+ \frac{n_i M_i^2}{\Sigma\, n_i\, M_i}$$

$$\overline{M}_w = \frac{n_1 M_1^2}{n_i\, M_i} \qquad (8.3)$$

For all macromolecules $\overline{M}_w$ is always greater than $\overline{M}_n$, When $\overline{M}_w$ is equal to $\overline{M}_n$, the polymer sample is perfectly homogeneous,. i.e., all fractions will have same molecular weight. Therefore, the ratio $\overline{M}_w$ / $\overline{M}_n$ gives a measure of a inhomogeneity or **polydispersion** of polymer. $\overline{M}_w$ is particularly sensitive to the presence of high-molecular-weight species whereas $\overline{M}_n$ is influenced more by species with lower-molecular-weights.

8-4 DETERMINATION OF MOLECULAR WEIGHT

Molecular weight of macromolecules is usually determined by i) Osmotic pressure method, ii) Ebulliometry, iiii) Cryoscopy, iv) Vapor-phase osmometry, v) Light scattering method, vi) Sedimentation method, vii) Viscometry, viii) End group analysis, ix) Ultra-centrifugation.

Of all these methods first four methods and end group analysis method are colligative in nature and hence give $\overline{M}_n$. End group anaylysis method is purely a chemical method. Rest of the methods give weight average molecular weight $\overline{M}_w$.

We are here under discussing only Osmotic pressure method, also popularly called **membrane osmometry**.

8-4-1 Membrane Osmometry

(a) Principle: This method is based on the phenomenon of osmosis, a colligative property. According to this phenomenon, when two solutions of different concentrations or a solution and its solvent are separated by a semipermeable membrane, always solvent migrates from less concentrated (or solvent) to more concentrated solution. The hydrostatic pressure developed as a result of this migration opposes the solvent migration and a dynamic equilibrium is set up. The hydrostatic pressure corresponding to this dynamic equilibrium is called **osmotic pressure**.

According to Thermodynamic theory of high polymer solutions, at any given temperature the osmotic pressure π is related to the **concentration, c, of polymer in gL⁻¹** through the expression

$$\frac{\pi}{c} = \frac{ST}{\overline{M}_n}(1 + Ac + Bc^2 + \ldots\ldots\ldots) \qquad (8.4)$$

where S is the solution constant and $\overline{M}_n$ is number average molecular weight of the solute, and A,B and so on are constants. The quantity π/c is called **Reduced osmotic pressure.**

For dilute solutions, c^2 and other higher concentration exponent terms are negligibly small and may be neglected. Therefore expression (8.4) may be written as,

$$\frac{\pi}{c} = \frac{ST}{\overline{M}_n}(1 + Ac) \qquad (8.5)$$

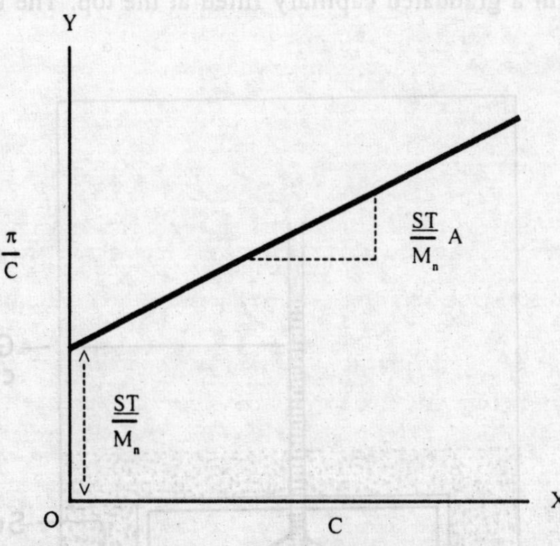

Fig. 8-1 Determination of molecular weight of polymers by osmotic pressure method

Equation (8.5) suggests that a plot of π/c versus c should give a stright line with a slope equal to $\frac{ST}{\overline{M}_n}$ A and intercept, obtained by extrapolation, equal to ST / $\overline{M}_n$

The number average molecular weight $\overline{M}_n = \frac{ST}{\text{intercept}}$ (8.6)

In fact, for very dilute solutions equation (8.5) may be simplified and written as

or

$$\pi = CST \qquad \text{where } C = \frac{c}{\overline{M}_n}$$

which is the famous van't Hoff osmotic pressure equation for dilute solutions. The non-applicability of van Hoff's equation for high polymer solutions is due to trapping of a large number of solvent molecules within the folds of macromolecules and there by rising the effective concentration of the solutions.

(b) Method: Osmometers have undergone many changes to accommodate the changing needs and availability of time. Here we give a brief account of two of the osmometers.

i) Conventional osmometry:

The osmometer illustrated in Fig. 8-2 consists of a glass or stainless steel funnel with a graduated capillary fitted at the top. The bottom of the

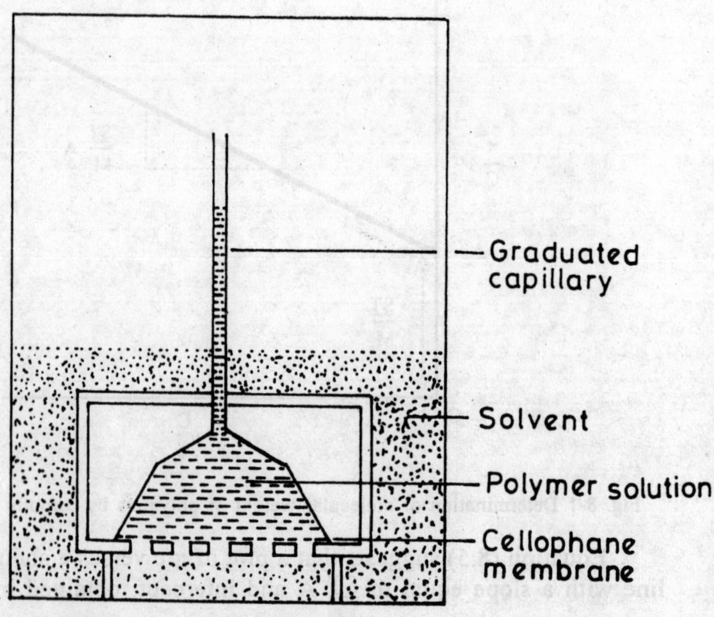

Fig. 8-2 Conventional Osmometry

chamber is perforated and fixed with a cellophane membrane. Now the chamber is filled with the experimental polymer solution, and placed in the solvent till it touches the lower end of the capillary and allowed to equilibrate. After 10 to 20 hours, when there is no change in the level of capillary, the rise in height is noted.

Then Osmotic pressure $\pi = hdg$

where h is the rise in height of capillary.

ii) High speed Membrane Osmometry (HSMO) :

This method cuts the experimental times drastically from 10-20 hours to 5-7 minutes. The apparatus consists of two cells, the upper cell are separated by membranes such as gel cellophane, collodion, nitrocellulose (having 11-13% nitrogen) and dinitrocollodion.

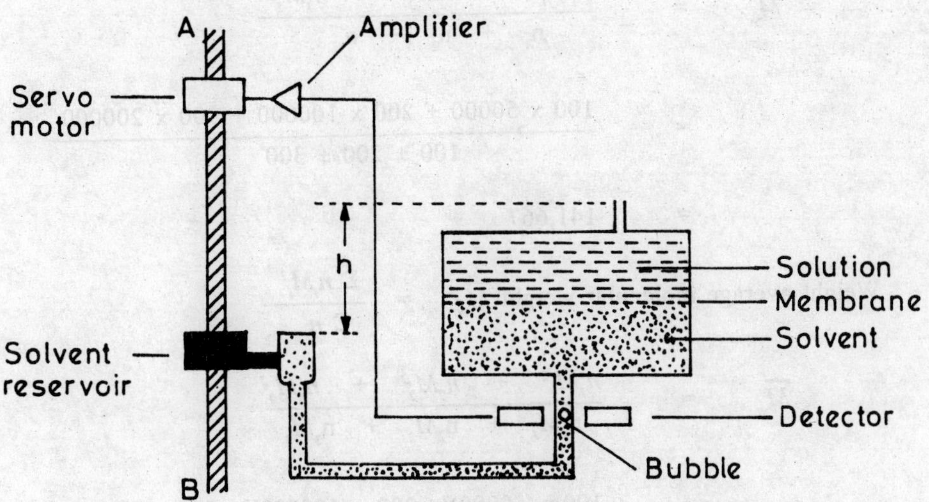

Fig. 8-3 High speed membrane osmometer (HSMO)

The short experimental times are due to rapid - equilibration on account of automatic maintenance of hydrostatic pressure with sensitive pressure sensing devices and closed solvent compartment. In the HSMO apparatus, illustrated by Fig 8-3 a bubble trapped in a capillary placed just below the solvent level and monitored by a photo-detector serves as a pressure sensing device. Any upward movement of air bubble is immediately detected by photodetector and the servo-motor is simulated to lift the solvent reservoir upwards on the threaded arm AB. This height difference, h, between solvent reservoir and solution level gives the osmotic pressure.

The only limitation of osmometry lies in the nonavailability of true semipermeable membranes. Molecular fragments having molecular weights below 10,000 generally pass through most of the semipermeable membranes. Osmometry is satisfactory only for macromolecules with molecular weight in the range 30,000 to 10^6.

8-5 NUMERICAL PROBLEMS

Problem 1: An irradiated sample of DNA was found to have 100 molecules of molecular weight 50,000 each, 200 molecules of molecular weight 100,000 each, 300 molecules of molecular weight 200,000 each. Calculate the number average and weight average molecular weights of the sample.

Number average molecular weight $\overline{M}_n = \dfrac{\Sigma\, n_i M_i}{\Sigma\, n_i}$

For the mixture of three polymer fragments

$$\overline{M}_w = \frac{n_1M_1 + n_2M_2 + n_3M_3}{n_1 + n_2 + n_3}$$

$$= \frac{100 \times 50000 + 200 \times 100000 + 300 \times 200000}{100 + 200 + 300}$$

$$= 141{,}667$$

Weight average molecular weight $\overline{M}_w = \dfrac{\Sigma\, n_i M_i^2}{\Sigma\, n_i}$

$$\overline{M}_w = \frac{n_1M_1^2 + n_2M_2^2 + n_3M_3^2}{n_1M_1 + n_2M_2 + n_3M_3}$$

$$= \frac{100 \times (50000)^2 + 200 \times (100000)^2 + 300 \times (200000)^2}{100 \times 50000 + 200 \times 100000 + 300 \times 200000}$$

$$= \frac{1.425 \times 10^{13}}{8.5 \times 10^7} = 167{,}647$$

Problem 2: A polymer sample has a composition of 20% by weight of fragments with molecular weight 50000 and 80% by weight of fragments with molecular weight 250,000. Find the $\overline{M}_n$ and $\overline{M}_w$ of the polymer sample.

$$\overline{M}_w = \frac{\Sigma\, n_i M_i 2}{\Sigma\, n_i}$$

but $n_iM_i = m_i$ the weight of ith fragment

where m_1, m_2, etc., represent individual weights of each fragment. If X is the total weight of the polymer

$$m_1 = \text{weight of 1st fragment} = \frac{20}{100}X = 0.2X$$

$$m_2 = \text{weight of 2nd fragment} = \frac{80}{100}X = 0.8X$$

$$\overline{M}_w = \frac{(0.2X) \times 50000 + (0.8X) \times 250000}{0.2X + 0.8X}$$

$$= \frac{10000\ X + 200000\ X}{X} = 210{,}000$$

$$\overline{M}_w = 210{,}000$$

we know $n_i M_i = m_i$

$$\frac{m_1}{m_2} = \frac{n_1 M_1}{n_2 M_2}$$

or

$$\frac{n_1}{n_2} = \frac{m_1 M_2}{m_2 M_1}$$

$$\frac{\text{No.of molecules with molecular weight 50000}}{\text{No.of molecules with molecular weight 500000}} = \frac{(0.2X)}{(0.8X)} \times \frac{250000}{50000}$$

$$= \frac{5}{4}$$

It means, for every 5 molecules of molecular weight 50,000 there are 4 molecules of molecular weight 250,000

Therefore,

$$\overline{M}_n = \frac{5 \times 50000 + 4 \times 250000}{5 + 4} = 138{,}889$$

Problem 3: A polymer sample is composed of 70% of fragments with M.wt. = 50,000 and 30% of fragments with M.wt = 500,000. Calculate $\overline{M}n$ and $\overline{M}w$. (Concept and working is same as given in earlier problem).

$$\overline{M}_w = \frac{0.7 \times 50000 + 0.3 \times 500000}{0.6 + 0.4}$$

$$\overline{M}_w = 185{,}000$$

$$\frac{\text{No.of molecules with M.wt 50000}}{\text{No.of molecules with M.wt 500000}} = \frac{(0.7) \times 500000}{(0.3) \times 50000}$$

$$= \frac{350000}{15000} = \frac{70}{3}$$

$$\overline{M}_n = \frac{70 \times 50000 + 3 \times 500000}{3 + 70} = \frac{5 \times 106}{73}$$

$$= 68493$$

Problem 4: Polymer A is composed of equal number of particles of molecular weight 50,000 and 90,000 and polymer B is composed of equal number of particles of molecular weight 70,000 and 250,000. Calculate $\overline{M}_w$ and $\overline{M}_n$ for both the polymers and comment on their polydispersity.

For polymer A:

$$\overline{M}_n = \frac{n_1 M_1 + n_2 M_2}{n_1 + n_2}$$

Given $n_1 = n_2 =$ say n

$$\overline{M}_n = \frac{n \times 50000 + n \times 90000}{n + n} = 70{,}000$$

$$\overline{M}_w = \frac{\Sigma\, n_i M_i^2}{\Sigma\, n_i M_i^2} = \frac{n \times (50000)^2 + n \times (90000)^2}{n \times 50000 + n \times 90000}$$

$$= \frac{1.06 \times 10^{10}}{1.4 \times 10^5} = 75{,}714$$

$$\text{Polydispersity of polymer A} = \frac{\overline{M}_w}{\overline{M}_n} = \frac{75714}{70000} = 1.08$$

for polymer B :

$$\overline{M}_n = \frac{n \times 70000 + n \times 250000}{n + n}$$

$$= \frac{320000n}{2n} = 160,000$$

$$\overline{M}_w = \frac{n \times (70000)^2 + n \times (250000)^2}{n \times 70000 + n \times 250000}$$

$$= \frac{6.74 \times 10^{10}}{3.2 \times 10^5} = 210,625$$

$$\text{Polydispersity of polymer B} = \frac{210625}{160000} = 1.32$$

The polydispersity values for polymer A and polymer B are 1.02 and 1.32 respectively. The polydispersity of polymer A is close to 1 suggesting that polymer A is more homogeneous than polymer B.

Problem 5: A plot of π/c versus c of solutions of polystyrene in toluene at 25°C gave an intercept of 0.15 atm cc g^{-1}. Calculate the molecular weight of polystyrene.

$$\text{Intercept} = \frac{\pi}{c_o} = 0.15 \text{ atm cc g}^{-1}$$

$$= \frac{0.15}{1000} = 1.5 \times 10^{-4} \text{ atm L g}^{-1}$$

$$\overline{M}_n = \frac{ST}{\pi/c}$$

$$= \frac{0.0821 \times 298}{1.5 \times 10^{-4}} = 163,105 \text{ g mole}^{-1}$$

Osmotic pressure being a colligative property the molecular weight obtained is the number average molecular weight.

Therefore,

$$\overline{M}_n = 163,105 \text{ g mol}^{-1}$$

Problem 6: $\overline{M}_n$ and $\overline{M}_w$ of a polymer containing equal number of molecules of molecular weights 50000, 100000, 1500000, 200000.
Let there be n moles of each polymer.

Therefore,

$$\overline{M}_n = \frac{\Sigma\, n_i M_i}{\Sigma\, n_i} = \frac{n \times 50000 + n \times 100000 + n \times 150000 + n \times 200000}{n + n + n + n}$$

$$= 125000$$

$$\overline{M}_w = \frac{\Sigma n_i M_i^2}{\Sigma n_i M_i} = \frac{nx(50000)^2 + nx(100000)^2 + nx(150000)^2 + nx(200000)^2}{n \times 50000 + n \times 100000 + n \times 150000 + n \times 200000}$$

$$= \frac{7.5 \times 10^{10}}{5 \times 10^5} = 150{,}000$$

Problem 7 : Calculate $\overline{M}_n$ and $\overline{M}_w$ of a polymer mixture of two polymers where polymer A with M.wt. = 80,000 has twice the number of molecular of polymer B with molecular weight 200000.

$$\overline{M}_n = \frac{\Sigma n_i M_i}{\Sigma n_i} = \frac{2 \times 80000 + 1 \times 200000}{2 + 1} = \frac{360000}{3}$$

$$\overline{M}_n = 120000$$

$$\overline{M}_w = \frac{\Sigma n_i M_i^2}{\Sigma n_i M_i} = \frac{2 \times (80000)^2 + 1 \times (200000)^2}{2 \times 80000 + 1 \times 200000} = \frac{5.28 \times 10^{10}}{3.6 \times 10^5}$$

$$= 146{,}667$$

8-6 SELF TEST QUESTIONS

8-6-1 For all polymers the following is true

A) $\overline{M}_n = \overline{M}_w$ B) $\overline{M}_n > \overline{M}_w$

C) $\overline{M}_n < \overline{M}_w$ D) $\overline{M}_n / \overline{M}_w = 1$

8-6-2 The weight molecular weight is given by

A) $\frac{\Sigma nM^2}{\Sigma nM}$ B) $\frac{\Sigma nM^2}{\Sigma Mn}$

C) $\frac{\Sigma nM^2}{\Sigma n}$ D) $\frac{\Sigma nM}{\Sigma n}$

8-6-3 Unlike colloidal solutions polymers solutions

(A) carry electric charge (B) carry no electric charge

(C) enhance electric charge (D) remove electric charge

8-6-4 The number average and weight average molecular weights of a polymer containing equal number of fragments with molecular weights 70,000 and 30,000

(A) 40000 and 58000 (B) 40000 and 1000000

(C) 58000 and 50000 (D) 50000 and 58000

8-6-5 In case of condensation polymers

(A) High molecular weight polymers are formed all at once

(B) Lower molecular weight polymers are formed all at once

(C) Molecular weight of polymer rises throughout the reaction

(D) Have no specific relation to their molecular weights.

8-6-6 Linear polymers have greater

(A) tensile strength (B) elasticity

(C) thermoplasticity (D) thermosetting

8-6-7 .... polymers soften on heating and regain original shape on cooling.

8-6-8 A typical example of inorganic polymer is ....

8-6-9 When the repeat unit and molecular formula of monomer are same, then it is said to be ....polymer.

8-6-10 Tensile strength is due to presence of .... groups in a polymer.

8-7 SHORT AND ESSAY TYPE QUESTIONS

8-7-1 Differentiate between number average molecular weight and weight average molecular weight of a polymer.
(O.U.,Apr.1987), (A.U.,Apr.1992), (N.U.,Oct.1991), (O.U.,Mar.1996)

8-7-2 Describe in detail the osmotic pressure method for the determination of molecular weight of a polymer. (A.U.,Apr.1992), (N.U.,Oct.1991)

8-7-3 What are thermosetting polymers? Explain with examples?

8-7-4 Discuss briefly the different types of macromolecules.
(O.U.,Oct.1990), (A.U.,Apr.1992)

8-7-5 Why is that mol.wt of a macromolecules is only an average molecular weight ? What is meant by number average molecular weight.
(O.U., Mar.1992), (N.U.,Sept.1995), (O.U.,Mar.1995)

8-7-6 State how molecular weight can be determined with an osmometer.
(A.U., Jun.1990), (SKU., Sept.1990) ,(O.U.,Oct.1991), (N.U.,Apr.1996), (N.U.,Sept.1995)

8-7-7 What are addition and condensation polymer? (N.U.,Apr.1991)

8-7-8 A 10g sample of polymer dissolved in 1 L of water, shows an osmotic pressure of 2.0×10^{-3} atm at 300K. Calculate the molecular weight of the polymer.
(A.U., Mar.1991)

8-7-9 What is number molecular weight of a polymer? Name the method used to determined it. (A.U.,Mar.1991)

8-7-10 What are macromolecules? (O.U., Apr.1987), (N.U.,Mar.1995)

8-7-11 A polymer has equal number of molecules of molecular weights 10,000, 20,000 and 40,000 daltons. Calculate the number average and weight average molecular weights of the polymer.

ANSWERS

8-6-1 **C**	8-6-2 **A**	8-6-3 **B**	8-6-4 **D**
8-6-5 **C**	8-6-6 **C**	8-6-7 **Thermoplastic**	8-6-8 **Silicones**
8-6-9 **Addition**	8-6-10 **polar**	8-7-8 **123, 150.**	8-7-11 23333; 30000

❋

Part IV

Practical Chemistry

1

SEMI-MICRO ANALYSIS

1-1 INTRODUCTION

The technique of semimicro analysis does not differ very greatly from that of macro analysis. As the volume of the solutions are of the order of 1 ml or less than that in the semi micro analysis, the apparatus should also match that scale, instead of the usual large glassware used for the macro analysis. The following points need to be kept in mind before proceeding to actual analysis.

I. Test tubes or semi micro tubes should be of 4 ml or 8 ml volumes for performing the tests. A centrifuge tube of 3ml volume and a boiling tube of 20ml volume should be used. The centrifuge tube should not be used for direct heating, but can be kept in a water bath.

II. Droppers with long endings should be used for handling the liquids. These droppers used in semimicro analysis are known as capillary droppers. The medicinal droppers are the best recommended ones. These droppers are essential, because a reaction at the most requires 30 to 40 drops of the solution. If an ordinary dropper is available, then 8 to 10 drops are sufficient to carry out the test.

III. In semimicro technique the precipitate is never separated by the filtration process, instead centrifuged in a hand or motor driven centrifuge. The heavy precipitate particles settle at the bottom of the centrifuge tube which is termed as residue. The supernatant liquid over it is called the centrifugate. The centrifugate is removed from the residue using a long stem dropper without disturbing the residue.

IV. Washing of the precipitate is another important technique of semimicro analysis. After removing the centrifugate from the residue (precipitate) a little water (1 to 2 ml) is added to the residue and is mixed well with a glass rod pointer/needle,centrifugation is done and the centrifugate is removed. This can be repeated atleast three times to get the precipitate of the required purity. The precipitate can be removed from the centrifuge tube using a small spatula of either glass or made of some inert metal. The need of removal of the precipitate is often not required, but, if it is essential and if the precipitate is very small in quantiry, then we can add the reagent with which the precipitate is to be tested directly to it in the centrifuge tube and the observation be recorded.

V. The volatile group gases can be detected in gas evolution tubes only. These tubes should essentially be of semimicro scale only. The reaction is performed with great care and observation, or else the evolved gas will escape the sight.

VI. Another very important and interesting technique which one should learn is the technique of spot tests. These are employed mainly for the confirmaroty tests which result in colored precipitates or solutions. The spot tests are performed on a spot plate having a number of circular cavities. A drop of the reactant and a drop of the reagent will give the result. If the spot plates are not available in the laboratory filter paper strips can be used in their place. Whatmann No 120 filter paper is best for this purpose as it has minimum spreading of the reagent solutions. The spot test technique is the most sensitive and highly sophisticated and the colored spots can be preserved as record of the test. The spot tests for the different anions and cations are given below.

1-2 SPOT TESTS

1. ACETATE : To a drop of neutral ferric chloride taken on a filter paper add a drop of the solution of the mixture and warm it on the bunsen flame, carefully. A red spot confirms acetate.

2. NITRATE : To a drop of diphenyl amine indicator taken on a porecelain tile or spot plate, add a drop of the test solution of the mixture. A blue color confirms nitrate.

3. CARBONATE : Place a drop of phenolphthalein indicator on a spot plate along with a small quantity of the mixture. A pink spot develops. Decolorization of this color by the addition of a drop of dil HCl solution confirm the presence of carbonate.

4. SULPHITE : A spot of potassium permanganate decolorizes on adding a drop of the solution of the mixture, confirming the sulphite.

5. SULPHIDES : A spot of acidified potassssium dichromate turns green by the addition of test solution confirms the presence of sulphide in the given salt mixture.

6. IODIDES/COPPER : To a drop of copper sulphate soultion add a drop of the solution of the salt mixture, yellow brown spot confirms iodide. The reverse of this test, i.e., development of brown spot by the addition of iodide to the salt solution confirms copper.

7. LEAD : A spot of the solution of the mixture turn yellow on adding a drop of potassium iodide confirming lead.

8. MERCUROUS : Change of a spot of the solution of the mixture to yellowish-green, by adding a drop of potassssium iodide confirms mercurous ion.

9. MERCURIC : If the spot of the solution of the mixture turns red on adding a drop of potassium iodide, it indicated the presence of mercuric ion.

10. BISMUTH : A spot of the solution of the mixture turns brown on adding potassium iodide drop, suggests the presence of bismuth.

11. IRON : Add to a drop of the test solution potassium ferrocyanide. A blue coloration suggests the presence of ferrous and/or ferric in the mixture.

12. ALUMINIUM : To a drop of aluminium hydroxide precipitate taken on a spot plate, add a drop each of dil. HCl and ammonium acetate followed by a drop of aluminon reagent. A red precipitate confirms aluminium.

13. NICKEL : To a drop of alkaline demethylglyoxime add a drop of the solution of the mixture on a fliter paper, a red spot confirms nickel.

14. MANGANESE: To a drop of the solution of the mixture add either sodium hydroxide or ammonia solution, a flesh colored precipitate confirms manganese.

15. BARIUM : To a spot of the mixture of the solution add a drop of potassium chromate, a yellow ppt confirms barium.

16. AMMONIUM : To a spot of the mixture of the solution on a filter paper add a drop of Nessler's reagent, an orange coloration confirms ammonium.

Inspite of these spot tests being very simple and attractive, the results should be taken with caution as the presence of any interfering radical may mislead us. So it is essential to separate the ions adopting standard techniques as discusssed below and then carry out these tests giving only a confirmatory status.

1-3 SEMIMICRO ANALYSIS

Similar to normal qualitative analysis even in semimicro analysis also the anions or acid radicals and cations or basic redicals are analysed separately. The semimicro analysis of the salt mixture can be performed in the following sequence.

1-3-1 Preparation of test solution

Test the solubility of the given salt mixture first in cold distilled water, if it is insoluble warm,it, even then if it is insoluble add a few drops of nitric acid **(Note: the solubility should not be tested with hydrochloric acid as it is a group reagent and moreover it reacts with many radicals).**

1-3-2 Sodium Carbonate extract

To test the acid radicals which are either sparingly soluble or insoluble in water, prepare the sodium carbonate extract of the given mixture. To about 0.5 mg (one pinch) of the given mixture add an equal quantity of sodium carbonate in a boiling tube, add to this 2 ml of distilled water and boil. If the solution is not clear, centrifuge, use the centrifugate as the test solution. If the solution is clear there is no need for centrifugation and can be used directly.

1-4 ANALYSIS OF ACID RADICALS

Acid radicals are divided according to their reaction with different reagents. The acid radicals to be studied are acetate (CH_3COO^-), nitrate (NO_3^-), phosphate (PO_4^{3-}), chloride (Cl^-), bromide (Br^-), iodide (I^-), carbonate (CO_3^{2-}), sulphite (SO_3^{2-}), sulphide (S^{2-}), sulphate (SO_4^{2-}) and borate (BO_3^{3-}).

1-4-1 Water Soluble Group

Radicals present: Acetate and Nitrate

Test for ACETATE

a) To about 0.5 mg of the mixture add about 1ml of distilled water if it dissolves then add to the solution four drops of neutral ferric chloride solution.	If a red coloration is obtained which upon heating forms a red precipitate indicates the presence of acetate.
b) To about 0.5 mg of the mixture add a few drops of conc. sulphuric acid and a few drops of either ethyl alcohol or isoamyl alcohol heat and pour the hot solution in a small quantity of water or sodium carbonate solution.	a pleasant odour due to formation of ester confirms the acetate.

$$FeCl_3 + 3NaCH_3COO \longrightarrow Fe(CH_3COO)_3 + 3\,NaCl$$

$$C_2H_5OH + NaCH_3COO \xrightarrow{H_2SO_4} NaOH + C_2H_5(CH_3COO)$$

Test for NITRATE

a) To another sample of 0.5 mg. of mixture add a few drops of conc.sulphuric acid and cool it under tap. Add from the sides of the test tube a few drops of freshly prepared ferrous sulphate solution.	A brown ring formation at the junction of the two liquids shows the presence of nitrate.
b) Add a few drops of conc. sulphuric acid to the given mixture along with a few copper turnings, heat it.	Golden fumes evolves which indicates the presence of nitrate.
c) A solution of the nitrate salt when poured carefully down the sides of the test tube containing diphenyl amine indicator.	A blue ring if formed confirms nitrate.

$$NaNO_3 + H_2SO_4 \longrightarrow NaHSO_4 + HNO_4$$

$$6FeSO_4 + 2HNO_2 + 3H_2SO_4 \longrightarrow 3Fe_2(SO_4)_3 + 2NO + 4H_2O$$

$$FeSO_4 + NO \longrightarrow [Fe(NO)]SO_4 \text{ Brown Ring}$$

$$H_2SO_4 + NaNO_3 \longrightarrow NaHSO_4 + HNO_3$$

$$Cu + 4HNO_3 \longrightarrow Cu(NO_3)_2 + 2NO_2 + 2H_2O$$

$$3Cu + 8HNO_3 \longrightarrow 3Cu(NO_3)_2 + 2NO + 4H_2O$$

$$2NO + O_2 \longrightarrow 2NO_2$$

1-4-2 Silver NItrate Group (Nitric Acid Group)

Radicals present: Phosphate, Chloride Bromide and Iodide.

To about 1 mg of the mixture add about 5ml of conc. nitric acid and boil till the fumes ceases. Add a few drops of distilled water to this solution. Divide the solution into two portions.

Test for PHOSPHATE

a) To one portion add ammonium molybdate solution	If a canary yellow crystalline precipitate is formed which is soluble in liqour ammonia or a conc. solution of either sodium hydroxide or KOH, the presence of phosphate is indicated.
b) To the solution of the mixture add a few drops of magnesium nitrate reagent.	A white crystalline precipatate confirms phosphate.

$$Na_2HPO_4 + 12(NH_4)_2 MoO_4 + 23 HNO_3 \longrightarrow (NH_4)_3[PMo_{12}O_{40}] + 2NaNO_3 + 21NH_4NO_3 + 12H_2O$$

$$(NH_4)_3PO_4 + Mg(NO_3)_2 \longrightarrow Mg(NH_4)PO_4 + 2NH_4NO_3$$

Test of HALIDES

To the second portion of the solution add silver nitrate solution.

a) White precipitate soluble in ammonium hydroxide and reappears with conc. nitric acid.	Chloride is present
b) Pale yellow precipitate which is sparingly soluble or completely insoluble in ammonia solution.	Bromide or iodide may be present
c) If chloride is present perform the chromyl chloride test as follows:	
To the mixture add conc. H_2SO_4 and solid potassium dichromate crystals, heat gently.	Colorless gas evolves which turns red on heating
Pass the above gas in NaOH or lead acetate solution.	Slow formation of a yellow ppt. or solution confirms chloride
d) If bromide or iodide is present add conc. sulphuric acid, a few drops of potassium dichromate solution and carbon tetrachloride then shake the mixture well.	Orange layer confirms bromide Violet layer confirms iodide.

$$NaX + AgNO_3 \longrightarrow AgX + NaNO_3 \ (X=Cl,Br,I)$$

$$AgCl + 2NH_3 \longrightarrow [Ag(NH_3)_2]Cl$$

$$[Ag(NH_3)_2]Cl + 2HNO_3 \longrightarrow AgCl + 2NH_4NO_3$$

$$K_2Cr_2O_7 + 4NaCl + 6H_2SO_4 \longrightarrow 2CrO_2Cl_2 + 2KHSO_4 + 4NaHSO_4 + 3H_2O$$

$$CrO_2Cl + 4NaOH \longrightarrow Na_2CrO_4 + 2NaCl + 2H_2O$$

$$K_2Cr_2O_7 + 6KBr + 7H_2SO_4 \longrightarrow 3Br_2 + Cr_2(SO_4)_3 + 4K_2SO_4 + 7H_2O$$

$$2Br_2 + CCl_4 \longrightarrow CBr_4 + 2Cl_2\uparrow$$

$$K_2Cr_2O_7 + 6KI + 7H_2SO_4 \longrightarrow 3I_2 + Cr_2(SO_4)_3 + 4K_2SO_4 + 7H_2O$$

$$2I_2 + CCl_4 \longrightarrow CI_4 + 2Cl_2\uparrow$$

1-4-3 Volatile Group (HCl Group) :

Radicals present : Carbonate, Sulphide, Sulphite and Sulphate.

Take about 0.5 mg of the mixture in a gas evolution tube, in the side tube of the apparatus add baryta water and then add dilute HCl to the mixture.

a) Colorless gas evolved which turns baryta water turbid.	Carbonate is present
b) Colorless gas with an odor of rotten eggs evolves turning lead acetate brown gray or black	Sulphide is present
c) Colorless gas with an ordor of burning sulphur evolves turning acidified potassium permanganate colorless and potassium dichromate green	Sulphite is present

To the solution of the salt mixture or sodium carbonate extract add barium chloride solution

a) White ppt. soluble in dil HCl	Carbonate is confirmed
b) White ppt. insoluble in dil HCl	Sulphate is confirmed

To the solution of the mixture or to the sodium carbonate extract add lead acetate solution.

a) white ppt. soluble in conc.NaOH and reappears with conc.HCl	Sulphate is confirmed
b) White ppt. highly soluble in any mineral acid or organic acid	Carbonate is confirmed

$$Na_2CO_3 + 2HCl \longrightarrow 2NaCl + CO_2 + H_2O$$

$$Na_2S + 2HCl \longrightarrow 2NaCl + H_2S$$

$$Na_2SO_3 + 2HCl \longrightarrow 2\ NaCl + SO_2 + H_2O$$

$$CO_2 + Ba(OH)_2 \longrightarrow BaCO_3 + H_2O$$

$$H_2S + (CH_3COO)_2\ Pb \longrightarrow PbS + 2CH_3COOH$$

$$Na_2SO_3 + 2HCl \longrightarrow 2NaCl + H_2SO_3$$

$$2\ KMnO_4 + 5H_2SO_3 \longrightarrow K_2SO_4 + 2MnSO_4 + 2H_2SO_4 + 3H_2O$$

$$K_2Cr_2O_7 + 3H_2SO_3 + H_2SO_4 \longrightarrow K_2SO_4 + Cr_2(SO_4)_3 + 4\ H_2O$$

$$BaCO_3 + 2HCl \longrightarrow BaCl_2 + H_2O + CO_2$$

$$Pb(C_2H_3O_2) + Na_2SO_3 \longrightarrow PbSO_3 + 2NaC_2H_3O_2$$

$$PbSO_3 + 2NaOH \longrightarrow Na_2SO_3 + Pb(OH)_2$$

1-5 ANALYTICAL CLASSIFICATION OF THE BASIC RADICAL

The basic radicals are classified into six groups, depending on their reactions with a certain group reagent, for the purpose of qualitative analysis, as shown below

Groups	Group Reagent	Radicals	Precipitate Formed	Distinctive Property
I Silver Group	Dil HCl	Pb^{++}, Hg^{++} & Ag^{+}	$PbCl_2$, Hg_2Cl_2, AgCl,	Chlorides insoluble in dil. HCl (cold)
IIA Copper Group	Dil HCl + H_2S	Hg^{++}, Bi^{+++}, Cd^{++}, Cu^{++}	HgS, Bi_2S_3, CdS, CuS	Sulphides insoluble in KOH:
IIB Arsenic Group	Dil HCl + H_2S	As^{+++}, Sb^{+++} Sn++	As_2S_3, Sb_2S_3 SnS	Sulphides soluble in KOH.
III Iron group	NH_4Cl + NH_4OH	Fe^{++}, Fe^{+++} Mn,$^{++}$, Cr^{+++} Al^{+++}	$Fe(OH)_2$, $Fe(OH)_3$ $Mn(OH)_2$, $Cr(OH)_3$ $Al(OH)_3$	Hydroxides of Fe & Mn insoluble in NaOH, that of Cr & Al are soluble in NaOH.
IV Zinc group	NH_4Cl + NH_4OH + H_2S	Co++, Ni++ Mn^{++}, Zn^{++}	CoS, NiS, MnS, ZnS	Sulphides of Co & Ni are HCl insoluble that of Mn & Zn are soluble in HCl.
V Alkaline earth group	NH_4Cl + NH_4OH + $(NH_4)_2CO_3$	Ba^{++}, Ca^{++} Sr^{++}	$BaCO_3$, $CaCO_3$ $SrCO_3$	Soluble in dil. acetic acid.
VI Alkaline group	Miscellaneous reagents	NH_4^{+}, K^{+}, Na^{+}, Mg^{++}	Precipitates formed according to the reagent used.	---

1-6 AMMONIUM RADICAL TEST

Before proceeding to the Basic radical analysis test for ammonium radial should be performed. Take 2 ml of the solution of the mixture and Nesslers reagent - orange ppt confirms ammonium radical.

To the mixture add NaOH solution & boil, colourless gas evolves becoming dense when a glass rod dipped in HCl is brought near it confirming ammonium radical.

1-7 ANALYSIS OF BASIC RADICALS

To the salt solution prepared either in distilled water or in dil. nitric acid add dil HCl and centrifuge. If a precipitate is fomed I group is present. Centrifuge and preserve the centrifugate for subsequent groups.

<table>
<tr><td>Residue:
1st group is present as $PbCl_2$, AgCl or Hg_2Cl_2.
(Group I)</td><td colspan="2">Centrifugate/Clear solution: Pass hydrogen sulphide gas, if ppt. occurs 2nd group is present. Centrifuge or else reject the solution. The ppts. may be HgS (black) PbS (black), Bi_2S_3 (brown), CuS (black) CdS (yellow), As_2S (yellow), Sb_2S_3 (orange), SnS_2 (white), Treat the ppt. with KOH and boil, centrifuge.</td></tr>
<tr><td>Residue:
Contains HgS, PbS, Bi_2S_3 CuS or CdS
(Group II A)</td><td>Centrifugate:
Contains $KAsS_2$, $KAsO_2$, $KSbO_2$, or K_2SnO_3
(Group II B)</td><td>Take a Fresh sample of reaction solution, if I & II group elements are absent, add solid NH_4Cl and NH_4OH solution. If precipitation occurs III group is present. centrifuge.</td></tr>
<tr><td colspan="3">If I or II group basic radicals are present, pass H_2S through the fresh sample, collect the centrifugate and boil it strongly to expel the hydrogen sulphide gas completely and then add solid ammonium chloride and ammonium hydroxide solution in excess.</td></tr>
<tr><td rowspan="2">Residue:
Contains $Al(OH)_3$, $Fe(OH)_3$, $Cr(OH)_3$, $MnO.2H_2O$
(Group III)
white gelatinous - Al Dirty green/ brown Ferrous/ Ferric, Green-Cr; pink or flesh or light brown- Mn^{2+}
(Group III)</td><td colspan="2">Centrifugate: Pass hydrogen sulphide gas into it, if ppt. is formed 4th group is present, centrifuge.</td></tr>
<tr><td>Residue:
contains ZnS, NiS, MnS or CoS
(Group IV)</td><td>Centrifugate: It is heated strongly till all the hydrogen sulphide gas is expelled out test with lead acetate, add ammonium chloride in excess then ammonia solution and ammonium carbonate. If ppt. occurs it is. carbonate of Barium, Calcium or Strontium. (Group V)</td></tr>
</table>

1-7-1 I Group Analysis

To about 1 ml of the solution prepared in distilled water or with a drop of HNO_3 acid add a few drops of dil. HCl. A white precipitate of either $PbCl_2$, $HgCl_2$ or AgCl is formed. Centrifuge the precipitate. Preserve the centrifugate.

Boil the precipitate with little water, keeping the test tube in a hot water rack. centrifuge.

<table>
<tr>
<td rowspan="2">Centrifugate: If the precipitate dissolves on boiling it is of Pb^{2+}.

Divide the solution in two portions to one protion add KI solution–yellow precipitate soluble in hot water is obtained and on cooling it, golden spangles are obtained, confirming Pb^{2+}.

To second portion add $KCrO_4$ solution, yellow precipitate confirms Pb^{2+}</td>
<td colspan="2">Residue: May contain Hg_2Cl_2 and AgCl & some undissolved $PbCl_2$. To remove lead chloride add 2 ml of water and boil again. centrifuge the hot solution, rejecting the centrifugate. To the residue add dilute NH_3 solution and centrifuge.</td>
</tr>
<tr>
<td>Residue: Blackness confirming mercurous Hg_2^{2+}</td>
<td>Centrifugate: May contain $(Ag(NH_3)_2Cl)$ add dilute nitric acid the white precipitate reappers. confirming Ag+</td>
</tr>
</table>

$$Pb(NO_3)_2 + 2HCl \longrightarrow PbCl_2\downarrow + 2HNO_3$$

$$PbCl_2 + HCl \overset{hot}{\rightleftharpoons} H\,(PbCl_3)$$

$$PbCl_2 + 2HCl \overset{hot}{\rightleftharpoons} H_2[PbCl_4]$$

$$Pb(NO_3)_2 + 2KI \longrightarrow PbI_2\downarrow + 2KNO_3$$

$$PbI_2 + 2KI \longrightarrow K_2(PbI_4)$$

$$PbNO_3 + K_2CrO_4 \longrightarrow PbCrO_4\downarrow + K_2NO_3$$

$$Hg_2(NO_3)_2 + 2HCl \longrightarrow Hg_2Cl_2\downarrow + 2HNO_3$$

$$2Hg_2(NO_3)_2 + 4NH_3 + H_2O \longrightarrow 2Hg + Hg_2\ NH_2O\ NO_3 + 3NH_4NO_3$$

$$AgNO_3 + HCl \longrightarrow AgCl\downarrow + HNO_3$$

$$AgCl + 2NH_3 \longrightarrow [Ag(NH_3)_2]Cl$$

$$[Ag(NH_3)_2]\ Cl + 2HNO_3 \longrightarrow AgCl\downarrow + 2NH_4NO_3$$

1-7-2 II Group Analysis

<table>
<tr><td colspan="5">To the ppt obtained by passing hydrogen sulphide gas, add KOH solution, Centrifuge:</td></tr>
<tr><td colspan="3">IIA: Residue: To the residue add dil. nitric acid, boil and centrifuge.</td><td colspan="2" rowspan="2">II B : Centrifugate: Pass hydrogen sulphide in excess and discard any ppt. if formed and add dil. HCl. If insoluble centrifuge.</td></tr>
<tr><td colspan="2">Residue: If black Hg^{2+}, dissolve the ppt. in a conc. solution of sodium sulphide. The ppt. dissolves add $SnCl_2$, grey ppt. confirms Hg^{2+}</td><td rowspan="3">Centrifugate: It may contain nitrates of Pb^{2+}, Bi^{3+}, Cu^{2+} or Cd^{2+}. Add excess of conc. NH_3 solution and centrifuge.

Centrifugate: If colorless-Cd^{2+} if blue -Cu^{2+}. If colorless pass hydrogen sulphide yellow ppt. - Cd^{2+}. If blue add potassium ferro cyanide solution. Reddish brown ppt confirms - Cu^{2+}.</td></tr>
<tr><td colspan="2">Residuee: May contain bismuth hydroxide or lead hydroxide. Add NaOH and centrifugate:</td><td rowspan="2">Residue: If yellow- As^{3+} dissolve it in ammonia solution and add dil. HNO_3 yellow ppt. confirms As^{3+}</td><td rowspan="2">Centrifugate: Divide it into two portions. To one portion add three drops of ammonia solution and 0.3 gm oxalic acid. Pass hydrogen sulphide gas- Orange ppt. confirms Sb^{3+}.

To the second portion add Hg_2Cl_2 solution dropwise. Grey ppt. confirms Sn^{2+}</td></tr>
<tr><td>Residue: Add dil. nitric acid, ppt. dissolves pour it in water white ppt. confirms Bi^{3+}</td><td>Centrifugate: Add potassium chromate solution and a few drops of dilute aceticacid yellow precipitate confirms Pb^{2+}</td></tr>
</table>

$$HgCl_2 + H_2S \longrightarrow HgS + 2HCl$$
$$2HgCl_2 + SnCl_2 \longrightarrow SnCl_4 + Hg_2Cl_2$$
$$Hg_2Cl_2 + SnCl_2 \longrightarrow SnCl_4 + 2Hg$$
$$2Bi(NO_3)_3 + 3H_2S \longrightarrow Bi_2S_3 + 6HNO_3$$
$$BiCl_3 + H_2O \longrightarrow BiOCl + 2HCl$$
$$CdSO_4 + H_2S \longrightarrow CdS + H_2SO_4$$
$$CdS + 2NH_3 + 2H_2O \longrightarrow Cd(OH)_2 + (NH_4)_2S$$
$$Cd(OH)_2 + (NH_4)_2 SO_4 + 2NH_3 \longrightarrow [Cd(NH_3)_4]SO_4 + 2H_2O$$
$$2CuSO_4 + K_4 [Fe(CN)_6] \longrightarrow Cu_2[Fe_2(CN)_6] + 2K_2SO_4$$
$$2AsCl_3 + 3H_2S \longrightarrow As_2S_3 \downarrow + 6 HCl$$
$$2 Na_3 AsO_3 + 6HCl + 3H_2S \longrightarrow As_2S_3 \downarrow + 6NaCl + 6H_2O$$
$$As_2S_3 + 6NH_3 + 3H_2O \longrightarrow (NH_4)_3 AsO_3 + (NH_4)_3 AsS_3$$
$$3As_2S_3 + 20 HNO_3 + 4H_2O \rightleftharpoons 6H_3AsO_4 + 9H_2SO_4 + 20 NO$$
$$2 SbCl_3 + 3H_2S — Sb_2S_3 \downarrow + 6 HCl$$
$$2Sb_2S_3 + 4 KOH \longrightarrow KSbO_2 + 3KSbS_2 + 2H_2O$$
$$KSbO_2 + 3KSbS_2 + 4HCl \longrightarrow 2Sb_2S_3 \downarrow + 4KCl + 2H_2O$$
$$4SbCl_3 + 12NH_4OH \longrightarrow Sb_4O_6 + 12NH_4Cl + 6H_2O$$
$$Sb_4O_6 + 4NH_4OH \longrightarrow 4NH_4(SbO_2)\downarrow + 2H_2O$$
$$SnCl_2 + H_2S \longrightarrow SnS + 2HCl$$
$$SnCl_2 + Hg_2Cl_2 \longrightarrow SnCl_4 + 2Hg$$

1-7-3 III Group Analysis

To the solution of the mixture add solid ammonium chloride till saturation then add ammonium hydroxide solution. If precipitation occurs, centrifuge, preserve the centrifugate and add to the residue a solution of NaOH and centrifuge:	
Residue: May contain MnO_2 $2H_2O$, $Fe(OH)_3$ Dissolve the ppt. in conc. nitric acid and divide the solution into two portions.	**Centrifugate:** May contains $Na(AlO_2)$ or $Na(CrO_2)$
a) To one portion add potassium ferrocyanide solution dropwise, blue coloration confirms Fe^{2+} / Fe^{3+}. b) To the second portion add sodium bismuthate, boil and cool, formtion of violet color confirms Mn^{2+}.	a) If yellow - Cr^{3+} is present. Add a drop of acetic acid and drop of lead acetate solution. Yellow ppt confirms - Cr^{3+}. b) Acidify the centrifugate with dil. HCl, add excess of ammonia- white ppt. confirms- Al^{3+}. Dissolve this ppt. in dil. HCl and add three drops of aluminon reagent and a drop of ammonium carbonate solution. Pink precipitate confirms Al^{3+}.

$$FeSO_4 + 2NH_4OH \xrightarrow{NH_4Cl} Fe(OH)_2\downarrow + (NH_4)_2\ SO_4$$

$$Fe(OH)_2 + 2HNO_3 \longrightarrow Fe(NO_3)_2 + 2H_2O$$

$$Fe(NO_3)_2 + K_4[Fe(CN)_6] \longrightarrow K_2Fe[Fe(CN)_6] + 2KNO_3$$

$$MnSO_4 + 2NH_4OH \xrightarrow{NH_4Cl} Mn(OH)_2 + (NH_4)_2SO_4$$

$$5NaBiO_3 + 2MnSO_4 + 16HNO_3 \longrightarrow 2HMnO_4 + 5Bi(NO_3)_3 + NaNO_3 + 2Na_2SO_4 + 7H_2O$$

$$CrCl_3 + 3NH_4OH \xrightarrow{NH_4Cl} Cr\ (OH)_3 + 3NH_4Cl$$

$$Cr(OH)_3 + NaOH \longrightarrow Na(CrO_2) + 2H_2O$$

$$Na_2CrO_4 + (CH_3COO)_2\ Pb \longrightarrow PbCrO_4 + 2CH_3COONa$$

$$AlCl_3 + 3NH_4OH \xrightarrow{NH_4Cl} Al(OH)_3 + 3NH_4Cl$$

$$Al(OH)_3 + NaOH \longrightarrow Na(AlO_2) + 2H_2O$$

$$Na(AlO_2) + HCl + H_2O \longrightarrow Al(OH)_3 + NaCl$$

1-7-4 IV Group Analysis

<table>
<tr><td colspan="3">If no precipitate is formed in the third group pass hydrogen sulphide in the same solution or to the centrifugate of third group. If ppt. is formed add dil. HCl (1:20) and centrifuge.</td></tr>
<tr><td rowspan="2">Residue: If black it, may be CoS or NiS. Dissolve the ppt. in dil. nitric acid or dil.HCl and boil. If the solution is green Ni^{2+} is present. If it is pink Co^{2+} is present. Add excess of NaOH-Green precipitate confirms-Ni^{2+}. Pink ppt. slowly turning brown-confims-Co^{2+}.</td><td colspan="2">Centrifugate: It may contain soluble $MnCl_2$ or $ZnCl_2$. Boil the solution to expel out all the hydrogen sulphide gas. Add NaOH in excess. If precipitation occurs, centrifuge.</td></tr>
<tr><td>Residue: Pink or brown ppt.indicates the presence of Mn^{2+}. Add sodium bismuthate and conc. nitric acid and boil. Violet color-confirms-Mn^{2+}.</td><td>Centrifugate: May contain Na_2ZnO_2. Pass hydrogen sulphide. Dirty white precipitate if obtained, divide it into two portions, To one portion add NaOH and to the other add dil. HCl. If it dissolves in both these reagents, Zn^{2+} is confirmed.</td></tr>
</table>

$$NH_4OH + H_2S \xrightarrow{NH_4Cl} NH_4HS + H_2O$$

$$Co(NO_3)_2 + NH_4HS + NH_3 \longrightarrow CoS \downarrow + 2NH_4NO_3$$

$$3\,CoS + 8HNO_3 \longrightarrow 3Co(NO_3)_2 + 2NO + 3S + 4H_2O$$

$$NiSO_4 + NH_4HS + NH_3 \longrightarrow NiS \downarrow + (NH_4)_2SO_4$$

$$3NiS + 8HNO_3 \longrightarrow 3\,Ni(NO_3)_2 + 2NO + 3S + 4H_2O$$

$$Co(NO_3)_2 + 2NaOH \longrightarrow Co(OH)_2 + 2NaNO_3$$

$$Ni(NO_3)_2 + 2NaOH \longrightarrow Ni(OH)_2 + 2NaNO_3$$

$$MnSO_4 + NH_4HS + NH_3 \longrightarrow MnS + (NH_2)_2\,SO_4$$

$$ZnCl_2 + NH_4HS + NH_3 \longrightarrow ZnS + 2NH_4Cl$$

$$ZnS + 2NaOH \longrightarrow Zn(OH)_2 + Na_2S$$

$$Zn\,(OH)_2 + 2NaOH \longrightarrow Na_2(ZnO_2) + 2H_2O$$

$$Na_2(ZnO_2) + H_2S \longrightarrow ZnS + 2NaOH$$

1-7-5 V Group Analysis

<table>
<tr><td colspan="3">To a fresh sample of the salt mixture add solid ammonium chloride, ammonium hydroxide and ammonium carbonate solutions. If a white crystalline ppt is formed 5th group is present. If precipitation occurs on the addition of solid ammonium chloride and ammonium hydroxide, centrifuge, reject the residue and to the centrifugate add ammonium carbonate solution.

To the ppt. obtained from ammonium carbonate addition, add dil. acetic acid. Divide the solution into three portions.</td></tr>
<tr><td>To the first portion add potassium chromate solution - yellow ppt. confirms Ba^{2+}.

Flame Test: Apple green - Ba^{2+} - confirmed.</td><td>To the second portion add ammoniumoxalate solution. White ppt. confirms Ca^{2+}.

Flame Test: Brick red-Ca^{2+} confirmed.</td><td>To the third portion add excess of nitric acid drop wise if ppt. forms- it confirms Sr^{2+}.

Note: This test is always decissive. If Ba^{2+} and Ca^{2+} are absent but precipitate is formed with ammonium carbonate solution, then it confirms the presence of Sr^{2+}.</td></tr>
</table>

$$BaCl_2 + (NH_4)_2\, CO_3 \xrightarrow[NH_4Cl]{NH_4OH\ \&} BaCO_3\downarrow + 2NH_4Cl$$

$$CaCl_2 + (NH_4)_2\, CO_3 \xrightarrow[NH_4Cl]{NH_4OH\ \&} CaCO_3\downarrow + 2NH_4Cl$$

$$SrNO_3 + (NH_4)_2\, CO_3 \xrightarrow[NH_4Cl]{NH_4OH\ \&} SrCO_3\downarrow + 2NH_4\uparrow NO_3$$

$$BaCO_3 + 2CH_3COOH \longrightarrow (CH_3COO_2)\, Ba + CO_2 + H_2O$$

$$(CH_3COO)_2Ba + K_2CrO_4 \longrightarrow BaCrO_4\downarrow + 2CH_3COOK$$

$$(CH_3COO)_2Ca + (NH_4)_2C_2O_4 \longrightarrow CaC_2O_4\downarrow + 2CH_3\, COONH_4$$

$$(CH_3COO)_2Sr + 2HNO_3 \longrightarrow Sr(NO_3)_2 + 2CH_3\, COOH$$

$$Sr(NO_3)_2 + (NH_4)_2\, CO_3 \longrightarrow SrCO_3\downarrow + 2NH_4NO_3$$

1-7-6 VI Group Analysis

It is a miscellaneous group with no particular group reagent. The radicals are potassium, sodium and magnessium and ammonium ions. The reactions of ammonium are already mentioned.

Reactions of K^+	(a) To the salt mixture solution, if tartaric acid is added, white crystalline ppt. is formed. (b) Flame Test: The flame appears violet when seen through blue glass.
Reactions of Na^+	(a) To the salt mixture solution, if uranyl magnesium acetate is added it forms a yellow ppt. (b) Flame Test: Golden flame is observed.
Reactions of Mg^{2+}	(a) To the solution of mixture, if ammonium hydroxide solution is added, it forms a white gelatinous precipitate. (b) Sodium hydroxide solution also give a white ppt. which is insoluble in excess of the reagent. (c) Salt mixture solution forms a white crystalline ppt. with sodium or ammonium phosphate solutions. (d) To the salt mixture solution add a little ammonium chloride and 8-hydroxy quinoline (oxime) reagent followed by ammonia solution slowly. Yellow ppt. confirms magenisum ion. (e) To potassium iodide solution add sodium hydroxide solution and then iodine solution till the solution turns yellow, now add the test solution, a yellow brown ppt. confirms magnesium.

$$NH_4Cl + NaOH \longrightarrow NH_3\uparrow + H_2O + NaCl$$

$$KCl + C_4H_4O_6H_2 \longrightarrow C_4H_4O_6HK + HCl$$

$$NaCl + 3UO_2(C_2H_3O_2)_2 + Mg(C_2H_3O_2)_2 + CH_3COOH \longrightarrow NaMg(UO_2)_3(C_2H_3O_2)_9 + HCl$$

$$MgSO_4 + 2NH_4OH \longrightarrow Mg(OH)_2\downarrow + (NH_4)_2SO_4$$

$$MgSO_4 + 2NaOH \longrightarrow Mg(OH)_2\downarrow + Na_2SO_4$$

$$MgSO_4 + Na_2HPO_4 + NH_3 \longrightarrow Mg(NH_4)PO_4\downarrow + Na_2SO_4$$

NOTE: The reporting of the radical is always recommended in their ionic symbols and not as their names. The name of the radical can be given in brackets.

For example;

REPORT:	Acid radicals	:	Cl^- (Chloride)	&	CO_3^{2-} (Carbonate)
	Basic radicals	:	NH_4^+ (Ammonium)	&	Cu^{2+} (Cupric)

1-8 IMPORTANT POINTS TO BE REMEMBERED

1-8-1 About Acid Radicals

The radicals which should not be given in the presence of each other in a mixture are

a) Nitrate and Bromide
b) Nitrate and Iodide
c) Bromide and Iodide
d) Chloride and Bromide
e) Chloride and Iodide
f) Carbonate and Sulphate / Sulphite / Sulphide
g) Phosphate and Chloride / Bromide / Iodide.

1-8-2 Possible combinations of Acid Radicals

a) Acetate and Chloride / Bromide / Iodide / Carbonate / Sulphate
b) Nitrate and Chloride / Carbonate / Sulphate
c) Chloride and Acetate / Nitrate / Carbonate / Sulphate
d) Bromide and Acetate / Carbonate / Sulphate
e) Iodide and Acetate / Carbonate / Sulphate
f) Carbonate and Acetate /Nitrate /Chloride / Bromide / Iodide
g) Sulphate and Acetate / Nitrate / Chloride / Bromide / Iodide

1-8-3 About Basic Radicals

No two radicals of the same group should be given eg. Aluminium and Iron Barium and Calcium, Cadmium and Tin. Usually ammonium ion is a common choice in every mixture.

❋

2

GRAVIMETRIC ANALYSIS

2-1 A BRIEF THEORY OF GRAVIMETRIC ANALYSIS:

The weight content of a given element or its ion in a substance is usually found by gravimetric analysis from the weight of a precipitate formed when the element or its ion is converted into an insoluble compound. In this analytical method the amount of a given component is found from the-results of weighing.

Usually a weighed sample of the substance to be analysed is brought into solution by a suitable method and the element to be determined is then precipitated as an insoluble compound. The precipitate is filtered off, washed thoroughly, ignited or dried and weighed accurately. The content of the element is calculated from the weight of the precipitate and its formula and then expressed as a percentage of the sample taken.

The choice of precipitant is very important in the gravimetric analysis to avoid the formation of a colloidal solution or sometimes impurities get coprecipitated, leading to incorrect results. Excess precipitant should be avoided.

Hence for the correct precipitation to take place apart from the precipitant, pH has to be maintained, as certain precipitates surface out in basic media such as those of iron or aluminium, whereas barium or chloride or silver, gets precipitated in acidic media and nickel precipitates out as its dimethyl glyoximate in almost neutral or acidic media.

The precipitate formed should be usually water insoluble which will make it easy for washing.

The temperature maintainance is also very essential, some precipitates occur in cold condition, but the best results are obtained in hot condition.

The weights used should be of a high calibre quality or else the results will not be correct.

2-2 ESTIMATION OF BARIUM AS (a) CHROMATE (b) SULPHATE.

Barium can be estimated as either chromate or as sulphate. These reactions and later the estimations can also be used for the estimation of chromate as barium chromate or estimation of sulphate as barium sulphate.

a. Estimation of Barium as chromate:

In a 500 ml. beaker take 200 ml of a solution containing not more than 400 milligram of barium at a neutral pH. To this solution add 1.0 ml of 6N acetic acid and 10 ml of neutral 3N ammonium acetate solution. Boil the solution, to this hot solution add dropwise a solution of ammonium chromate with vigrous strring. Transfer the beaker to a boiling water bath and allow it to remain on the flame the precipitate gets digested. Test the solution for complete precipitation by adding a few drops of the precipitant from the sides of the beaker, if it is not complete add a few milliliters of the precipitant to the reaction solution & digest further.

Filter the precipitate through a sintered glass crucible, which has been previously washed, dried and weighed. After transfering the precipitate completely wash it with hot water, till there is no precipitate of chloride in the filterate. This can be tested by transfering, a few drops of the filterate, directly from the crucible in a test tube containing silver nitrate. (the chloride is from the $BaCl_2$ solution).

Dry the precipitate first by suction and later in an hot air oven at 120°C. After cooling, the precipitate is weighed alongwith the crucible. The difference of weight of the empty crucible & this weight gives the weight of the precipitate.

Multiply the weight obtained by the respective chemical factor to get the amount of either barium or chromate as the case may be.

$$BaCl_2 + (NH_4)_2\,CrO_4 \longrightarrow 2NH_4Cl + BaCrO_4\downarrow$$

Calculation by the Chemical Factor method :

Chemical factor for Barium = 0.5421

Chemical factor for Chromate = 0.4579

The weight of the precipitate is considered to be x gms

Barium in Barium Chromate

Molecular weight of $BaCrO_4$ = 253.37

Atomic weight of Barium = 137.36

Amout of Ba in x gms of precipitate

will be : $\frac{137.36 \text{ x } x}{253.36}$ = 0.5421 x gms

Chromate in Barium Chromate

Molecular weight of $BaCrO_4$ = 253.37

Ionic weight of CrO_4^{-2} = 116.01

Amout of CrO_4^{-2} in x gms of precipitate

will be : $\frac{116.01 \text{ x } x}{253.36}$ = 0.4579 x gms

(b) Estimation of Barium as Sulphate:

In a 500 ml beaker, take 100 ml of the barium solution, which should not contain more than 0.15 gm of barium in the form of $BaCl_2$. The hydrochloric acid used to make the solution should not be more than 1% by volume. Heat the solution to boiling, to the hot solution add a slight excess of hot 1N sulphuric acid slowly with constant stirring. Digest the precipitate obtained, on a water bath, wash the precipitate carefully with hot water through a Whatman No.42 filter paper, slowly adding a small quantity of the precipitate everytime to the filter paper. Test for the complete removal of chloride with the filterate reacting it with $AgNO_3$ solution. Washing should be continued till no chloride precipitates.

Dry the precipitate alongwith the filter paper in an hot air oven at 120°C. Keeping the precipitate in the folds of the filter paper ignite it in a previously weighed silica crucible and incinirate it strongly. Cool the crucible and add a drop of sulphuric acid to convert any sulphide formed to sulphate. Heat the crucible strongly till fumes disappear, cool and weigh. Take atleast three weights by heating & cooling the crucible to get a constant weight.

Multiply the weight of the precipitate with the chemical factor.

$$BaCl_2 + H_2SO_4 \longrightarrow BaSO_4\downarrow + 2\ HCl$$

Calculation by the Chemical Factor method :

Chemical factor for barium = 0.5886

Chemical factor for sulphate = 0.4114

The weight of the precipitate is considered to be x gms

Barium in Barium Sulpahte

Molecular weight of $BaSO_4$ = 233.36

Atomic weight of Barium = 137.36

Amount of Ba in x gms of precipitate

will be : $\frac{137.36 \times x}{233.36}$ = 0.5886 x gms

Sulphate in Baraium Sulpahte

Molecular weight of $BaSO_4$ = 233.36

Ionic weight of SO_4^{-2} = 96

Amount of SO_4^{-2} in x gms of precipitate

will be : $\frac{96 \times x}{233.36}$ = 0.4114 x gms

2-3 ESTIMATION OF SILVER OR CHLORIDE

In a 500ml beaker take 200ml of a solution of silver nitrate dissolved in a 1% solution of nitric acid. The silver content should not be more than 1 milligram. Heat the solution to about 70°C, i.e., till a bubble appears in the solution, now add slowly a 0.2N hydrochloric acid solution with vigrous stirring.

(Care should be taken that the beaker is wrapped in a carbon paper or the reaction should not be expossed to light directly)

Allow the precipitate to digest on a water bath. Test the supernatant liquid with a drop of hydrochloric acid for observing the complete precipitation. Allow the precipitate to settle down in a dark cupboard overnight. Decant the precipitate through a previously weighed sintered glass crucible, wash the precipitate till it is free from the unreacted chloride. Dry the precipitate at 120°C, allow to cool in a desiccator and weigh. Repeat the heating, cooling and weighing process atleast thrice to get a constant weight.

$$AgNO_3 + HCl \longrightarrow AgCl \downarrow + HNO_3$$

Calculation by the Chemical Factor method :

Chemical factor for Silver = 0.7524

Chemical factor for Chloride = 0.2476

The weight of the precipitate is considered to be x gms

Silver in Silver Chloride

Molecular weight of silverchloride = 143.38

Atomic weight of silver = 107.88

Amount of silver = $\frac{107.88 \times x}{143.38}$

= 0.7524 x gms

Amount of Chloride = 0.2476 x gms

2-4 ESTIMATION OF IRON AS IT'S OXIDE.

In a 500 ml beaker take 100 ml of the ferrous ammonium sulpahte solution containing about 0.8 gms of it. Add 10 ml of 1:1 dilute hydrochloric acid to the solution and about one to two millilitres of dilute nitric acid is also added. The solution is boiled; keeping a policeman glass rod inside and a watch glass over the beaker; till the solution becomes clear yellow. Dilute the solution to 200 ml and boil the solution again, to the boiling solution add 1:1 ammonia solution dropwise till the smell of ammonia persists in the beaker and complete orange brown precipitation occurs. Digest the solution, the supernatant liquid should be colourless, test for complete precipitation by adding a few more drops of ammonia. Now decant the solution through a whatmann filter paper 42, allowing as little as possible amount of precipitate to flow. On complete decantation wash the precipitate with hot 1% ammonium nitrate solution and decant the washings through the filter paper, atleast 4 to 5 washings are required. Remove all the particles from the beaker with the help of the policeman glass rod. Test for any chloride if present with silver nitrate solution removing all the chloride by the process of washing the precipitate with hot ammonium nitrate solution. Let the whole liquid drain out completely through the funnel meanwhile heat a silica crucible to red heat, cool it in a desiccator to bring its temperature down to room temperature and weigh it acurately. Keep the filter paper containing the precipitate in the silica crucible with utmost care that no precipitate comes out of it. Heat the crucible till the filter paper and the precipitate dries. Now ignite the filter paper, when it gets brunt completely, incinitrate it till a brown residue is left in their crucible. Cool it to room temperature and weigh accurately to get the difference of weight as that of ferric oxide. Repeat the procedure of heating the crucible to red heat, cooling atleast thrice to get constant weight. Multiply the weight of ferric oxide with the chemical factor to get the amount of iron.

$$FeSO_4(NH_4)_2\,SO_4\,.6H_2O \longrightarrow FeSO_4 + (NH_4)_2\,SO_4 + 6H_2O$$

$$6FeSO_4 + 2HNO_3 + 3H_2SO_4 \longrightarrow 3Fe_2\,(SO_4)_3 + 2NO + 4H_2O$$

$$FeSO_4 + HNO_3 + 3HCl \longrightarrow FeCl_3 + H_2SO_4 + NO_2 + H_2O$$

$$Fe_2(SO_4)_3 + 6NH_4OH \longrightarrow 2Fe(OH)_3\downarrow + 3(NH_4)_2\,SO_4$$

$$FeCl_3 + 3NH_4OH \longrightarrow Fe(OH)_3\downarrow + 3NH_4Cl$$

$$2\,Fe\,(OH)_3 \xrightarrow{\text{heat}} Fe_2O_3 + 3H_2O\uparrow$$

Calculation by the Chemical Factor method :

Chemical factor of Iron = 0.69944

The weight of the precipitate is considered to be *x* gms

Iron in Iron oxide

Molecular weight of Fe_2O_3	=	159.70
Atomic weight of Iron	=	55.85
Amount of Iron	=	$\frac{111.7 \times x}{159.7}$
	=	0.69944 *x* gms

2.5 ESTIMATION OF NICKEL AS NICKEL DIMETHYL GLYOXIMATE

In a 500 ml beaker dissolve about 4 milligrams of nickel ammonium sulphate in about 100 ml of distilled water and add 5 millilitres of 1:1 dilute hydrochloric acid, dilute the solution to 200ml keep a policeman glass rod in the beaker and heat the solution to 70°C (till a bubble appears) keeping a watch glass over it. To the hot solution, add solution of dimethyl glyoxime dropwise, about 5 ml. immediately with vigrous stirring add 1:1 dilute ammonia solution directly in the solution and not down the walls of the beaker taking care that it does not spurt on the sides. Ammonia should be added till its smell persists. Digest the red precipitate on a steam bath for about half an hour. Test the supernatant liquid for complete precipitation, allow the precipitate to settle leave it for cooling and settle for an hour. Meanwhile weigh a sintered glass crucible, which has been heated & cooled previously and accurately. Filter the cold solution through the crucible by suction. Wash the precipitate with cold water until free from chloride, testing the washings with silver nitrate solution. Dry the precipitate in an hot air oven at 120°C for about an hour. Cool the crucible in a desiccator to bring it to room temperature and weigh as nickeldimethylglyoximate. Repeat the drying in the oven,cooling and weighing to get a constant weight. The weight of the precipitate when multiplied with nickel's chemical factor gives the amount of nickel present.

$$NiCl_2 + 2C_4H_8N_2O_2 \longrightarrow Ni(C_4H_7N_2O_2)_2 + 2HCl$$

Calculation by the Chemical Factor method :

Chemical factor of nickel = 0.20335

The weight of the precipitate is considered to be *x* gms

Nickel in nickel dimethyl glyoximate

Molecular weight of $Ni(C_4H_7N_2O_2)_2$	=	288.71
Atomic weight of nickel	=	58.71

Amount of Nickel $= \frac{58.71 \times x}{288.71}$

$= 0.20335\ x$ gms

2-6 ESTIMATION OF ALUMINIUM AS ALUMINIUM OXINATE

In a 250ml beaker dissolve about 4 milligrams of aluminium sulphate or better aluminium ammonium sulphate in 100 ml of distilled water, heat the solution to 70°C, (till a bubble appears), to the hot solution add about 10ml of 8-hydroxyquinoline (oxine) immediately add 25ml of 2N ammonium acetate solution. The yellow precipitate appeared is now heated gently with stirring to ensure complete precipitation. Test the solution for complete precipitation after allowing the precipitate to settle for about an hour. Meanwhile a sintered glass crucible is weighed which has been heated & cooled previously. Filter the precipitate by suction, dry the crucible containing the precipitate in a hot air oven for an hour at 120°C cool it in a desiccator to bring it to room temperature then weigh. Repeat the procedure of heating, cooling and weighing till constant weight is obtained. Multiply the weight of the precipitate with aluminium's chemical factor.

$$AlCl_3 + 3H\,(C_9H_6NO) \longrightarrow Al(C_9H_6NO)_3 + 3HCl$$

Calculation by the Chemical Factor method :

Chemical factor of Aluminium = 0.05878.

The weight of the precipitate is considered to be *x* gms

Aluminium in aluminium oxide

Molecular weight of Al $(C_9H_6NO)_3$ = 458.98

Atomic weight of aluminium = 26.98

Amount of Aluminium $= \frac{26.98 \times x}{458.98}$

$= 0.05878\ x$ gms

APPENDIX

APPENDIX - A

NAMES, SYMBOLS AND DEFINITIONS OF CERTAIN SI DERIVED UNITS

Physical Quantity	*Name*	*SI Unit*	
		Symbol	***Definition***
Force	Newton	*N*	$J\ m^{-1} = Pa\ m^{-2} = kg\ m\ s^{-2}$
Pressure	Pascal	*Pa*	$N\ m^{-2} = kg\ m^{-1}\ s^{-2}$
Energy	Joule	*J*	$N\ m = Pa\ m^{3} = kg\ m^{2}\ s^{-2}$
Power	Watt	*W*	$J\ s^{-1} = kg\ m^{2}\ s^{-3}$
Electronic charge (Q)	Coulomb	*C*	A. s
Electrical resistance	Ohm	Ω	$V\ A^{-1} = kg\ m^{2}\ s^{-3}\ A^{-2}$
Electrical conductance	Siemens	*S*	$\Omega^{-1} = kg^{-1}\ m^{-2}\ s^{3}\ A^{2}$
Electriç potential difference (E)	Volt	*V*	$J\ A^{-1}\ s^{-1} = kg\ m^{2}\ s^{-3}\ A^{-1}$
Frequency (ν)	Hertz	*Hz*	s^{-1} or cycle per second
Dipolemoment (μ)	Debye	*D*	C m

APPENDIX - B

GENERAL PHYSICAL CONSTANTS AND THEIR VALUES

Constant	*Symbol*	*Value*#	*Multiplication factor*	
			SI Units	*cgs Units*
Speed of light	*c*	2.997 923	$10^{8}\ m\ s^{-1}$	$10^{10}\ cm\ s^{-1}$
Elementary charge	*e*	1.602 189 2(46)	10^{-19} C	$10^{-20}\ cm\ g^{-1/2}$
Avogadro constant	*N*	6.022 045 (31)	$10^{23}\ mol^{-1}$	$10^{23}\ mol^{-1}$
Atomic mass unit	*u*	1.660 565 5(86)	10^{-27} kg	10^{-24} g
Electron rest mass	m_e	9.109 534 (47)	10^{-31} kg	10^{-28} g
Proton rest mass	m_p	1.672 648 5(86)	10^{-27} kg	10^{-24} g
Neutron rest mass	m_n	1.674 954 3(86)	10^{-27} kg	10^{-24} g
Gravitational acceleration	*g*	9.806 65	$m\ s^{-2}$	$10^{2}\ cm\ s^{-2}$
Gravitational constant	*G*	6.672 0(27)	$10^{-11}\ m^{3}\ kg^{-1}\ s^{-2}$	$10^{-8}\ cm\ g^{-1} s^{-2}$
Molar volume of an ideal gas at NTP	V_m	2.241 383 (70)	$10^{-2}\ m^{3}\ mol^{-1}$	$10^{4}\ cm\ mol^{-1}$
Faraday constant	*F*	9.648 456 (27)	$10^{4}\ C\ mol^{-1}$	$10^{3}\ cm\ g^{-1}\ mol^{-1}$
Planck constant	*h*	6.626 176 (36)	10^{-34} J s	10^{-27} erg s
Boltzmann constant	*k*	1.380 662 (44)	$10^{-23}\ J\ K^{-1}$	$10^{-16}\ erg\ K^{-1}$
Gas constant	*R*	8.314 41(26)	$J\ K^{-1}\ mol^{-1}$	$10^{7}\ erg\ K^{-1}\ mol^{1}$ $1.989717\ cal\ K^{-1}\ mol^{-1}$ $0.082056\ L\ atm\ K^{-1}\ mol^{-1}$

The numbers in parentheses represent the standard deviation in the last significant figure(s) cited. Values based on J.Phys.Chem.Ref.Data,2,663-734 (1973) and Pure and Appl.Chem;51, 1-41(1979).

APPENDIX - C

DATA ON THE STRENGTH OF AQUEOUS SOLUTIONS OF THE COMMON ACIDS AND OF AQUEOUS AMMONIA

Reagent	Specific gravity	Approximate Normality	Volume in mL required to make 1L of approx.1 N solution
Hydrochloric acid	1.18	11.3	89
Nitric acid	1.42	16.0	63
Sulphuric acid	1.84	36.0	28
Perchloric acid	1.66	11.6	86
Hydrofluoric acid	1.15	26.5	38
Phosphoric acid	1.69	41.1	23
Acetic acid	1.05	17.4	58
Aqueous ammonia	0.90	14.3	71

APPENDIX - D

DISSOCIATION CONSTANTS OF ORGANIC ACIDS AND BASES

1. Dissociation constants of carboxylic acids at 298K

Acid	K_{a_1}	K_{a_2}	K_{a_3}
Acetic acid	1.8×10^{-5}		
Benzoic acid	6.5×10^{-5}		
Chloroacetic acid	1.4×10^{-3}		
Citric acid	3.5×10^{-4}	1.7×10^{-5}	4.0×10^{-7}
Formic acid	1.8×10^{-4}		
Oxalic acid	5.9×10^{-2}	6.4×10^{-5}	
Phenol	1.3×10^{-10}		
Tartaric acid	1.0×10^{-3}	4.6×10^{-5}	

2. Dissociation constants of amines at 298K

Amine	K_b
Aniline	4.3×10^{-10}
Diethyl amine	5.4×10^{-4}
Ethyl amine	6.4×10^{-4}
Methyl amine	4.4×10^{-4}
Pyridine	1.7×10^{-9}
Trimethyl amine	6.4×10^{-5}

QUESTION PAPERS

OSMANIA UNIVERSITY

CHEMISTRY (New Syllabus) PAPER II
B.SC 2ND YEAR EXAMINATION, MARCH 1996.

ENGLISH VERSION

Time: 3 Hours Max Marks: 100

N.B.: Answer all questions. Answer all bits of a question together.

SECTION - A (Essay Type) (4x5=60)

1. • Compound - A(C_7H_6O) when treated with 40% aqueous NaOH gave compound-B($C_7H_6O_2$), which gave CO_2 with $NaHCO_3$, and compound-C(C_7H_8O), an alcohol. Compound-A when reacted with acetic anhydride in the presence of sodium acetate at elevated temperatures gave Compound-D ($C_9H_8O_2$) which also evolved CO_2 with $NaHCO_3$. Formulate the sequence of reactions and name compounds A, B, C and D. Write mechanisms and mention the name reactions involved in the reactions. What type of isomerism does compound-D exhibit and draw their structures. OR (4+8+3)

   • Describe with suitable examples the mechanisms and stereo chemistry of SN_1 and SN_2 reactions. (3+6+6)

2. • How ethyl acetoacetate is prepared by Claisen Condensation?

   • How a) n-butyric acid and b) z-butyric acid butanone are synthesised from ethyl acetoacetate. Write mechanisms involved in these reactions. (5+5+5)

   OR

   • Write notes, along with mechanisms, on the following reactions (a) benzion condensation (b) Friedel Craft acylation and (c) Kolbe-Schmidt reaction. (5+5+5)

3. • Mention two ores of Uranium. How is the metal extracted from its oxide ore. OR (2+7+6)

   • Discuss briefly "Valence Bond theory (VBT) of metals" and explain the high electrical conductivity and thermal conductivity of metals on the basis of this theory. (8+4+3)

4. • Explain number average and weight average molecular weights of macro-molecules.

   • A polymer has equal number of molecules of molecular weights 10,000; 20,000 and 40,000 daltons. Calculate the number average and weight average molecular weights of the polymer. OR (5+5+5)

   • Draw a neat labled phase diagram of lead-silver system. Explain briefly different parts in it. (3+4+4+4)

SECTION B (Short Answer Type) 8x15=40

5. • Which nitro compounds exhibit tautomerism? Draw the structures of an example and name them. OR (1+3)

- Among allyl chloride and vinyl chloride which one hydrolyses fast and explain why. (1+2+2)

6. • Write the structure of picric acid. Why is it called acid? (2+3)

OR

• Name the smallest cyclic ether. Write a reaction to prepare it. (2+3)

7. • What type of compounds is obtained by Williamsons Synthesis ? Give one example. (2+3)

OR

• Draw five canonical structures of benzene (5)

8. • Arrange the following in the increasing order of their basicity. a) Aniline b) dimethyl amine c) p-nitroaniline d) ammonia and e) methyl amine. (5)

OR

• State, Sytzeff's rule and explain with one example. (2+3)

9. • Write three different properties to distinguish transition elements from non-transition elements. (5)

OR

• Describe how the first transition series differs from the second and the third series. (5)

10. • Give one example of the artificial transmutation of elements and explain K- Capture. (2+3)

OR

• Explain in brief lanthanide Contraction. (5)

11. • What are point defects? Write brief notes on Schottky and Frankel defects in crystals. (1+2+2)

OR

• What is coagulation of Colloids? State Hardy-Schulze Law. (2+3)

12. • Draw lattice planes of a simple cubic crystal system and mention their respective Miller indices. (3+2)

OR

• What is the degree of freedom at triple point in water system. Explain using phase diagram. (2+3)

TELUGU VERSION

సమయం : 3 గంటలు అత్యధిక మార్కులు : 100

N.B. అన్ని ప్రశ్నలకు సమాధానము వ్రాయుము.
ఒక ప్రశ్నలోని అన్ని భాగములు కలిపి సమాధానము వ్రాయుము.

విభాగము - ఎ (వ్యాసరకము) 4 X 15 = 60

1. • 40% NaOH జలద్రావణముతో సమ్మేళనము $-A(C_7H_6O)$ను చర్య జరిపినపుడు ఏర్పడిన సమ్మేళనము $-B(C_7H_6O_2)$, $NaHCO_3$ తో కలిపి CO_2 మరియు ఒక ఆల్కహాల్ సమ్మేళనము $-C(C_7H_8O)$ ఇచ్చును. సోడియ ఎసిటేట్ సమక్షములో సమ్మేళనము –A, ఎసిటిక్ ఎన్‌హైడ్రైడ్ తో ఉన్నత ఉష్ణోగ్రతల వద్ద చర్య జరిపినపుడు ఏర్పడిన సమ్మేళనము - $D(C_9H_8O_2)$ గూడా $NaHCO_3$తో కలిసినపుడు CO_2 వెలువడును. చర్యల వరుసక్రమమును సాంకేతికముగా తెలిపి, A, B, C మరియు D సమ్మేళనాల నామములను పేర్కొనుము. చర్యా విధానమును వ్రాసి చర్యలలో ఏవి నామకరణ చర్యలో తెలుపుము. సమ్మేళనము –D ఏ రకమైన అణు సాదృశ్యమును ప్రదర్శించును మరియు వాటి నిర్మాణములను వ్రాయుము. లేదా (4 + 8 + 3)

● SN_1 మరియు SN_2 చర్యల యొక్క చర్యా విధానములు మరియు ప్రాదేశిక రసాయన శాస్త్ర పరిజ్ఞానమును సరియైన ఉదాహరణలతో వర్ణింపుము. (3+6+6)

2. ● క్లైసన్ కండెన్సేషన్ పద్ధతిలో ఈథైల్ ఎసిటోఎసిటేట్‌ను ఎట్లు తయారు చేయుదురు. ఈథైల్ ఎసిటోఎసిటేట్ నుండి (a) n-బ్యూటిరిక్ ఆమ్లము (b) 2-బ్యూటేనోన్‌లను ఎట్లు సంశ్లేషణము జరుపుదురు. ఈ చర్యలలో గల చర్యా విధానములను వ్రాయుము. (5+5+5)

లేదా

● ఈ క్రింది చర్యల పై చర్యా విధానముల తో కలిపి లఘు వ్యాఖ్య వ్రాయుము.

a) బెంజోయన్ కండెన్సేషన్
b) ఫ్రీడల్ (క్రాఫ్ట్) ఎసైలేషన్
c) కోల్బే–ష్మిట్ చర్య. (5+5+5)

3. ● రెండు యురేనియం ఖనిజములను తెలుపుము. దాని ఆక్సైడు ఖనిజము నుండి ఈ లోహమును ఎట్లు నిష్కర్షణము చేయుదురు. లేదా (2+7+6)

● ''లోహాల వేలెన్సీ బంధ సిద్ధాంతము (VBT)''ను గురించి క్లుప్తముగా వివరించుము మరియు ఈ సిద్ధాంతము నాధారముగా లోహాల అధిక విద్యుద్వాహకత మరియు ఉష్ణవాహకతలను వివరించుము. (8+4+3)

4. ● బృహదణువుల సంఖ్యా సగటు అణుభారము మరియు భారసగటు అణుభారము లను వివరించుము.

● ఒక పోలిమెర్ (అణుపుంజము)లో సమాన సంఖ్యలో గల అణువులు వాటి అణుభారములను 10,000; 20,000 మరియు 40,000 డాల్టన్‌లు కలిగియున్నవి.

● పోలిమెర్ యొక్క సంఖ్యా సగటు మరియు భారసగటు అణుభారాలను గణించుము. (5+5+5)

లేదా

● సీసము – వెండి వ్యవస్థ యొక్క చక్కని పావస్థ పటము (Phase diagram) గీయుము. దానిలో వేరు వేరు భాగాలను క్లుప్తముగా వివరించుము. (3+4+4+4)

విభాగము - బి (లఘు సమాధానరకము) 8 x 5 = 40

5. ● ఏనైట్రో సమ్మేళనాలు టాటోమెరిజమ్‌ను ప్రదర్శించును. ఒక ఉదాహరణమునకు నిర్మాణములను వ్రాసి మరియు వాటి పేర్లను తెలుపుము. లేదా (2+3)

● ఎలైల్ క్లోరైడ్ మరియు వైనైల్ క్లోరైడులలో ఏది వేగముగా జల విశ్లేషణము చెందును. మరియు ఎందుకో వివరించుము. (1+2+2)

6. ● పిక్రిక్ ఆమ్లము నిర్మాణమును వ్రాయుము. దానిని ఆమ్లమని ఎందుకు పిలుతురు. (2+3)

లేదా

● అతి చిన్న సైక్లిక్ ఈథర్ పేరును తెలుపుము. దానిని తయారుచేయు ఒక చర్యను వ్రాయుము. (2+3)

7. ● విలియమ్ సన్స్ సంశ్లేషణము పద్ధతిలో ఏ రకమైన సమ్మేళనము వెలువడును. ఒక ఉదాహరణమునివ్వుము. లేదా

● బెంజీన్ యొక్క ఐదు కేనోనికల్ నిర్మాణములను గీయుము. (5)

8. ● క్రింది పదార్థమును వాటి క్షారత పెరుగుదల క్రమాంకములో తెలుపుము.

(ఎ) ఎనిలిన్ (బి) డైమిథైల్ ఎమిన్ (సి) p-నైట్రో ఎనిలిన్ (డి) అమ్మోనియం మరియు (ఇ) మిథైల్ ఎమిన్. లేదా

- సెట్ జెట్స్ నియమమును తెలిపి ఒక ఉదాహరణతో వివరించుము. (5)

9. - పరివర్తన మూలకముల నుండి పరివర్త నేతర మూలకముల భేధమును తెలుపుటకు మూడు వేరు వేరు ధర్మాలను వ్రాయుము. లేదా
   - ద్వితీయ మరియు తృతీయ శ్రేణినుండి ప్రథమ పరివర్తన శ్రేణి ఎట్లు బేధములు కలిగియున్నదో వర్ణించుము. (2+3)

10. - కృత్రిమ మూలకాల పరివర్తనము నకు ఒక ఉదాహరణ నిచ్చి K-నిర్బంధము (K-Capture) ను వివరింపుము. లేదా
    - ''లాంథనైడ్ సంకోచము'' ను గురించి క్లుప్తముగా వివరించుము. (2+3)

11. - బిందులోపాలు (Point defects) అనగా నేమి. స్పటికాలలో షాట్కీ మరియు ఫ్రాంకెల్ లోపాల గురించి క్లుప్తమైన వ్యాఖ్య వ్రాయుము. లేదా (1+2+2)
    - కొలాయిడ్‌ల స్కందనము అనగానేమి. హార్డీ – స్కూల్జ్ నియమును తెలుపుము.

12. - ఒక సాధారణ ఘనాకార స్పటిక వ్యవస్థ యొక్కలాటిస్ తలాలను గీయుము. మరియు వాటి సరియైన మిల్లర్ సూచికలను తెలుపుము. లేదా
    - నీటి వ్యవస్థలో త్రిక బిందువు వద్ద స్వాతంత్ర్య పరిమితి ఏమి. ప్రావస్థా పటమునుపయోగించి వివరించుము. (2+3)

NAGARJUNA UNIVERSITY

B.Sc. DEGREE EXAMINATION, MARCH 1996

(Examination at the end of Second Year)

PART II - CHEMISTRY

Paper II - INORGANIC, ORGANIC AND PHYSICAL CHEMISTRY

Time: Three hours Maximum : 100 Marks

PART A - (5 x 12 = 60 marks)

Answer any FIVE questions selecting THREE from Section I and ONE each from Section II and Section III.

SECTION I
(ORGANIC CHEMISTRY)

1. (a) Give any two general methods of preparation of alkynes.

   ఆల్కైన్‌లను తయారు చేయు రెండు సాధారణ పద్ధతులను తెలుపుము.

   (b) What happens when propyne reacts with dil.H_2SO_4 in presence of $HgSO_4$?

   $HgSO_4$ సమక్షంలో సజల H_2SO_4 తో ప్రొపైన్ చర్య జరిపినప్పుడు జరుగు చర్య ఏమిటి?

   (c) What happens when ozone reacts with propyne?

   ప్రొపైన్ ఓజోన్‌తో ఎట్లు చర్య జరుపును.

—

2. (a) Give an account of Baeyer's strain theory and its limitations.

బేయర్ ప్రయాస సిద్ధాంతమును మరియు దాని పరిధులను గురించి వ్రాయండి.

(b) Give an account on Sachse-Mohr's theory.

సాక్స్, మోర్ సిద్ధాంతము గూర్చి తెలుపుము.

3. Discuss the structure of Benzene.

బెంజీన్ నిర్మాణమును చర్చింపుము.

4. (a) Write any two important methods of preparation of phenols.

ఫినాల్‌లను తయారు చేయు రెండు ముఖ్య పద్ధతులను వ్రాయండి.

(b) Starting from phenol how do you obtain the following

(i) Picric acid (ii) Phenyl acetate and (iii) Salicyl aldehyde.

ఫినాల్ నుండి క్రింది వాటిని ఎట్లు తయారుచేస్తారు.

(i) పిక్రిక్ అమ్లము (ii) ఫినైల్ ఎసిటేట్ (iii) శాలిసల్ ఆల్డిహైడ్.

5. Explain the mechanism involved in (a) Aldol condensation (b) Claisen condensation.

క్రింది వానిలో చర్యా విధానాన్ని వివరించండి.

(a) ఆల్డాల్ సంఘననము (b) క్లైయిజన్ సంఘననము

6. (a) How is Acetic acid prepared from ethyl alcohol?

ఈథైల్ ఆల్కహాల్ నుండి ఎసిటిక్ అమ్లాన్ని ఎట్లు తయారు చేస్తారు?

(b) What is the action of the following on it?

(i) $SOCl_2$ (ii) ROH in presence of conc.H_2SO_4 (iii) $LiAlH_4$.

అది క్రింది వాటితో ఎట్లా చర్య జరుపును.

(i) $SOCl_2$ (ii) గాఢ సల్ఫ్యూరిక్ అమ్ల సమక్షంలో ఆల్కహాల్‌తో (iii) $LiAlH_4$.

SECTION II
(INORGANIC CHEMISTRY)

7. Name the important ore of Thorium. Explain its extraction from the ore.

థోరియం యొక్క ముఖ్య ఖనిజమేది? దాని నుండి థోరియం నిష్కర్షణమును వివరింపుము.

8. What are transition elements? Explain (a) Magnetic character (b) Oxidation states and (c) Color of ions of transition metals.

పరివర్తన మూలకాలు అనగా ఏవి? వాటి (a) అయస్కాంత ధర్మాలు (b) ఆక్సీకరణ స్థితులు మరియు (c) రంగులను గూర్చి వివరించుము.

9. What is artificial radioactivity? Explain the phenomenon with examples. Give uses of radioactive isotopes in Agriculture, Medicine and Industry.

కృత్రిమ రేడియోధార్మికత అనగానేమి? ఉదాహరణలతో వివరించుము. వ్యవసాయం, వైద్యరంగం మరియు పరిశ్రమలలో రేడియోధార్మిక స్వస్థానియుల ఉపయోగాలను తెలుపుము.

SECTION III
(PHYSICAL CHEMISTRY)

10. What are the differences between Adsorption and Absorption? Derive Langmuir adsorption isotherm. What are its limitations?

అధిశోషణము, శోషణములకు గల తేడాలను తెలుపుము. లాంగ్‌మ్యూర్ అధిశోషణ సమోష్ణోగ్రత రేఖను ఉత్పాదించి దాని పరిమితులను తెలుపుము.

11 Explain the following terms :

(a) Plane of symmetry.

(b) Axis of symmetry.

(c) Centre of symmetry and

(d) Space lattice.

క్రింది వాటిని వివరింపుము.

(a) సౌష్ఠవతలము

(b) సౌష్ఠవాక్షము

(c) సౌష్ఠవ కేంద్రము మరియు

(d) ప్రాదేశిక జాలము.

12. What is a condensed system? Define condensed phase rule. Explain Silver-Lead system with the help of condensed phase rule.

సంగ్రహ వ్యవస్థ అనగానేమి? సంగ్రహ ప్రావస్థా నియమమును నిర్వచింపుము. దీనిని ఉపయోగించి సిల్వర్-లెడ్ వ్యవస్థను వివరింపుము.

PART - B (10 x 4 = 40 marks)

Answer any TEN of the following

13. Give an account of the conformations of *n*-butane

n-బ్యూటేన్ యొక్క అనురూపకాల గురించి వ్రాయుము.

14. Williamson's synthesis - explain.

విలియంసన్ సంశ్లేషణ - వివరించుము.

15. Iodoform reaction - Explain

అయొడో ఫారమ్ చర్య - వివరించుము.

16. Explain Hell-Volhard-Zelinsky reaction

హెల్-వోల్‌హర్డ్-జెలిన్‌స్కీ చర్యను వివరించుము.

17. Write briefly on Friedel-Crafts reaction.

ఫ్రీడెల్-క్రాఫ్ట్ చర్యను గూర్చి లఘువ్యాఖ్య వ్రాయండి.

18. What is peroxide effect? Explain.

పెరాక్సైడ్ ప్రభావము అనగానేమి? వివరించుము.

19. How is Nickel extracted by Mond's process?

మాండ్ పద్ధతిలో నికెల్‌ను ఎట్లు సంగ్రహించెదరు?

20. Explain briefly about Nuclear fission. Give its important applications.

కేంద్రక విచ్ఛిత్తి చర్య గురించి క్లుప్తంగా చర్చించుము. ఈ చర్య యొక్క అనువర్తనములను తెలుపుము.

21. Explain free electron theory of metallic bonding.

లోహాలలోని బంధాలను వివరించే స్వేచ్ఛా ఎలక్ట్రాన్ సిద్ధాంతమును వివరించుము.

22. Explain Lanthanide contraction.

లాంథనైడ్ కంట్రాక్షన్ గురించి వివరించుము.

23. Explain how the molecular weight of macromolecules determined by osmotic pressure method.

ద్రవాభిసరణ పీడన విధానము ద్వారా బృహదణువుల అణుభారమును కనుగొను విధమును వివరించుము.

24. Write a short note on the following:

(a) Tyndal effect.
(b) Gold number.

క్రింది వాటిపై లఘువ్యాఖ్య వ్రాయుము

(a) టిండాల్ ప్రభావము.

(b) గోల్డ్ సంఖ్య.

25. Explain the various defects in crystals.

స్ఫటికాలలో ఉండే వివిధ లోపాలను వివరించుము.

26. Calculate the number of degrees of freedom at the triple point for one component and two component systems.

ఏకఘటక వ్యవస్థకు, ద్విఘటక వ్యవస్థకు త్రికబిందువు వద్ద ప్రావస్థానియమమును ఉపయోగించి స్వతంత్ర పరిమితులను కనుగొనుము.

NAGARJUNA UNIVERSITY

B.Sc. DEGREE EXAMINATION, SEPTEMBER 1995

(Examination at the end of Second Year)

PART II - CHEMISTRY

Paper II - INORGANIC, ORGANIC AND PHYSICAL CHEMISTRY

Time: Three hours Maximum : 100 Marks

PART A - (5 x 12 = 60 marks)

Answer any FIVE questions selecting THREE from Section I and ONE each from Section II and Section III.

SECTION I

(ORGANIC CHEMISTRY)

1. Write any two general methods of preparation of Alkanes and discuss the Mechanism of electrophilic addition in propylene with examples.

   ఆల్కీనులు తయారీకై ఏవేని రెండు సామాన్య పద్ధతులను పేర్కొని ప్రొపిలీనులు ఎలక్ట్రోఫిలిక్ సంకలన చర్య యొక్క చర్యా విధానమును ఉదాహరణలతో సహా చర్చించుము.

2. Discuss the Mechanism and stereochemistry of SN_1 and SN_2 reactions taking suitable examples.

   SN_1 మరియు SN_2 చర్యల యొక్క క్రియా విధానము త్రిమితీయ రసాయన శాస్త్రము ఉదాహరణలతో సహా చర్చింపుము.

3. Write the general methods of preparations of Alcohols. How can you distinguish inter molecular and intra molecular Hydrogen bond with two examples?

   ఆల్కహాలులను తయారుచేయు రెండు సాధారణ పద్ధతులను వ్రాయుము. అంతరు అణుక, అంత్రాణుక హైడ్రోజను బంధాలను ఎటుల భేదపరిచెదరు. రెండు ఉదాహరణలిమ్ము.

4. How can you prepare cyclo Pentanone from esters of dicarboxylic acid, and explain Bayers Strain theory. Write the products formed by the addition of Hydro Iodic acid on Cyclopropane, Cyclo Butane and Cyclo Pentane.

   డైకార్బాక్సిలిక్స్ ఎస్టర్ల ద్వారా సైక్లో పెంటెనేన్‌ను ఎటుల తయారు చేయుదురు? బేయర్స్ స్ట్రైన్ థియరీను విశదీకరింపుడు. సైక్లోప్రోపెన్ సైక్లోబ్యూటెన్ సైక్లోపెంటన్‌లపై హైడ్రోఅయొడిక్ ఆమ్లం సంకలనం వల్ల వచ్చు ఉత్పన్నాలను తెలుపుము.

5. Write the Mechanisms of

   (a) Benzoin condensation

   (b) Claisen condensation

   (c) Haloform reaction

ఈ క్రిందివాని చర్యాత్మకతను వివరించుము :

(a) బెంజోయిను సంఘననము

(b) క్లైజన్ సంఘననము

(c) హెలో ఫారమ్ చర్య

6. How can you prepare propyne from acetylene and the following products from acetylene :

(a) Acetaldyhyde

(b) Oxalic Acid

(c) Acrilonitride

ప్రొపైన్‌ను ఎసిటిలీను నుండి ఎటుల తయారు చేయుదురు? ఈ క్రింది సంయోగపదార్థములను ఎసిటిలీను నుండి ఎటుల తయారు చేయుదురు :

(a) ఎసిటాల్‌డిహైడు

(b) ఆక్జాలిక్ ఆమ్లం

(c) ఎక్రిలోనైట్రైడ్

SECTION II
(INORGANIC CHEMISTRY)

7. Write any two methods of preparations of organometallic compounds and any Four properties.

ఆర్గానోమెటాలిక్ సమ్మేళనాలు తయారుచేయు పద్ధతులను ఎవైనా రెండింటిని మరియు ఏవైనా నాలుగు ధర్మాలను వ్రాయండి.

8. What are the important ores of Nickel? How is it extracted? Write its uses.

నికెల్ యొక్క ముఖ్య ఖనిజాలేవి? నికెల్‌ను ఏవిధంగా సంగ్రహించెదరు? దాని ఉపయోగాలను వ్రాయండి.

9. Write an account of application of Nuclear Isotopes.

కేంద్రక ఐసోటోపుల యొక్క అనువర్తనాలను గురించి విశదీకరించండి.

SECTION II
(INORGANIC CHEMISTRY)

10. Define phase rule and explain the terms involved in it. Apply phase rule to water system.

ప్రావస్థా నియమమును నిర్వచించి, అందలి పదములను వివరించండి. ప్రావస్థా నియమమును జల వ్యవస్థకు అనువర్తించండి.

11. Explain the mechanical, electrical and optical properties of colloids.

కొల్లాయిడ్ల యాంత్రిక, విద్యుత్, కాంతి ధర్మాలను వివరించండి.

12. What is Number average Molecular weight of a Polymer? How can it be experimentally determined by osmotic pressure method?

పాలిమెర్ యొక్క సంఖ్య సగటు అణుభారమనగానేమి? దానిని ఆస్మాటిక్ పీడన పద్ధతిలో ఎటుల ప్రయోగాత్మకంగా నిర్ణయించెదరు?

PART - B (10 x 4 = 40 marks)

Answer any TEN of the following

13. Explain Markoni Koff's rule with examples.

మార్కోనికాఫ్ నియమమును ఉదాహరణలతోసహా వివరించుము.

14. What is orientation of Benzene?

బెంజీను యొక్క స్థాన నిర్దేశకత అనగానేమి?

15. Explain Cannizaro reaction.

కెన్నిజర్ చర్యను వివరించుము.

16. How can you convert Phenol to Salcylic Aldehyde?

ఫినాల్‌ను సాల్‌సిలిక్ ఆల్డిహైడుగా ఎటుల మార్చుదురు?

17. How can you prepare glycerol from propylene?

ప్రొపిలీను నుండి గ్లిసరాల్‌ను ఎటుల తయారు చేయుదురు?

18. Write the principle behind Diels Alder reaction and explain.

డీల్సు ఆల్డర్ చర్య యొక్క ముఖ్య ఉద్దేశ్యమేమి? వివరించుము.

19. What are functions of Iron in biological system?

జీవ వ్యవస్థలో ఐరన్ యొక్క అనువర్తనాలేవి?

20. Explain metallic band on the basis of free electron theory and valency band theory.

స్వేచ్ఛా ఎలక్ట్రాన్ సిద్ధాంతపరంగాను, వేలెన్సీబంధ సిద్ధాంతపరంగాను లోహబంధాన్ని వివరించండి.

21. Mention important uses of Thorium.

థోరియం యొక్క ముఖ్య ఉపయోగాలను వ్రాయండి.

22. Explain Solvolysis.

సాల్వొలిసిస్ ను వివరించుము.

23. What is Braggs equation? Write down its application in crystal structure.

బ్రాగ్సు సమీకరణ అనగా నేమి? స్ఫటిక నిర్మాణములో దీని ఉపయోగమును తెలుపుము.

24. What is a EUTECTIC Mixture? Explain with an example.

సులభ ద్రవీకరణ మిశ్రమము అనగా నేమి? ఒక ఉదాహరణతో వివరించుము.

25. Explain the phenomenon of protection of colloids with an example.

కొల్లైడ్‌ల రక్షణ ప్రక్రియను ఉదాహరణతోసహ వివరించుము.

26. Define weight average Molecular weight of a Polymer.

పాలిమెర్ యొక్క భార సగటు అణు భారమును నిర్వచింపుము.

ANDHRA UNIVERSITY, SEPTEMBER 1994
PART II - CHEMISTRY
(Common Core Scheme)

Paper II – INORGANIC, ORGANIC AND PHYSICAL CHEMISTRY

Time: Three hours　　　　Maximum : 100 Marks

PART I - (7 x 10 = 70 marks)

Answer SEVEN questions, choosing any TWO from each Section A and C and any THREE from Section B. All questions carry equal marks.

SECTION A

1. **Mention two important ores of zirconium with formula. How is zirconium extracted? What are its uses?**

   జిర్కోనియం ముఖ్య ఖనిజాలు రెండింటి పేర్లను సంకేతనాలను తెలపండి. జిర్కోనియంను ఏ విధంగా దాని ఖనిజాల నుండి సంగ్రహణమును జరుపుతారు? దాని ఉపయోగాలను తెల్పండి.

2. **Describe two chemical methods for the preparation of inert gases from atmospheric air. Explain the structures of XeF_2 and XeF_4.**

   జడవాయు మూలకాలను, గాలి నుండి తయారు చేసే రెండు రసాయనిక పద్ధతులను తెల్పండి. XeF_2, XeF_4 నిర్మాణాలను గురించి వివరించండి.

3. **Discuss the following properties of transitional elements.**

   (a) variable oxidation states　(b) formation of coloured ions
   (c) paramagnetism　(d) formation of complexes.

   పరివర్తన మూలకాల యొక్క ఈ క్రింది ధర్మాలను వివరించండి.
   (a) ఆక్సీకరణ స్థితులు　(b) రంగు గల అయాన్‌ల ఏర్పరుచుట
   (c) పారా అయస్కాంత ధర్మము　(d) సంశ్లిష్ట సమ్మేళనాలను ఏర్పరుచుట.

SECTION B

4. **Discuss the mechanism of the following reactions of aldehydes and ketones:**

   (a) Addition reaction with sodium bisulphite (b) Reaction with Grignard reagent (c) Aldol condensation.

   ఆల్డిహైడ్‌లు-కీటోన్‌ల యొక్క ఈ క్రింది చర్యా విధానాలను వ్రాయండి.
   (a) $NaHSO_3$ తో సంకలన చర్య (b) గ్రిగ్‌నార్డ్‌కారకముతో సంకలన చర్య
   (c) ఆల్డాల్ సంఘటన చర్య.

5. **How is glycerol synthesised? How does it react with (a) Oxalic acid (b) Conc. HNO_3 (c) Potassium bisulphite.**

   గ్లిసరాల్‌ను ఏ విధముగా సంశ్లిష్టచర్య ద్వారా తయారు చేస్తారు. అది ఈ క్రింది పదార్థాలతో ఏ విధముగా చర్య జరుపుతుంది.

   (a) ఆక్సాలిక్ అమ్లం (b) గాఢ HNO_3 (b) $KHSO_4$

6. **Give three general methods for the preparation of alkynes. Give their reactions with (a) Hydrogen bromide (b) Ozone (c) Tollen's reagent.**

ఆల్కైన్‌లను తయారు చేయుటకు మూడు సాధారణ పద్ధతులను వివరించండి. అవి ఈ క్రింది వాటితో ఏ విధంగా చర్యను జరుపుతాయి.
(a) హైడ్రోజన్ బ్రోమైడ్ (b) ఓజోన్ (c) టాలిన్ కారకము.

7. Discuss the structures of benzene. Give the mechanism of electrophilic substitution of benzene.
బెంజీన్ యొక్క నిర్మాణమును గురించి చర్చించండి. బెంజీన్ యొక్క ఎలక్ట్రోఫిలిక్ ప్రతిక్షేపణ చర్యా విధానాన్ని వివరించండి.

SECTION C

8. State and explain distribution law. What are its limitations ? How is the law modified in case of (a) Association (b) Dissociation of the solute?
వితరణ నియమాన్ని నిర్వచించి వివరించండి. దాని యొక్క అవధులను తెలపండి. ఈ నియమము ద్రావణిలో (a) వియోజనము చెందినపుడు (b) సహచరితాణువులుగా మారినపుడు ఏ విధంగా మార్పు చేస్తారో తెల్పండి.

9. Explain plane of symmetry and centre of symmetry. Discuss the defects in crystals.
సౌష్టవ తలం మరియు సౌష్టవ కేంద్రంలను వివరించండి. స్ఫటికాలలో లోపాలను గూర్చి చర్చించండి.

10. Define critical constants. Derive the relation between critical constants and van der Waal's constants.
సందిగ్ధ స్థిరాంకాలను నిర్వచించండి. సందిగ్ధ స్థిరాంకాలకు వాన్‌డర్ వాల్ స్థిరాంకాలకు మధ్యగల సంబంధాన్ని ఉత్పాదించండి.

PART II - (10 x 3 = 30 Marks)
Answer ALL questions.

11. What are clathrate compounds? Give two examples.
పంజర సమ్మేళనాలు అంటే ఏమిటి? రెండు ఉదాహరణలు తెల్పండి.
12. Explain Hume-Rothery rules.
హ్యూమ్ -రూథరీ నియమాలను వివరించండి.
13. Name two important ores of nickel. Give their composition. Mention two uses of nickel.
నికెల్ యొక్క రెండు ముఖ్య ఖనిజాలను తెల్పి వాటి సంఘటనను వ్రాయండి. దాని యొక్క రెండు ఉపయోగాలను తెలపండి.
14. Explain the reactivity of acid chlorides.
ఆసిడ్ క్లోరైడ్‌ల చర్యాశీలతను వివరించండి.
15. What is haloform reaction?
హాలోఫారమ్ చర్యను వివరించండి.
16. Explain Clemmensen's reduction reaction.
క్లిమన్‌సన్ క్షయకరణ చర్యను గురించి వ్రాయండి.
17. Explain the acidic nature of acetylene.
ఎసిటిలీన్ ఆమ్ల స్వభావాన్ని వివరించండి.
18. What are azeotropic mixtures? Give two examples.
ఎజియోట్రోపిక్ మిశ్రమము అంటే ఏమిటి? రెండు ఉదాహరణలను తెలపండి.
19. Define the terms (a) unit cell and (b) crystal lattice.
యూనిట్ సెల్ మరియు స్ఫటిక జాలములను నిర్వచించండి.
20. What are cryohydrates? Give examples.
క్రయోహైడ్రేట్‌లు అంటే ఏమిటి? ఉదాహరణ ఇవ్వండి.

INTRODUCTORY POLYMER CHEMISTRY

This introductory text on Polymer Chemistry does not presuppose a high level of proficiency in the background subjects and is written for students of an average calibre who have studied in an Indian University for a B.Sc. (Pass) degree. This text would, therefore, be useful to those students who wish to be initiated in the subject at the grass-root level. For this end in view, special feature of the book is the introduction of elementary questions in the text based on the previously discussed matter and furnishing their answers at the end of the chapter.

In order to add to the usefulness of the book, ten specimen laboratory exercises, which can be performed in an ordinary Organic Chemistry laboratory, have been given in an Appendix of the book. A glossary of the terms used in the text has been given at the very beginning of the book.

A chapter on "Degradation and Stabilization of Polymers" has been added to enhance the utility of the book to the technologist and the practitioner. The recent Merrifield solid phase synthesis of polypeptides, to emphasize the usefulness of the solid supports in the synthesis of useful macromolecules has been taken note of in the book.

G.S. MISRA obtained his B.Sc. and M.Sc. degrees from the University of Allahabad and his Ph.D. from the University of London in 1948, working under the guidance of Prof. D.H. Hey FRS, on the subject of Addition Polymerization. He also attended a course in Plastics Technology at the Borough Polytechnic and was awarded a Bronze Medal by the City & Guilds as the first prize. He joined the Chemistry Department of Lucknow University in 1949 as an Assistant Professor and worked there till 1963. He was Director, Indian Lac Research Institute, Namkum, Ranchi (1963-69), Professor and Head of the Department of P.G. Studies and Research in Chemistry, University of Jabalpur (1969-79) and worked in the same capacity at the Jammu University (1979-83). He also spent a year (1954-55) at the University of Munich.

Prof. Misra has published about 80 research papers in the fields of Organic and Polymer Chemistry.

PUBLISHING FOR ONE WORLD

NEW AGE INTERNATIONAL (P) LIMITED, PUBLISHERS

New Delhi • Bangalore • Calcutta • Guwahati • Hyderabad
Lucknow • Madras • Mumbai • Pune • London • Bangkok

81-224-0471 pp 274 1993 Paper Rs. 95